ETHNOBOTANY
Evolution of a Discipline

Dedicated to
JOHN W. HARSHBERGER
1869–1929
A prolific floristic and taxonomic botanist,
who in 1895 first coined the word *ethnobotany*

ETHNOBOTANY
Evolution of a Discipline

Edited by
Richard Evans Schultes
and
Siri von Reis

DIOSCORIDES PRESS
Theodore R. Dudley, Ph.D., General Editor
Portland, Oregon

The publisher wishes to acknowledge the contribution of the late Theodore R. Dudley, General Editor of Dioscorides Press, in the preparation of this volume.

ISBN 0-931146-28-3

Printed in Hong Kong

DIOSCORIDES PRESS
The Haseltine Building
133 S.W. Second Avenue, Suite 450
Portland, Oregon 97204, U.S.A.

Library of Congress Cataloging-in-Publication Data

Ethnobotany : evolution of a discipline / edited by Richard Evans Schultes and Siri von Reis.
    p.   cm.
    Includes bibliographical references and index.
    ISBN 0-931146-28-3
    1. Ethnobotany. I. Schultes, Richard Evans, 1915–   . II. von Reis, Siri, 1931–   .
GN476.73.E84   1995
581.6′1—dc20   94-15515
                                                                     CIP

Contents

Contents

Foreword

A popular catch phrase of our time claims that history is dead. What exactly is meant by this phrase, I am not sure, but it strikes me that perhaps it refers to the history of peoples, politics, and princes. With increasing migrations and commingling of peoples, the collapse of East-West political confrontation, and the obsolescence of some princes, perhaps the pundits are indeed onto something. But a more important history—that of people and plants—is far from dead. Like a ball of callus in a tissue culture, it is just emerging from its undifferentiated state and is putting out roots and shoots that give it definition and a chance at breathing, living, and contributing to the world of tomorrow. A vital part of this newly developed aspect of history is ethnobotany.

Plants have always been more important than politics—both to human daily living and to history. Even today, millions of subsistence farmers have no idea who their nominal political leaders are, but they know a great deal about their plants—sometimes even more than scientists. Furthermore, plants have had a greater historical impact than have politicians. An excellent example is provided by the few wheat seeds of an unknown Mennonite who stuffed his (or more probably, her) pockets before leaving the Russian steppes in the 1870s. Those seeds grew into plants so tolerant of cold that they made winter wheat a practical reality for the United States and Canada. By 1919, Turkey red, resulting from the gift of those destitute foreigners, accounted for 98 percent of U.S. winter wheat. Its contribution to generations of North Americans, including farmers, millers, freight-train operators, bakers, accountants, capitalists, and consumers, has been far greater than the contributions of the presidents of the 1870s, not to mention the thousands of lesser legislators of that decade.

Mennonite wheat is just one of dozens of plants, many of them from primitive aboriginal societies, that directed and changed history since that era. The 1880s and 1890s saw the international development of bananas, African oil palm, rubber, the cola nut, the chocolate tree, quinine, and pineapple, to mention only a few. The effects of these plants far transcends the results of any political decisions of the time. Southeast Asia's banana so transformed Central American republics that their economies now rely on this fruit for their survival, and South America's pineapple changed Hawaii from a sleepy, exotic way station to an outpost of hard-driving enterprise from which it has never recovered. In less than a century the Amazon's rubber tree revolutionized industries and lifestyles around the world, exerting an influence greater than any political upheaval. Africa's oil palm probably changed the landscape of large parts of Southeast Asia more than World War II did, and quinine from Java's cinchona plantations affected the overall tropics in a way that no political ruler could ever match: it has saved thousands of lives in the tropics.

9

These few plants transformed more than just the tropics. The United States, Britain, France, and dozens of affluent nations have come to eat, drink, or otherwise rely on products from tropical plants including bananas, oil palm, cola nut, cacao, rubber, and others. Giant companies like Lever Brothers, Cadbury, Hershey, Dole, United Fruit, Goodyear, Michelin, Proctor and Gamble, and Colgate Palmolive rose to dominate life and public outlook in the Western world.

Strangely, though, in the twentieth century people began to forget the plant world's significance to themselves and their history. This has been the century of steel, machines, and the laboratory—all powered mainly by resources from below the soil rather than from it. Plastics, polyesters, pesticides, and manufactured pharmaceuticals along with planes, personal computers, and Polaris missiles have changed the modern industrial world possibly more than any plant. But now the plants and their age-old relationship with humans are returning once again to prominence. With increasing doubts about petroleum supplies, with environmental concerns over persistent materials such as plastics and certain pesticides, with sagging interest in synthetic textiles and manufactured products in general, and with the rising fears of ozone holes, global warming, toxic wastes, and other fearsome side effects of the twentieth century, the connections between people and plants are starting to tighten once again.

History, then, is not dead. Rising up like a phoenix of the forest and farm is a vibrant new branch that is in tune with the most basic issues facing a planet under stress: food, forestation, biodegradable resources, atmospheric carbon dioxide levels, and renewable agriculture, for example. When historians and the public at large wake up to this fact (and they will sooner or later), history will in larger and larger measure come to be based in great part on ethnobotany, the precious knowledge of the properties of plants among indigenous peoples amassed over millennia.

This volume, therefore, can be seen as a kind of prehistory that tells much about our past, but also much about our present and future. It outlines and details the evermore vital links that have connected people to their plants and that still connect them. And ethnobotany can give us a glimpse of the world to come.

NOEL D. VIETMEYER
National Academy of Sciences
National Research Council on International Relations
Washington, D.C.

Preface

Ethnobotany is a very old discipline. Knowledge of useful plants must go back to the beginning of human existence; even today there is evidence that simian animals seek out plants useful to their purpose. But primitive men and women must of necessity have been something approaching ethnobotanists: they certainly had to classify plants into those without any use, those from which they could obtain nourishment or stimulation, those which could alleviate ills or even cure sicknesses, those with psychoactivity, and those which, if ingested, could kill a person. Imagine their astonishment to discover that a few plants, the hallucinogens, put them, through visual and other hallucinations, in contact with the supernatural world, the ancestors, and the malevolent or benevolent spirits who controlled all the affairs of the human race on earth.

In every culture, and probably very early, there were men or women especially knowledgeable in the properties of plants. These individuals often rose to an exalted rank. To this day, in many unacculturated and superficially acculturated societies, the medicine men and women or shamans still hold a high position in many phases of tribal life. These individuals diagnose and "treat" illnesses, determine the causes of death, assure good harvest through rituals, explain climatic and natural phenomena, control methods of hunting and fishing, are the repository of tribal mythology and history, and usually are the specialists who manipulate and control hallucinogenic drugs that are held to be sacred.

Over millennia, the human race's dependence on the Plant Kingdom for most of its needs increased. Eventually agriculture, the ability of humans to domesticate certain plants, was discovered, first in the Old World, slightly later in the New World, leading to a different kind of dependence, particularly on food plants. The advent of agriculture had sundry effects on the old-fashioned ethnobotany, an effect which today is evident in the extremely significant knowledge of biological diversity found in many primitive societies.

Only in the twentieth century, however, has ethnobotany assumed the status of a distinct branch of natural science. Ethnobotany today has numerous definitions. The most widely employed explanation is "the use of plants in primitive societies," as opposed to the term *economic botany*, which is normally meant to indicate the study of plants used in advanced agroindustrial societies. Another definition maintains that ethnobotany comprises the complete registration and understanding of the classification, uses, and practical, religious, and superstitious concepts concerning plants in primitive or unlettered societies. Our own belief is that a broader definition is desirable, and we propose to consider the term to mean "the study of human evaluation and manipulation of plant materials, substances, and phenomena, including relevant concepts, in primitive or unlettered societies." It is obvious from such inclusive definitions that ethnobotany must be highly interdisciplinary, drawing from aspects of botany, anthropology, archaeology, phytochemistry, pharmacology, medicine, history, religion, geography, and numerous other tangentially pertinent sciences and arts.

The study of ethnobotany and research has undergone changes and has been significantly amplified in the century since the term was first used in 1895. Its constant growth resulted in the necessary proliferation of terms to describe sundry specialized subdivisions of the study: ethnobiology, ethnopharmacology, archaeoethnobotany (often called paleoethnobotany), ethnoecology, ethnomedicine, ethnomycology, socioethnobotany, and a number of others.

The first half of the twentieth century has seen a rapid development of interest in ethnobotany, as an admittedly incomplete enumeration of events and activities will illustrate. In the United States and Canada more research is being carried out each year; many more university courses in ethnobotany are now being offered and the number of doctoral dissertations based on ethnobotanical field work has proliferated. The discipline is alive and prospering in many other countries around the globe: Mexico, Colombia, Brazil, Germany, Spain, Italy, the United Kingdom, and especially France, India, China, and Kenya—to name only a few. National and international societies dedicated to ethnobotany or one of its specialized subdivisions have been formed. The Society of Ethnopharmacology, for example, was founded in 1990 during an international symposium held in Strasbourg, France. Symposia and meetings, national and international, are being held with more frequency; Ethnobotanica '92, for example, was held in Córdoba, Spain, in celebration of the 500th year of Columbus's arrival in the Western Hemisphere. National and international conservation organizations are devoting evermore attention to ethnobotanical conservation to salvage some of the remaining knowledge possessed by still aboriginal societies in various parts of the world. There is growing interest in the pharmaceutical industry in a number of countries, and in the United States an organization was established exclusively to study plants used in primitive societies.

Publishers are offering each year more books in ethnobotany. Numerous scientific journals are devoted to articles in this discipline: *Journal of Ethnopharmacology, Journal of Ethnobiology, Planta Medica, Ethnobotany, Social Pharmacology, Diversity, Environmental Conservation,* and *Environmental Awareness,* to name but a few. Sundry long-established botanical and anthropological publications are now accepting ethnobotanical papers of various kinds; *Bulletin d'Agriculture Tropical et Colonial* has had a long history of publishing ethnobotanical studies; *American Anthropologist* readily accepts articles in this field; and the German journal *Curare* is mainly ethnobotanical. Numerous journals formerly devoted exclusively to taxonomy or other fields of botany have realized the tangential importance of ethnobotanical research to many more specialized areas of research in the plant sciences. Even widely read popular publications (e.g., *National Geographic* and *Smithsonian*) welcome material on the uses of plants and their importance among indigenous peoples; and newspapers and the television are devoting more space and time to ethnobotanical conservation.

In the main, two kinds of goals influence ethnobotanical research. One—of great interest and significance to anthropologists, archaeologists, sociologists as well as to students of mythology and the origin of religions and related fields—concerns the psychological aspects of the ways that aboriginal peoples interpret and treat their useful plants. This aspect is of deep interest from an academic viewpoint. The other goal concerns the possibility of finding new species valuable for agriculture or industry or of discovering and salvaging new chemicals from the wild floras of the world.

Modern medicine or industry may never use some of the many new chemical compounds isolated from plants, but literally thousands of new secondary constituents of plants have been reported in the twentieth century: in 1950, for example, fewer than 5000 alkaloids were known, but in 1990 more than 10,000 were known, a majority from tropical species. From the chemical viewpoint, the world's flora—estimated to consist of up to 500,000 higher plants—has hardly been touched; it has been stated that fewer than 10 percent of the calculated 80,000 species of Amazonia, for example, have been superficially analyzed. Phytochemists would undoubtedly find it well nigh impossible to collect enough material from such a remote region to analyze 80,000 species. Concentrating on those plants that Amazonian tribal peoples by experimentation over the centuries have found to be bioactive could provide chemists with a kind of short cut.

The value of learning from the perspicacity of many aboriginal peoples about the biodiversity of plants—those which they use and those which they know only as elements of the forest—should not be overlooked. Ethnobotanical studies devoted to this aspect of field research should be given greater and immediate attention in view of the unchecked and wanton destruction of vast areas of the world's forests.

The use of plants usually requires some kind of technology. It may be simple, such as cooking, or complex, such as the elaboration of curare. Two general types of technology can be recognized. One concerns the technological treatment by indigenous peoples; this type might be termed *aboriginal technology*. The second type is the *advanced* or *complex technology* of modern industry and science. Although it is not generally realized, modern, advanced technology has to a great extent utilized and built upon some of the discoveries which were made in primitive societies over the centuries and which still comprise present-day aboriginal technology.

More and more has the practical value of ethnobotanical conservation been appreciated for the many contributions that it has offered to science. One of the most significant contributions is the intricate knowledge of minute and often hidden differences in species—that is, biodiversity—of such extreme importance to modern genetics. Aboriginal peoples usually possess an uncanny ability to appreciate diversity, even within a single species, often when trained botanists cannot distinguish the characters that indigenous peoples use to differentiate the strains, races, or ecotypes. Protecting areas of virgin vegetation believed to have an unusually high density of biological diversity has become one of the major aims of environmentalists, and attention to the ability of indigenous peoples to distinguish the aboriginally recognizable differences will be one of the most important contributions of ethnobotanists in the future.

A basic need—and one that has already begun to be met—is the training of more young people for ethnobotanical research and field work. A concurrent need is the availability of research funds from a variety of sources to support this urgent research. Along with these is the need to put international pressure on countries with government-supported or government-condoned programs that involve the invasion or systematic destruction of regions where defenseless indigenous peoples have lived for centuries—programs that promote reckless burning of forests, poisoning of rivers, leveling of indigenous housing, disintegration of cultures, and, not infrequently, the physical annihilation of whole tribes.

It is in the interest of crystallizing the urgency of intensifying ethnobotanical research that we have prepared this volume. While it has not been possible to publish material representative of all the many phases of ethnobotany nor from all the outstanding specialists, we have tried to present a broad spectrum from numerous geographical regions. Most of the contributions are original, written for this book, but we have reprinted a few because (1) they consider aspects not otherwise covered in this book, (2) they were published in obscure journals, (3) they are botanically significant though they are in fields not commonly associated with ethnobotany, or (4) they are simply outstanding.

Why explore the status of ethnobotany today? Because the time has come to evaluate this discipline and its position in the sciences and in environmental conservation, particularly from the point of view of the welfare of future generations. It is our sincere hope that publication of this collection of essays will help in increasing the preparation of more ethnobotanists, a distinct necessity for all humanity in view of the rapid disappearance of native knowledge with acculturation and the disappearance of the precious ethnobotanical knowledge of so-called primitive cultures.

Siri von Reis   Richard Evans Schultes

Acknowledgments

It is an immeasurable pleasure to express our deepest appreciation to the many colleagues and friends who, busy as they all are, willingly devoted time and energy to enrich this collection of essays with the wealth of their experiences in the sundry aspects of ethnobotany. Several of the authors are no longer with us, but their contributions remain.

Our first thanks must go to the very numerous preliterate societies whose people have cooperated wholeheartedly in the various fields of ethnobotany represented by the essays published in this book.

For many courtesies and practical suggestions during the preparation of the book, we express our appreciation to Mark J. Plotkin. We owe a debt of gratitude to Kathleen Horton, who with the greatest care and understanding typed much of the final manuscript. Our thanks go to Mary Gaudet for astute help and suggestions and for assistance in proofreading many of the original articles. Gustavo Romero aided us significantly in translating the several contributions that were submitted in Spanish.

We likewise thank the several publishers of the books and journals from which selected articles are reprinted here because of their pertinence to understanding the development and growth of ethnobotany.

It is a pleasure to express our debt to the libraries and librarians of Harvard University for the many courtesies in locating rare books or other difficult-to-locate sources: the Ames Economic Botany Library, the Gray Herbarium Library, the Arnold Arboretum Library, and the Tozzer Library of Archaeology and Ethnology.

For financial assistance during the preparation of this volume, we express our gratitude to Conservation International, the World Wildlife Fund, and Peter Thomson of Boston for their generosity.

We are sincerely appreciative of the patience, cooperation, and understanding of the editors and staff of Dioscorides Press during the preparation and publication of this volume.

Contributors

Janis B. Alcorn
Biodiversity Support Program
World Wildlife Fund
1250 24th St. NW
Washington, DC 20037

Edward F. Anderson
Desert Botanical Garden
1201 N. Galvin Parkway
Phoenix, AZ 85008

W. Balée
Institute of Economic Botany
The New York Botanical Garden
Bronx, NY 10458

Michael J. Balick
Institute of Economic Botany
The New York Botanical Garden
Bronx, NY 10458

Norman G. Bisset (deceased)
Chelsea Department of Pharmacy
King's College London
University of London
Manresa Road
London, England SW3 6LX

B. M. Boom
Institute of Economic Botany
The New York Botanical Garden
Bronx, NY 10458

Jan G. Bruhn
Department of Toxicology
Karolinska Institutet
Box 60-208
Stockholm S-104, Sweden

Eduino Carbonó
Herbario de la Universidad Tecnológica
    del Magdalena
Apartado Aéreo 731
Santa Marta, Colombia

R. L. Carneiro
Institute of Economic Botany
The New York Botanical Garden
Bronx, NY 10458

E. Wade Davis
1073 Clyde Avenue
West Vancouver, BC
Canada V7T 1E3

Memory Elvin-Lewis
School of Dental Medicine
Washington University
4559 Scott Ave.
St. Louis, MO 63110

William A. Emboden, Jr.
Severin Wunderman Museum
3 Mason
Irvine, CA 92718

Peter T. Furst
Department of Anthropology
University of Pennsylvania
Philadelphia, PA 19104

Ashoke K. Ghosh
Department of Anthropology
University of Calcutta
Calcutta, India

Charles B. Heiser
Department of Biology
Jordan Hall 142
Indiana University
Bloomington, IN 47405

Albert Hofmann
CH-4117
Burg-i-L.
Rittimatte, Switzerland

Bo R. Holmstedt
Department of Toxicology
Karolinska Institutet
Box 60-208
Stockholm S-104, Sweden

John O. Kokwaro
Department of Botany
University of Nairobi
Nairobi, Kenya, East Africa

Weston La Barre
Department of Anthropology
Duke University
Durham, NC

Walter H. Lewis
Department of Biology
Washington University
1 Brookings Dr.
St. Louis, MO 63130

Jan-Erik Lindgren
Department of Toxicology
Karolinska Institute
Box 210
S-171 77 Stockholm
Sweden

Frank J. Lipp
The Foundation for Shamanic Studies
P.O. Box 1939
Mill Valley, CA 94942

L. E. Luna
Swedish School of Economics
Helsinki, Finland

J. K. Maheshwari
National Botanical Research Institute
H.I.G.-130, Sector E
Aliganj Extension, Aliganj
Lucknow-226 020, India

Dennis J. McKenna
Aveda Corporation
4000 Pleasant Ridge Drive
Blaine, MN 55449

George R. Morgan (deceased)
Geography Department
Chadron State College
Chadron, NE 69337

Plutarco Naranjo
Universidad Central
Quito, Ecuador

Ong Hean Chooi
51, Jaban SS20/11
Damansara Utama 47400
Petaling Jaya, Malaysia

Mark J. Plotkin
Conservation International
1015 18th St., N.W., Suite 1000
Washington, DC 20036

Ghillean T. Prance
Royal Botanic Gardens, Kew
Richmond
Surrey, England TW9 3AB

Siri von Reis
Botanical Museum
Harvard University
26 Oxford Street
Cambridge, MA 02138

Carl A. P. Ruck
Department of Classical Studies
Boston University
745 Commonwealth Avenue
Boston, MA 02215

Judith Grace Schmidt
Institute of Economic Botany
The New York Botanical Garden
Bronx, NY 10458

Richard Evans Schultes
Botanical Museum
Harvard University
26 Oxford Street
Cambridge, MA 02138

Priyadarsan Sensarma
Department of Botany
Bangabasi College
Calcutta, India

Peter A. G. M. de Smet
Royal Dutch Association for the
    Advancement of Pharmacy
Alexanderstraat 11
2514 JL The Hague, The Netherlands

C. Earle Smith (deceased)
Department of Anthropology
University of Alabama

Victor Manuel Toledo
Instituto de Biología
Universidad Nacional Autónoma de
    México
Apartado Postal 70-275
04510 México, D.F., México

G. N. Towers
Department of Botany
University of British Columbia
Vancouver, BC
Canada

Nancy J. Turner
Environmental Studies Program
University of Victoria
Box 1700
Victoria, BC
Canada V8W 2Y2

R. Gordon Wasson (deceased)
Botanical Museum
Harvard University
26 Oxford Street
Cambridge, MA 02138

Garrison Wilkes
Department of Biology
Harbor Campus
University of Massachusetts
Boston, MA 02125

Gordon R. Willey
Peabody Museum of Archaeology &
    Ethnology
Harvard University
11 Divinity Avenue
Cambridge, MA 02138

PART 1:

General Ethnobotany

Although ethnobotany has only recently come into its own as a distinct academic component of the natural sciences, interest in the uses, symbolism, ritualistic, and other aspects of the practical, everyday interrelationship between people and plants has long been a definite part of human thought and activity. It must have had its numerous births in primitive societies when people depended on wild plants for most of their needs, but when settled life and agriculture began and numerous records could be kept, the dependency on plants assumed even greater importance.

Archaeological studies indicate the importance of ethnobotanical knowledge even before rudimentary writing evolved. This knowledge was preserved first verbally in songs and poems. Later these early oral beliefs were preserved in the Indian vedas and the earliest Chinese herbal, *Pen ts'ao kang mu* (1596). Even the Egyptian scrolls preserve the oral traditions of medicinal and other plant uses.

According to the late Agnes Arber (*Herbals: Their Origin and Evolution*, Cambridge University Press, 1986), the great student of ancient herbals, early ethnobotanical records, as well as later ones, reflect a particular bias:

From the beginning, the study of plants has been approached from two widely separate standpoints—the philosophical and the utilitarian. Regarded from the first point of view, botany stands upon its own merits as an integral branch of natural philosophy; whereas, from the second, it is merely a by-product of medicine and agriculture. At different periods in the evolution of the science, one or other aspect has pre-

19

dominated, but from classical times onwards, it is possible to trace the development
of these two distinct lines of enquiry, which have . . . converged though they have
more often, to their detriment, followed unconnected routes.

Thus, the men who composed the scrolls of the Egyptian pharaohs were early literary
ethnobotanists who collected and put down for posterity beliefs about "medicinal" plants.
Similarly, various Greek authors, including Theophrastus and Aristotle, established the
idea that each plant has a "psyche" or soul, a belief that survived in European thinking
well into the sixteenth century.

Western ethnobotany stemmed from the teachings mainly of Aristotle through Arab
translations in great part based on Aristotle, whom we might honor as the earliest "father
of Western ethnobotany." Much of the belief in the medicinal efficaciousness of plants that
permeates European herbals in the medicinal years can be directly traced to Greek ethnobotanical writings. Yet many herbals of the Middle Ages recorded local folk uses in
medicine of European plants and must be classed as records of general ethnobotanical
beliefs then accepted, particularly in ethnomedicine.

The same can be said of the records of other cultures. Five hundred years ago and
often earlier, the Chinese, for example, recorded extensive series of medicinal works
which influenced later publications. The early records of plants of India and their utilitarian value are found primarily in the vedas, long poems which have preserved many
ancient beliefs current throughout the nation.

The arrival of the Spaniards in the New World led, especially in Mexico, to numerous
written records of various aspects of general ethnobotany. Many of the uses and beliefs
concerning the local floras have survived to modern times. Our knowledge of much of the
ethnobotanical lore in pre-Conquest Mexico and Guatemala, however, does not come
from written records. Paintings, monuments, statues, and carvings, as well as archaeological plant remains, have given us a relatively wide insight into much of the ethnobotany of these regions.

An early ethnobotanical record of Brazil, written by Guilherme Piso (*História Natural
do Brasil Ilustrada*), was published in Holland in 1648. An early flora of French Guiana, prepared by Fusée Aublet and published in 1775 (*Histoire des Plantes de la Guiane Française*),
contains many ethnobotanical notes. Very few ethnobiological data, apart from edible
food plants, are known about early Peruvian agricultural or medicinal plants. The first
real ethnobotanical report from Peru and Chile is the diary of Spanish botanist Hipólito
Ruiz López (*Historico del Viage que Hizo a los Reynos del Peru y Chile el Botánico*, Ed. J. Jara-
millo-Orango, Real Academia de España, Madrid, 1952), who spent 11 years (1777–1788)
in these countries; his diary is full of the native uses and value of the plants and animals
of these Andean realms. In regions of North America colonized by the French and English, early reports of plant uses among indigenous peoples are sparse and are scattered
especially as incidental notes in general writing.

In contrast to the meager historical records of general ethnobotany, the twentieth century has seen numerous isolated research efforts on tribal ethnobotany in South America,
especially in Brazil, Colombia, Ecuador, Venezuela, and Suriname. At least eighteen specialists carry out field work in various parts of Colombia. One Ecuadorian ethnopharmacologist, Plutarco Naranjo, has published extensively on medicinal and toxic plants
employed by the Indians. Foreign ethnobotanists also carry out rather extensive field
work in Ecuador, primarily in general ethnobotany.

Several European universities and other organizations have engaged in ethnobotanical research in Asia and Africa. Among them is the ethnobotanical laboratory at the Jardin
Botanique in Paris, which, under the direction of Jacques Barrau, has been particularly
active in Southeast Asia. France's ethnobotanical community also is extremely active. Al-

though general ethnobotanical research has languished in Africa, John Kokwaro is one of the continent's few ethnobotanists now active in research. A most extraordinarily complete study of a larger sector of Africa, titled *The Medicinal and Poisonous Plants of Southern and Eastern Africa*, was published in 1962 by J. M. Watt and M. G. Breyer-Brandwijk.

In North America, ethnobotanical studies as a separate field had its birth with the coining of the word *ethnobotany* in 1895 by John Harshberger. Melvin Gilmore, Volney Jones, Richard Ford, and their students of the ethnobotanical laboratory of the University of Michigan have carried out general ethnobotanical research in numerous North American tribes. Other institutions furthering research on the continent include the University of New Mexico (Edward Castetter), the University of Victoria (Nancy Turner), and the Harvard Botanical Museum (Oakes Ames, Paul C. Mangelsdorf, R. E. Schultes, Margaret Towle, Dorothy Kamen-Kaye, and Wilma Wetterstrom).

What we have designated as general ethnobotany in this volume—the discovery, enumeration, and evaluations of uses of plants in primitive societies—is often not considered "scientific enough." It is, nevertheless, the basis of any and all types of ethnobotany and will continue to be of both academic and practical value.

The five contributions included in this section concern themselves with the definition and scope of ethnobotany. Janis B. Alcorn examines the object of ethnobotanical enquiry and the aims with which that enquiry is carried out, emphasizing the application of ethnobotanical data to development. E. Wade Davis traces the growth of the new synthesis in ethnobotany to an interdisciplinary orientation and discusses its potential. Frank J. Lipp reviews the conceptual and logical basis of ethnobotanical research as a preliminary step towards synthesis of ethnobotanical method and theory. Ghillean T. Prance enumerates four developments other than cataloging that add to the scientific and global importance of ethnobotany and that will become the future of ethnobotany: an interdisciplinary approach, the addition of more ecology, the study of peasant agriculturists, and quantitative ethnobotanical inventory. Finally, Priyadarsan Sensarma and Ashoke K. Ghosh clearly distinguish ethnobotany from phytoanthropology and identify their connotations and respective areas of research.

Not included in this section are four significant articles published elsewhere. In 1941, Volney H. Jones, in "The Nature and Status of Ethnobotany" (*Chronica Botanica* vol. 6, no. 10, pp. 219–221), pointed out that ethnobotany, which may appear to infringe on other scientific disciplines, differs in that the latter focus primarily on the application of knowledge gained to problems involving the contact of people and plants whereas ethnobotany is concerned exclusively with the interrelation of people and plants and takes for its sole aim the illumination of this contact. The second article, by F. Raymond Fosberg in 1948, "Economic Botany—A Modern Concept of Its Scope" (*Economic Botany* vol. 2, no. 1, pp. 3–14) considered the broadening scope of economic botany; once limited to the descriptions of economically important plants, in 1948 it encompassed careful taxonomic work, the application of physiological principles, recognition of ecological relationships, and genetics. The third article by Albert F. Hill appeared in 1948; "Ethnobotany in Latin America" (*Chronica Botanica* 16:176–180) outlined a definition of ethnobotany and discussed methods of accumulating new material in this discipline. Finally, an article by Michael J. Balick in 1991, titled "Ethnobotany for the Nineties" (*The Public Garden*, vol. 6, no. 3, pp. 11–13), addressed the influence of molecular biology and other laboratory-oriented botanical studies, which, as they increase, tend to lead to less stress on field work.

The Scope and Aims of Ethnobotany in a Developing World

JANIS B. ALCORN

In its early days, ethnobotany was implicitly shaped by imperialist motives (e.g., Brockway 1979): collectors were sent to gather useful plants from areas occupied by traditional cultural groups and the collected plants were used for commercial exploitation by the modern world. In today's developing world, however, ethnobotany is shaped by an explicit concern for collecting data within a framework whereby those data will contribute to the development of all classes in all nations and especially to planned development in the region from which specific data are collected. Policy-makers need information about economically valuable natural resources and the ways in which those resources are used so they can predict the outcome of development programs and facilitate the development and introduction of new, locally adapted crops and agricultural techniques (Alcorn 1992). Local problems facing policy-makers often need local solutions; the general problems of Malaysia and Brazil, for example, may be similar, but the solutions that work in one region often fail in another. Policy-makers and development planners need information about specific countries to create solutions that work in each nation's context. Today's ethnobotany responds to these needs.

This chapter examines the object of ethnobotanical inquiry and the aims with which that inquiry is carried out, emphasizing the application of ethnobotanical data to development. While citing representative examples of ethnobotanical research, this chapter does not attempt a history of major contributions to the field. Instead, the process of ethnobotanical inquiry is used as a vehicle to demonstrate the complex interrelationships within the ethnobotanical matrix.

What is Ethnobotany?

The scope of a field of inquiry is reflected in the questions asked by its researchers. Specific questions asked by ethnobotanists include the following: What plants are available? Why are they available? What plants are recognized as resources? What social, political, biological, economic, and ecological factors cause particular plants to be perceived as resources? How does the use of a certain set of resources influence the use/availability of others? How is ethnobotanical knowledge distributed among the human population? What do people think about plants? How do they differentiate and classify elements of their natural environment? From what resource zones are plant products harvested? How are they used? What are the economic and financial benefits derived from plants? How are populations of resources maintained? What effect does their management have upon the structure of local vegetation? What effect does their management have upon the structure and functioning of local institutions? What factors influence resource management decisions and thereby affect local plant populations? How have human activities and their consequences influenced the evolution of local plant populations? For what purposes are resources needed? To what stressors are human populations adapting? Are human choices of particular resources adaptive? If so, to what are they adaptive? How are human adaptive strategies affected by change? What changes are presently occurring, and what changes have taken place in the past?

As can be seen from these questions, modern ethnobotany is concerned with the "totality of the place of plants in a culture" (Ford 1978). It is the study of plant-human interrelationships embedded in dynamic ecosystems of natural and social components. Put another way, ethnobotany is the study of contextualized plant use. Plant use and plant-human interrelationships are shaped by history, by physical and social environments, and by inherent qualities of the plants themselves. The object of ethnobotanical inquiry is actually a sort of "text" (sensu Ricoeur 1971), the meaning of which is derived partially from the natural, social, and cultural contexts in which that text is played out. An ethnobotanical text evolves around a human community's use and management of vegetation.

Both plant and human "actors" have roles in the script of the ethnobotanical text. The term *role* refers both to role characterization and role action, as the following simplified example from a limited section of ethnobotanical text demonstrates. Humans eat "plant X." They plant, cultivate, and harvest "plant X." Thus, the human role is that of farmer and consumer; the role of the plant is that of food and crop. In this particular example, the actions of humans as farmers actively influence the evolution of "plant X" in a process called *domestication*. Thus, "plant X" plays the interactive role of domesticand. It also plays the role of commercial item. It is sold in markets, and its harvest dates and storage needs influence the daily lives of those who play the role of traders.

In a given ethnobotanical text, then, plants are used and, as a consequence, plant communities are disturbed and human lifeways are modified. As the text is acted out, the contextual environment of activity changes. For example, trade in "plant X" may lead to deforestation that affects the production of "plant X." As a consequence, "plant Q," which is better adapted to agricultural systems not dependent on the forest, slowly replaces "plant X" in the role of food. "Plant X," however, still is used medicinally and persists at the edges of the agroecosystem. Its name changes, and its ceremonial usage shifts. In its new role as weed and medicine, "plant X" is managed differently. The human role in "plant X's" evolution also has changed, but genetic traces of past domestication action remain.

Changes in the use of plants influence human strategies, and new plant-human interactions are initiated as "using" changes. Change influencing the ethnobotanical text comes also from outside the immediate setting. New technologies, new plants, and new politico-economic exigencies modify the "stage," enter the text, precipitate new roles, and produce new directions in the text's "story."

The roles played by plants reflect the biological and physical properties of the plants, the biological and perceived needs of humans, the natural and anthropogenic communities of which the plants are a part, and the genetically limited responses of plants to human disturbance. Species that are elements of the vegetation disturbed by humans are each manipulated and used in different ways. At the same time, different members of the human community participating in this interactive text vary slightly in their behavior. Each person uses his or her personal knowledge of plants, agriculture, house building, and medicine in a slightly different manner, and each responds somewhat differently to changes in the environment. Nonetheless, variation plays around clear, central patterns that define the ethnobotanical text.

The aims of ethnobotany are twofold: (1) to document facts about plant use and plant management and (2) to elucidate the ethnobotanical text by defining, describing, and investigating ethnobotanical roles and processes. Through these aims, ethnobotanists seek to understand the dynamics of the system of which plant use and plant management are a part.

The pursuit of these two aims is contextualized by an applied goal: the development of new plant-derived products; new or improved cultigens; and natural-resource-conservative, sustained-yield agroecosystems adapted to meet local needs and conditions. To achieve their aims, ethnobotanical researchers apply methods of diverse disciplines. Botanists, anthropologists, geographers, chemists, pharmacologists, and soil scientists are among those pursuing research in the ethnobotanical realm. As in any ecologically oriented endeavor, input from numerous disciplines is necessary. Basic to good ethnobotanical research are solid data about plants, people, and the environment. An integrative focus, however, is also fundamental, if the full potential of ethnobotanical inquiry is to be realized. The burden of integrating data lies upon the ethnobotanical field worker.

The ethnobotanical field worker observes the living system of which plant use is a part. Careful observation is not, however, an easy job. It requires looking at human activities and organizations from a plant's viewpoint and looking at plants and plant communities from a culturally informed as well as a scientific perspective. Only by looking at the material in this way can the ethnobotanical field worker recognize links between the subunits investigated by researchers in different disciplines. Methods of different disciplines reveal different aspects of the system under study much as different stains of the same material prepared for viewing under a microscope reveal different biological structures. It is up to the field-experienced worker to integrate the data and spot new areas requiring investigation.

Coming from an industrial society where plants are primarily esthetic parts of the urban environment and secondarily known to provide food, raw materials, habitats for wildlife conservation, and genetic resources for bioengineering, the field worker must undergo a process of reeducation to see an environment shaped by plants and landforms, an environment where plants are an integral part of human lives. This reeducation is easier for those trained in field biology, but these same individuals must overcome a bias to ignore or to classify simplistically the human component of the ecosystem under study (Anderson 1952; Posey 1984).

As an outsider, the newly arrived, Western-educated field biologist in a nonindustrialized society sees an environment of individual plant and animal species grouped into general associations. If educated in something other than field biology, the Western outsider sees the environment in units of vague vegetation gestalt-types, such as jungle, desert, garden, or fallow. The outsider also sees waterways, topographic variation, and soil types isolated in time and isolated from each other.

The indigenous person, on the other hand, is an insider and sees an integrated association of biotic and abiotic elements interacting over time (see Alcorn 1989b). Thus, for the insider, a particular plant's chemistry, phenology, community associates, habitat, uses, cultural value, mythic associations, growth curve, physical properties, and usual management come to mind in a constellation of associated knowledge whenever the plant or its name is encountered in walking, thought, or conversation. The insider also is enculturated to his or her place in society, the culture, and the politico-economic setting of his or her people and family. This knowledge is reflected in the environment "seen" by the insider on daily walks, but it is an aspect of the environment invisible to the outsider. It constitutes an unspoken part of the insider's conversation, unheard by the outsider; it is one of the "languages" that the ethnobotanical field worker aims to learn.

In nonindustrialized regions of the world, the knowledge of indigenous people about their biological environment is extensive. For example, Berlin (1978) has estimated that among the two American Indian groups with whom he has worked, biological knowledge is greater than all other domains of knowledge combined. In such regions, plants are a very important part of an individual's day-to-day environment, and plant behaviors, habit, and parts are used as resources. In addition, plants, like rocks, soil, and weather, are objects for thought as well as for use. Some ethnobotanists have stressed this by defining plant use as "behavioral response to plants" (Hays 1974). As important objects in the immediate environment, plants and their "behaviors" (for plants are not unchanging objects) are named, classified, studied, interpreted, and responded to.

The plant classification of a particular group reflects human cognitive patterns (Berlin 1973), utilitarian concerns (Hunn 1982; Morris 1984), the cultural salience of particular plants and the subsistence concerns of the group classifying the plants (Brown 1985). Plant forms and life cycles lend themselves as material for symbol and metaphor. Plants and plant communities give structure to the environment experienced by humans, they serve as landmarks for present locales and past events, and they give their names to local areas. Plants are included in mythologies and legends according to their characteristics and their cultural importance.

Long-term changes in vegetation affect culture and language (e.g., Meggers 1977; Hebda and Mathewes 1984). Particular plants may exert dominant influences upon cultural beliefs and art, be they hallucinogen (e.g., Reichel-Dolmatoff 1971; Dobkin de Rios 1974) or major crop plant (e.g., Hanks 1972; Nigh 1976). Their phenology may shape the yearly round of festivals (e.g., Guyot 1975). Their behavior is read as portents of the weather, harvests, or community health, and their responses to manipulation serve for divination (e.g., Alcorn 1984b). Plants also provide the raw materials for material culture and for economic livelihoods.

Ethnobotanical knowledge of these matters is held by individuals and constitutes one of the more important types of information transferred along information networks. It also is encoded in subsistence systems and cosmologies (Netting 1974; Reichel-Dolmatoff 1976). While individuals may not be aware of specific information thus encoded, information important for survival is ensured passage from generation to generation. Ethnobotanical knowledge is not only learned from others; it also derives from personal investigations that reflect indigenous epistemological methods. It is ideal material for the study of how adaptively important knowledge is shared and generated (e.g., Moore 1981), but with few exceptions (e.g., Stross 1973; Messer 1975) researchers have not studied ethnobotanical knowledge for these ends. The investigation of a region's ethnobotany includes the study of all these things as well as the investigation of resources, resource zones, and resource maintenance.

All these factors impinge upon the ethnobotanical text, but the economic and ecological aspects of ethnobotany have more pressing importance for the purposes of development planning. These latter aspects of ethnobotany have shaped the thinking of ethnobotanists in developing countries as they have sought to answer the questions "What is ethnobotany?" and "For whom is it practiced?" (e.g., Barrera 1979; Gómez-Pompa 1982; Sarukhan 1985). The remainder of this chapter focuses on those aspects of ethnobotany of vital importance to a developing world.

Entering the Ethnobotanical Matrix: What Good is This Plant?

While there are many potential means of entering a region's ethnobotanical matrix, one of the best remains the original ethnobotanical focus on plant use: What good is this plant? This question is one the investigator shares with indigenous collaborators, thus making conversation easy, and it is one of interest to planners seeking an inventory of their nation's natural resources. This entry path begins at the stage of identifying for what purpose plants are being used, and which plants are being used. After learning which plants are resources, the investigator then can answer other, more complex questions, but even at that deeper level the same initial question shadows the investigator's steps. The following discussion looks at ethnobotany from the perspective of this question, and in the process of outlining the scope of ethnobotany also sketches some of the interrelationships that develop between different subplots of the ethnobotanical text.

One of the most important uses that plants are "good for" is as producers of unique and diverse chemical compounds. The investigation of the use of plant-derived chemicals has been an important part of ethnobotany (e.g., Schultes 1960; Schultes and Hofmann 1980). In seeking to document indigenous knowledge about smells, tastes, and biological effects of plants, ethnobotanists contribute to the development of new chemical products ranging from sugar substitutes (Compadre et al. 1985) to pharmaceuticals (e.g., Swain 1972). The ethnobotanical field worker aims to provide full information necessary for ethnopharmacological investigations (Bruhn and Holmstedt 1981; Croom 1983) as well as to describe the health beliefs and medicinal system that form the context in which plant-derived medicines are used.

Eighty percent of the world's population relies on traditional medicine to maintain its health (Weragoda 1980). In recognition of the political, economic, and social barriers slowing the delivery of modern biomedical health care to most of the world's population, the World Health Organization (WHO) has embarked upon an ambitious program to evaluate herbal medicines (WHO 1978; Penso 1980; Akerele 1985). This project ultimately hopes to circumvent the problems of developing and distributing appropriate pharmaceuticals (McDermott 1980) by encouraging the cultivation and use of locally adapted medicinal plants with proven empirical value.

While claims are made as to the breadth and accuracy of indigenous peoples' medicinal knowledge, only preliminary attempts have been made to assess the value of a particular cultural group's total pharmacopeia (see Domínguez and Alcorn 1985; Jiú 1966; Ortíz de Montellano 1975; Alcorn 1990a). Ackerknecht (1942) estimated that while 25 to 50 percent of the world's ethnopharmacopeia was composed of "active" drugs, a much smaller percent was used in a way that suggested its specific activity was empirically valuable. The evaluation of indigenous medicine's efficacy is difficult (Alcorn 1984b).

Ethnomedical investigations have grown, but they continue to be heavily anthropological in nature. One of the challenges facing ethnobotanists is to integrate the chemical, pharmacological, and ecological study of plant-derived remedies into ethnomedical studies that focus on the social and symbolic functions of folk medicine. This is an example of the way in which the coordination of ethnobotanical and anthropological studies can enhance the effectiveness of both and at the same time make unique contributions to understanding human ecology.

As the concern for food self-sufficiency grows in developing nations (e.g., George 1977; Toledo 1985), so does interest in indigenous foods. Although documentation of local foods has a long ethnobotanical tradition, researchers since the mid 1900s have also focused on collecting the genetic resources of major crops (e.g., Hawkes 1983) and have studied the classification, selection, and dissemination of cultivars (e.g., Brush et al. 1981; Boster 1984). Nonetheless, wild foods continue to make important contributions to local diets and local economies (e.g., Messer 1972; Fleuret 1979) and are, therefore, important to food self-sufficiency.

As Western farming practices are adopted, however, the availability of nutritionally important wild greens and minor crops diminishes, and the nutritional quality of rural diets deteriorates (e.g., Dewey 1981; DeWalt 1983). The Western-style agricultural practices accompanying a shift to commercial agriculture weaken plant-human inter-relationships by reducing the diversity of plants available in the environment. Food choice reflects many factors. Ethnobotanists provide the data on locally important foods—their availability, maintenance, and processing—that must be considered together with data on nutrition, social interactions, and economic contexts if human ecologists are to for-mulate and test meaningful hypotheses about the reasons for food choice (e.g., Tanaka 1976; Ellen 1979; O'Connell and Hawkes 1981; Pulliam 1981).

Construction materials also are documented by ethnobotanical workers. The best way to investigate these resources is to study construction as it is being done, observing the selection, harvest, and preparation of building parts (e.g., Villers et al. 1981). In this way, alternative resources are documented, the reasons for choosing particular species can be investigated, and the resource zone from which harvesting is done is witnessed. While knowledge that a plant is used as a construction resource may tell us something about the plant's physical properties, factors other than its resistance to rot, its flexibility, or its weight may enter into resource choice (e.g., Alcorn 1981a). All the factors considered in construc-tion material choice constitute important data about a region's ethnobotanical text.

Particular craft and tool requirements vary from culture to culture and region to region. Dyes, baskets, rope, traps, bark beaters, nets, and cloth are but a few of the items made from plants. Unless these items are made for sale or are of special ceremonial significance, resources for their production are rather minor, and plant-human inter-relationships with them are usually weak. Factors influencing land use decisions may have a particularly strong influence on the availability of alternative craft resources and thereby upon resource choices. On the other hand, ethnobotanists may uncover other interrelationships. For example, Weinstock (1983) found that the commercial exploita-tion of wild craft resources may disrupt agricultural systems in such a way as to disrupt production of both agricultural crops and the craft material itself.

Plants valuable for livestock or wildlife feed and shelter have received relatively lit-tle attention by ethnobotanists, although indigenous consultants often reveal details about animal habits and diet, if asked, especially if hunting is important locally. Because animal foods are not a direct human use of plants, consultants may not stress this aspect of a plant's value. Nonetheless, such information could be valuable for understanding eth-nobotanical linkages in the ecosystem, and ethnobotanists should attempt to document it. Game or livestock management considerations may influence vegetation management decisions in unexpected ways. In areas where game management is of interest, such plant use data can be invaluable for preservation of game stocks for sport, sale, or direct con-sumption. In addition, nonpasture plants traditionally used for livestock feed sometimes have great potential for development into new feedstocks (e.g., Peters and Pardo-Tejedo 1982).

Investigation of firewood resources has been intense in the 1970s and 1980s due to concern over deforestation and the deepening firewood crisis (Eckholm et al. 1984). Na-tional and international organizations have collected ethnobotanical and biological data on the world's firewood resources (e.g., National Academy of Sciences 1980) to guide the selection and promotion of fuelwood crops in developing countries, but since the rea-sons for the firewood crisis vary from region to region, the solutions will vary also. Eth-nobotanical field workers document the types of firewood preferred for particular tasks, the amount of firewood used locally, where and how firewood is being harvested, how firewood supplies have been maintained traditionally, and reasons for local firewood shortages. They thereby aim to provide basic information vital for the successful resolution of firewood problems in specific regions.

Economic and financial benefits derived from the use, sale, and maintenance of plants are receiving increasing interest as researchers investigate incentives to conserve biodiversity (e.g., McNeely 1989). For example, Peters et al. (1989) have documented the potential financial returns from commercial exploitation of the traditional, nontimber forest products of Amazonia and Indonesia. Others (e.g., Vasquez and Gentry 1989) have investigated the ecologically destructive impacts of increased extraction of wild forest products in response to expanding markets. Surprisingly few studies, however, have focused on the economic and financial returns that traditional communities derive from maintaining a biologically diverse environment (e.g., Alcorn 1989a; Anderson 1990). Such studies demonstrate that indigenous peoples include in their cost-benefit analyses of alternative land uses the existence value of biodiversity, the option value of biodiversity, and the economic value of the ecological services provided by biodiversity, in addition to the use value of biodiversity.

Ethnobotany of Agroecosystems

Gathering information about what plants are used for and why is just the first step in the investigation of a region's ethnobotany. It is the study of resource zones and plant management systems that moves us directly into the area of ethnobotanical dynamics. Traditional means of resource management promote rich ethnobotanical texts. Within traditional agroecosystems, human activities influence both the crops and the natural vegetation occupying the region (see Posey and Balée 1989). By studying these systems, we learn how the use of plants, and thereby factors affecting the use of plants, can alter plant populations. Such knowledge is of immediate interest for development efforts concerned with such things as weed control, development of crops for "marginal" lands, conservation of crop genetic resources (e.g., Oldfield and Alcorn 1987), and sustained agricultural production (e.g., Alcorn 1990b).

Ethnobotanists have long been interested in the effects of plant management upon plants. The study of domestication, the dispersal of cultigens, the evolution of weeds, and the dispersal of weeds have been the central foci of many ethnobotanical studies. The domestication of cultigens has depended upon plant responses to human activities as well as upon human reaction to plant qualities and responses to management (e.g., Rindos 1984), and not all plants are preadapted for domestication (e.g., Kupzow 1980).

In the tropics, the ranges of plant management activities (e.g., Harlan 1975; Alcorn 1981b) and responsive effects are particularly wide. While some plants associated with agroecosystems evolved into weeds in response to human disturbance patterns, crops and weeds are not the only plants bearing genetic traces of ethnobotanical processes. Wild populations of many plants often are part of agroecosystems and, as such, are affected by agricultural decisions and practices aimed at crop species (e.g., Bye 1979, 1981; Ford 1981; Chacón and Gliessman 1982; Alcorn 1984b). Weed associations common to a given region, for example, reflect the type of agricultural disturbance occurring in that region (e.g., Kellman 1980). Thus, choosing to grow a particular crop and the selection of the manner in which it is to be cultivated influence much more than the crop. Wild populations (weed and nonweed) are, in addition, often protected or subjected to direct selective pressures (e.g., Harlan 1975; Bye 1979; Gordon 1982; Zizumbo and García-Marín 1982; Alcorn 1984b).

Both agricultural and nonagricultural activities disturb animals, which act as pollinators, herbivores, or disseminators, and thus disturb the context and selective agents affecting plant population dynamics and community structure. For example, factors influencing the placement of trails through forested areas between agricultural fields may influence the traffic lines of fruit dispersers and hence species distribution (e.g., Palmeirim and Etheridge 1985). Village placement can affect vegetation years after the villagers have left (e.g., Lambert and Arnason 1978; Sobey 1978). Knowledge of the factors influencing the placement of agricultural and nonagricultural activities and structures can help locate ethnobotanical links in a particular local system. Studies of the effects of human activities upon "wild" plants have only begun to probe these interesting aspects of the ethnobotanical text.

The ethnobotanical field worker, knowing factors that affect plant population dynamics, is mentally equipped to recognize and classify plant manipulation behaviors from an "effects" perspective. Agricultural activities include the disturbance of existing plants *en masse*, as vegetation (ranging from deforestation and the drainage of marshes to weeding activities), as well as the planting and protection of selected, individual plants. By focusing on how people manipulate individual plants in the context of *en masse* vegetation management, the ethnobotanist can begin to identify ethnobotanical guild groups and focus upon the impact that human activities have upon particular species and communities (e.g., Bye 1978, 1981; Alcorn 1984b).

Techniques of *en masse* vegetation management and individual plant management (Alcorn 1981b) vary in the intensity of their effects upon plants. We are, however, only beginning to understand the range of effects these techniques have upon plants. For example, one of the commonest, oldest, and most obvious *en masse* vegetation management practices is the use of fire (Bartlett 1956), but while fire's general impact has been the subject of much study, we must investigate the specific impact of fire management and timing under local conditions (e.g., Yellen and Lee 1976; Smith 1977; Uhl et al. 1982) to understand the role of this type of *en masse* vegetation management in the local ethnobotanical text. In some areas of the world, "fire-stick farming" (Jones 1969) is still used to ensure the presence of wild biological resources.

Fire is more commonly used in the context of "slash-and-burn" (or swidden) agriculture, an old and complex but still widespread form of farming (Cox and Atkins 1979) that utilizes fire to trigger and harvest the products of natural regenerative processes. Swidden agriculturalists farm vegetation, not soil; traditional slash-and-burn subsistence systems utilize successional plants and the successional process as agricultural resources (Alcorn 1989b). Cycles of successional vegetation (cycles of fallows) are used to capture nutrients from deep subsoil and from atmospheric dust (Harcombe 1980; Kellman 1980; Williams-Linera 1983), and then controlled burning of the successional vegetation is used to release these nutrients for agricultural use. The impact of this use of plants upon successional and climax vegetation is of potentially great ecological and economic importance (e.g., Gómez-Pompa 1971; Alcorn 1981a) but is as yet seldom studied.

It has been said that pre-industrial people make their living as "applied ecologists" (Gordon 1982). Much of the ecological knowledge they apply, however, is encoded in the subsistence system that they inherited from others (Alcorn 1989b). The agroecosystem reflects an ecological wisdom of which individual farmers may be unaware and which can be fully appreciated only by soil scientists, agronomists, entomologists, and other biologists who study them. More plant ecology studies need to be done in the context of traditional agroecosystems for fuller understanding of any region's ethnobotanical text. Ethnobotanical studies done by those with backgrounds in geography have focused upon describing these systems and the structures they create (e.g., Kimber 1973; Denevan 1980). The field ethnobotanist's primary description and assessment of these systems delineates the system and provides the questions upon which specialists can focus.

The land managed under a traditional system of agriculture usually contains many different patches: gardens or fields of various ages containing different combinations of cultigens and wild plants according to the farmer's decisions and the needs of his or her family, as well as wild vegetation in fallows (also of varying ages and composition depending on the plot's history) and in woodland or grassland (e.g., de Schlippe 1956; Conklin 1957; Kunstadter 1978; Posey 1983; Alcorn 1984b; Denevan et al. 1984). The vegetational diversity created by traditional systems apparently functions to protect the agroecosystem from pest outbreaks (e.g., Altieri and Letourneau 1982; Risch et al. 1983; Speight 1983; Altieri 1984). Traditional agroecosystems provide systems within which entomologists can identify potential pest problems before they become major problems as agriculture is "modernized" (Altieri 1984).

Plant management practices also affect soil and water dynamics. In some cases, agricultural practices prevent soil loss; in others, they accelerate it. Management of specific plants or *en masse* vegetation management can cause soil changes (e.g., Furley 1975; Allen 1984) that feed back into effects upon vegetation, then upon its management.

Indigenous maintenance of wild resources is, in effect, indigenous "conservation" of wild species. Traditional farming systems often create fallows and other corridors or patches of natural vegetation that act as resource zones, biodiversity conservation units, and gene banks for the regeneration of natural communities (e.g., Alcorn 1984a; Denevan et al. 1984; Lynch and Alcorn 1991). The study of resource conservation is an area of ethnobotanical inquiry that reaches into the heart of the ethnobotanical text and contributes to reaching the applied goals of the ethnobotanical endeavor. By identifying ecologically adaptive aspects of traditional resource management, ethnobotanists contribute to the development of resource-conservative, sustained-yield agroecosystems (e.g., Hart 1980; Gliessman et al. 1981; Wilson and Kang 1981; Hanks 1984; Alcorn 1984a, 1990c).

Another ethnobotanically important aspect of agricultural systems with development applications is experimentation. Traditional agricultural systems usually include places for a process by which new agricultural methods or cultigens are tried out (e.g., Johnson 1971; Vermeer 1979), as well as room for individual variation between farmers (e.g., Alcorn 1984b). Experimental activities mediate a region's ethnobotanical dynamics by controlling what, where, and how certain plants or techniques are incorporated into the existing agroecosystem. Such pathways, if known, also can be used by agricultural extension agents as a means to achieve farmer participation in evaluating and accepting "new" recommendations (Vermeer 1979; Alcorn 1992).

The Human Ecology Connection

Ethnobotany can play a major role in facilitating the study of human ecology. This is perhaps best reflected in the role that paleoethnobotany or archaeoethnobotany has played within archaeology to elucidate human ecological relations of the past. Because archaeologists are forced to look carefully at the impact of humans on their physical and biotic environment and to reconstruct what social conditions and ideologies could have contributed to the environment, they have explored the gamut of interrelations between people and plants.

Ethnobotanists have a venerable record of contributing to the interpretation of archaeological sites (Dimbleby 1978; Ford 1979). Modern ethnobotanical data are often used as analogs for interpreting past plant-human interrelationships. Botanical data have served as linchpins for theories about the life ways, social and economic history, and health of past peoples. The structure and methods of "contextual archaeology" described by Butzer (1982) mirror that of modern ethnobotany. Archaeologists draw connections between isolated artifacts of past lives to infer the interrelationships between natural and social systems and "to understand the human ecosystem defined by that systemic intersection" (Butzer 1982). Ethnobotanists do not use their data to reconstruct a static environment isolated in time; they use data to reconstruct the dynamics of plant-human interrelationships and their ecological consequences (e.g., Folan et al. 1979; Ford 1979; Voorhies 1982; Pearsall 1983; Peters 1983; Wiseman 1983).

Whether ethnobotanists focus on the uses of a particular plant, plants employed for a special purpose, reasons for plant use, the integration of plants into agroecosystems, evolutionary effects of plant-human interactions, markets, dynamics of plant knowledge, or plant classification, they contribute to elucidating aspects of a text also of interest to human ecologists. Human ecology and ecological anthropology are concerned with human adaptation to natural, social, and physical environments through the use of knowledge and culture (Bennett 1976; Orlove 1980). If ethnobotanists use appropriate anthropological approaches to contextualize their research, the ethnobotanical text they describe will provide a structure that includes junctures around which data from anthropologists, economists, and others can be integrated. Ecological anthropology studies are often handicapped by a failure to differentiate important elements of the biological environment or are limited by lack of data about specific resources.

By providing data about the plants of the ecosystem under anthropological study, ethnobotanists can contribute to more sophisticated analyses of human ecological relations. By providing details about the people in plant environments, anthropologists can, in turn, add to our understanding of the ethnobotanical text and its dynamics. For example, anthropological information already has contributed to our understanding of how social and ideological systems have affected the evolution and maintenance of crop plant diversity (e.g., Brush et al. 1981; Boster 1984). As applied human ecologists, economic and social anthropologists have been granted an expanded role in rural development projects (e.g., Barlett 1980; Baker et al. 1983; International Potato Center 1984), but ethnobotanists have rarely been given the opportunity to demonstrate the value of their contribution as members of such development teams (e.g., Salick 1990).

Ethnobotany for Development and Policy Planning

Ethnobotanical knowledge is important for development planners and policy-makers who devise solutions to local and regional problems (e.g., Brokenshaw et al. 1980; Chambers 1983; Alcorn 1984a, 1990b, 1992; Groenfeldt et al. 1990). An understanding of ethnobotany is critical for assessing the ecological relations within human-manipulated ecosystems. Knowledge of resources and successful resource management/harvesting systems provides information about human adaptation to the social and natural environment: what people are having to adapt to, how they are adapting, and what the consequences of their adaptations are.

Ethnobotanical studies also identify the effects of public policy on plant resources and their uses (e.g., DeWalt 1982; Alcorn and Molnar 1990; Lynch and Alcorn 1991). They document the negative results of certain types of development activities, thus enabling policy-makers to predict the effects of new policies under consideration and to formulate programs to circumvent the predicted negative impacts. Innovations build upon existing systems, and the aim of ethnobotanical studies is to provide the understanding of extant beliefs and decision-making factors that policy implementors and development planners need to introduce changes successfully (Brokenshaw and Riley 1980). Ethnobotanically informed understanding of human ecology is therefore prerequisite for designing rural development programs that work.

It is the interdisciplinarian ethnobotanist who can translate the complexities of plant-human interrelationships into terms appreciated by policy-makers. The aim of elucidating the ethnobotanical text ultimately means identifying patterns at the local, regional, and global levels—the discrete levels upon which planners focus. Once these patterns are determined, a computerized program for policy planners could be developed from the general model of the ethnobotanical text. Such a program would prompt collection of necessary local, regional, and "external" data and guide their interpretation to predict developmental problems and to suggest new paths of development at local, regional, and national levels. Such a program would be of value for those interested in a variety of aspects from agriculture and food issues to health issues and industrialization. The program could be designed to interact with social statistics, economic, floristic, and plant use databases, and might best be built as an "expert systems" program based on indigenous knowledge systems.

Ethnobotanical field workers can, as individuals, play a critical role in participatory economic development (sensu Chambers 1983) because they are in a unique position to catalyze interactions between others and the people whose knowledge they document. For example, Gary Nabhan, by founding the Native Seeds conservation network and writing popular books about North American ethnobotany (e.g., Nabhan 1982), has facilitated and stimulated interactions between North American Indians and scientists to solve shared problems. As one of the few professionals whose job it is to link people of old cultures to people of modern culture, ethnobotanists also may be able to influence how modern culture negotiates or averts the impending "end of nature" (McKibben 1989). Modern culture bearers, alienated from nature, desperately need to hear and heed the wise council of ancient culture bearers who understand the interdependent relationships between humans and nature. This communication, or its failure, occurs among people of the current generation. Ethnobotanists face the challenge of stimulating this communication.

There are hopeful signs in the increased participation in ethnobotany by indigenous peoples (e.g., MFM/STC 1980; Berlin 1984; Posey 1990), by institutions funded by the governments of developing countries (e.g., Gómez-Pompa 1982; Posey 1984), and by international committees (e.g., the Ethnobotany Specialist Group of the International Union for Conservation of Nature and Natural Resources). Indigenous people have joined the new "global coalition for biological and cultural diversity" that emerged from the 1990 Kunming Congress of the International Society for Ethnobiology. UNESCO–MAB (Man and the Biosphere Program) and the United Nations Environment Program are collaborating with national governments and local farm communities to initiate pilot projects conserving and developing traditional agroecosystems within MAB biosphere reserves through the integration of scientific and indigenous ethnobotanical knowledge (UNESCO–MAB 1984). Ethnobotanists from developing countries are collecting ethnobotanical data for use in making policy recommendations (e.g., Toledo et al. 1976; Posey 1984; Toledo and Barrera-Bassols 1984; Toledo 1985).

Policy-makers are beginning to give overdue respect to indigenous and peasant farmers as appreciation grows for the methods by which traditional agroecosystems meet local needs by integrating the maintenance of diverse crops and wild resources (Caballero N. 1979; Posey 1983; Oldfield and Alcorn 1991). In these systems lie potential answers for solving the dilemma (Daly and Cobb 1989) of how to integrate economic development, biodiversity conservation, national food self-sufficiency, and sustained agricultural production.

Today's ethnobotany offers development planners more than useful information about economically valuable plants and ecologically sustainable agricultural techniques. Policy-makers are beginning to respect the authority of indigenous and peasant communities and are seeking ways to build successful partnerships with them. Communities and government are increasingly recognizing that strong, local, common property regimes backed by the power of the state play a critical role in sustainable resource management. Community organizations have historically regulated access to biological resources and thereby mediated ethnobotanical relationships. On the cutting edge of ethnobotanical research today are two meta-questions that focus on the relationships between plants and socio-political institutions: How are common property regulations modified by the commercialization of plant products for international markets, and how does that, in turn, affect ethnobotanical relationships? How do local and state authorities prevent, or contribute to, overexploitation of biological resources?

Just as the voices of indigenous organizations now reach the world media, international conservation organizations, bilateral donors, and multilateral development banks, these voices also are reaching the ears of ethnobotanical researchers. As those voices are heeded by researchers who respond to local needs, the ethnobotanical text, as published in academic halls and development agencies, will assume more policy-relevant dimensions.

LITERATURE CITED

Ackerknecht, E. H. 1942. Problems of primitive medicine. *Bull. Hist. Med.* 11: 503–521.

Akerele, O. 1985. The W.H.O. traditional medicine program: Policy and implementation. *International Traditional Medicine Newsletter* 1: 1,3.

Alcorn, J. B. 1981a. Factors influencing botanical resource perception among the Huastec: Suggestions for future ethnobotanical inquiry. *Journal of Ethnobiology* 1(2): 221–230.

———. 1981b. Huastec noncrop resource management: Implications for prehistoric rain forest management. *Human Ecology* 9: 395–417.

———. 1984a. Development policy, forests, and peasant farms: Reflections on Huastec-managed forests' contributions to commercial production and resource conservation. *Economic Botany* 38(4): 389–406.

———. 1984b. *Huastec Mayan Ethnobotany.* Austin: University of Texas Press.

———. 1989a. An economic analysis of Huastec Mayan forest management. In *Fragile Lands of Latin America: Strategies for Sustainable Development.* Ed. J. Browder. Boulder, CO: Westview Press.

———. 1989b. Process as resource: The traditional agricultural ideology of Bora and Huastec resource management and its implications for research. In *Resource Management in Amazonia: Indigenous and Folk Strategies.* Eds. D. A. Posey and W. Balée. Advances in Economic Botany Series, vol. 7. Bronx, NY: New York Botanical Garden. 63–77.

———. 1990a. Evaluating folk medicine: Stories of herbs, healing and healers. *Latin American Research Review* 25: 259–270.

———. 1990b. Indigenous agroforestry strategies meeting farmers' needs. In *Alternatives to Deforestation.* Ed. A. B. Anderson. New York: Columbia University Press.

———. 1990c. Indigenous agroforestry systems in the Latin American tropics. In *Agroecology and Small Farm Development.* Eds. M. Altieri and S. Hecht. Boca Raton, FL: CRC Press.

———. 1992. Ethnobotanical knowledge systems: Resource for meeting rural development goals. In *Indigenous Knowledge Systems: The Cultural Dimension of Development.* Eds. D. M. Warren, D. Brokenshaw, and L. J. Slikkerveer. London: Kegan Paul.

Alcorn, J. B., and A. Molnar. 1990. Deforestation and forest-human relationships: What can we learn from India? Paper presented at Tropical Forest Ecology, the Changing Human Niche, and Deforestation—A Symposium. American Anthropological Association meeting, November 1990, New Orleans.

Allen, J. C. 1984. Soil response to forest clearing in the United States and the tropics: Geological and biological factors. *Biotropica* 17: 15–27.

Altieri, M. A. 1984. Pest-management strategies for peasants: A farming systems approach. *Crop Protection* 3: 87–94.

Altieri, M. A., and D. K. Letourneau. 1982. Vegetation management and biological control in agroecosystems. *Crop Protection* 1: 405–430.

Anderson, A. B. 1990. Extraction and forest management by rural inhabitants in the Amazon estuary. In *Alternatives to Deforestation*. Ed. A. B. Anderson. New York: Columbia University Press.

Anderson, E. 1952. *Plants, Man and Life*. Berkeley: University of California.

Baker, D., E. Modiakgotla, D. Norman, J. Siebert, and M. Tjirongo. 1983. Helping the limited resource farmer through the farming systems approach to research. *Cult. and Agric.* 19: 1–8.

Barlett, P. F, ed. 1980. *Agricultural Decision Making: Anthropological Contributions to Rural Development*. New York: Academic Press.

Barrera, A., ed. 1979. *La Etnobotánica: Tres Puntos de Vista y Una Perspectiva*. Mexico: INIREB.

Bartlett, H. H. 1956. Fire, primitive agriculture, and grazing in the tropics. In *Man's Role in Changing the Face of the Earth*. Ed. W. L. Thomas, Jr. Chicago: University of Chicago Press.

Bennett, J. W. 1976. *The Ecological Transition: Cultural Anthropology and Human Adaptation*. New York: Pergamon Press.

Berlin, B. 1973. Folk systematics in relation to biological classification and nomenclature. *Ann. Rev. Ecol. Syst.* 4: 259–271.

———. 1978. Ethnobiological classification. In *Cognition and Categorization*. Eds. E. Rosch and B. B. Lloyd. Hillsdale, NJ: Lawrence Erlbaum Associates; distributed by John Wiley and Sons. 11–26.

———. 1984. Contributions of Native American collectors to the ethnobotany of the Neotropics. In *Ethnobotany in the Neotropics*. Eds. G. T. Prance and J. A. Kallunki. Advances in Economic Botany Series, vol. 1. Bronx, NY: New York Botanical Garden. 24–33.

Boster, J. S. 1984. Classification, cultivation, and selection of Aguaruna cultivars of *Manihot esculenta* (Euphorbiaceae). In *Ethnobotany in the Neotropics*. Eds. G. T. Prance and J. A. Kallunki. Advances in Economic Botany Series, vol. 1. Bronx, NY: New York Botanical Garden. 34–47.

Brockway, L. H. 1979. *Science and Colonial Expansion: The Role of the British Royal Botanic Gardens*. New York: Academic Press.

Brokenshaw, D., and B. W. Riley. 1980. Mbeere knowledge of their vegetation and its relevance for development: A case study from Kenya. In *Indigenous Knowledge Systems and Development*. Eds. D. W. Brokenshaw, D. M. Warren, and O. Werner. Lanham, MD: University Press of America.

Brokenshaw, D. W., D. M. Warren, and O. Werner, eds. 1980. *Indigenous Knowledge Systems and Development*. Lanham, MD: University Press of America.

Brown, C. H. 1985. Mode of subsistence and folk biological taxonomy. *Current Anthropology* 26: 43–64.

Bruhn, J. G., and B. Holmstedt. 1981. Ethnopharmacology: Objectives, principles, and perspectives. In *Natural Products as Medicinal Agents*. Eds. E. Reinhard and J. L. Beal. Stuttgart: Hippokrates. 405–430.

Brush, S. B., H. J. Carney, and Z. Huaman. 1981. Dynamics of Andean potato agriculture. *Economic Botany* 35: 70–88.

Butzer, K. W. 1982. *Archaeology as Human Ecology: Method and Theory for a Contextual Approach*. New York: Cambridge University Press.

Bye, R. A., Jr. 1978. Evolutionary and ecological effects of wild onion gathering in Northwest Mexico. Paper presented at the 19th annual meeting of the Society for Economic Botany, St. Louis.

———. 1979. Incipient domestication of mustards in Northwest Mexico. *Kiva* 44: 237–256.

———. 1981. Quelites—ethnoecology of edible greens—past, present and future. *Journal of Ethnobiology* 1: 109–123.

Caballero N., J. 1979. Perspectivas para el quehacer etnobotánico en México. In *La Etnobotánica: Tres Puntos de Vista y Una Perspectiva*. Ed. A. Barrera. Mexico: INIREB.

Chacón, J. C., and S. R. Gliessman. 1982. Use of the nonweed concept in traditional tropical agroecosystems of southeastern Mexico. *Agro-Ecosystems* 8: 1–11.

Chambers, R. 1983. *Rural Development: Putting the Last First*. London: Longman.

Compadre, C. M., J. M. Pezzuto, A. D. Kinghorn, and S. K. Klamath. 1985. Hernandulcin: An intensely sweet compound discovered by review of ancient literature. *Science* 227: 417–419.

Conklin, H. C. 1957. *Hanunoo Agriculture*. Rome: FAO.

Cox, G. W., and M. D. Atkins. 1979. *Agricultural Ecology: An Analysis of World Food Production Systems*. San Francisco, CA: W. H. Freeman and Company.

Croom, E. M., Jr. 1983. Documenting and evaluating herbal remedies. *Economic Botany* 37: 13–27.

Daly, H. E., and J. B. Cobb, Jr. 1989. *For the Common Good: Redirecting the Economy Toward Community, the Environment, and a Sustainable Future*. Boston: Beacon Press.

Denevan, W. M. 1980. Latin America. In *World Systems of Traditional Resource Management*. Ed. G. A. Klee. New York: Halstead Press.

Denevan, W. M., J. M. Treacy, J. B. Alcorn, C. Padoch, J. Denslow, and S. Flores P. 1984. Indigenous agroforestry in the Peruvian Amazon: Bora Indian management of swidden fallows. *Interciencia* 9: 346–357.

de Schlippe, P. 1956. *Shifting Cultivation in Africa: The Zande System of Agriculture*. London: Routledge & Kegan Paul.

DeWalt, B. 1982. The big macro connection: Population, grain, and cattle in Southern Honduras. *Cult. and Agric.* 14: 1–12.

DeWalt, K. M. 1983. *Nutritional Strategies and Agricultural Change in a Mexican Community*. Ann Arbor, MI: UMI Research Press.

Dewey, K. G. 1981. Nutritional consequences of the transformation from subsistence to commercial agriculture in Tabasco, Mexico. *Human Ecology* 9: 151–187.

Dimbleby, G. W. 1978. *Plants and Archaeology*. Atlantic Highlands, NJ: Humanities Press.

Dobkin de Rios, M. 1974. The influence of psychotropic flora and fauna on Maya religion. *Current Anthropology* 15: 147–164.

Domínguez, X. A., and J. B. Alcorn. 1985. Screening of medicinal plants used by Huastec Mayans of Northeastern Mexico. *Journal of Ethnopharmacology* 13: 139–156.

Eckholm, E., G. Foley, G. Barnard, and L. Timberlake. 1984. *Fuelwood: The Energy Crisis That Won't Go Away*. Washington: Earthscan.

Ellen, R. F. 1979. Sago subsistence and the trade in spices: A provisional model of ecological succession and imbalance in Moluccan history. In *Social and Ecological Systems*. Eds. P. Burnham and R. F. Ellen. New York: Academic Press.

Fleuret, A. 1979. The role of wild foliage plants in the diet: A case study from Lushoto, Tanzania. *J. Ecol. Food and Nutrition* 8: 87–93.

Folan, W. J., L. A. Fletcher, and E. R. Klintz. 1979. Fruit, fiber, bark, and resin: Social organization of a Mayan urban center. *Science* 204: 697–701.

Ford, R. I. 1978. Ethnobotany: Historical diversity and synthesis. In *The Nature and Status of Ethnobotany*. Ed. R. I. Ford. Anthropological Papers, no. 67. Ann Arbor, MI: University of Michigan Museum of Anthropology. 33–49.

———. 1979. Paleoethnobotany in American archaeology. *Advances in Archaeological Method and Theory* 2: 285–336.

———. 1981. Ethnobotany in North America: An historical phytogeographic perspective. *Canadian Journal of Botany* 59: 2178–2188.

Furley, P. A. 1975. The significance of the cohune palm, *Orbignya cohune* (Mart.) Dahlgren, on the nature and in the development of the soil profile. *Biotropica* 7: 32–36.

George, S. 1977. *How the Other Half Dies: The Real Reasons for World Hunger*. Montclair, NJ: Allanheld, Osmun.

Gliessman, S. R., R. Garcia E., and M. Amador A. 1981. The ecological basis for the application of traditional agricultural technology in the management of tropical agro-ecosystems. *Agro-Ecosystems* 7: 173–185.

Gómez-Pompa, A. 1971. Posible papel de la vegetación secundaria en la evolución de la flora tropical. *Biotropica* 3: 125–135.

———. 1982. La etnobotánica en México. *Biótica* 7: 151–161.

Gordon, B. L. 1982. A Panama Forest and Shore. Pacific Grove, CA: Boxwood Press.

Groenfeldt, D., J. B. Alcorn, S. Berwick, D. Flickinger, and M. Hatziolos. 1990. *Opportunities for Ecodevelopment in Buffer Zones: An assessment of two cases in Western India*. Report for USAID/India. Washington, DC: World Wildlife Fund–US.

Guyot, M. 1975. Le système cultural Bora-mirana. In *Culture sur Brulis et Évolution du Milieu Forestier en Amazonie du Nord-Ouest*. Eds. P. Centlivres, J. Gasche, and A. Lourteig. Bulletin de la Société Suisse d'Ethnologie, Geneva.

Hanks, J., ed. 1984. *Traditional life-styles, conservation, and rural development*. International Union for Conservation of Nature and Natural Resources (IUCN) Commission on Ecology Papers No. 7. Gland: IUCN.

Hanks, L. M. 1972. *Rice and Man: Agricultural Ecology in Southeast Asia*. Arlington Heights, IL: AHM Publishing Company.

Harcombe, P. A. 1980. Nutrient accumulation during the first year of recovery of a tropical forest ecosystem. In *Recovery and Restoration of Damaged Ecosystems*. Eds. J. Cairns, K. L. Dickson, and E. S. Herricks. Charlottesville: University of Virginia Press.

Harlan, J. R. 1975. *Crops and Man*. Madison, WI: American Society of Agronomy.

Hart, R. D. 1980. A natural ecosystem analog approach to the design of a successional crop system for tropical forest environments. *Biotropica* 12 (supplement, tropical succession): 73–82.

Hawkes, J. G. 1983. *The Diversity of Crop Plants*. Cambridge, MA: Harvard University Press.

Hays, T. E. 1974. *Mauna: Explorations in Ndumba Ethnobotany*. Ph.D. thesis, University of Washington. Ann Arbor, MI: University Microfilms.

Hebda, R. J., and R. W. Mathewes. 1984. Holocene history of cedar and native Indian cultures of the North American Pacific Coast. *Science* 225: 711–713.

Hunn, E. 1982. The utilitarian factor in folk biological classification. *American Anthropologist* 84(4): 830–847.

International Potato Center. 1984. *Breaking New Ground: Anthropology in Agricultural Research*. Lima, Peru.

Jiú, J. 1966. A survey of some medicinal plants of Mexico for selected biological activities. *Lloydia* 29: 250–259.

Johnson, A. W. 1971. *Sharecroppers of the Sertao*. Stanford, CT: Stanford University Press.

Jones, R. 1969. Fire-stick farming. *Australian Natural History* 16: 224–228.

Kellman, M. 1980. Geographic patterning in tropical weed communities and early secondary succession. *Biotropica* 12 (Special Edition Tropical Succession): 34–39.

Kimber, C. T. 1973. Spatial patterning in the dooryard gardens of Puerto Rico. *Geographical Review* 63: 6–26.

Kunstadter, P. 1978. Ecological modification and adaptation: An ethnobotanical view of Lua' swiddeners in northwest Thailand. In *The Nature and Status of Ethnobotany*. Ed. R. I. Ford. Anthropological Papers, no. 67. Ann Arbor, MI: University of Michigan Museum of Anthropology. 169–200.

Kupzow, A. J. 1980. Theoretical basis of plant domestication. *Theor. Appl. Genet.* 57: 65–74.

Lambert, J. D. H., and T. Arnason. 1978. Distribution of vegetation on Maya ruins and its relationship to old land-use at Lamania, Belize. *Turrialba* 28: 33–41.

Lynch, O., and J. B. Alcorn. 1991. *Empowering Local Forest Managers: Towards More Effective Recognition of Tenurial Rights, Claims, and Management Capacities of the People Occupying "Public" Forest Reserves (Paa Sangin) in Thailand*. Washington, DC: World Resources Institute Center for International Development and Environment.

McDermott, W. 1980. Pharmaceuticals: Their role in developing societies. *Science* 209: 240–245.

McKibben, B. 1989. *The End of Nature*. New York: Doubleday.

McNeely, J. A. 1989. *Economics and Biological Diversity*. Gland: International Union for Conservation of Nature and Natural Resources (IUCN).

Meals for Millions and Save the Children (MFM/STC). 1980. *O'odham I:waki: Wild Greens of the Desert People*. Tucson, AZ: MFM.

Meggers, B. J. 1977. Vegetational fluctuation and prehistorical cultural adaptations in Amazonia: Some tentative correlations. *World Archaeology* 8: 287–303.

Messer, E. 1972. Patterns of "wild" plant consumption in Oaxaca, Mexico. *J. Ecol. Food and Nutrition* 1: 325–332.

———. 1975. *Zapotec Plant Knowledge: Classification, Uses and Communication about Plants in Mitla, Oaxaca, Mexico*. Ph.D. thesis, University of Michigan. Ann Arbor, MI: University Microfilms.

Moore, J. A. 1981. The effects of information networks on hunter-gatherer societies. In *Hunter-Gatherer Foraging Strategies*. Eds. B. Winterhalder and E. A. Smith. Chicago: University of Chicago Press.

Morris, B. 1984. The pragmatics of folk classification. *Journal of Ethnobiology* 4: 45–60.

Nabhan, G. P. 1982. *The Desert Smells Like Rain*. San Francisco, CA: North Point Press.

National Academy of Sciences. 1980. *Firewood Crops*. Washington, DC: National Academy of Sciences.

Netting, R. McC. 1974. Agrarian ecology. *Ann. Rev. Anthrop.* 3: 21–56.

Nigh, R. B. 1976. *Evolutionary Ecology of Maya Agriculture in Highland Chiapas, Mexico*. Ph.D. thesis, Stanford University. Ann Arbor, MI: University Microfilms.

O'Connell, J. F., and K. Hawkes. 1981. Alyawara plant use and optimal foraging theory. In *Hunter-Gatherer Foraging Strategies*. Eds. B. Winterhalder and E. A. Smith. Chicago: University of Chicago Press.

Oldfield, M. L., and J. B. Alcorn. 1987. Conservation of traditional agroecosystems. *BioScience* 37: 199–208.

———, eds. 1991. *Biodiversity: Culture, Conservation and Ecodevelopment*. Boulder, CO: Westview Press.

Orlove, B. S. 1980. Ecological anthropology. *Ann. Rev. Anthrop.* 1980: 235–273.

Ortíz de Montellano, B. 1975. Empirical Aztec medicine. *Science* 188: 215–220.

Palmeirim, J., and K. Etheridge. 1985. Influence of man-made trails on foraging by tropical frugivorous bats. *Biotropica* 17: 82–83.

Pearsall, D. 1983. Evaluating the stability of subsistence strategies by use of paleoethnobotanical data. *Journal of Ethnobiology* 3: 121–137.

Penso, G. 1980. The role of WHO in the selection and characterization of medicinal plants (vegetable drugs). *Journal of Ethnopharmacology* 2: 183–185.

Peters, C. M. 1983. Observations on Maya subsistence and the ecology of a tropical tree. *Am. Antiq.* 48: 610–615.

Peters, C. M., M. J. Balick, F. Kahn, and A. B. Anderson. 1989. Oligarchic forests of economic plants in Amazonia: Utilization and conservation of an important tropical resource. *Conservation Biology* 3: 341–349.

Peters, C. M., and E. Pardo-Tejedo. 1982. *Brosimum alicastrum* (Moraceae): Uses and potential in Mexico. *Economic Botany* 36: 166–175.

Posey, D. A. 1983. Indigenous ecological knowledge and development of the Amazon. In *The Dilemma of Amazonian Development*. Ed. E. F. Moran. Boulder, CO: Westview Press. 225–257.

———. 1984. A preliminary report on diversified management of tropical forests by the Kayapó Indians of the Brazilian Amazon. In *Ethnobotany in the Neotropics*. Eds. G. T. Prance and J. A. Kallunki. Advances in Economic Botany Series, vol. 1. Bronx, NY: New York Botanical Garden. 112–126.

———. 1990. Intellectual property rights: What is the position of ethnobiology? *Journal of Ethnobiology* 10: 93–98.

Posey, D. A., and W. Balée, eds. 1989. *Resource Management in Amazonia: Indigenous and Folk Strategies*. Advances in Economic Botany Series, vol. 7. Bronx, NY: New York Botanical Garden.

Pulliam, H. R. 1981. On predicting human diets. *Journal of Ethnobiology* 1: 61–68.

Reichel-Dolmatoff, G. 1971. *Amazonian Cosmos*. Chicago: University of Chicago Press.

———. 1976. Cosmology as ecological analysis: A view from the rain forest. *Man* 11: 307–318.

Ricoeur, P. 1971. The model of the text: Meaningful action considered as text. *Social Research* 5: 529–562.

Rindos, D. 1984. *The Origins of Agriculture: An Evolutionary Approach*. New York: Academic Press.

Risch, S. J., D. Andow, and M. A. Altieri. 1983. Agroecosystem diversity and pest control: Data, tentative conclusions, and new research directions. *Environ. Entomol.* 12: 625–629.

Salick, J. 1990. Amuesha forest use and management: An integration of indigenous use and natural forest management. Paper presented at 31st annual meeting of the Society for Economic Botany, June 1990, Madison, WI.

Sarukhan, J. 1985. Ecological and social overviews of ethnobotanical research. *Economic Botany* 39: 431–435.

Schultes, R. E. 1960. Tapping our heritage of ethnobotanical lore. *Economic Botany* 14: 257–262.

Schultes, R. E., and A. Hofmann. 1980. *The Botany and Chemistry of Hallucinogens*. 2nd ed. Springfield, IL: Charles C. Thomas.

Smith, J. M. B. 1977. Man's impact upon some New Guinea mountain ecosystems. In *Subsis-*

tence and Survival: Rural Ecology in the Pacific. Eds. T. Bayliss-Smith and R. Feachem. New York: Academic Press.

Sobey, D. G. 1978. Anogeissus groves on abandoned village sites in the Mole National Park, Ghana. *Biotropica* 10: 87–99.

Speight, M. R. 1983. The potential of ecosystem management for pest control. *Agriculture, Ecosystems and Environment* 10: 183–199.

Stross, B. 1973. Acquisition of botanical terminology by Tzeltal children. In *Meaning in Mayan Languages.* Ed. M. S. Edmondson. The Hague: Mouton.

Swain, T., ed. 1972. *Plants in the Development of Modern Medicine.* Cambridge, MA: Harvard University Press.

Tanaka, J. 1976. Subsistence ecology of Central Kalahari San. In *Kalahari Hunter-Gatherers: Studies of the Kung San and Their Neighbors.* Eds. R. B. Lee and I. DeVore. Cambridge, MA: Harvard University Press.

Toledo, V. M. 1985. *Ecología y Autosuficiencia Alimentaria: Una Estrategia Basada en la Diversidad Biológica, Ecológica y Cultural de México.* Mexico City: Siglo XXI.

Toledo, V. M., A. Argueta, P. Rojas, C. Mapes, and J. Caballero. 1976. Uso múltiple del ecosistema, estrategias del ecodesarrollo. *Ciencia y Desarrollo* 11: 33–39.

Toledo, V. M., and N. Barrera-Bassols. 1984. Ecología y Desarrollo Rural en Pátzcuaro. Mexico City: Instituto de Biología, Universidad Nacional Autónoma de México.

Uhl, C., H. Clark, K. Clark, and P. Maquirino. 1982. Successional patterns associated with slash-and-burn agriculture in the Upper Rio Negro region of the Amazon Basin. *Biotropica* 14: 249–254.

UNESCO–MAB. 1984. Action plan for biosphere reserves. *Nature and Resources* 20: 1–12.

Vasquez, R., and A. H. Gentry. 1989. Use and misuse of forest-harvested fruits in the Iquitos area. *Conservation Biology* 3: 350–361.

Vermeer, D. E. 1979. The tradition of experimentation in swidden cultivation among the Tiv of Nigeria. *Applied Geography Conferences* 2: 244–257.

Villers R. L., R. M. López F., and A. Barrera. 1981. La unidad de habitación tradicional campesina y el manejo de recursos bióticos en el area Maya Yucatanense. *Biotica* 6: 293–322.

Voorhies, B. 1982. An ecological model of the Early Maya of the central lowlands. In *Maya Subsistence.* Ed. K. Flannery. New York: Academic Press.

Weinstock, J. A. 1983. Rattan: Ecological balance in a Borneo rain forest swidden. *Economic Botany* 37: 58–68.

Weragoda, P. B. 1980. Some questions about the future of traditional medicine in developing countries. *Journal of Ethnopharmacology* 2: 193–194.

WHO (World Health Organization). 1978. *Drug Policies and Management: Medicinal Plants.* WHO Document WHA31.33. Geneva: WHO.

Williams-Linera, G. 1983. Biomass and nutrient content in two successional stages of tropical wet forest in Uxpanapa, Mexico. *Biotropica* 15: 275–284.

Wilson, G. F., and B. T. Kang. 1981. Developing stable and productive biological cropping systems for the humid tropics. In *Biological Husbandry.* Ed. B. Stonehouse. Boston: Butterworths.

Wiseman, F. M. 1983. Subsistence and complex societies: The case of the Maya. *Advances in Archaeological Method and Theory* 6: 143–189.

Yellen, J. E., and R. B. Lee. 1976. The Dobe-/Du/da environment: Background to a hunting and gathering way of life. In *Kalahari Hunter-Gatherers: Studies of the Kung San and Their Neighbors.* Eds. R. B. Lee and I. DeVore. Cambridge, MA: Harvard University Press.

Zizumbo V. D., and P. C. García-Marín. 1982. *Los Huaves: La Appropriación de los Recursos Naturales.* Chapingo, Mexico: Depto. de Sociología Rural, Universidad Autónoma Chapingo.

Ethnobotany: An Old Practice,
A New Discipline

E. WADE DAVIS

Ethnobotany as an academic discipline has its roots in the numerous observations of explorers, traders, missionaries, naturalists, anthropologists, and botanists concerning the use of plants by the seemingly exotic cultures of the world. It was born as much as anything out of the coalescence of disparate field reports, and from its beginnings it has struggled to find the unifying theory innate to many more narrowly delimited scientific disciplines.

As a result, ethnobotany has at times suffered from a lack of orientation and integration, and its traditional task of cataloging the uses of plants has been criticized as lacking theoretical content. In part these criticisms have been valid. Ethnobotanical data collected without reference to an intellectual problem may be eclectic to the point of inutility. Yet critics of the practice of ethnobotany usually overlook two important considerations. First, the act of compiling raw information provides the foundation of any natural science, and without a basic inventory, theoretical formulations are not possible. Second, ethnobotany remains on one level what it has always been—a science of discovery. Its contributions to the welfare of the human race have not been trivial. In the field of medicine alone, between 25 and 50 percent of the modern drug armamentarium is derived from natural products, and most of these compounds were first used as medicines or poisons in a folk context (Farnsworth and Morris 1976; Holmstedt and Bruhn 1983). Today, in an era marked by the massive destruction of diversity not only of plants and animals but of human cultures as well, basic plant exploration remains a vital and essential contribution of the ethnobotanist.

Moreover, critics of ethnobotany often overlook the considerable theoretical advances that the field has made increasingly (Alcorn 1984; Johns 1990). The old practice of ethnobotanical discovery, of finding new plants, labeling them according to the rules of binomial nomenclature, and incorporating their usefulness into modern society has been augmented by an intellectual perspective that views both the plants and their utilization as but a metaphor for understanding the very cognitive matrix of a particular society. This trend has resulted in sophisticated studies which, depending on the problem at hand, may concurrently employ a range of methodologies derived from various academic disciplines; it is precisely this interdisciplinary orientation that allows the ethnobotanist to pose and answer questions that cannot be approached by narrower specialists. This chapter highlights the significance of this new synthesis in ethnobotany by offering an overview of its development and a discussion of its potential.

For much of Western intellectual history, botany and what we now know as ethnobotany were synonymous fields of knowledge. Indeed, at its inception, ethnobotany was less an academic discipline than a point of view, one perspective by which European scholars and plant explorers went about classifying the natural world. Botanical exploration was stimulated by a desire to systematize creation, but it was motivated by the promise of economic gain. Ethnobotany was a strategy that sought to satisfy an economic imperative by yielding new natural products of commercial potential. From the start, then, ethnobotany has been intimately linked to botanical exploration, and its history has run parallel to the evolution of both systematic and economic botany.

Although it was not until 1895 that the term was coined by University of Pennsylvania botanist John Harshberger (1896), ethnobotany as a practice goes back in the European intellectual tradition at least to the wanderings of Dioscorides, the Greek surgeon who traveled about the Mediterranean area at the behest of Emperor Nero. His *De Materia Medica*, written in 77 A.D., contained detailed descriptions of the botany and medicinal properties of some 600 plants. Dioscorides carefully noted the habitats of the plants, when and how they were to be gathered, which were edible, poisonous, and/or therapeutic, and he included in his compilation recipes and formulae for their use. He also took note of exotic plants—spices in particular—that had significant economic potential.

Dioscorides was also a good teacher, but as his work became canonized, the example of his field work was forgotten and the man himself became a text (Boorstin 1983). Dioscorides had studied plants, but for more than a thousand years after his death botanists studied Dioscorides. Botany throughout much of the Middle Ages became a process of embroidering his original work, searching through the fields of northern Europe for herbs that might mimic the plants he had gathered much earlier on the shores of the Mediterranean. As the artistry of the Medieval herbals flourished, numerous editions of Dioscorides' work came forth, and in each one the illustrations became ever more fanciful, straying further and further from the original.

It was not until the Renaissance that botanists moved out of the monasteries and into the fields. Artists led the way—notably Hans Weiditz who drew from life to illustrate Brunfels's otherwise traditional herbal *Herbarum Vivae Eicones* (1530). Leonhart Fuchs finally broke the hold of Dioscorides in 1542 by publishing in his lavishly illustrated *De Historia Stirpium* the names of some four hundred plants native not to the Mediterranean but to his own Germany. Fuchs's herbal was of historic significance. Its illustrations were scientifically accurate, and in turning attention to the local flora it acknowledged the work of Hieronymous Bosch, who in *Neu Kreutterbuch* (1539) had finally abandoned all attempts to match Greek or Latin names recorded from Dioscorides with elements of the northern flora. Without doubt, the most novel aspect of Fuchs's herbal was the appearance for the first time of some of the strange and exotic botanical arrivals from the Americas. More than any other force, these new discoveries would oblige botanists finally to follow the example set long ago by Dioscorides: they would have to become explorers of an entirely new world of plants (Boorstin 1983).

Reports from the Americas had both delighted and bewildered the Old World. Columbus turned to plants to prove that he had reached the Indies. He mistook the small inedible nut (*nogal de pais*) as the coconut described by Marco Polo. Any vaguely aromatic plant became evidence of the Spice Islands. His ship's surgeon dug up the roots of the common kitchen rhubarb (*Rheum raponticum*) and concluded that he had found a limitless supply of the valuable cathartic drug *Rheum officinale*, a native of China. These were understandable mistakes, to be sure, for the plants involved represented only the beginning of an outpouring of botanical marvels that stirred the foundations of Europe. Maize, cinchona, potatoes, tobacco, tomatoes, coca, manioc, chocolate, chilies, pineapples, and rubber would over the long term prove to be economically the most significant, but for the botanist they were but a handful of the thousands of novel plants that collectively raised immediate and significant intellectual questions. How were they to be labeled and arranged? How was one to know what was really new? What did it mean to be new? What, indeed, was the entity to be studied? If this avalanche of material called out for systematization, the first task of the botanist was to decide upon the units of classification.

The English naturalist John Ray, who offered the first species concept in *Methodus Plantarum* (1682), continued in his three-volume *Historia Plantarum* (1686–1704) to offer the first systematic treatment of all the plants then known to Europe. Ray studied plants as living organisms and, having considered the morphology of the entire plant, determined that a species comprised "a set of individuals who give rise through reproduction to new individuals similar to themselves" (quoted in Boorstin 1983, p. 434). This was a vital conceptualization for it provided all naturalists at the time with a building block from which a revolutionary taxonomic system could be constructed.

It was left to Carl Linnaeus to do that work, however, and with his arrival on the scene, the history of botanical and ethnobotanical exploration truly began. The genius of Linnaeus, and particularly his elegant concept of binomial nomenclature, provided the classificatory framework that could embrace the plethora of new discoveries coming into Europe from all parts of the earth. In *Species Plantarum* Linnaeus placed a binomial label on each of the 5900 plants known to European botanists. Then he began to look overseas, and in the midst of the Age of Discovery, he dispatched his students to every corner of the globe. In nearly every instance the ostensible purpose of the botanical expeditions was to obtain new plants for use as foods, textiles, and medicines that would benefit Europe.

Linnaeus evidently shared in the spirit of the mercantile age. Writing in a museum catalog in 1754 he presaged optimistically the role that plant exploration would play in the development of human society:

> Man, ever desirous of knowledge, has already explored many things; but more and greater still remain concealed; perhaps reserved for distant generations, who shall . . . make many discoveries for the pleasure and convenience of life. Prosperity shall see its increasing Museums, and all the knowledge of the Divine Wisdom, flourish together; and at the same time all the practical sciences . . . shall be enriched; for we cannot avoid thinking that what we know of the Divine works are far fewer than those of which we are ignorant.

Linnaeus's students not only brought back enormous quantities of specimens but also accounts of the cultures they had visited, the customs of the inhabitants, and, in particular, the way indigenous peoples used their plants—the very material of classical ethnobotany. Peter Kalm was dispatched to North America to find exotic forms of the mulberry, which his financial backers hoped would provide the basis of a new silk industry. To the Middle East went Frederick Hasselquist, and still further afield to China traveled Pehr Osbeck. Others sailed to the Spanish Indies, Tartary, Persia, Surinam, India, and what is now South Africa. Daniel Solander, another of Linnaeus's protégés, joined Cook on a three-year circumnavigation of the world and returned to Europe with some 1200 new species, including more than 100 new genera.

The nineteenth century saw botanical exploration at a peak. The voyages of Alexander von Humboldt and Aimé Bonpland electrified Europe and inspired the travels of numerous young naturalists. Alfred Wallace spent four years on the Amazon and another eight in the Malay archipelago. Joseph Hooker joined the Antarctic expedition of James Ross, a work that secured him later contracts to study the flora of the Himalayas and Ceylon and led indirectly to the establishment of the Royal Botanic Gardens at Kew, England. The self-taught naturalist Richard Spruce, perhaps the most important plant explorer of his era, spent seventeen years in the northwestern Amazon and adjacent regions of the Andean Cordillera. For months at a time these botanists were dependent on the fellowship of indigenous peoples, who in turn were dependent on the forest for virtually every aspect of their material culture. As a result, the travel journals that have come down to us contain a wealth of ethnobotanical lore.

By the latter part of the nineteenth century this rich repository of data that lay scattered throughout the writings of botanical explorers, frontier physicians, missionaries, and traders drew the attention of a number of academics who felt the examination of indigenous societies and their plants demanded systematic study. In 1874 Stephen Powers coined the term *aboriginal botany* to describe the study of "all forms of the vegetable world which the aborigines used for medicine, food, textile fabrics, ornaments, etc." (Powers 1875, p. 373). Following his example, botanists produced a series of descriptive compilations listing plants used cross-culturally for a particular purpose (Palmer 1871; Millspaugh 1892). These studies, in turn, illuminated the inherent limits of data based on anecdote or isolated reports and led to the first systematic studies of the ethnobotany of individual ethnic groups (e.g., Mooney 1889, 1891; Kroeber 1907, 1920; Stevenson 1915; Teit 1930; Robbins et al. 1916; Vestal and Schultes 1939).

This methodological reorientation marked what Ford (1978) identified as a major change in the theoretical focus of ethnobotany. Increasingly as ethnologists joined the field, the emphasis shifted from the raw compilation of plant names and uses to an intellectual perspective that viewed the character of a people's relationship with the plant world as but one means of approaching an understanding of the cognitive foundations of a culture. If the botanist remained wedded to a utilitarian mandate to obtain new sources of wealth for Western society, the anthropologist was far more interested in the interactions between humans and plants, the dynamic processes by which each influenced and held sway over the lives of the other.

As anthropologists working in ethnobotany became concerned with the "totality of the place of plants in a culture" (Ford 1978), the intellectual potential of the discipline began to be realized. The study of plants became a vehicle for addressing general issues of ethnological significance. Several themes emerged. The important concept of cultural relativism was reinforced by studies of folk classification, which revealed that aboriginal taxonomies, while not necessarily coinciding with Linnean concepts and categories, were equally complex and firmly rooted in biology (Conklin 1954; Berlin et al. 1974). Studies of hallucinogenic plants offered insights into the origin and character of complex religious beliefs (La Barre 1938; Reichel-Dolmatoff 1971, 1975). Work in medical anthropology highlighted the significance of non-Western concepts of health and healing and, in doing so, emphasized the elaborate connection between spiritual belief, psychological predisposition, and pharmacology that underlies all indigenous practices involving psychotropic preparations.

Perhaps most significantly, by emphasizing the ongoing dynamic interrelationship between humans and plants, anthropologists explicitly incorporated advances in ecology and thus redefined the human role as but one element in a complex and ever-changing equation. Plants ceased to be considered passive objects. The notion that humans were omnipotent was challenged. Plants and human societies became viewed as co-dependents, and the task of the ethnobotanist was transformed from one of compilation to one of understanding and evaluating in a biologically meaningful way their complex interactions. Critically, once viewed from an ecological perspective, the biological basis of the past and present use of plants could be ascertained and the diachronic changes in the relationship between plant and human societies could be measured both qualitatively and quantitatively by the ethnobotanical field worker (Bye 1976).

One consequence of this modern synthesis in ethnobotany is an integrative focus that draws upon the expertise of a host of specialists, including botanists and anthropologists as well as ecologists, geographers, soil scientists, pharmacologists, phytochemists, cultural ecologists, and environmental conservationists. Indeed, the interdisciplinary range of ethnobotany is so broad that it has become virtually impossible and certainly meaningless to define sectarian boundaries of the field. Contemporary ethnobotany may embrace pursuits as different as agroforestry and land management, ethnoscience and cognitive anthropology; studies in plant domestication, iconographic and ceramic interpretation; archaeological plant remains; the symbolic role of psychoactive preparations; the search for novel plant-derived medicines; and even the struggle for conserving the world's threatened rain forests. No single theoretical or methodological orientation can encompass the breadth of inquiry that is modern ethnobotany. Rather, the ethnobotanist must adopt a problem-solving approach and select a research team and a methodology appropriate to the specific task at hand.

Two major challenges mark the future of ethnobotany. First, the long-standing task of cataloging what is now known, of documenting which plants are and are not important to a society, and of sifting through the immense repositories of folk beliefs for plants that may serve the needs of human societies—all these must continue at an ever-accelerating pace. It is a tiresome yet tragically true admonition that the rate of destruction of biological and cultural diversity, particularly in tropical rain forest areas, promises to rob us within a single generation of the accumulated wisdom of millennia and of our ability to preserve even certain elements of this knowledge.

Ethnobotanists must record not only lists of plant uses but a vision of life itself. This is the second and much more difficult task—to understand not just how a specific group of people uses plants but how that group perceives them, how it interprets those perceptions, how those perceptions influence the activities of members of that society, and how those activities, in turn, influence the ambient vegetation and the ecosystem upon which the society depends.

Contemporary ethnobotanists are meeting these challenges. The task of documenting folk knowledge has been systematized as never before by the quantitative methodologies pioneered by researchers at the Institute of Economic Botany of The New York Botanical Garden (Boom 1985a, 1985b, 1987; Balée 1986, 1987), at the Harvard Botanical Museum (Laboratory of Economic Botany), at Colombia's Instituto de Ciencias Naturales, and at numerous other institutions in the Americas, Europe, Africa, and Asia. By establishing standard, one-hectare (2.5-acre) forest inventory plots and by carefully cross-referencing their data through subsequent interviews with informants, these researchers have shown that at least two indigenous societies in the Amazon utilize over 80 percent of the plant species found growing in the local forest.

Other researchers have concentrated on patterns of plant utilization. Kvist and Holm-Nielsen (1987) compared the pharmacopeia of seven indigenous groups in lowland Ecuador and selected plants likely to be pharmacologically active by identifying those species or genera used for the same purpose by tribes without cultural contact. Davis and Yost (1983a, 1983b, 1983c) studied the ethnobotany of the Waorani of Amazonian Ecuador, one of the later groups to be contacted in South America, and found that despite an extraordinary knowledge of plants and forest ecology, the Waorani use very few medicinal plants. Medical studies conducted at the time of contact indicate that the Waorani were an exceedingly healthy people. Within the overall framework of good health, however, the tribe suffered from a number of readily identifiable conditions, and the ethnobotanical data indicate that for each major affliction the Waorani went to some experimental effort to find cures.

The precise and limited materia medica of the Waorani stands in marked contrast to the vast pharmacopeias of their more acculturated neighbors and suggests that the plethora of plants used by these surrounding groups reflects at least in part the chaos of contact and the accelerated experimentation that occurred in post-contact times. This suggestion, while challenging the notion that indigenous pharmacopeias necessarily developed slowly over hundreds of years, in no way denigrates indigenous healing practices. On the contrary, it reveals traditional healers for what they are—active scientific experimenters whose work reflects social needs and whose laboratory happens to be the rain forest.

One significant trend in contemporary ethnobotany reflected in these and other studies is the active collaboration in the field between ethnographers and botanists (Vickers and Plowman 1984; Anderson and Posey 1986; Denevan and Padoch 1988) and between Western scientists in general and indigenous collectors (Berlin 1984; Baker et al. 1987). The advantages of such an interdisciplinary and intercultural approach are myriad. For one, anthropologists are becoming increasingly familiar with the fundamentals of plant taxonomy and aware of the importance of properly collected voucher specimens. Botanists, in turn, are learning to conduct field work with an enhanced awareness of the significance of the dynamic interplay between informant and investigator.

Indeed, in the process of successfully obtaining any ethnobotanical information, there is an ineffable quality to the interaction between researcher and informant, impossible to quantify but often critical to the legitimacy of the data and the overall value of the research. A sensitivity to cultural values is fundamental, but of equal importance is an awareness on the part of the investigator that the way a question is formulated and posed has a direct bearing on the way it is answered.

Another advantage of anthropologists and botanists working together is a practical one. Seldom do botanists have the opportunity or inclination to spend a year or more in a single location, and almost never do they have the time or training necessary to master indigenous languages. Yet the depth of one's knowledge of any aspect of a people's ethnobotany is directly proportional to one's overall knowledge of the culture. Data filtered through a language that neither informant nor researcher consider their native tongue and transmitted in a language foreign to the culture of the informant are clearly affected by the process and may well be suspect. The only means to circumvent this dilemma is the active collaboration of an ethnographer already fluent in the native language.

One example of such a collaboration is the aforementioned work Jim Yost and I undertook among the Waorani (Davis and Yost 1983a, 1983b, 1983c). This group of approximately 660 Indians inhabits a vast region of over 20,000 square kilometers (7000 square miles) south of the Río Napo and north of the Río Curaray in central eastern Ecuador. Like many small Amazonian groups, the Waorani are acephalous and highly egalitarian, with a political and social structure based on an extended network of kinship ties. Mutual hostility between the Waorani and all outside groups, together with a subsistence base and settlement pattern that linked the tribe to the interriverine forests, caused this group to be unusually isolated. Its language is unique, and at the time of contact in 1958 only two loan words were discovered. To date, no linguistic congeners have been found (Peeke 1973). Dr. Yost, a linguist and anthropologist by training, is one of perhaps three Westerners who speak this language fluently. Working together in the field for only a month, he and I were able to collect approximately 80 percent of the plants recorded by Yost during his eight years of living with the Waorani. In absolute numbers, the ethnobotanical collections compared favorably with the collections obtained by an ethnobotanist working alone for an entire year among a neighboring indigenous group (Pinkley 1973). More important, the proper taxonomic identification of the plants allowed us to reconsider biomedical data collected at the time of contact and to formulate the novel hypothesis concerning Waorani ethnomedicine (Davis and Yost 1983b). Without doubt, this idea would not have occurred to either of us had we not worked together, sharing the benefits of our different academic backgrounds.

Berlin (1984) has taken this interdisciplinary approach one step further by training and incorporating indigenous collaborators in the process of obtaining ethnobotanical data. The results are impressive. Indigenous collectors dramatically increase the range of coverage of a botanical survey, and their personal involvement virtually guarantees the gathering of invaluable information on the cultural significance of the plants. Moreover, considering local people as counterparts rather than as merely informants balances the relationship between outsider and insider, and encourages a local appreciation of traditional knowledge.

Perhaps the best examples of the interdisciplinary approach in ethnobiology and ethnobotany are to be found in the advancing field of ethnoecology (Bye 1976; Posey 1979, 1984; Alcorn 1981, 1984; Denevan and Padoch 1988). For several years Darell Posey and his team have worked among the Mebêngôkre-Kayapó, who currently inhabit a reserve of some 2 million hectares (5 million acres) in the Brazilian state of Pará (Posey 1983, 1984, 1985). Their studies document an indigenous system of integrated land management of remarkable sophistication, and their conclusions have transformed our understanding of lowland Amerindian agricultural practices.

Slash-and-burn (swidden) agriculture, with its cycle of clearing, burning, and planting leading to one or two good crop yields followed by seasons of diminishing returns and long periods of regenerative fallow, has long been recognized as the most adaptive means of farming in the lowland tropics. The analytical focus of most studies, however, has been on the principle cultigen (often manioc in the Amazon), and the tendency has been to measure the life span of the field in relation to that crop alone. Once that crop is no longer productively grown, the field is said to be abandoned, a new section of virgin forest is felled, and the cycle begins once more. Viewed from this perspective the agricultural system as a whole appears to be inherently wasteful and critically dependent on population densities.

Posey and his associates have shown that, on the contrary, the time devoted to the production of the primary carbohydrate is but a phase in the organic evolution of the agricultural setting. Old fields are, in Posey's words, "anything but abandoned" (Posey 1984, p. 114). Long after the peak two- or three-year period is past, these fields continue to yield domestic plant products and remain repositories of a plethora of other useful raw materials. One study revealed that 94 percent of the 368 plants collected in old fields had some medicinal purpose (Anderson and Posey 1986). Residual trees provide edible fruits and oils. Other useful products include fish poisons, dye plants, insect repellents, body cleansers, firewood, and materials for making thatch, rope, packaging, and crafts (Posey 1984).

Arguably the most compelling aspect of Posey's important work is the evidence he presents indicating that the Kayapó deliberately manage this complex agroforestry system and that they do so on a sustained yield basis. The biological use of insects, the manipulation of semidomesticated plants, and the deliberate encouragement or transplanting of wild trees and medicinal plants along trailsides and in fields are elements of a complex integrated system of management that stands in marked contrast to the crude and destructive patterns of modern land use in the tropics.

It is to the contemporary environmental crisis that ethnobotany promises to make what may turn out to be its most significant and historic contribution. Tropical forests are being destroyed at an alarming rate. The result is extinction not only of plants and animals but of human societies that have, over the course of thousands of years, developed an intimate knowledge of the forest and the natural products it contains. Largely responsible for this tragic destruction are misguided development programs initiated by government and international agencies struggling to deal with problems of massive foreign debt, population pressures, chronic poverty, and unemployment. Economic arguments will continue to dominate the public policy debate. Conservationists must seek an environmental policy consistent with these economic realities by showing that the long-term income-generating potential of the standing forest equals or exceeds the short-term gain resulting from its destruction.

Ethnobotany can contribute to this strategy in two ways. First, ethnoecological studies may provide models for profitable and environmentally sound multiple use land management programs. Second, ethnobotanists can invoke the considerable economic potential of as yet undiscovered or undeveloped natural products (Myers 1983; Balick 1985). Of an estimated 75,000 edible plants, for example, only 2500 have ever been eaten with regularity, a mere 150 have entered world commerce, and a scant 20, mostly domesticated grasses, stand between human society and starvation (R. E. Schultes, pers. com.). To diversify this resource base is one goal of ethnobotany, and numerous promising crops that can be exploited in ecologically sound ways have already been identified (National Academy of Sciences 1975; Balick 1984).

Possibly the greatest economic potential of ethnobotany lies in the area of folk medicine. Annual worldwide sales of plant-derived pharmaceuticals currently total over $20 billion, and a great many of these drugs were first discovered by traditional healers in folk contexts (Farnsworth 1982). The gifts of the shaman and the sorcerer, the herbalist and the witch, include such critical drugs as pilocarpine, digitoxin, vincristine, emetine, physostigmine, atropine, morphine, and reserpine (Farnsworth 1988). The forests of tropical America have yielded scopolamine, cocaine, quinine, and d-tubocurarine. An impressive 70 percent of all plants known to have antitumor properties have been found in tropical forests (Myers 1983).

This wealth of vital drugs has come from but a minor segment of the tropical flora. In the Amazon approximately 1 percent of the plants have been studied chemically and an astonishing 90 percent have not yet been subjected to even a superficial chemical analysis (Schultes 1988b). Any practical strategy for expanding our knowledge of this living "pharmaceutical factory" (Myers 1985, p. 210; Schultes 1987, 1988a) must include ethnobotanical research. To attempt to assay the entire flora without consulting Amazonian Indians would be logistically impossible and intellectually foolish.

Yet if ethnobotanists are to seize upon traditional knowledge as a means of rationalizing the preservation of threatened rain forests, they must do far more than search for new wealth. The tropical forest, with its thousand themes and the infinitude of form, shape, and texture, appears at times to mock the terminology of the Western scientist. Millennia ago men and women entered that forest, and through adaptation, cultures emerged, hundreds of them, the complexities of which rivaled even those of the dense vegetation out of which they were born. To stay alive, these men and women invented a way of life and, lacking the technology to transform the forest, they chose instead to understand it.

The intellectual achievements of Amazonian Indians suggest that the ultimate challenge of ethnobotany will lie not merely in the identification and extraction of natural products, but rather in the discovery and elaboration of a profoundly different way of living with the forest. Recall once again the Waorani of eastern Ecuador. Like many Amazonian groups, the Waorani identify both psychologically and cosmologically with the rain forest. Since they depend on that environment for a large part of their diet, it is not surprising that they are exceptionally skilled naturalists. It is the sophistication of their interpretation of biological relationships that is astounding. Not only do they recognize such conceptually complex phenomena as pollination and fruit dispersal, they understand and accurately predict animal behavior. They anticipate flowering and fruiting cycles of all edible forest plants, know the preferred foods of most forest animals, and may even explain where any particular animal prefers to pass the night. Waorani hunters can detect the scent of animal urine at forty paces in the forest and can accurately identify the species of animal from which it came.

Confronted by this awesome sensitivity to the surrounding forest one cannot help but reflect on whether the ethnobotanical community has really paused to consider the implications of such awareness. Ethnobotanists recognize the shaman as an intellectual peer and properly delineate the experimental nature of shamanistic practice. Yet when we attempt to account for these discoveries, the phrase that is inevitably employed is "trial and error." It is a reasonable term and may well account for certain processes and transformations, but at another level it is a euphemism disguising our ignorance of how Amazonian Indians come up with their insights.

Consider, for example, two well-known Amazonian preparations—the arrow or dart poison curare and the hallucinogen ayahuasca. The former is derived principally from a number of species in several genera of lianas (e.g., species of *Chondrodendron*, *Abuta*, and *Curarea*), and the latter is also a liana (*Banisteriopsis* species). In both instances the active principles are found in the bark.

What is fascinating about these preparations from an epistemological point of view is their elaboration, which involves a number of procedures that are either exceedingly complex or that yield a product the use of which would not have been inherently obvious to the inventor. In the case of curare, the bark is rasped and placed in a funnel-shaped leaf compress suspended between two hunting spears. Cold water is then percolated through and the drippings collected in a ceramic pot. This dark-colored liquid is slowly heated over a fire and brought to a frothy boil numerous times until the fluid thickens. It is then cooled and later reheated until a thick layer of viscous scum gradually forms on the surface. This scum is removed, the dart tips are spun in the viscid fluid, and the darts are then carefully dried by the fire. The procedure itself is mundane. The realization, however, that this orally inactive substance, derived from but a handful of the hundreds of forest lianas, could kill when administered intramuscularly is profound.

In the case of ayahuasca it is the sophistication of the actual preparation that is impressive. The drug may be prepared in various ways but usually the fresh bark is scraped from the stem and boiled for several hours until a thick, bitter liquid is produced. The active compounds are the ß-carbolines harmine and harmaline, the subjective effects of which are suggested by the fact that when first isolated they were known as telepathine.

Significantly, the psychoactive effects of ayahuasca are enhanced dramatically by the addition of a number of subsidiary plants. This is an important feature of many folk preparations and it is due in part to the fact that different chemical compounds in relatively small concentrations may effectively potentiate each other. In the case of ayahuasca, the usual admixtures are the leaves of two shrubs (*Psychotria viridis* and *P. carthaginensis*) and a scandent liana (*Diplopterys cabrerana*) (Schultes and Hofmann 1980). All three of these plants contain tryptamines that are orally inactive, unless monoamine oxidase inhibitors are present. The ß-carbolines found in *Banisteriopsis caapi* are inhibitors of precisely this kind, and thus they potentiate the tryptamines. The result is a powerful synergistic effect, a biochemical version of the whole being greater than the sum of the parts.

The experimental process that originally led to the manipulation and combination of these morphologically dissimilar plants, and the discovery of their unique chemical properties, is far more profound than the phrase "trial and error" suggests. The patterns that any researcher—and the shaman most certainly has earned that title—observes in nature depend on cognitive constructs and an intellectual synthesis, and reflect, in turn, culturally patterned thoughts and values. Sensitivity to nature is not an innate attribute of South American Indians. It is a consequence of adaptive choices that have resulted in the development of highly specialized perceptual skills. Those choices, in turn, spring from a comprehensive view of nature and the universe in which humans are perceived as but an element inextricably linked to the whole.

It is this unique cosmological perspective that has enabled the shaman to comprehend implicitly the intricate balance that is the Amazon forest. Another worldview altogether, one in which the human race stands apart, now threatens the forest with devastation. Perhaps the most important contribution of the new synthesis in ethnobotany will be its ability to promote actively a dialogue between these two worldviews such that folk wisdom may temper and guide the inevitable development processes that today ride roughshod over much of the earth.

LITERATURE CITED

Anderson, A. B., and D. A. Posey. 1986. Manejo de cerrado pelos índios Kayapó. *Boletim do Museu Paraense Emílio Goeldi, Botânica*, 2(1): 77–98.

Alcorn, J. B. 1981. Haustec noncrop resource management: Implications for prehistoric rain forest management. *Human Ecology* 9: 395–417.

―――. 1984. Development policy, forests, and peasant farms: Reflections on Haustec-managed forests' contributions to commercial production and resource conservation. *Economic Botany* 38(4): 389–406.

Baker, M., D. Neill, W. Palacios, and J. Zaruma. 1987. *Plant Resources of Amazonian Ecuador*. Second annual report of "Flora del Ecuador," Institute of Economic Botany. Bronx, NY: New York Botanical Garden.

Balée, W. 1986. Análise preliminar de inventário florestal e a etnobotânica Ka'apor (Maranhão). *Boletim do Museu Paraense Emílio Goeldi, Botânica*, 2(2): 141–167.

―――. 1987. A etnobotânica quantitativa dos índios Tembé (Rio Gurupi, Pará). *Boletim do Museu Paraense Emílio Goeldi, Botânica*, 3(1): 29–50.

Balick, M. J. 1984. Ethnobotany of palms in the Neotropics. In *Ethnobotany in the Neotropics*. Eds. G. T. Prance and J. A. Kallunki. Advances in Economic Botany Series, vol. 1. Bronx, NY: New York Botanical Garden. 9–23.

―――. 1985. Useful plants of Amazonia: A resource of global importance. In *Amazonia*. Eds. G. T. Prance and T. A. Lovejoy. Key Environment Series. New York: Pergamon Press. 339–368.

Berlin, B. 1984. Contributions of Native American collectors to the ethnobotany of the Neotropics. In *Ethnobotany in the Neotropics*. Eds. G. T. Prance and J. A. Kallunki. Advances in Economic Botany Series, vol. 1. Bronx, NY: New York Botanical Garden. 24–33.

Berlin, B., D. E. Breedlove, and P. H. Raven. 1974. *Principles of Tzeltal Plant Classification: An Introduction to the Botanical Ethnography of a Mayan-Speaking Community in Highland Chiapas*. New York: Academic Press.

Boom, B. M. 1985a. "Advocacy botany" for the Neotropics. *Garden* 9(3): 24–28.

―――. 1985b. Amazonian Indians and the forest environment. *Nature* 314: 324.

―――. 1987. *Ethnobotany of the Chácobo Indians, Beni, Bolivia*. Advances in Economic Botany Series, vol. 4. Bronx, NY: New York Botanical Garden. 1–68.

Boorstin, D. J. 1983. *The Discoverers*. New York: Random House.

Bye, R. A. 1976. *Ethnoecology of the Tarahumara of Chihuahua, Mexico*. Ph.D. thesis, Harvard University.

Conklin, H. C. 1954. *The Relation of Hanunoo Culture to the Plant World*. Ph.D. thesis, Yale University, New Haven, CT.

Davis, E. W., and J. A. Yost. 1983a. The ethnobotany of the Waorani of eastern Ecuador. *Botanical Museum Leaflets* (Harvard University) 29(3): 159–217.

―――. 1983b. The ethnomedicine of the Waorani of eastern Ecuador. *Journal of Ethnopharmacology* 9(2–3): 273–298.

―――. 1983c. Novel hallucinogens from eastern Ecuador. *Botanical Museum Leaflets* (Harvard University) 29(3): 291–295.

Denevan, W. M., and C. Padoch. 1988. *Swidden-Fallow Agroforestry in the Peruvian Amazon*. Advances in Economic Botany Series, vol. 5. Bronx, NY: New York Botanical Garden. 1–107.

Farnsworth, N. R. 1982. The consequences of plant extinction on the current and future availability of drugs. Paper presented at AAAS Annual Meeting, 3–8 January, Washington, DC.

————. 1988. Screening plants for new medicines. In *Biodiversity*. Ed. E. O. Wilson. Washington, DC: National Academy Press. 83–97.

Farnsworth, N. R., and R. W. Morris 1976. Higher plants—the sleeping giant of drug development. *American Journal of Pharmacy* 148(2): 46–52.

Ford, R. I. 1978. Ethnobotany: Historical diversity and synthesis. In *The Nature and Status of Ethnobotany*. Ed. R. I. Ford. Anthropological Papers, no. 67. Ann Arbor, MI: University of Michigan Museum of Anthropology. 33–49.

Harshberger, J. W. 1896. Purposes of ethnobotany. *Botanical Gazette* 21(3): 146–154.

Holmstedt, B., and J. G. Bruhn. 1983. Ethnopharmacology—a challenge. *Journal of Ethnopharmacology* 8: 251–253.

Johns, T. 1990. *With Bitter Herbs They Shall Eat It*. Tucson: University of Arizona Press.

Kroeber, A. L. 1907. *The Arapaho*. American Museum of Natural History, Bulletin 18.

————. 1920. Review of uses of plants by the Indians of the Missouri River region, by Melvin Randolph Gilmore. *American Anthropologist* 22: 384–385.

Kvist, L. P., and L. B. Holm-Nielsen. 1987. Ethnobotanical aspects of lowland Ecuador. *Opera Botanica* 92: 83–107.

La Barre, W. 1938. *The Peyote Cult*. Yale University Publications in Anthropology, no. 19. New Haven, CT: Yale University Press.

Millspaugh, C. F. 1892. *Medicinal Plants*. Philadelphia: John C. Yorston & Company.

Mooney, J. 1889. Cherokee plant lore. *American Anthropologist* 2(3): 223–224.

————. 1891. The sacred formulas of the Cherokee. Bureau of American Ethnology 7th Annual Report (1885–1886). Washington, DC. 301–397.

Myers, N. 1983. *A Wealth of Wild Species*. Boulder, CO: Westview Press.

————. 1985. *The Primary Source*. New York: Norton & Company.

National Academy of Sciences. 1975. *Underexploited Tropical Plants with Promising Economic Value*. Report of an Ad Hoc Panel of the Advisory Committee on Technology Innovation. Washington, DC: National Academy of Sciences.

Palmer, E. 1871. Food products of the North America Indians. U.S. Commissioner of Agriculture Report, 1870. Washington, DC. 404–428.

Peeke, M. C. 1973. *Preliminary Grammar of Auca*. Norman, OK: The Summer Institute of Linguistics.

Pinkley, H. V. 1973. *The Ethno-ecology of the Kofán*. Ph.D. thesis, Harvard University.

Posey, D. A. 1979. *Ethnoentomology of the Kayapó Indians of Central Brazil*. Ph.D. thesis, University of Georgia.

————. 1983. Indigenous ecological knowledge and development of the Amazon. In *The Dilemma of Amazonian Development*. Ed. E. F. Moran. Boulder, CO: Westview Press. 225–257.

————. 1984. A preliminary report on diversified management of tropical forests by the Kayapó Indians of the Brazilian Amazon. In *Ethnobotany in the Neotropics*. Eds. G. T. Prance and J. A. Kallunki. Advances in Economic Botany Series, vol. 1. Bronx, NY: New York Botanical Garden. 112–126.

————. 1985. Native and indigenous guidelines for new Amazonian development: Understanding biological diversity through ethnoecology. In *Change in the Amazon Basin*. Vol. 1, *Man's Impact on Forests and Rivers*. Ed. J. Hemming. Manchester, England: Manchester University Press. 156–181.

Powers, S. 1875. Aboriginal botany. *California Academy of Sciences Proceedings* 5: 373–79.

Reichel-Dolmatoff, G. 1971. *Amazonian Cosmos*. Chicago: University of Chicago Press.

————. 1975. *The Shaman and the Jaguar*. Philadelphia: Temple University Press.

Robbins, W. W., J. P. Harrington, and B. Freire-Marreco. 1916. *The Ethnobotany of the Tewa Indians*. Bureau of American Ethnology Bulletin no. 55. Washington, DC. 1–124.

Schultes, R. E. 1987. Ethnopharmacological conservation: A key to progress in medicine. *Opera Botanica* 92: 217–224.

————. 1988a. Conservation looks to the medicine man. *Soc. Pharmacol.* 2(1): 83–91.

————. 1988b. *Where the Gods Reign*. Oracle, AZ: Synergetic Press.

Schultes, R. E., and A. Hofmann. 1980. *The Botany and Chemistry of Hallucinogens*. 2nd ed. Springfield, IL: Charles C. Thomas.

Stevenson, M. C. 1915. Ethnobotany of the Zuñi Indians. Bureau of American Ethnology 30th Annual Report (1908–1909). Washington, DC. 35–102.

Teit, J. A. 1930. *The Ethnobotany of the Thompson Indians of British Columbia*. Bureau of American Ethnology 45th Annual Report (1927–1928). Washington, DC. 441–552.

Vestal, P. A., and R. E. Schultes. 1939. *The Economic Botany of the Kiowa Indians as It Relates to the History of the Tribe*. Cambridge, MA: Botanical Museum of Harvard University.

Vickers, W. T., and T. Plowman. 1984. *Useful Plants of the Siona and Secoya Indians of Eastern Ecuador*. Fieldiana: Botany, n.s., no. 15. Chicago: Field Museum of Natural History.

Ethnobotanical Method and Fact: A Case Study

FRANK J. LIPP

Ethnobotany is the study of the interactive relationships between nonindustrial societies and their floral environment. Although the term is of recent coinage, ethnobotanical research has a long and rich history, which includes the Egyptian queen Hatshepsut, who in 1495 B.C. sent an official to distant areas to collect living specimens of fragrant trees (Coats 1970, p. 243). Similarly, in the New World the Aztecs regularly dispatched envoys to distant provinces in search of new medicinal and ornamental plants, which were brought back wrapped in beautifully woven mantles, the roots being packed in balls of earth (Lipp 1976, p. 188). The earliest known ethnobotanical work, *The Condition of the Flora of the Southern Region*, was written at the end of the third century A.D. by Hi-Han and relates to the Chinese introduction and utilization of numerous plants from Southeast Asia (Millot 1968, p. 1741).

Although ethnobotanical studies have been commonly carried out by botanists (e.g., Melvin R. Gilmore) and by anthropologists (e.g., David P. Barrows, Ralph L. Roys), it is noteworthy that several scientists made important contributions to both botany and anthropology. Harley H. Bartlett and Joseph F. Rock, whose major field of research was botany, also made significant contributions to Asian ethnology (Rock 1947; Voss 1961; Sutton 1974). The eminent anthropologist Alfred L. Kroeber produced a statistical study of the interisland floral relationships of the Galápagos Archipelago (1916). Eduard Seler, the distinguished scholar and organizer of Middle American anthropology, collected numerous ethnobotanical specimens in Mexico, resulting in the description of 160 new species and genera; 60 species and two genera were named for him and his wife, Caecilie Seler-Sachs (Loesener 1922, pp. 321–324).

The fields of ethnobotanical research principally encompass the utilization of plants in non-Western societies; folk taxonomy; archaeological vegetal remains; the origins and domestication of cultivated plants; the ecological effects of human activity upon plant communities; and the symbolic role of plants in religion, folklore, and the monuments of early civilization. In the 1980s, ethnobotanical mitigation, which comprises the protection and transplanting of plant resources valuable to American Indians, became a major component of federally mandated vegetative management and environmental impact programs (Peri et al. 1983).

Even a cursory study of the vast literature dealing with the aforementioned topics indicates that the theoretical and methodological approaches involved depend primarily upon the investigator's principal discipline, whether it be botany, anthropology, agronomy, pharmacology, geography, or another science. Like psychoneuroimmunology and other new disciplines, it is the interdisciplinary and integrative nature of the field that vastly enriches it. At the same time, it should be clear that no single theoretical framework and methodology can encompass the diverse areas of study within the field of ethnobotany. Even a broadly based, ecologically oriented systems approach would be of limited use to, for example, linguists studying local plant names as a means of identifying the origins and movements of ancient peoples (Austerlitz 1962; Friedrich 1970).

This chapter is concerned primarily with the conceptual and logical basis of ethnobotanical research. Although research techniques cannot be entirely separated from the examination of their logic-in-use, ethnobotanical research methods have been adequately described by Coville (1895), Fosberg (1960), and Lipp (1989), among others.

Two aspects of ethnobotany are the search for plants with biodynamic properties and the study of the ongoing plant domestication process. Although some commonalities exist, the theories and methods employed in the two endeavors differ substantially. To elucidate the investigative processes involved in these two facets of ethnobotany, case material from the author's research will be used.

The search for bioactive plants with potential medicinal, nutritive, and other applications is carried out by a systematic search of the literature, herbaria, and ethnobotanical surveys in the field, or a combination of the three. Of a Brazilian specimen of *Dimorphandra parviflora* Spruce ex Bentham collected in 1851, Richard Spruce (*1465*) annotated: "From the seeds of this a snuff is made. *Paricá* Ling. Ger." This information was encountered in The New York Botanical Garden Herbarium on a photographic copy of a herbarium sheet located at the Royal Botanic Gardens, Kew. Since this species is taxonomically related to *Elizabetha princeps* Schomburgk ex Bentham, an ingredient in Amazonian narcotic snuffs (Schultes and Holmstedt 1968), chemical studies would appear to be warranted. The Lingôa Geral term *paricá* is commonly associated with *Anadenanthera peregrina*, although the genus does not occur in northwestern Amazonas where Spruce collected (Schultes 1954). The term *paricá* is often employed for any snuff (e.g., that prepared from *Virola*) (Schultes, pers. com.). The later numbers of Spruce were never named and listed by Bentham. Therefore other undescribed ethnobotanical specimens may be located among them.

Species of the genus *Cyperus* are taken separately or in combination with *Banisteriopsis*, *Datura*, and other hallucinogenic plants by the Yagua Indians of the Peruvian Amazon (Chaumeil and Chaumeil 1979, pp. 45–53). *Cyperus* species corresponding to the common name *pirípirí* are also recognized as being hallucinogenic by the Jívaro Indians of eastern Ecuador (Harner 1973, p. 137). Although the species involved are not identified, at least one, *C. articulatus* Schlechter, is reported to have "intoxicating" effects (von Reis and Lipp 1982, p. 15). Cascudo (1968, p. 62; 1971, p. 538) cited the root and bark of *Pithecellobium diversifolium* Bentham, along with *Acacia jurema* Martius and *Mimosa nigra* Huber, as an ingredient in the hallucinogenic potions and cigars widely used in the possession cults (*catimbó*) of northeastern Brazil.

In North America, several plants from the Pacific Northwest indicate physiological effects that warrant phytochemical analysis. A club-moss, *Lycopodium selago* L., was employed along the Pacific Coast to produce a "sort of intoxication" by chewing the stems and swallowing the juice (Gorman 1896, p. 80). Significantly, a Peruvian *Lycopodium* species is used as an additive to a mescaline cactus potion to increase its effects (Dobkin 1968, p. 191). Also, the bitter roots of *Lupinus nootkatensis* Donn were ingested by the Tlingit, Kwakiutl, and perhaps other American Indian tribes to induce a state of intoxication (Kurtz 1894, p. 351; Boas 1921, pp. 199, 551). Unfortunately, specimen sheets of this lupine deposited in the herbarium of the New York Botanical Garden by Franz Boas and George

Hunt do not contain any ethnobotanical notes of interest. The Tlingit also made an intoxicant by boiling the roots of the "dead persons berries," *Streptopus amplexifolius* (L.) DC. (De Laguna 1972, p. 409).

Although the medicinal and alimentary aspects of *Oplopanax horridus* Miquel were extensively reviewed by Turner (1982), it should be noted that Tlingit shamans subsisted solely on the root juices of this plant to obtain dream visions (Holmberg 1856, p. 349; Erman 1870, p. 370). The plant was also said to increase hypnotic powers and give the shamans "power over others" (Palmer 1871, p. 370; Gorman 1896, p. 73). A hypoglycemic substance found in the roots of this species, as well as food abstinence, causes blood glucose levels to fall (Large and Brocklesby 1938). Severe hypoglycemia may result in loss of consciousness and in convulsions, conditions not infrequently found in the literature relating to shamanic behavior (Layard 1930, p. 528; Lewis 1971, pp. 179–185; Bolton 1981, pp. 264–265).

Dennis Rosmini (pers. com. 1987) reported that during an influenza epidemic in which many people died in the Haida village of Skidegate, a gravely ill girl was administered teas made from devil's club (*Oplopanax horridus*). The girl claims to have experienced a significant dream vision, and she attributes her vision and subsequent recovery to the tea. On the Asiatic side of the Pacific Ocean, Tungus, Gilyak, and Ainu shamans induce a state of ecstasy by inhaling the smoke and chewing the roots of *Ledum palustre* L. (Knoll-Greiling 1959, p. 54; Ohnuki-Tierney 1980, p. 208n.).

Folk tales and myths frequently enable the researcher to shed light on the use of biodynamic plants in a particular culture (Lucas 1960; Miller 1963). In a tale of the Pima-Papago tribe of southern Arizona and northwestern Mexico, a cannibal monster is made "crazed, as if drunken" and put to sleep with the use of narcotic cigarettes (Curtis 1908, p. 19). One of the ingredients of these cigarettes was a lichen called *jevud hiosig*, "earth flower," which, when mixed with tobacco and smoked, has a dizzying, narcotic effect (Castetter and Underhill 1935, p. 27; Castetter and Bell 1942, p. 112).

Lenora Curtin's (1949, pp. 77–78) Piman informants likened the effects of this lichen to that of marihuana and noted that the smoking of this lichen "makes young men crazy." During ethnobotanical research among the Papago, samples of this lichen were obtained and later identified as *Parmelia conspera* (Ehrhart) Ach. Considerable magical potency is attributed to this plant by the Papago-Pima, since it is also used for success in hunting, in gambling, in love, and for disposing of an enemy. Whether these specimens account for the use of lichens with similar properties by the Mohave (Devereux 1949, pp. 111–112) and Kiowa (Vestal and Schultes 1939, p. 12) or whether other lichen genera are involved cannot be presently ascertained.

Eupatorium solidaginifolium A. Gray or *pihol* is also smoked by the Papago as a substitute for tobacco, producing a slight nervous tremor. It is said to make people "crazy," a property attributed also to *Nicotiana trigonophylla* Dun. or *wiyowpul*. This effect may be due to the high alkaline content of these two species and the method of smoking employed: deliberately swallowing the smoke and exhaling slowly from the nostrils. A related plant with similar properties, *Eupatorium berlandieri* DC., was used as a tobacco substitute by the Apache (Rothrock 1880, p. 65). Coury (1969, p. 221) stated that the genus *Eupatorium* has narcotic properties and identified it with the Aztec *quauxoxouqui*.

Bourke (1894, p. 125) noted that American Indians of southeastern Texas smoked the leaves and bulb of a "drago" to induce ecstatic visions. Utilized to prevent tooth decay and gum diseases, the only plant currently used by the name *drago* in the Texas-Mexico borderlands is *Jatropha dioica* Cer. (Clarissa Kimber, pers. com. 1975). In his work on the flora of Texas, Havard (1885, p. 513) identified *drago* as a different euphorbaceous plant, *Mozinna spathulata* Ortgies.

The foregoing body of data is an expression of the natural historical approach to field method, which exhaustively follows the details of a topic for its intrinsic interest, with a minimum of generalizations. At the same time there appears to be little or no connection with any specific scientific problem. Systematic data, once accumulated, will have value—sometimes in wholly new directions, and new hypotheses and insights into a problem will emerge or be suggested. Empirical data are kept as free from interpretation as possible and should enter the picture only after adequate data have been collected. Hypotheses are deduced from systematically ordered data rather than from preconceived, analytic, hypothesis-formulations.

The tradition of collecting and preserving data is deeply rooted in Western intellectual history. This tradition derives its force and stimulus from the utility of the information for the solution of practical concerns and from the awareness that the knowledge and products of human activity are rapidly being destroyed on the frontiers of an expanding civilization.

The notion, however, that data collection is itself value free, that facts are facts, oversimplifies the understanding of the scientific process and invests the extensive collection of "fact" with a false sense of truth. Consequently, natural scientific description has moved in the direction of theory construction that would not do violence to the characteristics of particular cases.

In this methodological approach, scientific activity is thought of primarily in terms of the framing and testing of hypotheses. The researcher conceives his or her first task to be the careful consideration of alternative hypotheses; data are then brought to bear on them so that they may be accepted or rejected, modified, or refined. The collection of data is then guided by their pertinence to the hypothesis to be verified. A field situation where this method was employed is the study of the domestication process among swidden agriculturists.

Several authorities on plant domestication, especially Darwin (1896, p. 338), Kempton (1936), and Weatherwax (1936, 1942) have regarded cultigens as a product of the plant-breeding skill of American Indians, who took useful wild plants into cultivation, acquired the efficiency of directed or artificial selection, and used this knowledge in the improvement of favorable varieties and the transformation of wild species of plants. Others, such as Manglesdorf (1963, p. 39), Darlington (1963), Simmonds (1962), and De Wet and Harlan (1975), have emphasized the role of population genetic factors in the domestication process.

To investigate the interaction and relative importance of bioevolutionary and cultural factors in the plant domestication process, a set of hypotheses was formulated and tested in a longitudinal study of the annual agricultural cycle among swidden farmers in highland Oaxaca, Mexico.[1] Methods of research consisted of participant observation, key informant and structured interviewing, and the collection of native texts. Field observations included measurements of sample kernel and ear characteristics; altitude, temperature, varietal flowering and maturation rates; mapping of locations and distances between corn fields; and soil and vegetative analyses of sample swidden fields at various stages of their crop-fallow cycle. Although the field data have not been completely analyzed, preliminary results are as follows.

Though the number of varieties and the annual agricultural cycle vary greatly in the region, the methods of artificial selection, and the maintenance and diffusion of the maize-dominated seed complex were quite similar in the villages studied. Maize varieties are classified into natural groupings according to ecological criteria and color; secondary criteria include size and weight, maturation period, and the presence or absence of dentation and starch. Microenvironmental adaptation for each variety is the primary selective criterion for maize as well as for beans (*Phaseolus*), and squashes and pumpkins (*Cucurbita*). Prior to and during planting, maize seeds are selected on the basis of viability of the embryo; rejection of butt and tip kernels is predicated on the absence of embryo viability.

When harvested, the maize ears are separated by size into groups for storage: the smallest and poorest ears are eaten first, and the largest ears are saved as seed corn. Periodically during the year, the stored corn is sorted and putrid ears are removed, exerting selective pressure for phenotypes resistant to rot disease. Seed for planting may be exchanged or bought from friends, neighbors, or adjacent villages. Excellence of seed overrides price or social factors in the acquisition of seed for planting.

Diffusion of maize is dependent also upon the swidden cycle, since cultivation of a new plot necessitates obtaining seed for an ecological zone different from the previous planting. Careful, selective breeding of maize and particular ecological and biological factors operate equally in the domestication process to maintain intravarietal homogeneity and intervarietal diversity.

When foreign varieties are introduced into a suitable microenvironment, however, these factors cannot prevent varietal hybridization, which in turn results in increased yields and vigor, but renewed selective breeding greatly diminishes the chances for the future derivation of new varieties. Although having a high degree of genetic stability, all native varieties (except for one corresponding to a high-altitude, early maturing form of Olotón) are the result of the past hybridization of sets of prehistoric mestizo races (Major Goodman, pers. com. 1981).

Cultural selection practices and cultigen diversity are affected only in part by the growth of nonfarming occupations, basic food imports, and integration into the regional market economy. Varieties of maize and other crops are grown in experimental plots to determine their viability under various ecological conditions. Hybridization between native varieties is recognized by changes in seed color and is avoided by planting varieties separate from each other. If hybridization has occurred, farmers recover their preferred color variety by removing from the harvest those colored kernels that will be used in planting.

Zea mexicana (Schrader) Kuntze, or *ca·mo·k* in Mixe, is interplanted with maize or grown separately and is processed for consumption by soaking in lime to remove the hard pericarp. It is not widely cultivated with maize, however, since it tends to grow amassed, which, according to the Mixe, prevents air from freely entering the fields, retarding maize growth and making the fields difficult to weed.

Artificial selection of maize prior to and during planting and at harvest times was ascertained by means of systematic observations of the agricultural cycle. That Johannessen (1982), who used only interviewing techniques in his study of maize domestication in Guatemala, did not uncover these determining elements of the process suggests that the catalytic transfer and integration of anthropological and botanical methods probably is the best means for the study of the plant domestication process.

Conclusion

This chapter briefly outlined two methodologies in use in ethnobotanical research. Although their differences have been emphasized, the total demands of both positions include data and generalization, hypotheses and observation. Even the seemingly atheoretical, fact-gathering approach involves theories, implicit or explicit, in the way in which the data are assembled and described.

The logic of theory construction includes, of necessity, a comparison and comprehensive analysis of empirical data, the elaboration and analysis of different hypothetical variants explaining the data, and comparison of these variants among themselves and with the data. This can be achieved only by using deductive, in addition to inductive, methods of reasoning, which may intertwine in a striking fashion. Given the plethora of data relating to the human-plant relationship, ethnobotany can successfully preclude itself from being overwhelmed in an ocean of facts only by carrying out investigations that are strengthened by a sound theoretical, as well as methodological, orientation, since the essence of science lies not in discovering facts but in finding new ways to think about them.

NOTES

[1]The field work upon which this paper is partially based was conducted over a 24-month period between 1978 and 1981 in the Distrito de Mixe, southeastern Oaxaca, Mexico.

ACKNOWLEDGMENTS

Financial support was provided by N.S.F. grant number 77-19159 and a Grant-in-Aid from the Wenner-Gren Foundation for Anthropological Research. Major M. Goodman, North Carolina State University at Raleigh, provided identifications of the maize varieties from winter plantings in Florida. I am also indebted to Richard Prince and Rupert Barneby of The New York Botanical Garden for providing identifications of the botanical specimens collected. Voucher specimens for the plants cited in this paper have been deposited at The New York Botanical Garden Herbarium.

LITERATURE CITED

Austerlitz, R. 1962. A linguistic approach to the ethnobotany of South-Sahalin. *Proceedings, 9th Pacific Science Congress, 1957*, Vol. 4: 302–303.
Boas, F. 1921. *Ethnology of the Kwakiutl.* Bureau of American Ethnology, 35th Annual Report 1913–1914, part 1. Washington, DC.
Bolton, R. 1981. Susto, hostility and hypoglycemia. *Ethnology* 20: 261–276.
Bourke, J. G. 1894. Popular medicine, customs, and superstitions of the Rio Grande. *Journal of American Folklore* 7: 119–146.
Cascudo, L. 1968. *Dicionário do Folclore Brasileiro.* Vol. 2. Rio de Janeiro: Edicôes de Ouro.
————. 1971. *Antologia do Folclore Brasileiro.* Sao Paulo: Biblioteca de Ciências Sociais, Martins.
Castetter, E. F., and W. H. Bell. 1942. *Pima and Papago Indian Agriculture.* Albuquerque, NM: University of New Mexico Press.
Castetter, E. F., and R. M. Underhill. 1935. Ethnobiological studies in the American Southwest. II. *The Ethnobiology of the Papago Indians.* University of New Mexico Bulletin 275, biological series, vol. 4, no. 3.
Chaumeil, J., and J. P. Chaumeil. 1979. Chamanismo yagua. *Amazonia Peruana* 2(4): 35–69.
Coats, A. M. 1970. *The Plant Hunters.* New York: McGraw-Hill.
Coury, C. 1969. *La Médicine d'l Amérique Précolombienne.* Paris: Les Editions Roger Dacosta.
Coville, F. V. 1895. Directions for collecting specimens illustrating the aboriginal uses of plants. *Bulletin, U.S. National Museum,* no. 39, part J.
Curtin, L. S. M. 1949. *By the Prophet of the Earth.* Santa Fe, NM: San Vincente Foundation.
Curtis, E. S. 1908. *The North American Indian.* Vol. 2. Cambridge: University Press.
Darlington, C. D. 1963. *Chromosome Botany and the Origins of Cultivated Plants.* New York: Hafner.
Darwin, C. 1896. *The Variation of Animals and Plants Under Domestication.* Vol. 1. New York: D. Appleton & Company.
De Laguna, F. 1972. *Under Mount Saint Elias: The History and Culture of the Yakutat Tlingit.* Smithsonian Contributions to Anthropology, vol. 7, part 1. Washington, DC: Smithsonian.
Devereux, G. 1949. Magic substances and narcotics of the Mohave Indians. *Brit. J. Med. Psychol.* 22: 110–116.
de Wet, J. M., and J. R. Harlan. 1975. Weeds and domesticates: Evolution in the man-made habitat. *Economic Botany* 29: 99–107.

Dobkin, M. 1968. Trichocereus pachanoi—a mescaline cactus used in folk healing in Peru. *Economic Botany* 22: 191–194.

Erman, A. 1870. Ethnographische Wahrnehmungen und Erfahrungen an den Küsten des Berings-Meeres. *Zeitschrift f. Ethnologie* 2: 295–327, 369–393.

Fosberg, F. R. 1960. Plant collecting as an anthropological field method. *El Palacio* 67: 125–138.

Friedrich, P. 1970. *Proto-Indo-European Trees: The Arboreal System of a Prehistoric People*. Chicago: University of Chicago Press.

Gorman, M. W. 1896. Economic botany of southeastern Alaska. *Pittonia* 3: 64–85.

Harner, M. J. 1973. *The Jívaro*. Garden City, NY: Anchor Press/Doubleday.

Havard, V. 1885. Report on the flora of western and southern Texas. *Proceedings, U.S. National Museum* 8: 448–533.

Holmberg, H. J. 1856. Ethnographische Skizzen über die Völder des Russischen Amerika. *Acta Societatis scientarum Fennicae*, vol. 4. Finska ventenskap-societeten, Helsingsfors. 281–422.

Johannessen, C. L. 1982. Domestication process of maize continues in Guatemala. *Economic Botany* 36: 84–99.

Kempton, J. H. 1936. Maize as a measure of Indian skill. In *Symposia on Prehistoric Agriculture*. University of New Mexico Bulletin 296: 19–28.

Knoll-Greiling, U. 1959. Rauschinduzierende Mittel bei Naturvolkern und ihre individuelle und soziale Wirkung. *Sociologus* 9: 47–60.

Kroeber, A. L. 1916. Floral relations among the Galapagos Islands. University of California Publications in Botany, vol. 6, no. 9: 199–220.

Kurtz, F. 1894. Die Flora des Chilacatgegietes im südöstlichen Alaska. *Botanische Jahrbücher f. Systematik, Pflanzengeschichte, u. Pflanzengeographie* 19: 327–431.

Large, R. G., and H. N. Brocklesby. 1938. A hypoglycemic substance from the roots of the devil's club (*Fatsia horrida*). *Canadian Medical Association Journal* 39(1): 32–35.

Layard, J. W. 1930. Shamanism: An analysis based on comparison with the flying tricksters of Malekula. *J. Royal Anthropol. Society of Great Britain and Ireland* 60: 525–550.

Lewis, I. M. 1971. *Ecstatic Religion: An Anthropological Study of Spirit Possession and Shamanism*. Baltimore, MD: Penguin Books.

Lipp, F. J. 1976. A heritage destroyed: The lost gardens of ancient Mexico. *Garden Journal* 26: 184–188.

———. 1989. Methods for ethnopharmacological field work. *Journal of Ethnopharmacology* 25: 139–150.

Loesener, T. 1922. Ueber Maya-Namen und Nutzwendung yucatekischer Pflanzen. In *Festschrift Eduard Seler dargebracht zum 70. Geburtstag von Freunden, Schülern u. Verehrern*. Ed. W. Lehman. Stuttgart: Strecker & Schröder. 321–343.

Lucas, E. H. 1960. Folklore and plant drugs. *Papers, Michigan Academy of Sciences, Arts, Letters* 45: 127–136.

Manglesdorf, P. C. 1963. The evolution of maize. In *Essays on Crop Plant Evolution*. Ed. J. Hutchinson. Cambridge: University Press. 23–40.

Miller, O. H. 1963. Search for new drugs in medical folklore. *J. Amer. Pharmacol. Association*, n.s., 3(3): 131–132.

Millot, J. 1968. L'ethnobotanique. In *Ethnologie Générale*. Ed. J. Poirier. Paris: Editions Gallimard. 1740–1766.

Ohnuki-Tierney, E. 1980. Shamans and *Imu*: Among two Ainu groups—toward a cross-cultural model of interpretation. *Ethos* 8: 204–228.

Palmer, E. 1871. Food products of the North American Indians. U.S. Commissioner of Agriculture Report, 1870. Washington, DC. 404–428.

Peri, D. W., S. M. Patterson, and J. L. Goodrich. 1983. *Ethnobotanical Mitigation, Warm Springs Dam—Lake Sonoma, California*. Prepared for U.S. Army Corps of Engineers, San Francisco District.

von Reis, S., and F. J. Lipp. 1982. *New Plant Sources for Drugs and Foods from The New York Botanical Garden Herbarium*. Cambridge, MA: Harvard University Press.

Rock, J. F. 1947. *The Ancient Na-khi Kingdom of Southwestern China*. Harvard-Yenching Inst. Monograph Series, vols. 8–9. Cambridge, MA: Harvard University Press.

Rothrock, J. T. 1880. Notes on economic botany of the western United States. *Pharmaceut. J. and Trans.* 107: 664–666.

Schultes, R. E. 1954. A new narcotic snuff from the northwest Amazon. *Botanical Museum Leaflets* (Harvard University) 16: 241–260.

Schultes, R. E., and B. Holmstedt. 1968. The vegetal ingredients of the myristicaceous snuffs of the northwest Amazon. *Rhodora* 70: 113–160.

Simmonds, N. W. 1962. Variability in crop plants, its use and conservation. *Biol. Rev.* 37. 422–465.

Sutton, S. B. 1974. *In China's Border Province: The Turbulent Career of Joseph Rock, Botanist-Explorer*. New York: Hastings House.

Turner, N. J. 1982. Traditional use of devil's club (*Oplopanax horridus*; Araliaceae) by native peoples in western North America. *Journal of Ethnobiology* 2(1): 17–38.

Vestal, P. A., and R. E. Schultes. 1939. *The Economic Botany of the Kiowa Indians as It Relates to the History of the Tribe*. Cambridge, MA: Botanical Museum of Harvard University.

Voss, E. G. 1961. Harley Harris Bartlett. *Torrey Botanical Club Bulletin* 88: 47–66.

Weatherwax, P. 1936. The origin of the maize plant and maize agriculture in ancient America. In *Symposia on Prehistoric Agriculture*. University of New Mexico Bulletin 296: 11–18.

———. 1942. The Indian as a corn breeder. *Proceedings, Indiana Academy of Science* 51: 13–21.

Ethnobotany Today and in the Future

GHILLEAN T. PRANCE

Ethnobotany has reached an exciting time in its history. For many years this important science has been concerned primarily with cataloging the numerous uses of plants by indigenous peoples around the world. This is the vital foundation for any biological science. Without a catalog or inventory of the interactions between people and plants, we cannot progress into the other areas that will characterize ethnobotany in the future. Even in the cataloging phase, ethnobotany has been a multidisciplinary study involving the interaction between botanists, anthropologists, chemists, pharmacognosists, and specialists in other fields.

Not only have the plants been described as, for example, in the many publications of Richard E. Schultes in the *Botanical Museum Leaflets* of Harvard University, but collaborating chemists have reported a host of new compounds based on these botanical collections. Many of these compounds have been proven to be biologically active, and, as a result of this work, some have been found in societies far beyond the single indigenous group within which they were discovered.

Without the vast store of basic knowledge about plant uses, plant chemistry, and indigenous cultures, the present phase of ethnobotanical research could not begin. It is tragic that the inventory is far from complete, when the plant knowledge of indigenous peoples built up over thousands of years of interaction with their environment is being lost at an ever-increasing rate.

Scientists will not be able to gather ecological, agroforestry, or soil management data when indigenous groups have been acculturated or exterminated. It is, therefore, all the more urgent to consider now what information we should gather from the remnants of the great indigenous cultures of the world, such as those of the South American Indians, the African bushmen, or the natives of New Guinea. It is imperative that we have a complete understanding of the way of life of these peoples, rather than merely a catalog of the plants that are useful cures, good foods, or the best fibers. Many of these useful plants may be lost forever if we do not learn quickly how they are cultivated and managed by the appropriate indigenous peoples.

The destruction of the world's tropical rain forests continues at an alarming rate. What is of even more concern is that much of this destruction obviously makes no economic sense in terms of results, for the costs outweigh the benefits for all concerned. The abandoned cattle pastures of the Trans-Amazon Highway, the ruined soils of Central Africa, or the devastated forests of Malaysia testify to this fact.

In most of these areas, prior to invasion by Western civilization, relatively high populations were sustained with much less damage to the environment. How were the Amazon Indians able to have such a large population in pre-Columbian times, yet still preserve an abundance of animals such as the now nearly extinct manatees and giant turtles of the Amazon, as well as the amazing diversity of the forest? Their management was obviously much more sound environmentally than that of the Westerners who have, in many regions, largely replaced them (see Roosevelt 1985).

It is, therefore, crucial to learn as much as we can about indigenous ecology and management systems before it is too late. It would be a pity to lose all this knowledge when the world is so desperately in need of information to better manage the vast area of its surface that lies between the tropics of Cancer and Capricorn. The challenge facing the ethnobotanist is to discover as much of this information as remains before it is too late and to pass it along to the people who can make decisions that may help to conserve both the peoples and the plants involved.

Although there is still considerable cataloging to be done in ethnobotany, and although this work is even more urgent today than in the past as indigenous groups are so rapidly losing their traditions or even becoming extinct (e.g., Davis 1977), many other developments are adding to the scientific interest and global importance of ethnobotany, four of which are discussed in this chapter.

An Interdisciplinary Approach

Ethnobotany has always been to some extent an interdisciplinary science. In particular, there has been considerable cooperation between botanists and chemists—for example, work on the various hallucinogenic plants of the Amazon (der Marderosian et al. 1968; Schultes and Holmstedt 1968; Schultes et al. 1969; Rivier and Lindgren 1972; Buckley et al. 1973; Schultes and Swain 1976) and work on the curare alkaloids (Marini-Bettalo et al. 1957). In general, however, ethnobotany has often lacked interaction between disciplines. Work has been carried out by botanists with very little training in ethnology and very little contact with anthropologists, or by anthropologists with little background in biology who usually collect poor specimens that are difficult to identify.

This separation of disciplines is encouraged, in part, by funding policies of various foundations. For example, some years ago I planned to collaborate with an anthropologist, Robert Carneiro, from a neighboring institution, the American Museum of Natural History. Our proposal to do a quantitative ethnobotanical inventory of a forest in the Kuikurú Indian territory was first rejected by the systematic biology section of the National Science Foundation because it was considered to be too anthropological. Later, our proposal was rejected by the anthropological section of the same organization because it was considered to be too biological! These rejections delayed the study by some eight years, but thanks to a private foundation (The John Edward Noble Foundation), this study has been completed (although with South American Indian groups other than the Kuikurú) with spectacular results (see "Quantitative Ethnobotanical Inventory" below). We have been able to have both a botanist and an anthropologist carry out quantitative inventories in indigenous areas of the Amazon.

One of the most important aspects for the future of ethnobotany is the maintenance of closer working relationships between different disciplines: botanists, anthropologists, chemists, soil scientists, and others. A good example of the efficiency of this team approach is the work coordinated by Darrell Posey with the Kayapó Indians of Gorotire, Brazil. Posey has been able to put together an impressive team to study the ethnobiology of the Kayapó. The new and interesting results about the ecology of the Indians, their management of the forests, their trailside plantations, and their use of many species of insects has been made possible by the interaction of entomologists, botanists, ecologists, zoologists, soil scientists, and even astronomers (see Posey 1984; Posey et al. 1984; Hecht and Posey 1989).

The Institute of Economic Botany of the New York Botanical Garden is finding similarly rewarding results in the work of a multidisciplinary team in the northeastern Peruvian Amazon. This research effort, which focuses on fruits native to the Peruvian Amazon, has included studies of the ecology of fruit trees in their natural environments, experiments in silvicultural management of fruit species, research into local traditional methods of production and management, and inquiries into the complexities of marketing systems involving fruits in the region. Members of the project team, which includes an anthropologist, ecologist, botanist, and forester, are getting results that not only are of interest to their specialized disciplines but which also provide a multifaceted view of the role of these plants in the societies and environments of the Peruvian Amazon, as well as an assessment of the economic potential of selected fruits.

One multidisciplinary project needing further work is the study of medicinal and stimulant plants of the tropical rain forest with *in situ* chemical analysis of the products actually used. To date, most chemical work has been done after botanists have collected and air-dried plants. The chemistry of dried plants is often quite different from that of fresh material or that of a mixture of several different plants (usually fresh) in a concoction: volatile compounds are easily lost through air-drying, and the analysis of individual plants ignores the reactions that may take place when chemicals from different plants are mixed together.

What is needed is a series of teams that consist of an ethnobotanist, an anthropologist, and a chemist working together in the field: the botanist to identify plants accurately, to collect them, to study their ecology and environment; the anthropologist to get the maximum amount of data about the uses of the plants, the way in which they are prepared and served, even the folklore surrounding them, and to distinguish magic plants from true medicinals; and the chemist to perform field tests, to fix the different concoctions by the method most appropriate and most likely to conserve the active ingredients, thus ensuring that these can be analyzed in the laboratory. Much chemical work done in the past yielded negative results because the analyses performed were not of the medicine used, but rather of dried plants. An interdisciplinary approach, however, would be likely to yield significant results in the future.

Ethnoecology

One of the developments in today's ethnobotany has been the inclusion of more ecology. Many of the exciting developments in tropical biology have been the results of a greater emphasis on ecological aspects, such as animal-plant interactions, coevolution, defense mechanisms, forest dynamics, tree demography, and light gap ecology. This has begun to carry over into ethnobotany and is already yielding results of considerable interest.

For example, it appears that the Amazon Indians have been quite successful at the management of areas of rain forest on a continuing, sustained yield basis. This is in marked contrast to many contemporary development and exploitation projects. There is a vast store of ecological information gathered by the Indians in more than 10,000 years of experience in the forest environment. Their ecological and management experience is in many respects probably much more useful than is the pure knowledge of plant uses. Indications from research suggest that Amazonian Indians have a phenomenal amount of ecological information to contribute and that it soon will be too late to glean even the remnant of that information which survives today.

Some of the best examples of such studies are again those of Posey and his team (e.g., Posey 1983a, 1984; Posey et al. 1984). The detailed study of the Kayapó Indians has shown many interesting aspects of ecology. A good example is the Kayapó habit of planting trees and herbs along forest trails. Trails that to the uninitiated appear to go through virgin forest are, in fact, carefully managed ecosystems with plantings that provide food, medicines, and other needs to hunters and travelers. Some shade-tolerant plants grow under this type of management, plants that would never thrive in open clearings of the field (Posey 1984). The Kayapó also have a phenomenal knowledge of the use of insects both as a defense mechanism for their crops and as a source of useful products such as honey (Posey 1983b).

Slash-and-burn agriculture systems often have been called wasteful because they use the fields for two or three years and then "abandon" them to regenerate. Studies have shown that in many indigenous groups this is far from the truth. Posey (1984) called these former agricultural areas *anything-but-abandoned fields* and showed that old fields are frequently revisited by the Indians and represent a continually managed resource that provides many items of importance to their life, including food plants, medicines, and game animals.

The Kayapó are not unique in their use and management of old fields. More and more studies are showing that these areas are consciously manipulated by indigenous agriculturalists. A good example is that of the Bora Indians in Peru, who have also been the subject of a multidisciplinary study (Denevan et al. 1984, 1985; Padoch et al. 1985; Denevan and Padoch 1988). In their study, Denevan and Padoch analyzed former fields from three to nineteen years old. These old fields, existing in varying stages of swidden fallow, continue to contribute important resources long after the staple annual crops have disappeared. The species-diverse forest that is allowed to regenerate contains a mixture of useful planted species (fruits such as uvilla, *Pourouma cecropiifolia* Martius, and umarí, *Poraqueiba sericea* Tul.) and numerous forest species, some of which also yield food, fiber, medicines, and other useful items.

Another ethnobotanical study that has gone far beyond cataloging useful plants is Janis Alcorn's work on the Huastec Mayans (Alcorn 1981, 1984, 1989). Focusing on plant-human interactions but concentrating on their ecological contexts and dynamic aspects, Alcorn has provided a broad and thorough picture of how one traditional Mexican population related to its vegetative environment. Among various interesting aspects of Huastec ethnobotany is the fact that this group also manages supposedly "abandoned" swidden fields.

The use of so-called fallow fields is by no means confined to the peoples of the New World. Kunstadter (1978) discussed this subject among the Lua' swiddeners in northeastern Thailand.

In the Kayapó study, Posey and Anderson (1985; Anderson and Posey 1989) found that 94 percent of the 368 plants collected in old fields were of medical significance to the Indians. This is quite apart from all the other useful products obtained from these areas: food, fish and bird baits, thatch, paints, oils, insect repellents, firewood, construction materials, fibers, and so on.

Another interesting aspect of the Kayapó study is the use of natural openings (light gaps) in the forest for the planting of crops. These clearings are created by natural tree falls or by the Kayapó when they fell a tree to collect something such as honey. Into such small openings in the forest the Kayapó plant a wide range of crops, including manioc, taro, cupa (*Cissus gongylodes* Burch ex Baker), yams, and beans. These crops thrive in such areas, where their productivity is greater than in open fields.

There are many other examples of the ecological knowledge of indigenous peoples (e.g., Salick 1989). As ethnobotany increasingly turns to this area, more exciting discoveries are being made. The Maué Indians have for many years cultivated the stimulant plant guaraná (*Paullinia cupana* var. *sorbilis* Ducke). What is poorly known is that the young leaves of this plant have around their margins a series of extra-floral nectaries that are much visited by protective ants. Since these nectaries disappear by the time the leaves are mature, they have been little studied scientifically. The Indians, however, know that the ants are friends and that the guaraná plants will grow better and produce more when infested by the ants that feed on the extra-floral nectaries of the young leaves. How different from the modern approach of using pesticides to kill off all insects, whether harmful or friendly!

Most modern farming in the Amazon suffers greatly from poor soils that are hard to manage. Much cherished for agriculture are the few patches of humus-rich, dark soil that occur in scattered areas throughout the Amazon, especially in areas surrounded by extremely poor, sandy soils. In Brazil this soil is termed *terra prêta dos indios* (Indian black soil), because it is an Indian-made soil dating back to former occupants of these areas (Smith 1980). These groups obviously had a technology to create what is the richest soil in the Amazon. They did not leave the trail of disaster and erosion that follows many contemporary agricultural projects on tropical soils.

New ecologically oriented papers are appearing increasingly (e.g., Carneiro's 1983 study of manioc use by the Kuikurú Indians), and the future of ethnobotany will certainly involve a great deal more emphasis on ecology. In some ways, the collection of this information is a more difficult and usually lengthier task than that set for themselves by earlier ethnobotanists. Such work is successfully carried out only when some deeper understanding of culture, not just that of plants, has been achieved; often long-term and, at times, tedious research is involved.

It is significant that most of the ethnobotanical studies that have provided ecological information are long-term and multidisciplinary. This means that in the future we must enable ethnobotanists to spend extended periods with the groups they are studying and we must encourage the participation of specialists from the many different fields that can contribute to ethnobotany, including anthropologists, geographers, chemists, soil scientists, entomologists, and others.

The Study of Peasant Agriculturists

Much ethnobotanical research is carried out among the indigenous tribal peoples of the world as it is somehow considered more adventurous or romantic to study a forest tribe than the peasant agriculturists who may live nearby. One of the most neglected aspects of ethnobotany today is the study of the *campesinos, caboclos, mestizos*, peasants, or whatever they are called locally. Many of these people also have an extensive knowledge of plants and of management of the environment in which they live (see Frechione et al. 1989; Parker 1989).

Although these people live in contact with modern medicines, they often still maintain large plant pharmacopeias and use native plant remedies. This occurs both from choice and necessity, as many are too poor to depend on purchased drugs. For example, van den Berg (1982, 1984) found over 600 medicinal plants used in the Ver-o-Peso market of Belém, in Pará, Brazil. These represent the plants known by and used by local peasant farmers (*caboclos*) and even city dwellers, not by remote Indian tribes.

A visit to the Belém market or to any other Amazonian market such as that of Iquitos at the other extreme of lowland Amazonia, quickly shows the variety of plants that are used by peasants and city dwellers. A wealth of fruits, tuber crops, medicinal plants, fibers, craft work, vegetables, and ornamental plants is found in these markets. These materials that are currently used by nontribal peoples merit study as much as do those used by the Indians. In fact, many of the plant uses, as well as management and processing techniques, have come from the Indians, some from tribes that are long extinct.

Research by members of the Institute of Economic Botany staff in Iquitos, Peru, has concentrated on the *campesinos* and is yielding interesting results. Directed by anthropologist Christine Padoch, these studies have been conducted both in the agricultural areas of the region and in the markets of Iquitos, and are being carried out in conjunction with an ecological study of locally underexploited fruits of further economic potential.

The study of *campesino* agriculture, especially in the village of Tamshiyacu, has indicated an agroforestry system that may be considered a market-oriented update of the traditional forms found among the Bora and other tribal groups of the region. After producing manioc and other annual crops, swidden fields gradually are turned into fruit orchards that with minimal labor required for management, remain productive for decades. Two of the plants central to this system are the umarí (*Poraqueiba sericea*) and the Brazil nut (*Bertholletia excelsa* Humboldt and Bonpland). The former is one of the most popular fruits in the Iquitos market, and its production yields a steady income to the farmers. The latter enters into production at a later stage. Into this system enter many other useful trees such as uvilla (*Pourouma cecropiifolia* Martius) and the oil-yielding *Couepia dolichopoda* Prance. After about thirty-five years of production, the secondary forest is felled to begin the cycle again. Unlike many other swidden systems, however, in this system not all the felled trees are burned; umarí trees are made into charcoal and sold, and the wood of some of the Brazil nut trees is used for construction. Under this type of management, the cycle can be repeated without undue loss of soil nutrients.

The ethnobotanical study of a system of this sort has considerable application to current research on methods of agroforestry. At the same time, the study of the market in Iquitos is providing economic and sociological data about the products of such villages as Tamshiyacu. It is also giving leads to the project as to what fruits should enter into the detailed ecological study of fruits that is now under way.

There are far too few data about such *campesino* ethnobotany and agricultural management systems. We need to encourage future students and researchers to carry out extensive work with these people who have learned much from the Indians while developing their own techniques and discovering new uses for plants. As with the Indians, the lives of the *campesinos* are centered on the plants they use for food, building materials, medicines, crafts, utensils, and many other necessities. Moreover, their ethnobotanical knowledge is also under threat from the appearance of huge cattle ranches, hydroelectric dams, aluminum factories (as in Bacarena, Brazil), and other development projects. The lifestyle and the inimitable knowledge of the *campesinos* is a fast-disappearing resource that could be used for more rational development of the region.

Quantitative Ethnobotanical Inventory

It is often stated that Amazonian Indians know uses for almost all the plants in the forest and that this is a way to learn how useful the standing rain forest could be. This type of statement frequently occurs in the conservation literature, and, indeed, a knowledge of indigenous use of the forest could be very useful for conservation. Until well into the twentieth century, however, there were few data, and certainly no quantitative data, to back the assertion that Amazonian Indians use most trees in the forest. For this reason, another project of the Institute of Economic Botany has been to make quantitative inventories of plant use in Amazonia (see Prance et al. 1987).

The first study, among the Panoan-speaking Chácobo Indians of Bolivia, has been completed by Brian Boom (1985a, 1985b, 1989). The method involves selecting a hectare of forest near an Indian village and making a standard inventory of the area, where every tree over 10 centimeters (4 inches) in diameter is measured, collected, and identified. We then seek to find out the uses the Indians have for each of the tree species, through informants both in the area of the inventory and through discussion in the village. The results are spectacular; the Indians use 82 percent of the species on the hectare (75 of 91 species), representing 95 percent of the individual trees (619 of 649 trees) (see Table 1).

Table 1. Results of a quantitative inventory of the Chácobo Indian use of rain forest, showing their use of the species on a 1-hectare (2.5-acre) plot (from Boom 1985a, 1985b).

Use category	Trees used		Species used	
	Number	Percent	Number	Percent
Commercial	5	0.8	1	1.1
Fuel	163	25.0	14	16.0
Medicinal	271	42.0	23	25.0
Construction & crafts	225	35.0	23	25.0
Food	264	41.0	33	36.0

A similar study has been carried out by anthropologist William Balée with the Tupí-Guaraní-speaking Ka'apor Indians of the extreme east of Brazilian Amazonia. It showed that the Ka'apor Indians use 100 percent of the trees and lianas over 10 centimeters (4 inches) in diameter, and data from subplots indicate that they use many of the other plants in the forest, including herbs, shrubs, and epiphytes (see Balée 1985, 1986).

Both botanist Boom and anthropologist Balée are continuing their studies with other tribes, the former with the Panaré of Venezuela and the latter with the Arawete in the Xingú River basin of Brazil. Data from these four widely separated tribes in different language groups, combined with those of primate specialist Katherine Milton on the Carib-speaking Arara of the lower Xingú, will give us the quantitative data needed to show how strong the relationship is between the Indians and the forest.

If these Indians, who are the remnants of much more extensive and diverse populations, have so much knowledge of the use of the rain forest, it is vital to discover this information before both the plant species and the Indian cultures become extinct. These data will also strengthen the case for conservation of the forest and of the indigenous cultures. In addition, data is needed on how the Indians manage their supplies of all these useful products, which products they simply extract from the forest and which they in some way manage as they do the plantings along the trails and in treefall gaps in the Kayapó region.

Conclusion

The kind of research discussed in this chapter will become the future of ethnobotany. It is research that expands on the already vast catalog both to quantify the data and to add a new ecological dimension through a multidisciplinary team approach. There is still much to do in ethnobotany, which could prove to be an even more exciting and rewarding field to the graduate students and young researchers of the future than it has already been to previous researchers. There is an urgent need to train more students in the field and to develop new positions for them.

ACKNOWLEDGMENTS

I am grateful to the staff members of The New York Botanical Garden Institute of Economic Botany, whose work I have cited here, for sharing the results of their research with me, principally William Balée, Brian Boom, Christine Padoch, and Charles Peters; to the Andrew W. Mellon Foundation and the John Edward Noble Foundation for support of the Institute and the work cited; to the National Science Foundation for support of much of my ethnobotanical work; and to Darrell Posey for helpful discussions. I thank William Balée, Michael Balick, and Christine Padoch for a critical reading of an earlier draft of this manuscript, and the many people in the countries discussed here who have collaborated with our research effort. This paper is Contribution No. IEB-8 of The New York Botanical Garden Institute of Economic Botany.

LITERATURE CITED

Anderson, A. B., and D. A. Posey. 1989. Management of a tropical scrub savanna by the Gorotire Kayapó of Brazil. Advances in Economic Botany Series, vol. 7. Bronx, NY: New York Botanical Garden. 159–173.

Alcorn, J. B. 1981. Haustec noncrop resource management: Implications for prehistoric rain forest management. *Human Ecology* 9: 395–417.

———. 1984. Development policy, forests, and peasant farms: Reflections on Haustec-managed forests' contributions to commercial production and resource conservation. *Economic Botany* 38(4): 389–406.

———. 1989. Process as resource: The traditional agricultural ideology of Bora and Haustec resource management and its implications for research. In *Resource Management in Amazonia: Indigenous and Folk Strategies*. Eds. D. A. Posey and W. Balée. Advances in Economic Botany Series, vol. 7. Bronx, NY: New York Botanical Garden. 63–77.

Balée, W. 1985. Ka'apor Indian forest management. Paper presented at the 84th Annual Meeting of the American Anthropological Association, December 1985, Washington, DC.

———. 1986. Análise preliminar de inventário florestal e a ethnobotânica Ka'apor (Maranhão). *Boletim do Museu Paraense Emílio Goeldi, Botânica*, 2(2): 141–167.

Boom, B. M. 1985a. "Advocacy botany" for the Neotropics. *Garden* 9(3): 24–28.

———. 1985b. Amazonian Indians and the forest environment. *Nature* 314: 324.

———. 1989. Use of plant resources by the Chácobo. Advances in Economic Botany Series, vol. 7. Bronx, NY: New York Botanical Garden. 78–96.

Buckley, J. P., R. J. Theobald, Jr., I. Cavero, B. A. Krukoff, A. P. Leighton, and S. M. Kupchan. 1973. Preliminary pharmacological evaluation of extracts of takini: *Helicostylis tomentosa* and *Helicostylis pedunculata*. *Lloydia* 36: 341–345.

Carneiro, R. 1983. The cultivation of manioc among the Kuikurú of the upper Xingú. In *Adaptive Responses of Native Amazonians*. Eds. W. Vickers and R. Hames. New York: Academic Press. 65–111.

Davis, S. 1977. *Victims of the Miracle*. New York: Cambridge University Press.

Denevan, W. M., and C. Padoch. 1988. *Swidden-Fallow Agroforestry in the Peruvian Amazon*. Advances in Economic Botany Series, vol. 5. Bronx, NY: New York Botanical Garden. 1–107.

Denevan, W. M., J. M. Treacy, J. B. Alcorn, C. Padoch, J. Denslow, and S. F. Paitan. 1984. Indigenous agroforestry in the Peruvian Amazon: Bora Indian management of swidden fallows. *Interciencia* 9: 346–357.

———. 1985. Indigenous agroforestry in the Peruvian Amazon: Bora Indian management of swidden fallows. In *Change in the Amazon Basin*. Vol. 1, *Man's Impact on Forests and Rivers*. Ed. J. Hemming. Manchester, England: Manchester University Press. 237–155.

Frechione, J., D. A. Posey, and L. F. da Silva. 1989. The perception of ecological zones and natural resources in the Brazilian Amazon: An ethnoecology of Lake Coari. Advances in Economic Botany Series, vol. 7. Bronx, NY: New York Botanical Garden. 260–282.

Hecht, S. B., and D. A. Posey. 1989. Preliminary results on soil management techniques of the Kayapó Indians. Advances in Economic Botany Series, vol. 7. Bronx, NY: New York Botanical Garden. 174–188.

Kunstadter, P. 1978. Ecological modification and adaption: An ethnobotanical view of Lua' swiddeners in northwest Thailand. In *The Nature and Status of Ethnobotany*. Ed. R. I. Ford. Anthropological Papers, no. 67. Ann Arbor, MI: University of Michigan Museum of Anthropology. 169–200.

der Marderosian, A., H. V. Pinkley, and M. F. Dobbins. 1968. Native use and occurrence of N,N-dimethyltryptamine in the leaves of *Banisteriopsis rusbyana*. *American Journal of Pharmacy* 140: 137–147.

Marini-Bettalo, G. B., et al. 1957. Nota 8. Gli alcaloidi della *Strychnos solimoesana* Krukoff. *Rend. Istituto Super. Sanita* 20: 242–357.

Padoch, C., J. Chota Inuma, W. de Jong, and J. Unruh. 1985. Amazonian agroforestry: A market-oriented system in Peru. Agroforestry systems 3: 47–58.

Parker, P. P. 1989. A neglected human resource in Amazonia: The Amazon Caboclo. Advances in Economic Botany Series, vol. 7. Bronx, NY: New York Botanical Garden. 249–259.

Posey, D. 1983a. Indigenous ecological knowledge and development of the Amazon. In *The Dilemma of Amazonian Development*. Ed. E. F. Moran. Boulder, CO: Westview Press. 225–258.

———. 1983b. Keeping of stingless bees by the Kayapó Indians of Brazil. *Journal of Ethnobiology* 31: 63–73.

———. 1984. A preliminary report on diversified management of tropical forests by the Kayapó Indians of the Brazilian Amazon. In *Ethnobotany in the Neotropics*. Eds. G. T. Prance and J. A. Kallunki. Advances in Economic Botany Series, vol. 1. Bronx, NY: New York Botanical Garden. 112–126.

Posey, D. A., and A. B. Anderson. 1985. Campo/Cerrado management by the Kayapó Indians. Paper presented at 24th meeting of the American Anthropological Association, Washington, DC.

Posey, D. A., J. Frechione, J. Eddins, L. F. da Silva, D. Meyers, D. Case, and P. MacBeath. 1984. Ethnoecology as applied anthropology in Amazonian development. *Human Organization* 43: 95–107.

Prance, G. T., W. Balée, B. M. Boom, and R. L. Carneiro. 1987. Quantitative ethnobotany and the case for conservation in Amazonia. *Conservation Biology* 1: 296–310.

Rivier, L., and J. E. Lindgren, 1972. Ayahuasca, the South American hallucinogenic drink: An ethnobotanical and chemical investigation. *Economic Botany* 26: 101–129.

Roosevelt, A. 1985. Resource management in the Amazon before the conquest: Beyond ethnographic projection. Advances in Economic Botany Series, vol. 7. Bronx, NY: New York Botanical Garden. 30–62.

Salick, J. 1989. Ecological basis of Amuesha agriculture, Peruvian Upper Amazon. Advances in Economic Botany Series, vol. 7. Bronx, NY: New York Botanical Garden. 189–212.

Schultes, R. E., and B. Holmstedt. 1968. The vegetal ingredients of the myristicaceous snuffs of the Northwest Amazon. *Rhodora* 70: 113–160.

Schultes, R. E., B. Holmstedt, and J. E. Lindgren. 1969. Phytochemical examination of Spruce's original collection of *Banisteriopsis caapi*. De plantis toxicariis e mundo novo tropicale commentationes. III. *Botanical Museum Leaflets* (Harvard University) 22: 121–164.

Schultes, R. E., and T. Swain. 1976. De plantis toxicariis e Mundo Novo tropicale commentationes. XIII. Further notes on *Virola* as an orally administered hallucinogen. *Journal of Psychedelic Drugs* 8: 317–324.

Smith, N. 1980. Anthrosols and human carrying capacity in Amazonia. *Ann. Assoc. Amer. Geog.* 70: 553–566.

van den Berg, M. E. 1982. Plantas medicinais na Amazônia. Conselho Nacional de desenvolvimento Científico e Tecnológico, Programa Trópico Umido, Belém, Brasil.

———. 1984. *Ver-o-Peso: The ethnobotany of an Amazonian market*. In *Ethnobotany in the Neotropics*. Eds. G. T. Prance and J. A. Kallunki. Advances in Economic Botany Series, vol. 1. Bronx, NY: New York Botanical Garden. 140–149.

Ethnobotany and Phytoanthropology

PRIYADARSAN SENSARMA AND ASHOKE K. GHOSH

Although the terms *ethnobotany* and *phytoanthropology* may appear to be synonymous, they are not; they are merely allied. To help readers understand the difference between the two terms, this chapter will clearly explain the connotations of both terms and identify their respective areas of research.

Ethnobotany

Etymologically, the term *ethnobotany* refers to the study of plants related to people. Harshberger (1896) applied this term to the study of plants used by primitive and aboriginal people. For some time few investigators were attracted to this new area of study, but from the fourth decade of the twentieth century, more and more people have begun investigations in various aspects of ethnobotany.

With the opening of new vistas of ethnobotanical studies, the scope of ethnobotany has now greatly enlarged. It is evident from the syllabus of ethnobotany suggested by Professor D. R. Nurez of the University of Murcia, Spain, that ethnobotany may be broadly divided into the following major subareas: introduction (basic concepts and relationship), general ethnobotany (concept and methodology, ethnotaxonomy, psychoethnobotany, socioethnobotany—plants and folklore), archaeobotany and paleoethnobotany, agroethnobotany, gastroethnobotany, technoethnobotany, pharmacoethnobotany, ethnotoxicology of plants, stimulating vegetals and hallucinogens, aromatic plants and perfumery, gardening, plants and environmental management, and plants and public education.

In ascertaining the scope of ethnobotany, Manilal (1988) stated, "Today, the term denotes the entire realm of useful relationship between plants and man." It has, therefore, become apparent that more emphasis is laid on ethnobotany with a view to understanding the usefulness of plants by studying their utilitarian value in different communities, especially among tribals or peoples in primitive societies.

Phytoanthropology

The word *phytoanthropology*[1] owes its origin to three Greek words: *phyton* (plant), *anthropos* (human), and *logia* (branch of knowledge). Hence, it may be said that the term refers to the branch of knowledge that studies humans, or communities of humans, from the plant perspective.

It may be observed that communities of people respond differently towards plants. All communities do not view the same plant with equal reverence or for the same uses. This is especially common in India, where people of different ethnic, linguistic, and religious groups, and even the same group at various cultural levels, live side by side, sometimes in small areas like villages or *tehsils*. Yet, human-plant relations of these communities are found to vary. Ten examples serve to illustrate this point.

Ficus religiosa L. is a holy plant to Buddhists and Brahmans, some of whom worship the plant and refuse to cut any part of it. On the other hand, Bhats use the leaves as an abortifacient.

Borassus flabellifer L. is very important economically, and most parts of the plant are used. In addition to the normal economic use of this plant, Santals take the ash of the male inflorescence as a contraceptive, while the Bhils use pulp of the heartwood for inducing abortion.

Santals apply powder of the seeds of *Butea monosperma* Taubert on sores, but Bhats use ash of the seeds as an ingredient in an abortifacient preparation.

Bhils chew the roots of young seedlings of *Bombax ceiba* Burman to increase sexual vigor, while Santals apply a paste of the flowers to ripen boils.

According to the *Agni Purana*, applying a mixture of the sap of *Anthocephalus chinensis* (Lamk) Richard ex Walp on sexual organs, along with honey and cane sugar, is beneficial for conjugal life. Baishnabs, however, consider the plant holy, rural Bengalees use a decoction of stem-bark to treat dyspepsia, and Santals take juice of the leaves for stomach trouble and make a decoction of the leaves for washing throat wounds.

Lodha women use pedicels (about 20 centimeters/8 inches long) of *Sagittaria sagittifolia* L. as an abortifacient in the first trimester of pregnancy, while Santal women take a decoction of the plant with black pepper for retention of the fetus. Rural Bengalees, Oraons, and Mundas eat the boiled tubers as vegetables.

Rural Bengalees, Oraons, and Mundas eat the green and ripe fruits of *Spondias pinnata* (L. f.) Kurz, but Santals apply a paste of root-bark with mustard oil to treat muscular pain.

Santals apply a decoction of the leaves of *Sida acuta* Burman f., with a paste of black pepper and lime, on swollen scrota, while rural Bengalees, Lodhas, and Oraons use stem-bark as cordage.

Rural Bengalees and Oraons apply powder of the seeds of *Nyctanthes arbortristis* L. with mustard oil to remove dandruff, chew stem-bark with betel leaves as an expectorant, and take a decoction of the leaves mixed with black pepper to treat intermittent fever. Santals eat fried leaves of this plant to treat influenza, and Lodhas eat boiled leaves as a vegetable.

One final example is *Ocimum sanctum* L. To Baishnabs in particular and to Hindus in general, this plant is holy. Lodhas also consider it to be holy. Rural Bengalees and Oraons, however, give juice of the leaves with common salt and honey to children for colds and coughs, apply a paste of the leaves with lime on ringworm, and mix the juice of leaves with mustard oil to treat earaches. Santals apply a paste of the leaves along with a paste of chilies to treat malarial fever.

These examples, which illustrate the diverging attitudes to uses of the same plant by different communities living in the same or similar environment, cause us to ask, Why do divergent attitudes occur even when communities live in the same or similar environments? Is it due to religious beliefs, social taboos, and prejudices, or is it lack of communication between communities of the same habitat? An investigator of phytoanthropology should attempt to learn the answers to these questions. Appropriate answers might help trace the cultural evolution of sundry communities.

It should be pointed out in this connection that studies on the cultural history of India have revolved around the contributions of hermitages and courts and that scant attention has been paid to the cultural history of pre-Aryan tribals and their contribution to the main stream of Indian culture. It is needless to mention that in India, an area with more than 15,000 phanerogamic species and approximately four tribes, the human-plant relationship has been very close from ancient times. Even today, rural Indians and tribal societies are intimate with plants and forests, but the relationship differs from community to community. Phytoanthropologists should try to find out the nature, extent, and reasons for differences among communities in their responses to the plants of their surroundings.

Conclusion

It appears, then, that while ethnobotany aims to discover the human-plant relationship, both utilitarian and esthetic, phytoanthropology studies the extent of similarities and dissimilarities in the responses of various communities to their plant neighbors, and the reasons thereof. Ethnobotany and phytoanthropology are interrelated and at times overlap, but they are not identical. Investigators in these two areas can contribute to one another and cooperate frequently, but the basic training of an ethnobotanist should concentrate on phytotaxonomy, while anthropology should form the basis of a phytoanthropologist's preparation.

Although some preliminary research in phytoanthropology has commenced, phytoanthropological investigations have not yet become extensive in India. In other countries, numerous anthropologists are working on human-plant relations, but few countries have many tribes in the same ecohabitat. India, with its rich floral wealth and varied tribal populations, offers opportunities for phytoanthropological investigations.

NOTES

[1]This term was perhaps used for the first time in a Ph.D. thesis submitted to Calcutta University by A. Das who worked under the supervision of Professor Ashoke K. Ghosh.

LITERATURE CITED

Harshberger, J. W. 1896. Purposes of ethnobotany. *Botanical Gazette* 21(3): 146–154.
Manilal, K. S. 1988. Linkages of ethnobotany with other sciences and disciplines. *SEBS News Letter* 7(1–3): 1–2.

PART 2

Socioethnobotany

One of the newer developments in ethnobotany deals with the question of how indigenous peoples can be compensated for sharing their ethnobotanical knowledge with the industrialized world. Many researchers, especially anthropologists, have come to believe that when a commercial company markets a product discovered originally from primitive societies, some kind of recompense should be made to the society from which the information was obtained. For lack of a more descriptive short name, we may call this new development "socioethnobotany."

There seems, fortunately, to be wide acceptance of this new approach to ethnobotany. The question, however, that often remains is this: "How can this approach be efficiently administered?" For example, in many regions where ethnopharmacological research is under way, money would either be useless or of little value and difficult for unacculturated people to handle. Numerous other avenues, however, may be open. It has been suggested that visits by doctors, nurses, and dentists might be arranged. In regions where these societies have had preliminary education in missionary schools, scholarships for further study of promising students might be possible. The basic desire is that indigenous peoples receive some tangible return for sharing their knowledge of the ambient flora with researchers from far-off regions.

Pharmaceutical and industrial organizations may soon set up methods to facilitate financial or other kinds of aid to indigenous peoples from whom information on native useful plants results in the discovery of marketable products in developed nations. Several pharmaceutical firms of the United States have already taken preliminary steps in this direction, including Shaman Pharmaceuticals, Inc., a California-based company focused on the discovery and development of novel pharmaceuticals from higher plants, usually those employed in primitive societies. The company donates a percentage of its research budget to local communities or their functional "governments" and has promised to donate a percentage of any profits. In 1989, it organized The Healing Forest Conservancy, a non-profit foundation committed to conserving biocultural and biological diversity, the development and management of natural and biocultural resources, and helping indigenous people participate in and share responsibility for the process of plant collection. The Conservancy is dedicated to conserving tropical forests and promoting the welfare of the people who inhabit those forests. Two other pharmaceutical companies, Sandoz and Schering-Plough, are waiting to see if any plant used by a non-Western healer makes it to the commercial market (none has so far) before making a commitment, and the National Cancer Institute has promised a "large" but unspecified percentage of royalties to native healers and/or their governments. Brigham Young University ethnobotanist Paul Cox, working with village healers in Samoa, has stated his intention to donate one-third of any patent royalties he receives to the healer who shared the information. In 1984 Cox learned of a plant used to treat yellow fever on the island of Upolu. The National Cancer Institute isolated the antiviral compound prostratin from the plant, which now is a candidate for clinical trial as a possible AIDS therapy.

The single contribution in this section, written by one of the most enthusiastic exponents of this new approach to ethnobotany, Victor Manuel Toledo of the Universidad Nacional Autónoma de México, examines the forces that have influenced ethnobotanists in Mexico to become involved with social change, national economic self-determination, and Indian rights.

New Paradigms for a New Ethnobotany: Reflections on the Case of Mexico

VICTOR MANUEL TOLEDO

Every science is, at the same time, a goal, a methodology, and an aggregate of knowledge. These are, however, neither immutable truths nor permanent or static entities. On the contrary, as new advances are made, every science changes its goals, methods, and contents. Kuhn (1962) clarified the phenomena that produce "scientific revolutions," namely, radical changes of *paradigms* or, as Kuhn would say, the step from a "normal science" to an "anomalous science."

Ethnobotanical research in Mexico has experienced marked changes in what could be considered the orthodox practice of this discipline. These changes may be noted in three areas:

1. Ethnobotanical research expanded greatly in the 1970s and 1980s. For example, the number of ethnobotanical papers presented in the four Mexican Botanical Congresses from 1975 to 1984 went from seventeen in 1975 to thirty-one in 1978, to eighty-one in 1981, and to seventy-four in 1984 (Table 1). Adding to the total the forty-five papers presented in the first ethnobotanical symposium held in 1976, a total of 248 papers were presented in only eight years (1976–1984).

2. Ethnobotanical research has attracted and united the most dissident, heterodox, and radical investigators, generally young scientists in disciplines such as biology, agronomy, anthropology, medicine, and pharmaceutical sciences.

3. Ethnobotanical research has become a discipline preoccupied with social change, technological innovation, the country's economic self-determination, and the struggle of Indian peoples.

This chapter examines the new paths taken by ethnobotanical research in Mexico. We will explore the causes that have motivated researchers to adopt an attitude of questioning and change, and show how ethnobotany in Mexico is practiced under new paradigms: that is, it has become a discipline that tends to be practiced with new goals, methodologies, and approaches.

Table 1. Number of papers presented at the Mexican National Congress of Botany (1960–1984).

Field	1960 (I)	1963 (II)	1966 (III)	1969 (IV)	1972 (V)	1975 (VI)	1978 (VII)	1981 (VIII)	1984 (IX)
Taxonomy & floristics	11	15	16	28	13	25	31	83	125
Ecology	9	20	20	21	26	52	61	130	132
History, bibliography, & collections	—	2	6	3	2	4	7	13	26
Physiology	5	3	2	7	9	23	14	35	33
Ethnobotany	11	19	9	17	10	17	31	81	74
Education	2	—	—	—	2	6	5	4	0
Morphology & anatomy	1	2	5	3	13	21	14	37	30
Miscellaneous	30	36	17	23	0	24	22	21	0
Total	69	97	75	102	75	172	185	404	420

The Questioning of Normal Ethnobotany

In its orthodox or "normal" version (to use Kuhn's term), ethnobotany has been and is either economic botany (when practiced by botanists) or ethnoscience (when practiced by ethnologists or linguists). In the first case, ethnobotany is a discipline oriented towards the exploration of new plant resources able to be converted into new raw materials for industry (food, textile, chemical, pharmaceutical, etc.). In the second case, ethnobotany becomes an instrument for understanding the role played by plants in the material culture. These two directions seem to be the main forces behind ethnobotanical research in, for example, North America (Ford 1978). In Mexico, however, neither of these two objectives guides most of the current research, but rather, as we will see, both have been questioned as legitimate forms of scientific endeavor.

Before we explore how the questioning of normal ethnobotany has come about in Mexico, it is convenient to ask why this questioning took place. Commonly, the transferring of certain scientific traditions from a developed country to one less developed takes place as a simple imitation of methods, styles, and approaches in research. In Mexico this happened in agriculture, animal husbandry, biochemistry, and other fields, but the introduction and rapid development of modern ethnobotany took place parallel to the radical questioning of the style in its conduct. This interesting phenomenon is inseparably linked to the following factors:

1. The notable increase of new professionals in the fields of biology, agronomy, and anthropology during the last half of the twentieth century, which has generated a large body of young investigators.

2. The strong politicalization that since the popular student movement of 1968 has marked teaching in the educational centers of biology (Facultad de Ciencias de la Universidad Nacional Autónoma de México), agronomy (Universidad Autónoma de Chapingo), and anthropology (Escuela Nacional de Antropología e Historia) in Mexico. This movement has left behind a core of new professionals with critical and inquisitive minds and with a legitimate concern about the social role of science.

3. The enormous ecological and floristic wealth of Mexico, resulting from the great size of its territory, its complex orography, and its biogeographical location. This wealth is shown by forty-five types of vegetation and 30,000 species of flowering plants, of which at least 3500 are endemic to Mexico (Lamlein, pers. com.).

4. The cultural and linguistic richness of the country expressed by the presence of fifty-five ethnic groups, which are distributed in all ecogeographic units of the country.

5. The profound Mexican social crisis that in rural areas results in (a) an increasing destruction of natural resources (vegetation, flora, fauna, soils), which advances at an annual rate of 1 to 2 million hectares (2.5 to 5 million acres) as a result of agricultural expansion and above all cattle ranching, industrial pollution, and the over-exploitation of forests (Toledo 1985a, 1986), and (b) the impoverishment, marginality, and exploitation of the large numbers of farmers, especially among Indian groups (Esteva 1983).

6. The country's marked Indian struggle in the 1970s that has given voice and conscience to the dominated cultures, not only in reference to Indian economic and political oppression, but also to the destruction of Indian natural resources and cultures (including their botanical knowledge).

Particularly important for the discussion and for the theoretical, conceptual, and political reflection of ethnobotany, four symposia were organized (in 1976, 1978, 1981, and 1984), the last three parallel with the National Botanical Congresses. These symposia served as interdisciplinary fora for open discussion of the significance and application of ethnobotany, its social and ideological importance, and its political role. Minutes of the first symposium were published (see Barrera 1979; Barcena et al. 1982), as well as some of the papers from the third and fourth symposia (see Toledo 1982; Caballero 1984a). While the National Botanical Congresses and other events (like some meetings on traditional medicine) have served to present the innumerable ethnobotanical studies that have been carried out, the four symposia have played the role of "philosophical fora" about the activities of this discipline.

The objections to normal ethnobotany in its two forms (i.e., as economic botany or ethnoscience) made during such events refer to two aspects: the ultimate use of ethnobotanical studies and the relationship that ethnobotany establishes with members of the indigenous ethnic groups that are studied. The first objection deals with a radical questioning of the ideological validity of the universal character and political neutrality of scientific work, that is, of the commonly accepted idea that the work of the researcher and the knowledge produced can be placed above the economic interests of the country and its social classes (Rose and Rose 1980a, 1980b; Gorz 1980). This is the dimension in which the ethnobotanist's work as searcher for new products is questioned.

In its origins, the ethnobotanist's work can be seen as an honorable activity and almost somewhat heroic, given that its findings constitute true contributions to the well being of humanity (and this was, perhaps, the case of the plant hunters of the eighteenth and nineteenth centuries). Nonetheless, as the development of capitalism accentuated the process of privatization and commercialization of manufactured goods, the new discoveries remained trapped in the economic interests of countries and diverse social sectors. An example is seen in the role played by economic botanists in the expansion of imperial England during the nineteenth century (Brockway 1979).

In this process, an apparatus of commercialization, each time more complex, has been placed between the investigator and/or discoverer of new plant products and the users of those products in such a way that the supposed universal benefit of ethnobotanical research has become almost a myth. In truth, and as a consequence of a marked process of monopolization and transnationalization of the capital on which the world economy lives, a good part of the industry that feeds on raw materials of plant origin tends to convert the new products into commercial products, the distribution of which remains restricted by patents, and the consumption of which remains reduced by the purchasing capacity of the users.

This is especially true of the pharmaceutical industry, currently dominated by approximately fifty transnational companies. The annual per capita consumption of pharmaceuticals in developed countries is estimated to be from fifty to seventy dollars, while in developing countries it barely reaches six dollars. This means that 80 percent of the world consumption of medicines occurs in developed countries, and only 15 percent occurs in developing countries (see Fattorusso 1983; *World Development*, vol. 11, no. 3, 1983). This development is paradoxical considering that an important part of the raw materials that serve as the basis of the modern pharmaceutical industry comes from tropical countries. The same pattern of inequality is seen between the social classes of any given country since, given their high cost, pharmaceuticals cannot be acquired by classes with lower economic resources.

Similar tendencies are observed in other industries, such as the food and seed industries, which are dominated by corporations (Mooney 1979). Whether we want it or not, orthodox botany has little by little been converted into a supplier of raw materials for a monopolistic and transnational industry so that the results of investigations increasingly imply less and less universal profit.

The second aspect of the objection to normal ethnobotany refers to the way the discipline approaches Indian informants. In utilitarian or academic ethnobotany, Indian peoples become, like plants, mere objects of research. Between the scientific researcher and the Indian informant, therefore, a partial and asymmetric relationship is established, which results from an encounter between a dominant and a dominated culture. For the orthodox researcher, furthermore, issues such as the social situation of the Indian community that he/she studies, and that community's future in a regional and national context remain outside his/her interests and universe of knowledge.

The same can be said of the economic exploitation to which the Indians have been subjected as members of a marginal social segment, or the role played by plants in their cultural resistance. In orthodox ethnobotany, Indians are of interest only as "suppliers" of new raw material or of new cognitive or linguistic structures. For that reason scientists and others tend to deny the double character of Indians as subjects of contemporary history and as citizens of the current world. In the same context, it is difficult to recognize Indian knowledge as another science different from academic science and in danger of extinction, but not (for that reason) less valid in practical terms, an issue that was suggested by Levi-Strauss in the mid 1900s and that has been shown by ethnoscientific investigations.

In summary, the questioning of these two central aspects of the normal practice of ethnobotany has kept new Mexican ethnobotanists from reproducing orthodox patterns of the work in this discipline. This questioning has obligated them to seek new objectives and consequently new ways to conceive and practice ethnobotany. The remainder of this chapter reviews the three principal currents that have emerged as opposing options to normal or orthodox research.

Ethnobotany, Rural Economy, and New Ecotechnology

The so-called Indian cultures, non-Western peoples, or agrarian societies with which ethnobotanists work comprise the vast rural sector not integrated into the modern world and are represented by the farming community, and on a lesser scale, by tribal societies. Cox and Atkins (1979, p. 111) estimated that 60 percent of the planet's surface in use by humans is still worked by this traditional sector. This is particularly true in Mexico where

farmers dominate rural spaces and comprise the dominant sector in agricultural and for-
estal production.

From an economic point of view, the farming community presents a format for
achieving production that is different from the format presented by the modern market
economy, which is based on the generation of goods and accumulation of capital. For this
reason, investigators have come to recognize the existence of the farmer's means of production (MCP). MCP is defined as a format in which production is fundamentally oriented towards autoconsumption, even though a part of the product is designated for sale
in the market. The farmer thus constitutes an economy in which there is a relative predominance of useful value over exchange value.

To this fundamental definition, other factors are added, such as the one-class character of the producer who is at the same time a property owner who works and a worker
who owns, the simple technological level of the productive processes, the tendency neither to buy nor to sell work force, and the predominantly family or community character
of social relationships. MCP is, furthermore, a marginal, subordinate, and secondary
economy in which the producers are permanently exploited through unequal interchange
of products, money, and work.

Examination of MCP from an ecological perspective reveals a particularly notable
phenomenon: the tendency to achieve production in harmony with nature's laws (Toledo
1980). In fact, the entire farm economy tends to satisfy a large part of its needs from the
ecosystems that are the basis of its production process, and not from the social sector with
which it is only partially and relatively connected. Given that farm producers satisfy their
most elemental material needs through ecological interchange with nature and not
through interaction with the market, they tend to achieve a production capacity that
maintains the renewing capacity of the ecosystems.

This trait of farm economy rationality has caused traditional cultures to become attractive to investigators (i.e., searchers for new technologies). In the face of the limitations and the inefficiency of modern technological models of an economy directed towards the accumulation of capital, and the models that predominate in agriculture
(monocultures driven by the green revolution), cattle ranching (extensive exploitation),
and forestry (plantations of a single species and the extraction of a few species), ecological theory finds innumerable promising alternative models in the farmer's means of appropriation from nature. This new ecotechnology that has for approximately a decade
dedicated itself to the discovery, inventory, and evaluation of the productive farmer has
been focusing on the strategies of multiple use (like polyculture) and on integrated systems of production used by the farming communities.

From this perspective, the sum of empirical knowledge that the diverse Indian cultures possess about plants, animals, soils, climate, and ecogeographic units serving their
production strategies becomes enormously important in deciphering and comprehend-
ing traditional technological models. One must not forget that each method of production
(or economic rationality) takes us not only to a form of organization of producers, a strategy of appropriation from nature, and a mix of technologies, but also to the sum of knowledge about the elements of the ecosystem and its interrelations. For that reason, ethnobotanical research, together with ethnozoology, ethnoedaphic studies, and ethnoecology
are converted to disciplines helpful in the search for new technological options, without
which the new ecotechnology is not possible (e.g., Posey 1983; Alcorn 1984).

In Mexico, numerous ethnobotanical investigators and research groups have concentrated their efforts within the new ecotechnology perspective. Particularly notable are
the studies by agronomists and biologists following the current agroecology lead of the
late E. Hernández-Xolocotzi. This group has been conducting ethnobotanical research on
the phenomenon of small-scale or *campesino* agriculture and has become an alternative to

current agricultural investigation backed by institutions of the Green Revolution (Centro Internacional de Mejoramiento de Maíz y Trigo, Instituto Nacional de Investigación Agrícolas, Colegio de Postgraduados, and others). Standing out in this new trend are studies by Zizumbo and Colunga (1982) among the Huaves of Oaxaca, by Illsley (1983) and Vara-Morán (1980) among the Mayas from north of the Yucatán, by Martínez-Alfaro and collaborators among the Nahuas and Totonacos of la Sierra Norte de Puebla, and by González (1984, p. 242) and Romero (1984, p. 236) in Tabasco.

The contribution by ethnobotanical research in the search for appropriate models for use of the rich tropical rain forests has become particularly noteworthy (Gómez-Pompa 1982). In Mexico, the study of Indian use of the ecosystems of the humid tropics has become a priority task because so much land has been destroyed by the expansion of agriculture and animal husbandry; Rzedowski (1978) estimated tropical forests exist at 10 percent of their original distribution.

One example of the help provided by ethnobotanical research in the search for new technological options in the humid tropics is the multiple use model generated by Toledo et al. (1978) within the botanical-ecological project in the region of Uxpanapa, Veracruz. This model used the botanical (and zoological) knowledge of thirteen native cultural groups from the Mexican tropics (Caballero et al. 1978) to reveal the potential usefulness of 1128 inventoried species of plants and vertebrate animals found in an area of approximately 1200 hectares (3000 acres) with primary and secondary forests. The results of this study were very encouraging, and detailed information was obtained about the usefulness of 244 plant species from seventy-one families occupying primary and secondary forest. The main conclusions derived from this study were as follows:

1. The predominance of plant products (82.3 percent) over animal products (17.7 percent).

2. The great variety of products available: 703 products from 332 useful plant and animal species in a total of 1128 inventoried species (Figure 1), or a ratio of 0.62 products for each inventoried species.

3. The importance of the tropical rain forest as a source of three main products: medicines (210 products or 29.9 percent of the total), foods (200 products or 28.4 percent), and woods (lumber, fibers, and others) (124 products or 17.6 percent). In addition, the remaining 24 percent of the products were distributed among firewood (3.9 percent), drugs (3.6 percent), stimulants (1.8 percent), forages (1.8 percent), and others (gums, resins, dyes, tannins, flavorings, sweeteners, and domestic uses).

4. The great importance of nonwood over wood products obtained from the rain forest as reflected by the number of goods (82.4 percent versus 17.6 percent). A similar figure is again revealed by the detailed list of plant products (Table 2): only 25.6 percent (148 out of 579) of the plant products are obtained from trunks. This rejects the idea of the rain forest as a predominant timber source.

5. The economic importance of both the primary and the secondary forest, which in our analysis produced almost the same amount of plant products (283 versus 296; Table 2).

6. The overwhelming importance of secondary forests as the main source of forages, firewoods, and chemical substances (medicines, drugs, stimulants and dyes) (Table 2).

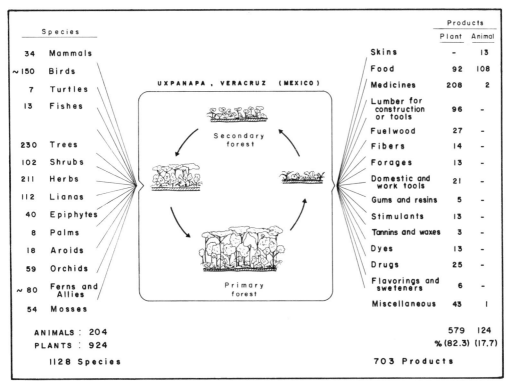

Figure 1. When compared with animal species, plant species in primary and secondary forests yield more useful products to indigenous peoples. Data based on Toledo et al. (1978).

Another aspect of ethnobotanical research in Mexico is the family garden, principally from the humid and subhumid tropics. Table 3 shows the principal ethnobotanical studies of this traditional agricultural system. The family garden is enormously rich in species and has great potential as a provider of needs (foods, medicines, and forage, principally). Other interesting studies include contributions by Lot et al. (1979) about the ethnobotany of aquatic plants and their role in the important traditional agricultural system of the Valle de México, known as Chinampa, and by F. Espinoza, who demonstrated that the mix of plants accompanying farm crops and considered by modern science to be weeds is in reality species with great utility for the farm producer.

Ethnofloristic Inventories and the Nationalization of Knowledge

With a base approximating 1500 published ethnobotanical papers (currently under bibliographic investigation), studies carried out in more than twenty Indian groups, and preliminary revision of the country's principal herbaria, it has been estimated that at least one-sixth of the flora of Mexico, some 5000 species, should have some use (Caballero 1984a; Caballero et al 1985). Of these, 2237 species have been registered by Díaz (1976) as medicinal, and more than 700 as foods (Mapes, pers. com.). It is unknown, however, what percent of the potential has been inventoried, which species can contribute to the satisfaction of the basic needs of the country's population, which species should be studied or protected, and finally, how ethnobotany can contribute to a project of national development.

Table 2. Available plant products of primary and secondary forests in Uxpanapa, Veracruz, Mexico (after Caballero et al. 1978).

Products	Part of plant used[1]												
	Roots	Stems	Trunks	Exudates	Barks	Spines	Leaves	Flowers	Fruits	Seeds	Buds	Entire	Total
Foods	0/6	6/1	-/-	2/0	-/-	-/-	2/6	5/1	29/18	5/6	2/1	0/2	51/41
Medicines	11/18	3/8	6/1	7/11	7/12	1/0	21/42	4/5	6/10	6/3	1/1	4/20	77/131
Woods	-/-	1/2	57/13	-/-	0/1	-/-	0/1	-/-	-/-	-/-	-/-	0/1	58/18
Tools	-/-	2/0	11/7	-/-	-/-	-/-	-/-	-/-	1/0	-/-	-/-	-/-	14/7
Lumber	-/-	-/-	15/5	-/-	-/-	-/-	-/-	-/-	-/-	-/-	-/-	-/-	15/5
Fuelwood	-/-	-/-	12/14	-/-	0/1	-/-	-/-	-/-	-/-	-/-	-/-	-/-	12/15
Domestic tools	3/0	5/1	5/0	1/0	3/0	-/-	3/9	2/0	2/4	2/1	-/-	1/1	27/16
Forages	-/-	1/0	-/-	-/-	-/-	-/-	3/4	-/-	0/1	-/-	-/-	0/4	4/9
Gums & resins	-/-	-/-	-/-	4/1	-/-	-/-	-/-	-/-	-/-	-/-	-/-	-/-	4/1
Fibers	-/-	-/-	-/-	-/-	3/8	-/-	-/-	-/-	2/0	-/-	-/-	-/-	5/9
Tannins & waxes	-/-	0/1	-/-	-/-	0/1	-/-	0/1	-/-	0/1	-/-	-/-	-/-	0/3
Dyes	0/1	0/2	1/0	-/-	-/-	-/-	1/3	-/-	0/4	1/0	-/-	-/-	3/10
Stimulants	0/2	0/2	-/-	2/1	-/-	-/-	1/0	0/1	0/2	1/1	-/-	-/-	3/10
Flavorings & sweeteners	0/1	-/-	-/-	-/-	-/-	-/-	2/0	1/0	2/0	-/-	-/-	-/-	5/1
Drugs	0/1	1/3	1/0	-/-	1/1	-/-	0/2	-/-	1/10	-/2	-/-	0/2	4/21
Subtotals	14/29	19/20	108/40	16/13	14/24	1/0	33/68	12/7	43/50	15/13	3/2	5/30	283/296
Total	43	39	148	29	38	1	101	19	93	28	5	35	579

[1]First number indicates "from primary forest"; second number indicates "from secondary forest."

Table 3. The main ethnobotanical studies on kitchen gardens in Mexico with some data on plant richness and plant uses.

Ecoclimatic zone	Peasant community or region	References	No. of plant species	Main plant uses
Hot & humid	1. Ejido Benito Juárez, Oaxaca	Martínez-Alfaro 1970	—	Food, medicines, lumber
	2. Ejido Augustín Melgar, Veracruz	Toledo et al. 1978	75	Food, medicines, domestic uses
	3. F. I. Madero, Tabasco	González 1984	202	Food, lumber, fuelwood
	4. Ejido Habanero, Tabasco	Romero 1984	219	—
	5. Ejido Mantilla, Tabasco	Romero 1984	218	—
Hot & subhumid	6. North of Yucatán	Barrera 1981	92[1]	Food, medicines, lumber
	7. Ejido Yaxcabá, Yucatán	Vara-Morán 1980	100	Food, other, medicines
Temperate & subhumid	8. La Libertad, Coatepec, Veracruz	Montes et al. 1976	151	Other

[1]Only trees and shrubs

These questions point to a second trend in ethnobotanical research in Mexico, a trend that has been developing around the idea that the country's flora constitutes a national heritage that must be protected from commercial exploitation. To inventory, study, protect, and utilize the plants are high-priority goals that will permit control of this national property. This is the reason for the creation of ethnobotanical computer databases capable of filing all the information on the country's useful plants obtained from bibliographic review, from the herbaria, and from work directly in the field (including farmers' markets). Such a database should be the starting point for the completion of research capable of allowing massive productive use of these potential resources.

Currently, three projects exist within this trend: the database on useful plants of Mexico from the Instituto Nacional de Investigaciónes Sobre Recursos Bióticos (INIREB), the bank of information on medicinal plants from the Unidad de Investigación Biomédica en Medicina Tradicional y Herbolaria (the old "Instituto Mexicano Para el Estudio de Plantas Medicinales"), and the Banco de Información Etnobotánico of the botanical garden of the Universidad Nacional Autónoma de México (see Caballero et al. 1985). To these one must add the Etnoflora Yucatanense, the first inventory of a region of the country dedicated as much to the taxonomic location and cataloging of species as to the study of their utilitarian value for the Mayan Indian culture (see Sosa et al. 1985). This project also includes an ethnobotanical database (Zizumbo and Colunga 1984).

It is interesting to point out that the ethnofloristic inventories have concentrated on the study of two plant resources of enormous importance for the country: food plants (Arellano et al. 1984, p. 226; Caballero 1984a; Tena-Flores and González 1984, p. 225) and medicinal plants (Amo and Anaya 1982; Lozoya and Lozoya 1982; González-Elizondo 1984). Mexico lacks a chemical-pharmaceutical industry, so that almost all the medicines that are consumed in the country are supplied by about thirty transnational companies, which import 75 percent of the raw material that forms the active components (Wionczek 1983). On the other hand, the use of medicinal plants in folk medicine is normally practiced in wide sectors of the population. With regard to foods, there exist 19 million inhabitants in the country with nutritional deficiencies. Mexico imports large quantities of basic foods (8.1 million tons in 1985), such as corn, beans, and milk, and the food-producing industry has been strongly penetrated by transnational companies, as is also the case for fruits, legumes, and seeds (Rama and Vigorito 1979; Barkin and Suárez 1984).

These studies demonstrate that the rich flora of the country and the knowledge of it that Indian cultures have accumulated could serve as the basis for a policy to nationalize the industry, which would include the nationalization of knowledge; that is to say, the creation of a body of basic information managed by Mexican investigators able to achieve what Bates (1985) called the passage of third group species (plants only locally useful) to the secondary and primary groups (plants with widespread use). This process of "neo-domestication" of wild plants with potential usefulness has already been initiated in the Universidad Autónoma de Chapingo (Estrada 1984, p. 257) and above all in the botanical garden of the Universidad Nacional Autónoma de México (Caballero et al. 1985). In the latter, ethnobotany forms the first link in a chain of research (genetic, tissue culture, horticulture, etc.) tending to make viable new plant resources of importance for the development of the country. In both cases, ethnobotanical research is allied to a project for national development: to provide basic information about raw materials to create new active beginnings, or to substitute imports, or to diversify food patterns. The great richness of the useful flora could return the whole country to self-sufficiency in foods (Toledo 1985b). Caballero (1984a, p. 10) summarized the situation thus:

Without forgetting the intrinsic importance that ethnobotany has as a scientific discipline, as a generator of knowledge, and also without forgetting its role of strengthening the cultural affirmation of Mexico's ethnic groups, ethnobotanical research plays a very important role as a model of development of the country, opposite the model of imminent destruction that now prevails. Thought of in this way, ethnobotany returns to a point between modern science and technology and the long experience of the country's farm population that allows the creation of optimal choices to make use of plant resources.

Ethnobotany and the Struggle of Indian People

The idea of a political ethnobotany formulated as a discipline for helping in the struggle of ethnic minorities arose with force during debates of the first and second ethnobotanical symposia held in 1976 and 1978 (Caballero 1979; Argueta 1982) and may be considered a response by the academic community to two phenomena: the intense politicalization of Mexican anthropology and the increasing Indian struggles in Mexico and all Latin America (see *Indianidad y Descolonización* 1979; Bonfil-Batalla 1981). The critique of the functional studies made by Anglo-Saxon anthropologists regarding Mexico's farm communities produced a Mexican anthropology committed to the study and the struggle of the indigenous populations, and it made many changes to the approach and methodologies of anthropological research in rural areas (see Hewitt de Alcantara 1985). Influenced by this new trend, some Mexican ethnobotanical researchers have tried to place scientific work at the service of the rural communities studied or to make Indians researchers of their own botanical wisdom.

> I keep thinking that the best ethnobotanist would be a member of a cultural minority and, trained as a botanist and as an ethnologist, would study, from within and as part of it, the traditional knowledge, the cultural significance, and the traditional management and use of the flora. And it would be even better if his studies could bring economic and cultural benefit to his own community (Barrera 1972, p. 22).

Although no complete experience exists, the idea that ethnobotany should strengthen the Indian struggle has been implicit in some research projects, such as the ethnobiology of Pátzcuaro, Michoacán, or that of San Mateo del Mar in Oaxaca. These projects were conceived with the idea that the first beneficiaries of the studies should be the Indian communities that were studied. From my experience with the Pátzcuaro project, I also concluded that the reversion of knowledge is a central and ultimate end of ethnobotanical research (Toledo 1982).

Another interesting contribution in the struggle of Indian people was the inclusion of ethnobotany (and ethnobiology) in the education (technical and university) of many young Indians sponsored by the Public Education Secretariat (Secretaría de Educación Pública) through the Office of Popular Cultures (Dirección de Culturas Populares). This trend has entered a period of maturity, leaving behind its romantic and utopian phase. Almost ten institutions and techno-scientific groups for the study and assistance of the Indian struggle have ethnobotanical investigators. Among these are the Group of Assistance to Ethnic Development (Grupo de Apoyo al Desarrollo Étnico, GADE) and the Center of Interdisciplinary Research for Integral Rural Development (Centro de Investigaciónes Interdisciplinarias para el Desarrollo Rural Integral, CIIDIR), both in Oaxaca; the Mexican Institute of Traditional Medicine "Tlahuilli" (Instituto Mexicano de Medicinas Tradicionales Tlahuilli) in Morelos; The Michoacán Center of Ecological and Social Studies (Centro de Estudios Ecológicos y Sociales de Michoacán); the Project on the Natural

Resources of the Guerrero Mountain (Projecto sobre los Recursos Naturales de la Montaña de Guerrero) (Viveros and Casas 1985); and diverse groups of the Sierra Norte de Puebla.

Towards a New Ethnobotany

Everything seems to indicate that ethnobotany in Mexico lives by a process similar to that which Kuhn (1962) called a "scientific revolution": that is, the replacement of an academic tradition with a new way to conceive and carry out research. This trend, however, will have to be thoroughly corroborated by a detailed sociological analysis that obviously goes beyond the limits of this chapter. What can be assured is that as a result of the intense discussion about the inner workings of Mexican ethnobotany during the 1980s, the way in which research in this field is conceived involves new approaches, new methods, and new goals. These phenomena are now evidenced in the literature and the researchers.

There is the proliferation in the literature not only of papers with philosophical, theoretical character dedicated to reflection on the meaning of this discipline (Barrera 1979; Caballero 1979; Argueta 1982; Toledo 1982), but also of studies of methodological character. For example, papers exist that discuss the methodology of the study of medicinal plants (Martínez-Alfaro 1976), the exploration of farming communities (Gispert et al. 1979), the analysis of markets (Bye and Linares 1983), and the system of plant use categories (Argueta et al. 1982). These papers seem to indicate that Mexican ethnobotany has entered a new dimension in which it is no longer isolated and a discipline enclosed within itself, but has become part of a new interdisciplinary trend loosely tied to the problems of production and politics. This trend makes mere academic contributions (i.e., knowledge for knowledge itself) or basic science cease to be the main objective(s) of research; instead they become intermediate products along the way to goals of applied character. An example is the Mexican contribution to the knowledge of folk taxonomy of the Maya (Barrera et al. 1976) and the P'urhepecha (Mapes et al. 1981).

The second evidence of a new ethnobotany is seen in the nature of the researcher. Just as a new literature is arising, a new type of ethnobotanist seems to be arising, one that is less specialized, less politically naive, and more conscious of his or her social role. Two factors have played a critical role in this metamorphosis: the participation of these new ethnobotanists in multidisciplinary research groups, and their recognition that the Indian groups with which they work are the most exploited and marginal sector of Mexican society. In this way, the new ethnobotanists become professionals knowledgeable not only in botany and in the Indian culture they study but also in the policies of rural development, farm economy, food problems, and mechanisms of social exploitation. To this we add their authentic concern for the destiny of their research and for the future of the Indian communities where they work, as well as their willingness to become part of multidisciplinary teams.

Examples of this new ethnobotany can be found in the research of Caballero and Mapes (1985) among the P'urhepecha of Michoacán, of Zizumbo and Colunga (1982) among the Huaves of Oaxaca, and of Viveros and Casas (1985) and De Avila-Blomberg (1985) among the Mixtecs of Guerrero and Oaxaca. They have left behind the old paradigms in a "natural" way, because their way of perceiving reality and of conceiving their own work are as much the result of acquiring a social and political consciousness as of rigorous and academic training and a philosophical reflection on scientific endeavors.

The new ethnobotanists believe that classical or normal ethnobotany has already suffered a breakdown or rather a reconstruction. As Kuhn (1962, p. 139) wrote:

The transition of a paradigm in crisis to a new one from which emerges a new tradition of normal science is far from being a process of accumulation reached through an articulation or broadening of the old paradigm. It is rather a reconstruction of the field, based on new fundamentals, reconstruction that changes some of the most basic theoretical generalizations as well as many of the methods and applications of the paradigm.

ACKNOWLEDGMENTS

The following colleagues provided unpublished or difficult-to-obtain materials, and almost all listened to, commented upon, and discussed the main ideas in this essay: Javier Caballero, Cristina Mapes, Patricia Moguel, Benjamin Ortíz, and Narciso Barrera-Bassols. I am thankful to Felipe Villegas for the drawings in Figure 1. Originally written in Spanish, this chapter was translated into English by P. L. and G. A. Romero.

LITERATURE CITED

Alcorn, J. B. 1984. Development policy, forests, and peasant farms: Reflections on Huastec-managed forests' contributions to commercial production and resource conservation. *Economic Botany* 38(4): 389–406.

Amo, S. D., and A. L. Anya. 1982. Importancia de la sistematización de la información sobre plantas medicinales. *Biotica* 7(2): 293–304.

Arellano, J., et al. 1984. Plantas comestibles del sureste de México. Res. 9o. Congreso Mex. Bot.

Argueta, A. 1982. Historia etnobotánica y situación indígena. In *Memorias del Simposio de Etnobotánica*. Eds. A. Barcena et al. Mexico: Instituto Nac. de Antropología e Historia (INAH). 274–279.

Argueta, A., et al. 1982. Análisis de las categorías antropocéntricas empleadas en los estudios etnobotánicos. In *Memorias del Simposio de Etnobotánica*. Eds. A. Barcena et al. Mexico: Instituto Nac. de Antropología e Historia (INAH).

Barcena, A., et al., eds. 1982. *Memorias del Simposio de Etnobotánica*. Mexico: Instituto Nac. de Antropología e Historia (INAH).

Barkin, D., and B. Suárez. 1984. *El fin del Principio: Las semillas y la seguridad alimentaria*. Entro de Ecodesarrollo. Edit. Océano. 187 pp.

Barrera, A. 1972. La etnobotánica. In *La Etnobotánica: Tres Puntos de Vista y Una Perspectiva*. Ed. A. Barrera. Mexico: INIREB. 19–26.

———. 1981. La unidad de habitación tradicional campesino y el manejo de los recursos bióticos en el área maya yucatanense. *Biotica* 5(3): 115–128.

Barrera, A., et al. 1976. *Nomenclatura Etnobotánica Máya*. Col. Científica No. 36. Mexico: Instituto Nac. de Antropología e Historia (INAH).

Bates, D. M. 1985. Plant utilization: Patterns and prospects. *Economic Botany* 39(3): 241–265.

Bonfil-Batalla, G., ed. 1981. *Utopia y revolución: El Pensamiento Político Contemporáneo de los Indios en América Latina*. Edit. Nueva Imagen.

Brockway, L. H. 1979. *Science and Colonial Expansion: The Role of the British Royal Botanic Gardens*. New York: Academic Press.

Bye, R. A., and E. Linares. 1983. The role of plants found in the Mexican markets and their importance in ethnobotanical studies. *Journal of Ethnobiology* 3(1): 1–13.

Caballero, J. 1979. Perspectivas para el quehacer etnobotánico en México. In *La Etnobotánica: Tres Puntos de Vista y Una Perspectiva*. Ed. A. Barrera. Mexico: INIREB.

———. 1984a. La etnobotánica: Base para el desarrollo de nuevos recursos vegetales. In *Memorias del Simposio de Etnobotánica*. Ed. E. Hernández-Xolocotzi. UACH (en prensa).

———. 1984b. Recursos comestibles potenciales. In *La Alimentación en México*. Ed. T. T. Reyna. Instituto de Geografía, Universidad Nacional Autónoma de México (en prensa).

Caballero, J., and C. Mapes. 1985. Gathering and subsistence patterns among the P'urhepecha Indians of Mexico. *Journal of Ethnobiology* 5(1): 31–47.

Caballero, J., et al. 1978. Flora útil o el uso tradicional de las plantas. *Biotica* 3: 102–186.

———. 1985. La unidad de investigación sobre recursos genéticos del Jardín Botánico de la

UNAM. In *Memorias de la 1a. Reunión Nac. de Jardínes Botánicos*. SEDUE–Asoc. Mex. de J. B. (en prensa).

Cox, G. W., and M. D. Atkins. 1979. *Agricultural Ecology: An Analysis of World Food Production Systems*. San Francisco, CA: W. H. Freeman and Company.

De Avila Blomberg, A. 1985. *Etnobotánica de dos comunidades Mixtecas de la sierra sur de Oaxaca*. Projecto de Investigación CIIDIR–Instituto Politecnico Nacional.

Diaz, J. L. 1976. *Índice y Sinonimia de las Plantas Medicinales de México*. Monografía Científica No. 1. Instituto Mexicano de Plantas Medicinales.

Esteva, G. 1983. *The Struggle for Rural Mexico*. South Hadley, MA: Bergin and Garvey.

Estrada, E. 1984. Avance en las investigaciónes sobre plantas medicinales en la Universidad Autónoma de Chapingo y el Colegio de Postgraduados. Res. 9o. Congreso Mex. Bot.

Fattorusso, V. 1983. Essential medicaments for the third world. *World Development* 11(3): 180–184.

Ford, R. I. 1978. Ethnobotany: Historical diversity and synthesis. In *The Nature and Status of Ethnobotany*. Ed. R. I. Ford. Anthropological Papers, no. 67. Ann Arbor, MI: University of Michigan Museum of Anthropology. 33–49.

Gispert, M., et al. 1979. Un nuevo enfoque de la metodología etnobotánica en México. *Medicina Tradicional* 2(7): 41–53.

Gómez-Pompa, A. 1982. La etnobotánica en Mexico. *Biotica* 7(2): 151–162.

González-Elizondo, M. 1984. *Las Plantas Medicinales de Durango. Cuadernos de Investigación Tecnológica*. CIIDIR–Instituto Politécnico Nacional, Universidad de Durango.

González, T. 1984. Descripción del uso, manejo y ecología de los huertos familiares de la Ranchería F. I. Madero, Tabasco. Res. 9o. Congreso Mex. Bot.

Gorz, A. 1980. On the class character of science and scientists. In *Ideology of/in The Natural Sciences*. Eds. H. and S. Rose. Cambridge: Schenkman Publishing Company. 34–46.

Hewitt de Alcantara, C. 1985. Anthropological Perspectives on Rural Mexico. Routledge & K. P.

Illsley, C. G. 1983. *Vegetación y Milpa en el ejido de Yaxcabá Yucatán*. Thesis, Escuela de Biología, Universidad Michoacana de San Nicolás Hidalgo.

Indianidad y Descolonización en América Latina. 1979. Documentos de la Segunda Reunión de Barbados. Edit. Nueva Imagen.

Kuhn, T. S. 1962. The *Structure of Scientific Revolutions*. Chicago: University of Chicago Press.

Lot, A., et al. 1979. The Chinampa: An agricultural system that utilizes aquatic plants. *Journal of Aquatic Plants Management* 17: 74.

Lozoya, X., and M. Lozoya. 1982. *Flora Medicinal de México*. Vol. 1, *Plantas Indígenas*. Mexico: IMSS.

Mapes, C., G. Guzmán, and J. Caballero. 1981. Elements of the P'urhepecha mycological classification. *Journal of Ethnobiology* 1(2): 231–237.

Martínez-Alfaro, M. A. 1970. Ecología humana del Ejido. B. Juárez, Oaxaca. *Publ. Esp. Inst. Nac. Invs. For. México* 7: 1–156.

———. 1976. Posible metodología a sequir en el estudio de las plantas medicinales mexicanas. In *Estudios sobre Etnobotánica y Antropología Medica*. Ed. C. Viesca. 75–83. IMEPLAM.

Montes, J. M., et al. 1976. Los huertos familiares, su importancia desde el punto de vista etnobotánico. In *Memorias del Simposio de Etnobotánica*. Eds. A. Barcena et al. Mexico: Instituto Nac. de Antropología e Historia (INAH). 196–214.

Mooney, P. R. 1979. *Seeds of the Earth: A Public or Private Resource?* Ottawa: Inter Pares.

Posey, D. A. 1983. Indigenous ecological knowledge and development of the Amazon. In *The Dilemma of Amazonian Development*. Ed. E. F. Moran. Boulder, CO: Westview Press. 225–257.

Rama, R., and R. Vigorito. 1979. *El Complejo de Frutas y Legumbres en México*. ILET–Edit. Nueva Imagen.

Romero, C. 1984. Etnobotánica de los huertos familiares en dos ejidos de la Región de la Chantalpa, Tabasco. Res. 9o. Congreso Mex. Bot.

Rose, H., and S. Rose. 1980a. The incorporation of science. In *Ideology of/in the Natural Sciences*. Eds. H. and S. Rose. Cambridge: Schenkman Publishing Company. 16–33.

———. 1980b. The myth of neutrality of science. In *Science and Liberation*: 17–32. Eds. Arditti et al. Boston: South End Press.

Rzedowski, J. 1978. *Vegetación de México*. Edit. Limusa.

Sosa, V., et al. 1985. La flora de Yucatán. *Ciencia y Desarrollo* 60: 37–46.

Tena-Flores, J., and M. Gonzalez. 1984. Plantas silvestres comestibles de Durango. Res. 9o. Congreso Mex. Bot.

Toledo, V. M. 1980. La ecología del modo campesino de producción. *Antropología y Marxismo*. 3: 35–55.

―――. 1982. La etnobotánica hoy: Reversión del conocimiento, lucha indígena y projecto nacional. *Biotica* 7(2): 141–150.

―――. 1985a. La crisis ecológica. In *México ante la Crisis*. Ed. H. Aguilar-Camin and P. González-Casanova. Tomo II. Siglo 21 Eds.

―――. 1985b. *Ecología y Autosuficiencia Alimentaria: Una Estrategia Basada en la Diversidad Biológica, Ecológica y Cultural de México*. Mexico City: Siglo XXI.

―――. 1986. La guerra de las Reses: Porque la ganadería es causa primera de la destrucción biológica y ecológica de México. In *Ambiente y Desarrollo en Mexico*. Ed. E. Leff. (en prensa).

Toledo, V. M., et al. 1978. El uso múltiple de la selva basado en el conocimiento tradicional. *Biotica* 3: 85–101.

Vara-Morán, A. 1980. La dinámica de la milpa en Yucatán: El Solar. In *Seminario sobre Producción Agrícola en Yucatán*. Eds. E. Hernández-Xolocotzi and R. Padilla y Ortega. Gob. Edo. de Yucatán-SPP-SARH. 305–342.

Viveros, J. L., and A. Casas. 1985. *Etnobotánica Mixteca: Alimentación y subsistencia en La Montana de Guerrero*. Tesis, Facultad de Ciencias, Universidad Nacional Autónoma de México.

Wionczek, M. 1983. Research and development of pharmaceutical products in Mexico. *World Development* 11(3): 255–261.

Zizumbo, D., and P. Colunga. 1982. Aspectos etnobotánicos entre los Huaves de San Mateo del Mar, Oaxaca. *Biótica* 7(2): 223–270.

―――. 1984. El Banco de datos etnobotánicos de Yucatán. Publ. Restringida. Mexico: INIREB.

PART 3

Historical Ethnobotany

Consideration of the oldest historical aspects of ethnobotanical records up to relatively recent accounts and their significance to modern ethnobotany provide us with the basis of the evolution of a discipline. While we must consider today's reports and discoveries in view of our modern scientific interpretations, the wealth of information in ancient written records cannot be taken lightly and often can be of incredible cultural and practical value.

Much of the most ancient "literature" lies in the realm of archaeoethnobotany in the study and interpretation of plant remains or in the interpretation of petroglyphs or other records that pre-date any written languages. Nonetheless, the written accounts are even more significant. The Egyptian scrolls, for example, are replete with the uses of plants, especially medicinal species, and indicate knowledge not only of the flora of Egypt but of neighboring regions. Perhaps the most important Egyptian medicinal record, now known as the Ebers Papyrus, was compiled approximately 1550 B.C. from earlier sources. It offers 700 formulas and folk medicines. Even earlier Babylonian records are heavily ethnobotanical.

In India, ethnobotany is likewise of great age. The earliest reports, which associated religion and medicinal plants, have led to the establishment of a system of medicine—ayurvedic medicine—much of which has survived in modern India. One of the most important of these early Indian sources is the *Rig Veda*, which has been useful in the attempt to identify the source of the sacred hallucinogenic plant—a narcotic plant which became a god (Soma) in ancient India. Other records of the ancient ethnobotanical values of plants go far back in Indian history and, incidentally, have contributed to the encouragement of modern Indian ethnobotanical research, an effort which is not equalled in its enthusiastic support in many other areas of the world. India has established a society of ethnobotany and the journal *Ethnobotany*. Among the country's numerous, very active ethnobotanists are J. K. Maheshwari, G. M. Oza, and S. K. Jain, author of the magistral *Dictionary of Indian Folk Medicine and Ethnobotany* and one of the leading ethnobotanists in the world today.

The ancient Chinese literature is a rich source of ethnobotanic data and, especially, ethnopharmacological information. From very early times in Chinese culture, medicinal and especially hallucinogenic plants were recorded in herbals such as Li Shih-chen's encyclopedic work, *Pen ts'ao kang mu*, which, although not published until 1596, is based on many centuries of verbally preserved, traditional knowledge. Even much earlier sources are known, including an anonymous collection from the Nuttan Dynasty (206 B.C. to 220 A.D.) and several important works from the Chin Dynasty (265 to 420 A.D.).

In the New World, "written" records go back to the pictorial writing of the Aztecs and Mayas, which have been studied and deciphered. These records indicate extensive ethnobotanical familiarity with the properties of plants of the rich floras and the importance of many species, especially in the magico-religious contexts of both cultures. At the time of the Spanish arrival in the Americas, the Aztecs already had a flourishing botanical garden in Mexico. The first herbal in the New World, published in 1542 and known as the Badianus manuscript, is the product of the Aztec physician Martín de la Cruz, who worked with elderly medicine men to document "first-hand" information. The volume is profusely illustrated with accurate colored drawings of the major medicinal plants of the Aztecs, and has their Nahuatl names and descriptions of their therapeutic values. The species of these plants can be identified easily, and the information has, in great part, survived to this date in Mexican ethnomedicine.

At about the same time in the fifteenth century, the King of Spain sent his personal physician to live with the Aztecs and study their medicines. Dr. Francisco Hernández spent several years with the Indians and wrote an extraordinary, fully illustrated volume in Latin, *Rerum Medicarum Novae Hispaniae Thesaurus, seu Plantarum, Animalium, Mineralium Mexicanorum Historia*, which was not published until 1651. Unfortunately, it has not been translated into English, although a modern Spanish edition exists.

Other than these two studies, nothing of similar historical importance to ethnopharmacology appeared in the Americas in this very early period. The few incidental references in the primarily religious literature of post-Conquest Mexico are annotations by non-medical writers.

Interest in ethnobotany has continued in Mexico with local and foreign ethnobotanists following up the rich heritage of nearly 500 years of numerous incidental references to plant uses among the many tribes of that nation. Anthropologists and botanists are paying more attention to ethnobotanical research, and interest in the wealth of information still existing in tribes has attracted many younger, well-trained specialists in numerous aspects of ethnobotanical investigation.

European ethnobotanical records claim as their written beginnings the work of Dioscorides, a Greek botanist and physician. Travelling widely in Greece, Italy, and Asia, Dioscorides produced in the first century B.C. *De Materia Medica*, in which some 500 plants were described. For fifteen centuries thereafter, this work was considered *the* authority in Europe. Following the invention of printing, however, numerous herbals appeared. Among the most influential herbals of the Middle Ages and Renaissance are Brunfels' *Herbarium Vivae Eicones*, Gerard's *The Herball or Generall Historie of Plantes*, Leonhart Fuchs' *De Historia Stirpium*, and Parkinson's *Paradisi in Sole Paradisus Terrestris* and *Theatrum Botanicum*.

During the Middle Ages, the "Doctrine of Signatures" permeated ethnopharmacology. It taught that God had placed plants on earth for the exclusive use of humanity and, on each, had placed a sign or signature to indicate its therapeutic value. Thus, the bloodroot was considered to be a blood tonic; the lungwort, with its dotted leaves resembling tubercular lung tissue, was considered useful against pulmonary troubles; and the mandrake, paralleling the human form, was naturally considered a panacea. Although this doctrine was widely accepted in Europe mainly because it conformed with the religious and philosophic authoritarianism of the period, many of the herbals produced during the Middle Ages influenced the development of ethnobotany for centuries and still are repositories of interesting ethnobotanical data, some of which might still be worthy of serious consideration.

Three contributions to historical ethnobotany are presented in this section. William A. Emboden, Jr., demonstrates how the art and artifacts of ancient Near Eastern civilizations are rich repositories of ethnobotanical information, heretofore neglected in favor of written records from Egypt during the same time period. Peter T. Furst examines the secondary role of some of the sacred inebriants of Indian Mexico as medicinal plants, using the Badianus manuscript of 1552 as his primary source of information about these plants. Carl A. P. Ruck looks at the use of plants in Greek mythology. Readers wanting a general overview of historical aspects of North American ethnobotanical reports and their significance to the modern development of ethnobotany are referred to Richard I. Ford's previously published article, "Ethnobotany: Historical Diversity and Synthesis" (in *The Nature and Status of Ethnobotany*, 1978, ed. by R. I. Ford, Anthropological Papers, no. 67, Ann Arbor, Michigan, University of Michigan Museum of Anthropology, pp. 33–49).

Art and Artifact as Ethnobotanical Tools in the Ancient Near East with Emphasis on Psychoactive Plants

WILLIAM A. EMBODEN, JR.

While there is considerable research on agriculture of the ancient Near East, most of the writing has dealt with sustenance crops such as grains, pulses, dates, and others. Various practices in agriculture at an early date in this area have been discussed by modern writers, but reliance on silting still seems to be the most plausible explanation for a stabilized agriculture practice. Gathered plants are not excluded by the progressive movement into new agricultural modalities.

Most neglected, and still very controversial, are the several kinds of psychoactive plants employed by early peoples. It is suggested that art and artifact have been sources often overlooked in determining the ethnobotanical content of any early civilization. The suggestion is made that early civilizations in the area of the Fertile Crescent employed *Datura, Cannabis, Claviceps, Mandragora, Nymphaea, Vitis,* and possibly *Papaver* as medicaments and ritual narcotics. They are well revealed in the remaining art and artifacts of these civilizations. As many of the images are imprecise in their execution, identification must be made in the context in which they are represented and is therefore often conjectural.

In the sciences, ethnobotany is one of the most recent and rapidly expanding disciplines, as exemplified by two relatively new and important publications, the *Journal of Ethnopharmacology* and the *Journal of Ethnobiology*. These publications were born as it became abundantly evident that ethnobotanical articles were being buried in journals devoted to other disciplines; and, yet, more and more scientists were turning to this multifaceted study. Unlike the more conventional branches of science, ethnobotany relies upon a greater database than does any other scientific discipline. Anthropology, archeology, pharmacology, biochemistry, and the many areas that comprise biology are all components of the broad discipline of ethnobotany. The earlier term *economic botany* was something of a misnomer in that it implied only economic considerations of the plant sciences, although it often was used to cover courses and writings in the area now defined as ethnobotany.

At a time when many molecular biologists have almost forgotten the uses of the past, the ethnobotanist is assembling historical art and artifact as part of the database. Further

verification of assertions and hypotheses deriving from these sources may come from analytical chemistry or electron microscopy, but it is mandatory to have individuals well
trained in diverse areas to bring together, and make sense of, the pieces left to us by former civilizations and contemporary civilizations threatened with extinction. To this end,
we need ethnomusicologists, ethnomycologists, pharmacologists, analytical chemists,
computer scientists, biochemists and, primarily, botanists who are well grounded in biology and have an interest in history.

It is important to note that the discipline of ethnobotany has been able to advance on
the basis of two primary developments, both from the first half of the twentieth century:
first, the evolution of a technology that allows rapid analysis of materials and data; and second, the discovery of ancient civilizations as something more than curiosities. The origins
of Western civilization can be traced to the ancient Near East, which was unknown to the
greatest of the Greek writers of antiquity and was regarded until late as a false start in the
development of both art and science. We have preferred to think of our own civilization as
having developed from Hellenic thought and values. Despite the enormous accumula-
tions of art and artifacts that were garnered during the nineteenth century, interpretation
of these materials was not forthcoming. While the picture is still fragmentary, we now are
able to interpret correctly much of what was formerly regarded as merely decorative.

It was the Neolithic revolution that initiated the ordered life of social stratification
within settled communities. Following this, more complicated federations of communities rose in a city-state plexus that required political systems, agricultural priesthood, currency, and economic systems, and that permitted the development of the luxuries of art
and writing. All of this happened about the fourth or the fifth millennium B.C. in what has
been called the Fertile Crescent of arable land surrounding the wilderness of Arabia. The
area included Egypt, Palestine, Syria, and Mesopotamia, which, until the end of the fourth
millennium, remained an unassimilated complex of simple villages with primitive agriculture and tribal principles rather than any ordered governments.

In a few generations the greatest transformation in the history of any people took
place. Writing appeared, monumental architecture rose from rubble, agriculture underwent revolutionary changes, governments replaced less-than-feudal states, and religion
and science made their appearances. Like an estivation period followed by germination,
a succession of cities grew, blossomed, and reached fruition. Egypt and Mesopotamia
were the luminaries of this great period of art, architecture, science, and engineering.
Long before any written record appeared, these civilizations had produced sculpture and
painting so imbued with information that they codify thought in many ways similar to
written language. Syria, Palestine, Sumeria, Anatolia, and the Levant were perhaps lesser
luminaries, but they blazed a trail like a comet and, in regard to ethnobotanical data,
made contributions as important as those of Egypt and Mesopotamia.

It is important to understand that early civilizations tied artistic expression to religion
and that this religion was based upon magic—the magic that comes from grain, from
brewing, from states of elevated consciousness associated with plants, from pain reliev-
ers, from healing herbs, and from resinous plants for embalming the body. In brief, the art
of this early period and of these cultures is a revelation of riches for the ethnobotanist.

In this chapter I would not propose to do the work of the anthropologist, archeologist,
or theologian. Instead, I will suggest, from their discoveries, thematic materials that either
have been neglected or have been subjected to alternative interpretations or to an extension of ideas that have been only partially formulated.

Egypt and Sumeria share the trait of being two great river valleys in which agriculture
could flourish through silting. No civilization develops in the absence of a stable agricultural base. In Egypt, the progressive aridity of North Africa and the Levant drove peoples
into the Nile Delta and thus became the determining factor as to where a settled existence

might emerge. Likewise in Mesopotamia, the Sumerians, a non-Semitic and non-Indo-European people, at the beginning of the third millennium B.C. gave birth to an independent art that reflected their preoccupations during this preliterate period. It would be yet another seven centuries before writing would appear in clay tablature, but the vessels of this epoch are revealing in form, in floral patterns, and in the animals depicted. These independent creations are in many ways similar to the vessels found in the escarpments and caves that escaped the periodic inundation of the Nile Delta.

Mesopotamia was settled by people who left the increasingly arid Persian Gulf region and who had earlier made their homes in the Iranian highlands. In their new homeland, they settled on islands and banks around marshes and relied upon annual flooding of the rivers, as well as on their own irrigation ditches, in which they probably cultivated small fish and edible aquatics such as the boiled rhizomes of *Nymphaea* species. The silting of vast areas allowed the planting of barley and wheat. Reed-clay huts rapidly gave way to an astonishing originality in terra-cotta and in brick architecture. From this period, we have two art forms: pottery that goes beyond ornamentation (meaningfully painted and incised) and figurines. Since the discovery of these two forms, weavings with designs have been found at Eridu in a grave dating to 3500 B.C.

Persepolis and Sus provide the first example of brush painting on pottery that goes beyond ornamentation. By the end of the second millennium B.C., the vessels are of exceedingly varied form. The designs are plant and animal motifs, and occasionally a human figure. It is hard to agree with Lloyd (1961, p. 302) that the appearance of the human form, such as a hunter with a bow, has the sole function of filling a gap in a design. The compositions are too well conceived to permit such "gaps" and, as Coe (1973) has said of the Maya, these people did not decorate; rather they depicted reality in all their art and artifacts. Certainly, the hunter with his bow becomes a focal figure in a civilization that is still hunting-gathering; his vocation is pivotal to the survival of the civilization. Likewise, Lloyd (1961, p. 302) referred to depictions of beasts and birds as being "irrelevant, since the painter himself, concerned only with its decorative value, may well have been ignorant of its traditional significance." My response is this: we need not concern ourselves with traditional themes in such an early civilization; rather we must see these people as depicting their reality and certainly not merely decorating.

Fish were a primary food source. The ibis may be viewed in the context in which birds have been viewed in early civilizations: as shamanic manifestations. The concept extends from the raven of the early Eskimos to the dove–Holy Ghost theme in early Christian iconography and hagiography. Horses, birds, dogs, fish, and floral motifs are treated by Lloyd (1961, p. 302) as "mere hieroglyphs." Such hieroglyphs are precursors of language and were the embodiment of thought of the time; the facile dismissal of such themes is unfortunate. Lloyd's assertion that "the great majority (of animal figures) are crudely made playthings for children and of no artistic interest" contrasts with his statement that those which are based upon human forms place them in the category of cult objects.

If ethnobotanists have made one great contribution, it is, I believe, to rethink these things that previously have been considered unworthy of serious consideration. It is not necessary to give excessively plastic expression to these abstracted ideas to find them meaningful. In the "ur-language" of this early period of ideograms, we find the genesis of a language that speaks to us of hunters with horses, of the activities of fishing, of the use of dogs in the hunt, and of the technology of the bow and arrow as an extraordinary leap beyond clubs. In one polychrome bowl of Tell Halaf ware from northern Mesopotamia, a central floral motif of anthers surrounded by numerous petals suggests a water lily or *Nymphaea*. While such identity is tentative, the water lily is a prime contender for any illustration of marshland plants. Further, the narcotic qualities of the flowers (Emboden 1981, 1982a, 1982b) and the edibility of the rhizome after boiling and leaching would make it a floral emblem par excellence, telling us much about life and religion, as opposed to the writing of the "Al Ubaid" phase of Mesopotamian development. This tablature was reserved specifically for the purpose of inventory of goods and administration, but in it we are able to identify aspects of ethnobotany, such as payment in grain, bread, and beer, which give us a glimpse of the role of plants in an emerging civilization.

The earliest records from the ancient Near East indicate that healing was accomplished by incantations and plants; both were seen as ridding the ailing body of demonic possession. Persons capable of eliciting in themselves and others states of hypnosis, delirium, or psychological transcendence made up a caste of shamans who mediated the journey of the spirit from the realm of the seen to that of the unseen and whose powers to cast out demons resided in numerous plants. Each plant was known by a name that more often characterized power than it described plant morphology or attributes.

Babylonian medicine most probably was carried into eastern Mesopotamia and Assyria by caravan routes. We know that a number of the plants mentioned in Assyrian tablature are Sumerian. One Babylonian record dating to 2250 B.C. indicates that Babylonia and Egypt then had a trade in drugs and that most of these were oils, gums, and resins. Oils provided a matrix for the carriage of several kinds of drugs.

Babylonian and Assyrian medicine is known primarily through an assemblage of clay tablets from the library of the palace of Assurbanipal, the Assyrian king who ruled Nineveh from 668 to 626 B.C. About 800 pieces of these tablets consist of medical texts that are believed to be of Babylonian origin in thought and that refer to a period between 2000 and 3000 B.C. Evidence to that end, in the absence of a written language, must come from art and artifacts: pottery shards, textile patterns, paintings, implements, stelae, statuary, and even architectural layout. The ethnobotanist must regard all these as tools to the unlocking of the complex pattern of plant use in these earliest of civilizations. The alternative is to use derivative texts, which in some instances may be misleading. The union of art and artifact with later writings provides a base for understanding the earliest uses of plants and plant products. Chemistry may further validate finds, such as residues in unguent jars.

Thompson (1924) cited about 250 drugs of vegetable origin as present in the Assurbanipal tablature. These are, however, compounded and often represent diverse combinations of important plants. Some can be identified; others must remain unknown, owing to the absence of relevant figures and morphological data. It should be noted that only 120 mineral substances are mentioned as medicaments, thus placing plants in the forefront of early medicine. According to the translations of R. Campbell Thompson, these plants include almond (oil), asafoetida, calendula, chamomile, ergot, fennel, henbane, myrrh, liquorice, lupine, mandrake, opium poppy, pomegranate, saffron, and turmeric. *Cannabis*, which figures prominently in healing in China and India, also would have been a major element of barter along the early trade routes leading into and out of Assyria.

Poultices of plant substances were common, and there is every reason to believe they were efficacious. Turnips were kneaded with milk to make a poultice paste, as were barley and wheat flours. It is worth noting that "rotten grain" was prescribed. While this may have been a practice to help conserve fresh grain, it also served to introduce fungus-infected grains to the areas of wounds. These fungi undoubtedly produced some antibiotics that assisted the healing process of the poultice.

Stomach pains seem to figure high on the list of common complaints, and to this end the family Apiaceae is most commonly recommended for such ailments. Herbs, seeds, roots, and resins frequently were macerated and put into beer or wine as a method of dispersing oils and other components that might not have been soluble in water. Oils, honey, and herbs were mixed with wine to be administered by clyster. Enemas, both warm and cold, are registered in this early tablature.

Disease is called "the hand of the spirit demon" in this codified material from Assurbanipal, and one of the common ways of driving out the spirit is fumigation. All oil-producing plants may be placed on hot coals to produce fragrant smoke. In this context, it is significant to note that the majority of plant oils investigated have bactericidal or bacteriostatic properties, as well as fungistatic virtues, and that as fumigants they no doubt served the purpose for which they were used. This ancient ritual has come down to the censers of the contemporary Catholic church and the spicers that are found in Orthodox synagogues. Wherever crowds of people gathered, ritual purification of the air was conducted to drive out demons. In reality, the custom served to discourage the dissemination of disease-causing organisms. In the same manner, fragrant plants laden with volatile oils were placed upon the floors of temples and houses to be crushed under foot, thus releasing these oils into the air—shades of the aerosols that dominate Western homes!

It is worth noting that the healers were not all men. One of the earliest Babylonian tablets from Nippa, dating from the kings of the Babylonian dynasty of circa 2000 B.C., mentions a shaman as Pir-Napistum of the school of healing of Eridu. He is called to heal but designates the task to his wife, a healing priestess. For a comatose patient she makes a vegetable and herbal poultice and cooks it in water. The mass placed upon the patient's head causes him to awaken, and he is then instructed to eat the medicinal concoction. This "magic food" recalls the practices of Native Americans who did not distinguish between food and medicine but were concerned with plants as "power" (Vogel 1970, p. 583).

By contrast, Egyptian hieroglyphs correspond to the emergence of the dynastic periods and virtually explode with a wealth of information about religion, politics, predynastic periods, and most important, agriculture. The information is incised in stone in Egyptian pre-dynastic art and only later is found in papyri.

Of this period, one of the most exquisite and informative intaglio pieces is an enormous, presumably votive mace-head from Hierkonopolis (in the Ashmolean Museum) celebrating the Scorpion King of Upper Egypt. Anthropologists have concentrated on the activities of this Scorpion King in opening a canal and consolidating the Upper Egyptian kingdoms, and on the mace as an emblem of power akin to the scepter of contemporary coronations. Apart from this overt interpretation and signature of high office, attention should be drawn to the form of the piece, its relief, and its intaglio iconography. True to the typical mace-head, it is obovoid and drilled; however, the concentric rings (the first to cap the mace-head) have between them an entire circle of horizontal lines in relief. Below are successive layers of depictions interrupted by stately palms. Plants are shown having globose heads but no leaves.

Is it possible that this mace-head is a modified opium poppy capsule? It has circumscissile dehiscence as a motif and the aforementioned stylized plants. If this be the case, the appearance of an opium poppy in this area would be in opposition to the studies of Krikorian (1975), who claimed that if we are to validate the opium poppy as a plant of the ancient Near East, we should have evidence in the form of the seeds or capsules themselves. Seeds are so small as to disappear easily; both seeds and capsule probably would deteriorate over thousands of years. Perhaps the finest records are those incised in stone. Any such assertion, however, must be taken as a hypothesis requiring further evidence for validation.

Papaver somniferum is widely accepted to be the result of domestication of the wild *P. setigerum*, but a time-scale for such an event is imperfectly known. Poppy seeds and pods have been found in bogs and lakes of the third millennium B.C. in north central Europe. Gongora (1868) wrote of finding 55,000-year-old poppy capsules and seeds in a limestone cave near the area of Spain now known as Granada, but he saw these only as symbols and neglected their psychoactive properties. The seeds are nutritious and a fine source of both proteins and oils, while the unripe fruit latex provides opium.

Thompson (1924) asserted that the Akkadian word *irru* was used to designate the opium poppy. While his philological reasoning was sound, his errors in the identification of other plants led to a rather generalized criticism of his scholarship. Thompson believed that there were cognates in the Sumerian texts, but Krikorian (1975) refuted this suggestion on philological grounds. Kritikos and Papadaki (1967) cited R. Dougherty, former curator of the Babylonian Collection at Yale University, to support the contention that the opium poppy was known to the Sumerians in the third millennium B.C. Those who take exception to the *Hul Gil* ideograms must contend, however, with the union of these glyphs with associated quotations, cited by Thompson, in which this plant is presented with such connections as this: "Early in this morning old women, boys and girls collect the juice, scraping it off the notches (of the poppy capsule) with a small iron blade and place it within a clay receptacle." Sonnendecker (1962) asserted that the *Hul Gil* ideograms may be found in Sumerian tablets of the fourth millennium B.C. Marinatos (1948, p. 85) cited the goddess Nisaba of Babylonia and Assyria as portrayed with opium poppies growing out of her shoulders!

In the absence of any seed or capsules from this early date, we are left with conjecture and debate as to the real antiquity of the opium poppy in this region. Nonetheless, the plant is mentioned forty-two times in tablature of the temple of Assurbanipal.

Ancient Uruk or Warka has given us an extraordinary stone vase upon which plants are figured in the lower register (Figure 1). This rare find from the alluvial plains of southern Iraq depicts a religious scene incised in alabaster, the figures being left in relief. The plant forms are varied; all have leaves and flowers or fruits, but the level of stylization precludes identification. Given, in the vase, the tendency toward the depiction of mythical and monstrous beasts, we may well expect to encounter equally mythical plants, which have no counterpart in the real world. On the other hand, the very nature of the technique limits the details that can be successfully expressed. One wonders what important plants these might be.

Any evidence that might be used to link Egypt and Mesopotamia during the final centuries of the fourth millennium B.C. must be found by way of resemblances in art, architecture, pictographic writing, and the conventions expressed in all these. It is very difficult to account for all the similarities, and evidence for physical contact between these anthropologically unrelated peoples is equivocal at best. Also, we are overwhelmed by the amount of tomb intaglio and murals from Egypt, which far exceeds the entire corre-

Figure 1. Stone vase from Uruk (Warka) with grain plants in the lowest register. Iraq Museum.

sponding anthropological and archeological remains of all the other collective lands of the ancient Near East.

In the third millennium B.C., Sumerian civilization was flourishing in the alluvial basin at the head of the Persian Gulf as an amalgamation of city-states (much like the Classical and post-Classical Maya civilization), each ruled by an oligarchy. This is the early dynastic period of Sumeria, with Ur, Erech, and Kish as three of its leading cities. Agriculture progressed as the result of irrigation canals that served to move people and objects, as well as rich muck and water, to parched lands. There can be little doubt that the small fish of these canals were netted as food and that the aquatic rhizomes of *Nymphaea*, when boiled and leached, could provide a crude carbohydrate. The same rhizomes, when raw, or the flower buds, when macerated, could provide a psychoactive decoction (Emboden 1982a).

A series of cylinders or rollers cut in intaglio and rolled upon pitch or clay leaves records daily life and religious activities. On one such cylinder, there is incised the figure of a plant with three giant flower buds emerging from five mounds (rhizomes or bulbs?) and protected by crouching, masked bulls. Human figures are on either side of the bull-minotaurs, protecting them from the attack of some monstrous bird. The three flowers strongly suggest the repeated early dynastic symbol of ancient Egypt in which this sacred trinity of flowers is found with a very high frequency (Rands 1953; Emboden 1981).

One of the great treasures of early dynastic Ur is a headdress of leaves of beaten gold and three large flowers with eight petals and a center of carnelian. The leaves are obovate with acuminate tips and may not relate to the flowers. The diameter of the flowers being approximately 12 to 15 centimeters (5 to 6 inches) suggests that they may represent the water lily or perhaps the opium poppy with the carnelian being interpreted as the capsule contained within the corolla. With this headdress, found by the Woolley Expedition, was a necklace of leaves that are obovate with acuminate tips and another rank of leaves in beaten gold that strongly suggest *Cannabis* (Figure 2). Others (Lloyd 1961, p. 302) have suggested willow (*Salix*), but the venation is more like that of *Cannabis*, as is the leaf morphology.

In early dynastic Egypt the flower of *Nymphaea* is regularly found in the headdresses

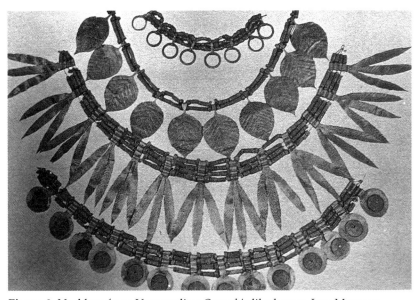

Figure 2. Necklace from Ur revealing *Cannabis*-like leaves. Iraq Museum.

of figures in tomb murals. At Ur, the famous "Ram in a Thicket" (Figure 3) is an emblem of the strength of Tammuz. The goat was an emblem of virility, and the "thicket" is a highly formalized plant of oppositely branched dichotomies of eight appendages. Two of these bear the same eight-petaled, beaten gold flowers with golden centers. The formality of the presentation makes it difficult to visualize the goat as simply a goat. Aspects of a deity mark the animal in its stance before the sacred plant upon which it rests its hooves. This goat-man is akin to Hellenic satyrs and centaurs. It is clearly a hybridized motif, and the portrayal is of a priestly order.

Figure 3. Ram (Tammuz) representation. The branched plant has two remaining flowers that suggest *Papaver*. British Museum.

The ruling class of Mesopotamia was altered by a seemingly peaceful ascendancy of Semitic Akkadians into the ruling classes. Unfortunately, the amount of extant Akkadian art is limited, but the few pieces known to art historians are of extremely high esthetic and technological merit. The Guti tribes from Iran's mountains overran all Sumeria except Lagash, where the Sumerian tradition seemed to have continued its development toward progressive refinement and naturalistic depiction. During this same time, Egypt was undergoing serious changes as the pharaohs were overthrown by a feudal noble class that brought with it war and anarchy.

Not until the final years of the third millennium was there an Egyptian revival; it appeared with the initiation of the Middle Kingdom. The cult of Ra was replaced by the Osirian tradition of death and resurrection not unlike the Mesopotamian legends of Tammuz. The importance of this to the anthropologist, archeologist, and ethnobotanist is that with the Osirian tradition came the establishment of refined tomb art in Upper Egypt. Events of ordinary life frequently are portrayed in these tombs, and arid conditions have left much of the tempera painting intact. Brewing of beer, viticulture, oenology, grain harvest, and preoccupation with the sacred water lily are all strongly evident.

According to the mycological researches of R. Gordon Wasson (1970), a people calling themselves Aryans descended from the north through Afghanistan to occupy the Indus valleys. These were Iranians (Aryan being a cognate). In the land from Palestine to Mesopotamia and Iran, settled in succession by the Sumerians, the Hittites, the Mitannians, and finally by the Indo-Iranians or Aryans, there is an amalgamation of shared legend and mythology. It is Wasson's contention that the Gilgamesh legend of the quest for the miraculous herb that is taken from him by a serpent is a common legend reaching throughout Eurasia from as far back as the Stone Age. Wasson asserts that the Soma-Haoma myth relates to the fly-agaric mushroom (*Amanita muscaria*) and that this related to sacred-tree mythology in that the birch tree supports fly-agaric in a mycorrhizal relationship in which fly-agaric is, in a secondary way, the "fruit" of that sacred tree.

Amanita muscaria contains ibotenic acid which breaks down to muscimole and muscazone, psychoactive agents that would serve as a fine adjunct to shamanic ecstasis. The fungus does not grow in the areas of the ancient Near East but may have been brought by the invaders as the sacrament that figures in their sacred books, collection of poems, and the *Rig Veda*, and around which much music, liturgy, and philosophy is centered. The Aryan invasion of 3500 years ago, in the second millennium, presumably introduced this plant that subsequently was lost or forgotten about 3000 years ago when its use was abandoned by the priesthood. We are left to wonder why there are no remains or records of this sacred mushroom in any of these areas of contact. Wasson's thesis is most intriguing and possibly correct, but ethnobotanists must continue to look for more clues in every aspect of art and artifact to validate such assertions. Certainly it would be most interesting to add this plant to those many already suggested.

The richness of the Egyptian Middle Kingdom is paralleled by an astonishing period of Mesopotamian art little recoverable from other than a limited cache of rollers or cylinders. Fragments of sculpture, the famous Hammurabi stela, and the great walled temples of Uruk, Kish, Eridu, and Nippur remain. One area, Mari, on the Middle Euphrates, leaves us with the finest remaining statuary and a small number of wall paintings showing palms and trees with enormous, presumably mythical, flowers. The rest of Mesopotamia fell into Kassite (Iranian) hands and inherited the impoverished vestige of the Guti rule. Imagination became stereotyped. In the second millennium, Assyrian merchants brought Mesopotamia's civilization to Anatolia, and clay tablets indicate that the Anatolian plateau had in it elements of an Indo-European culture, that of the Hittites who would later unite Anatolia in the east.

If the opium poppy had not previously entered Near Eastern civilization, it certainly would have done so at this time. Such a contention may be based in part upon a famous Assyrian fertility seal in which Athirat or Ashera (the Mesopotamian Ishtar) is shown with a magical plant, two priests, and two winged demons with "pollen bags." It is the last that has led to a suggestion that the magical tree is the date palm, which the Assyrians knew to be divided into sexes. Pollination was practiced as an essentially magical act. The form of the plant, however, is that of either a poppy or a pomegranate, neither of which require deliberate pollination. The pomegranate, because of its copious seed, is an obvious emblem of fertility, while the poppy is emblematic of power, for reason of its conquest over pain and grief. It is for this reason that the poppy goddess of Knossos wears a corona of three poppy capsules. Ishtar is associated more often with magical plants than with fertility, and the temptation is to make the association with the poppy, rather than with the pomegranate.

In both Sumerian and Akkadian myths, Ishtar (Inanna) descends into the underworld (much like the later stories of Persephone), where she undergoes shamanic death, to be rescued three days later by two sexless creatures (the cult of Ishtar-Inanna involved eunuchs) who finally return her to her own city of Erech. Thus, she is associated with the shamanic sleep of death that is the essential quality of the opium poppy. Hence, it is not unrealistic to associate the "Ishtar cylinder" with the poppy. Further evidence derives from the intense preoccupation of Ishtar worship with magic involving plague amulets, exorcism, and later, dream divination.

Plants also figure in these practices. The onion, *Allium*, was used in medicine, divination, and exorcism. Possession, as the result of misdeeds and broken taboos, could lead to the offender being given an onion bulb by a shaman. Each successive layer of the onion represented a misdeed that could be obliterated by peeling the layer and throwing it into a magical fire in which it was burnt to oblivion. This manner of voiding transgressions also symbolized the removal of skin and flesh from the skeleton, a common shamanic theme in diverse cultures.

In a similar manner, wheat and barley made into bread not only were eaten or made into poultices but, when moldy, served a magical purpose. A loaf of bread placed upon the head of an ailing child would draw the sickness, by magic, into the bread. The loaf then would be rubbed down the child's body from head to foot. Ultimately, the loaf would be eaten or thrown to a dog who magically would take up the child's ailments.

Plants figure in the earliest Sumerian and Babylonian mythology, namely, the legend of Gilgamesh. The antiquity of this legend is attested to be about the identification of the father of Agga, king of Kish and foe of Gilgamesh, on an alabaster bowl, dating to circa 2700 B.C., from the Diyala Valley. Many later versions exist. Fragments of the narrative are dated to nearly the end of the third millennium B.C. In the course of his journey in search of the plant of immortality, the protagonist passes through a magical garden in which grapes of carnelian and lapis lazuli produce a magical wine. In the eleventh tablet, Gilgamesh descends into the sea to find the plant of immortality. On the journey back to his homeland, the plant is taken from him by a serpent. In the last (twelfth) tablet is a fascinating episode involving a willow tree (*Salix*) guarded and coveted by Inanna (Ishtar). Of the many plants that might have figured in such a legend, it is curious that the willow tree should happen to be the most common source of the most popular analgesic medicine. Salicin, found in its young twigs and leaves, removes pain and inflammation and is the antecedent of modern aspirin.

From the Sumerian triumph of architecture, agriculture, art, and cuneiform script, we have by circa 2000 B.C. the barbarian movement into this civilization to shifting some of its finest populace into the Aegean and Anatolian civilizations under Greeks and Hittites. The question still remains, why were these people so vulnerable? Had their civilization already peaked, and, if so, what was the reason for the decline? As with other civilizations, we may look to agriculture.

Prior to 8000 B.C., wheat was a grass common to the Fertile Crescent that was to become Mesopotamia and the cradle of civilization. This wild wheat crossed with a "goat grass," and the progeny subsequently formed a hybrid known as emmer (*Triticum dicoccum*), the plump grains of which could support a civilization. A subsequent hybridization produced even more heterozygosis and eliminated a brittle rachis that had necessitated hand harvesting. The grain would not shatter from the plant through the action of wind as had the wild form before it.

The Sumerians learned how to cultivate this wheat and also barley, native to that area in diverse forms, and soon were trading grain for lapis lazuli and carnelian. In the Ubaid and Uruk periods, wheat was widely cultivated and was almost a monopoly, but as early as early dynastic times, wheat production began to decline owing to improper and excessive irrigation, as well as to a corresponding salinization of the land. This led to the predominance of barley, which is much more tolerant of excess salt in the soil. Thus, by 3000 B.C., trade routes between Mesopotamia and Iran had to be opened and negotiated through mountain passes and river valleys. By the end of the early dynastic period, a transit trade route was established between south Mesopotamia and the Indus valley.

While art historians tend to speak of the embellishments in art, architecture, and artifacts, ethnobotanists may find considerable other ramifications. Many writers have ignored the extensive agricultural diversity of crop plants and medicinal plants that entered and exited by these extensive land-sea routes. Crops of grain and other food stuffs alone will not suffice to sustain populations for any considerable period of time. The role of medicinal and drug plants is perhaps as important as that of food plants and may have been as significant a factor in the establishment and maintenance of the first civilizations as was the more evident grain.

The Uruk period saw the first appearance of open cast copper chisels. An axe of this period was found in Susiana, and it was bound in linen, attesting to the knowledge of the growing, retting, and removal of fibers of linen from *Linum*. The production of linen sacks allowed grain to be stored under circumstances in which it was less likely to rot (e.g., animal skins or pottery). Fabric came to replace animal skin as the exclusive clothing of these early civilizations. Linen provided the ideal container for grain and artifacts to be carried along difficult trade routes. The ropes tying the sacks shut were sealed with resins or pitch in which cylinders were rolled to leave an impression giving specific categories of information. Cylinder imprints also served to identify early wines and vineyards. Rolled in pine pitch used to seal the amphoras of wine in early Egypt and subsequently in Hellenic civilizations, these cylinder impressions gave the area of cultivation, the cultivator, and the date.

The years surrounding 2500 B.C. are important in that they mark what one might characterize as the end of several early periods. Egypt at this time had just entered the Old Kingdom period, which followed the proto-dynastic period. Mesopotamia was well into its early dynastic phases. Anatolia was in the Second Early Bronze Age. A new kind of sophistication was appearing in most areas of the ancient Near East. Architecture was beginning to have a certain grandeur, ritual was becoming stratified by hierarchies of priests, trade routes were becoming well established, and the luxuries of life were coming into evidence in increasingly large numbers of nonessential items such as jewelry, nonritual artifacts, elaborate furniture, and all the ornamentation that can be afforded only by an advanced civilization free from subsistence patterns.

As castes grew in hierarchies of priests, the shaman-priest was concerned with prophesy or divination. Egypt had to build diverse centers of religious activity to accommodate the division of labor among the elite. Magical and medical papyri grew in number. Agriculture was well established as a result of inundation and limited crop irrigation by canals. Emmer, flax (*Linum*), two kinds of barley (Upper Egyptian and Lower Egyptian), and, after the Ptolemaic period, wheat became the principal crops. The extraordinary amounts and varieties of beer produced were made not from grain, but from barley bread that was fermented—a far less wasteful practice, since unused bread would be recycled as beer and the leftover mash fed to domestic animals.

The various papyri are often diverse in content. Thus, in the *Book of the Dead*, which presents concepts of death and resurrection, numerous scenes of domestic life give an extraordinarily fine view of all aspects of Egyptian agriculture, viticulture, and oenology. We find in the papyri that large amounts of beer and wine were poured as sacred libations at the time of a death or upon the completion of a monument. The same kind of information is codified in numerous tomb paintings. During the Middle Kingdom and thereafter, every vase, chair, musical instrument, sarcophagus, textile, and weaving contained information concerning either daily life or the afterlife. Most of this was in the form of depictions, and, even in the absence of a Rosetta stone and hieroglyphic understanding, there is a great deal that we could comprehend regarding all thirty dynasties.

One product that is often neglected in discussions of ancient Egypt is oil from crops of the castor bean (*Ricinus communis*). Olive oil was imported, as the tree was not grown there successfully until the Ptolemaic period. Alternative oils for commercial use were the fruit oils of the moringa tree (*Moringa drouhardi*), linseed oil from seeds of *Linum*, oil from the balanos tree (*Balanites aegyptiaca*) as well as sesame and saffron oils. Castor oil was especially important in that, when mixed with natron, it produced a smokeless flame that could be used in homes for lighting and in tombs to allow painters to produce murals. Alternative light sources would have covered the extraordinary art with a coating of soot.

Considerable debate exists over the probability of the opium poppy existing in early dynastic Egypt as well as in Assyria. Gabra (1956) suggested that the word *shepen* refers to poppy and *shepenen* to the opium poppy. These words appear in most medical papyri and in some papyri devoted to magic, notably the *Ebers Papyrus*. Those who argue against a number of "exotics" in early Egypt must take into account that from the late pre-dynastic period, trade in timber and numerous other commodities was accomplished by intercourse with the Levant and the montane regions of Lebanon. Also, as early as the fifth dynasty, expeditions to Punt are recorded, and we have reason to believe that these originated at an even earlier date. While the "land of Punt" has never been firmly located, we know from commodities such as sandalwood, ebony, giraffes, baboons, ivory, leopard skin, and gold that the area or region was most certainly a part of the Somaliland coast. Harbors and ships for import were stationed along the Red Sea. Donkeys brought items of barter from Nubia to the south as far as equatorial Africa. The tomb of Ramses II, 1304–1237 B.C., presents us with a complete depiction of tributes that came from conquered Nubia and places south.

It is a temptation for the ethnobotanist to find psychoactive plants in early dynastic Egypt, and to that end many have tried to place *Cannabis* in this context. The contention that *smsm t*, mentioned in both the *Berlin Papyrus* and *Ebers Papyrus*, corresponds to *Cannabis* is highly unlikely. No mummy has been found wrapped in hemp fiber; no rope from the base of *Cannabis* exists in Egypt from this period. No residues of hashish have been found in any lipid matrices from funerary jars or unguent containers. It was not until the third century A.D., when the Roman emperor Aurelian imposed a tax on an Egyptian fiber, that we can identify hemp. The same may be said for Babylonia (unless one accepts the Waterman thesis advanced in 1930 that *qu-nu-bu*, mentioned during the reign of Esarhaddon in circa 680 B.C., is translatable as *Cannabis*).

By contrast, Gabra (1956) identified opiates in a residue from an "unguent vessel" of the eighteenth dynasty. Both the narcotic *Mandragora autumnalis*, *M. vernalis*, and fruits of *Papaver somniferum* figure in tomb paintings and vessels of early dynasties and become quite common by the eighteenth dynasty (Emboden 1979). These are found in frequent conjunction with the sacred blue water lily of the Nile marshes, *Nymphaea caerulea*, which has been determined to have narcotic properties much as do its relatives *N. alba*, *N. ampla*, and the related *Nuphar lutea*. The union of three genera, all psychoactive and all involved in a healing presentation, is to be seen in the colored limestone intaglio of Meritaton and Semenkhara.

The full exegesis of psychoactive plants in the context of dynastic Egypt is discussed by Emboden (1979, 1981). The argument is made that these plants and their psychoactive constituents were adjuncts to the state of ecstasis among the priestly castes of ancient

Egypt and that they lead us to a very new way of viewing Egyptian art and artifacts, as well as those of other ancient civilizations. A limestone relief of the Amarna period circa 1350 B.C. shows us the healing of King Semenkhara by his consort Meritaton using *Mandragora* and *Nymphaea*; this is a fine example of the specific context of these plants. These plant motifs appear again in the eighteenth dynasty portrait of Tutankhamen on his throne with his queen (Figure 4).

Figure 4. Throne chair of Tutankhamen (Post Amarna) depicting his queen healing him using *Mandragora* and *Nymphaea*. Cairo Museum.

One bit of iconography that still puzzles Egyptologists is the depiction of "Lady Tuth-Shena" on the stela in which she is before the god Horus. Emanating from the sun disc on Horus's head are five "rays" of tubular flowers that strongly suggest *Datura* (Figure 5). Since *Datura* is pantemperate and pantropical, the genus could not be considered scarce in any region. It is also a genus with easily identifiable virtues. It has been used in every area in which it is known, in rites of passage and in diverse forms of shamanism. Its psychoactive properties are extraordinary, and one of the usual modalities in the *Datura* experience is that of mystical flight, an out-of-the-body sensation.

This explanation, like so many others relating to Tuth-Shena and Horus, might seem specious were it not for the other, associated plants that have psychoactive properties: the central flower and leaf of *Nymphaea caerulea*; at the foot of Horus, the unguent jar wrapped with the narcotic water lily bud; the strand of grapes and their leaves hanging from the opposite side of the supporting pedestal upon which offerings rest; the four repeated representations of a cleft water lily leaf in the series of glyphs at the right-hand margin. It is the realm of the dead, evidenced by the resin cone on the head of Tuth-Shena. The light is the light of Horus, realized in the psychoactive flowers of *Datura* which "illuminate" Tuth-Shena in allegorical fashion. It is the power of Horus before which she throws up her hands in awe. *Vitis*, *Nymphaea*, and *Datura* are the intoxicating elements portrayed in this scene of shamanic manifestation.

Figure 5. Tuth-Shena in awe of the god Horus, who emits rays of *Datura* from the sun disc on his head.

It is perhaps by coincidence that the frequency of portrayal of psychogenic plants is correlated with the level of development of ancient civilizations, but I do not think so. A shamanic caste appears and, subsequently, there is further shamanic stratification, the adjuncts to these priestly offices—that is to say, psychoactive plants—increase. The length of the associated rituals is progressively increased, and the litanies or magical incantations become hypertrophied. We can see the same thing among the Maya. It parallels the complexity of medicine and medicinal practices, for all these are inseparable at a certain level. They are manifestations of belief systems that are enhanced by altered states of consciousness.

In conclusion, rather than to try to elucidate the complexity of ancient agricultural practices, it may be more appropriate to comment on certain categories of plants important to a settled state of existence, and on the inclusion of greater numbers of plants with psychoactive properties as a civilization evolves. The best place to find these plant species may not be in cuneiform script or in the hieroglyphics of papyri, but in the art and artifacts of the civilization. This is especially necessary for the ancient Near East, where ethnobotanical evaluation has been virtually absent.

LITERATURE CITED

Coe, M. 1973. *The Maya Scribe and His World*. New York: The Grolier Club.

Emboden, W. A. 1979. The sacred narcotic water lily of the Nile: *Nymphaea coerulea* Sav. *Economic Botany* 33(1): 395–407.

———. 1981. Transcultural use of narcotic water lilies in ancient Egyptian and Maya drug ritual. *Journal of Ethnopharmacology* 3: 39–83.

———. 1982a. The mushroom and the water lily: Literary and pictorial evidence for *Nymphaea* as a ritual psychotogen in Mesoamerica. *Journal of Ethnopharmacology* 5: 139–148.

———. 1982b. The water lily and the Maya scribe: An ethnobotanical interpretation. *New Scholar* 8: 103–127.

Gabra, S. 1956. *Papaver* species and opium through the ages. *Bulletin de l'Institut d'Égypte* 37: 39–46.

Gongora, M. 1868. *Antigüedades Prehistóricas de Andulucía*. Madrid: C. Moro.

Krikorian, A. D. 1975. Was the opium poppy and opium known in the ancient Near East? *Journal of the History of Biology* 8(1): 95–114.

Kritikos, P. G., and S. P. Papadaki. 1967. The history of the poppy and of opium and their expansion in antiquity in the eastern Mediterranean area, part I. *Bulletin of Narcotics* 19(3): 17–38.

Lloyd, S. 1961. *The Art of the Ancient Near East*. New York: Frederick A. Praeger.

Marinatos, S. 1948. *The Creto-Mycenaean Religion*. Athens.

Rands, R. L. 1953. The water lily in Maya art: a complex of alleged Asiatic origin. Anthropological Papers, *Bureau of American Ethnology Bulletin* 151: 75–153.

Sonnendecker, G. 1962. Emergence of the concept of opiate addiction. *Journal Mondial de Pharmacie* 3: 277.

Thompson, R. C. 1924. *The Assyrian Herbal . . . A Monograph on the Assyrian Vegetable Drugs*. London: Luzac. 8: 42–47.

Vogel, V. J. 1970. *American Indian Medicine*. Norman: University of Oklahoma Press.

Wasson, R. G. 1970. Soma of the Aryans: An ancient hallucinogen. *Bulletin of Narcotics* 22(3): 25–30.

"This Little Book of Herbs": Psychoactive Plants as Therapeutic Agents in the Badianus Manuscript of 1552

PETER T. FURST

It is not generally appreciated that some of the same psychoactive plants that played such an important role in religious ritual and divination among the Aztecs were also used as medicinal herbs in the treatment of disease and other physical afflictions. One primary source on this additional role is an early Aztec herbal, the so-called *Codex Badianus* of 1552; another is Book 11 of Fray Bernadino de Sahagún's encyclopedic *General History of the Things of New Spain*, better known as *The Florentine Codex*.

The Aztecs' knowledge of the healing properties of plants and their therapeutic use of intoxicating—and, in other contexts, sacred or even divine—members of the Plant Kingdom in the treatment of disease should not be lightly dismissed. As Schultes (1990, p. 8) observed:

> If a plant has any physiological effect on the body, it indicates that it has at least one bioactive chemical. We should know this chemical: it may not be used in our pharmacopeias; it may be used for a completely different application; occasionally it may be valuable for the same purpose; or some of the chemical constituents may provide the synthetic chemist with an unknown base on which to create new compounds.

That indigenous peoples are keen observers of the natural world and that their shamans have frequently proved to be excellent empirical chemists is well documented in the ethnobotanical literature. Though only a fraction of their medicinal plants has been tested scientifically, the number of species employed by Native Americans alone is staggering. Moerman's two-volume compendium, *Medicinal Plants of Native America* (1986), has 17,000 entries! We have no comparable list for Middle and South America, but we can be sure that in light of the wealth of plant life in the Central American and Amazonian rain forests, the number may well be substantially higher.

Indigenous peoples of Mesoamerica had, and still have, extensive knowledge of healing plants. No estimates are available for Mexico as a whole, but judging from the few modern studies and the great array of medicinal plants still available in local markets, they probably number in the thousands. It is also a fact that apart from the common shamanic techniques of "psychosomatic" curing, including sucking and the retrieval of lost, strayed, or stolen souls, the extensive native *materia medica* of herbal remedies was—as it still is today—sometimes reinforced or augmented by nonempirical therapy, including objects of mineral and animal origin, as well as magical spells and incantations.

With respect to a secondary or additional role of some of the sacred inebriants of Indian Mexico as "real" medicine, which is the subject of this chapter, I am reminded of claims made to me by Huichol shamans that their divine hallucinogenic peyote cactus, *Lophophora williamsii*, is also an effective medicament. I saw it being used as such in 1966 and 1968 when Huichols on the peyote pilgrimage to the north-central high desert of San Luis Potosí rubbed peyote juice on cuts and scratches that briars and cacti had left on their arms and legs. Peyote, they explained, would keep these wounds from becoming infected. Magic? No. Unbeknown to them and, at the time, to me, modern science had fully confirmed their assertions some time earlier when researchers at the University of Arizona isolated from the sacred magical cactus of Mexican Indians and the Native American Church of North America a chemical that exhibited antibiotic activity against a wide spectrum of bacteria and at least one fungus, including strains of the penicillin-resistant *Staphylococcus aureus* (McLeary et al. 1960, pp. 247–249; P. T. Furst 1976, pp. 112–113). As Schultes (1990) pointed out:

> Modern medicine has found therapeutic or experimental uses for some of the psychoactive principles or for semi-synthetic analogues of them which are now valued in Western medicine for a variety of applications.

History of the Badianus Manuscript

By any measure, the sixteenth-century Aztec herbal published in English translation in 1940 as *The Badianus Manuscript* is a remarkable document. Above all it is testimony to the advanced state of empirical indigenous medicine before the European invasion.

Although the idea of illustrated herbals obviously originated in Europe, where such books date back to antiquity, and the text, translated into Latin from Nahuatl (Aztec), occasionally appears to reflect a European rather than an indigenous medical tradition, the Badianus manuscript is the oldest known medical text from the Americas composed entirely by indigenous—in this case Aztec—scholars. In 184 opaque watercolor illustrations rendered stereotypically, in a style derived from pre-Hispanic codex painting rather than with the naturalism that would facilitate botanical identification, the herbal depicts many of the most important plants employed in Aztec medicine. The descriptions of their preparation and use in the treatment of disease are the earliest we have from post-Conquest Mexico or anywhere else in the Americas. In addition, nearly seventy more plants are discussed but not illustrated. Probably most of these medicinal herbs are still in use in the Mexican countryside; of those that have been tested by modern science, many have proved to have therapeutic effects.

The herbal also reveals much about the therapeutic and prophylactic uses of bird and animal blood and body parts, as well as earths and bezoar stones and other nonherbal substances. If these were mostly magical, lacking the pharmaceutical underpinnings of the herbal *materia medica* of sixteenth-century Mesoamerica, they were probably no less effective than, say, crushed spiders stuffed into walnut shells that were still being used much more recently in England against gout, or the widespread, blind-faith use of bleeding with leeches or scalpels that was the fashion in European and Euroamerican medicine as recently as the nineteenth century. In fact, the nonherbal substances of the Aztec physicians were probably more effective, because they were generally only additives to therapeutic herbs. Indeed, as Cortés, Sahagún, Hernández, and other writers of the early colonial period affirmed, Aztec medicine was in many respects superior to its European counterpart.

So, certainly, was personal hygiene. For Europeans, bathing was anathema, a certain source of illness. As the hero in Clavell's best-selling novel *Shogun* put it, in his granny's opinion a man should bathe only twice in his life—at birth and at death. In contrast, like other American Indians, the Aztecs made frequent use of the therapeutic sweat bath (*temascal* in Nahuatl), a practice as integral to native medicine as it was to spirituality. As this herbal and other sources make clear, medicinal plants before, during, and after were indispensable to the sweat bath's restorative effects.

Before turning to the topic of the title of this chapter, it should be noted that the Badianus manuscript, the name by which this fascinating early colonial document has become known, is a misnomer. Its real author was not Juan Badiano, or, to use the Latin form, Juannes Badianus, but an Aztec physician named Martín de la Cruz. Badiano, too, was an Aztec, born in Xochimilco, but he was only the recorder and translator of de la Cruz's prescriptions into Latin. Both were associated with the Franciscan College of Santa Cruz de Tlatelolco, an institution of higher learning for Indian children.[1] De la Cruz, a physician at the College, may have taught Aztec medicine; Badiano was reader (lecturer) in Latin. The most famous teacher at the College (and, for a time, also its Father Superior), was, of course, Fray Bernadino de Sahagún, author of that indispensable sourcework on Aztec Mexico, *The Florentine Codex*. The herbal was actually composed in the years when Sahagún himself was absent from the Colegio. Yet Emmart (1940, p. 25) was surely right when she called him its real inspiration:

> While Juannes Badianus gives Jacobo de Grado (then the Superior at Tlatelolco) the credit for "laying the task upon his shoulders," she writes in her commentary to the 1940 facsimile edition, "it rightly belongs in greater measure to Fray Bernadino, true father of the history of medicine in Mexico, for he awakened the interest in native medicine and gathered together the group of Aztec physicians who gave instruction in their indigenous pharmaceutical remedies."

De la Cruz identified himself as author at the beginning of the book, in the dedication to Don Francisco de Mendóza (the son of Antonio de Mendóza, first Viceroy of New Spain, who had been a strong supporter and financial backer of the college at Tlatelolco):

> A little book of Indian medicinal herbs composed by a certain Indian, physician of the College of Santa Cruz, who has no theoretical learning, but is well taught by experience alone. In the year of our Lord Savior 1552.

In signing it at the end with equal self-deprecation, Badiano affirms that he is the translator, not the author:

> Juannes Badianus, the translator, to the fair-minded reader, Greetings. I beg again and again, most excellent reader, that you consider favorably the work I have put into this translation, such as it is, of this little book of herbs.

Though he was raised in the Christian faith and given a classical education, one cannot help but wonder what Badiano might really have been thinking when he went on to describe himself and his learned Aztec physician-colleague as "poor unhappy Indians

. . . inferior to all mortals," their poverty and "insignificance implanted in us by nature" and hence meriting the recipient's indulgence for a work so unworthy that it "will not surprise me, that you cast it out where it deserves."

In size the original is modest, measuring 20 centimeters (8 inches) long, 15 centimeters (6 inches) wide, and 2 centimeters (0.75 inch) thick, with velvet covers that were once held together by metal clasps. These have long since disappeared, but otherwise this interesting manuscript from the early Spanish colonial period, sent to Europe in the early 1550s as a gift for Charles V, is in remarkably good shape.

The title page bears the inscription, "ex libris didaci Cortauilae," who has been identified as Diego Cortavila y Sanabria, a pharmacist active in Madrid in the seventeenth century (Garibay 1964, p. 5). The herbal later passed in some unknown way into the library of Cardinal Barberini, which in 1902 became part of the Apostolic Library of the Vatican. In its catalog it is listed as *Codex Barberini, Latin 241.*

The existence of the herbal had remained unknown until Charles Upson Clark, a professor of classical languages, Smithsonian researcher, diplomat, and World War I military intelligence officer, chanced upon it in 1929 in the Vatican Library. Coincidentally, another scholar, Lynn Thorndyke, saw it at almost the same time while making an inventory of the Barberini collection. To add to the coincidences, in the same year an Italian publisher brought out an edition of a sixteenth-century Italian copy of the Badianus manuscript in the Royal Library of Windsor entitled *Erbe Medicinali del Messico o Libellus de medicinalibus Indorum herbis, quem quidam Indus Collegii Sanctae Crucis medicus composuit anno Domini 1552.* Except for a few British scholars, the existence of this copy, too, was unknown until the Librarian of the Corsiniana Library, G. Gabrieli, brought it to light. The herbal remained in the Vatican until May 1990, when Pope John Paul took the manuscript back to Mexico and presented it to President Carlos Salinas de Gortari as a gift from the Vatican to Mexico. Thus, nearly 450 years after it traveled to Europe as a gift to the Spanish crown, "the little book of herbs" had come full circle, not, of course, to the former convent of Tlatelolco, but to a new home, the rare book collection of the library of the National Museum of Anthropology.

These and other facts about the history and content of the Aztec herbal we owe to Emily Walcott Emmart, who translated the herbal from Aztec rather than from the Latin text and published it in full color with explanatory introductory chapters in 1940 as *The Badianus Manuscript.* Born around the turn of the century, Emmart was trained not in botany or Mesoamerican linguistics but in cytology, having received her Ph.D. in that discipline in 1930 from Johns Hopkins University. She spent seven years on the research and translation, completing the manuscript for publication in 1938. Just analyzing the Aztec entries with the help of the Molina and Simeón dictionaries of the Aztec language and the sixteenth-century Olmos grammar took her almost a year. In 1935, while she was dividing her time between academic duties at Johns Hopkins and the Aztec herbal, the Smithsonian sent her to Madrid as a delegate to the Tenth International Congress for the History of Medicine. It was during this trip that she got her first look at the original manuscript in the Vatican Library and the chance to compare the original with Professor Clark's page-by-page photographs.

In Rome, Emmart also met Cardinal Tisserant, at the time Pro-Prefect of the Biblioteca Apostolica Vaticana, and through him his niece, Marie Therese Missionier-Vuilleman, a graduate of the Academy of Fine Arts in Rome. It was the latter who executed the watercolor copies of the botanical illustrations in the manuscript. "They are," she wrote (Emmart 1940, p. xxii), "a labor of infinite care and exactness covering a period of more than two years." The Smithsonian paid for them, but considerably greater funds were needed if the final manuscript was to see the light of day in a full-color edition, something on which she insisted from the beginning.

Funding the publication during the years of the Great Depression was a problem. Of crucial importance, Emmart wrote, was the Amateur Gardeners of Baltimore, her home city, and then, through members of that organization, the Garden Club of America. The story is of interest because, with the steady decline of government funding, researchers are increasingly compelled to piece together support from a variety of sources. In that respect, not much seems to have changed since the late 1930s.

As Emmart tells it, while she was still working on the translation, Elizabeth L. Clark, a friend with a strong interest in plants, ethnobotany, and the preservation of ancient works of art, chanced to see it. Clark presented the project to the Baltimore club. They, in turn, not only pledged support but took it to the Garden Club of America. As it happened, that organization had recently established a Founders' Fund in honor of its first president, Mrs. J. Willis Martin, the awards to be used to assist worthy projects in the field of botany. It was particularly fitting, wrote Emmart, that this fund's first award should have gone toward the publication of the earliest herbal in the New World. A complete facsimile in full color was even then enormously expensive, and it took the combined efforts of more than fifty prominent members of the Garden Club of America, led by Mrs. Harold Irving Pratt, the first chairperson of the Founders' Fund, to complete the funding for an edition that would do justice to the original, as indeed the Johns Hopkins edition did.

Finally, a 1964 Mexican facsimile edition, with the Latin text rendered into Spanish, should be mentioned. It gives recognition to the Aztec physician Martín de la Cruz as the herbal's author, with his name above the title, rather than to the translator. This fine full-color edition bears the title of the Latin translation, *Libellus de Medicinalibus Indorum Herbis* (A Little Book of Indian Medicinal Herbs). The 394-page work was published by the Instituto Mexicano del Seguro Social (IMSS), with an introduction and linguistic commentary by the noted Nahuatl scholar, Ángel María Garibay, and essays by other Mexican scholars on the significance of the herbal to botany, medicine, dentistry, zoology, mineralogy, and ethnohistory. Two thousand copies were printed in linen and another 215 were numbered and bound in leather. This edition also improves on the 1940 English version in that it uses the same folio numbers as the original, whereas the 1940 English edition organizes the illustrations into numbered plates. Unfortunately, like the Johns Hopkins Press edition, this version is long out of print.[2]

Four Psychoactive Plants with Therapeutic Value

As noted earlier, 251 plants and their therapeutic uses are discussed in the de la Cruz–Badianus herbal. Of these, 184 are illustrated. Another early colonial herbal, contained in Book 11 of Sahagún's *The Florentine Codex*, lists 255 plants, and yet, as Debra Hassig (1989, p. 32) has noted, they overlap in only fifteen instances, "attesting to the great variety of plants in central Mexico." Among the overlapping species are some that functioned not only as sacred intoxicants and stimulants in religious ritual but also for medicinal purposes to remove or alleviate the physical symptoms of disease. The indigenous worldview did not have the sharp dividing line that in Western culture would separate these functions or contexts from each other. Among these sacred intoxicants with medicinal uses are *Datura*, morning glory, tobacco, and cacao.

DATURA

Several species of *Datura* are listed as herbal remedies both in the de la Cruz herbal and in Book 11 of Sahagún's *The Florentine Codex*. Emmart (1940, p. 245) wrote that "numerous species of *Datura* were known and cultivated by the Aztecs. They were valued not only for their beauty but because of their narcotic properties. The narcotic drug stramonium is common to the species of this genus."

Plate 49 in the 1940 edition[3] depicts two different species side by side (Figure 1), with the legend beneath them reading, "For pain in the side. The application of the herbs called *tolohuaxihuitl* and *nexehuac*, ground in water and applied, takes away pain in the side." Emmart's (1940, p. 253) notes for this plate read as follows:

1. *Tolohuaxihuitl*—one of the most widely known of the narcotic remedies of the Aztecs is that composed of the extract of various species of *Datura*, commonly referred to in this manuscript as *tolohua* (Pl. 22, 42) or *tolohuaxochitl* (Pl. 49, 57, 64, 65, 86). Numerous other native names such as *toloache*, *toloatzin*, *nacazul*, and *tlapatl* (still known as *nacazul* and *tlapatl* in some areas) have been given to various species of *Datura*. The plants possessing the names *toloache* or *nacazul* have been identified as *Datura innoxia* Miller (W. E. Safford, *Daturas of the Old World and New*, Smithsonian Annual Report, p. 549, 1920), and the plant *tlapatl* as the jimson weed, *Datura stramonium*. In the preliminary description of the Badianus manuscript the author referred to the *tolohuaxochitl* as *Datura stramonium*. Dr. Blakeslee has suggested that the red seeds shown in the illustration of *tolohuaxochitl* are indicative of the species *Datura meteloides* Dunal. This species was used as a narcotic agent in the treatment of infected and abscessed glands (p. 42); in the cure of pain in the side (Pl. 49); for pain of the pubes (Pl. 57); in an ointment to cure cracks in the soles of feet (Pl. 86). From *Datura* is obtained the narcotic stramonium, which is composed of 4 percent of the alkaloids hyoscyamine, atropine, and scopolamine.

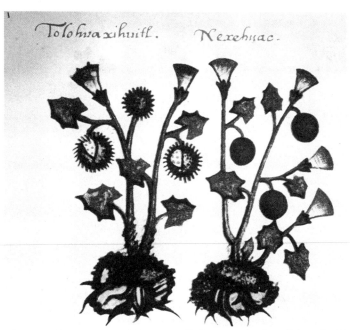

Figure 1. Two species of *Datura*, called, respectively, *tolohuaxihuitl* and *nexehuac*, as depicted in Plate 49 of the de la Cruz–Badianus manuscript. The species at the left is clearly *D. innoxia* Miller. Emmart (1940, 253) identified the other, whose name translates as "sleep plant," as *D. ceratocaula* Ortega.

About the plant the Aztecs called *nexehuac*, from *nexua* (to sleep), Emmart wrote that the artist "drew a picture which is similar to that of the *tolohuaxihuitl*, but with blue flower and dark blue smooth-skinned fruit. This character suggests that the plant may be *Datura ceratocaula* Ort."

In this connection it should be noted that Martínez (1959, pp. 327–328) listed eight species of *Datura* known as *toloache*, including *D. ceratocaula* Ortega, *D. innoxia* Miller, *D. meteloides* Dunal, *D. discolor* Bernh., *D. pruinosa* Greenman, *D. stramonium* L., *D. quercifolia* Humboldt, Bonpland, and Kunth, and *D. villosa* Fernandez-Villar (the latter particularly in Bolaños, Jalisco, and San Luis Potosí, all sharing similar chemistry.

Tolohuaxochitl appears again in the prescription for Plate 57 in the de la Cruz manuscript:

> CURE OF THE PUBES. When this part feels pain, it is to be anointed with a juice which you are to press out and make from the bark of the *macpalxochitl* tree, the briars, the herbs *toloaxihuatl* and *xiuhtontli*, Indian knife, flint, the fruit which we call *tetzapotl* and the *texoxoctli* stone, all ground in the blood of a swallow, a little lizard, and a mouse. But remember to heat this juice. And if the swelling or pain is vehemently burning, you should not hesitate to cut this part, clean the cut part and anoint it with a liquid of the herb *tlalhuaxin* ground in the yolk of an egg.

Emmart (1940, p. 262) identified the *macpalxochitl*, from *macpal* or *macpalli* (palm, hand) and *xochitl* (flower), as *Chiranthodendron pentadactylon*, one of the best-known Mexican trees, native to the mountains of Oaxaca and Guatemala and cultivated in Mocetuzuhma's gardens. The flowers of this tree were used as an antiepileptic, and the juice of the bark and leaves was pressed out and used to relieve pain. Martínez (1959, pp. 421–422) noted that the flowers of this 10-meter (30-foot) high tree were sold fresh or dried in herbal markets as "flor de manita" (little hand flower) for treatment of heart problems, epilepsy, eye inflammation, and hemorrhoidal pain.

Tetzapotl, or stone zapote, Emmart wrote, is probably *Calocarpum mammosum* (L.) Pierre, whose fruit is still employed in a lotion to relieve pain. Emmart quoted Standley to the effect that the Aztecs used the seed coat as a remedy for epilepsy, while the sap was believed to have vomitive and anthelmintic properties.

In Plate 41 of the de la Cruz herbal, *tolohuaxihuitl* is shown side by side with a plant called *tlahchinolpan yxhuaxiuitl* as remedies for struma (goiter) or scrofula (Figure 2). The prescription reads as follows:

> FOR STRUMA OR SCROFULA. One who has struma gets relief from the malady if you put on his neck a plaster of the herbs, which grow in a pleasure-garden or in a thicket or in a burned thicket of reeds, *tolohuaxihuitl* (Datura), *tonatiuhyxiuh* (sun herb) of the root of *tecpatl*, the leaves of a bramble-bush; grind them with the stone, which may be found in the intestines of a swallow, and with its blood.

In her commentary to this, Emmart identified the affliction as the swelling of the glands of the neck, which the Aztecs called *mazacocoliztli* and which was treated with plasters made of herbs mixed with the bezoar stone of a swallow. Of the three main ingredients of the plaster, *toloa* was generally used to relieve pain, *tonatiuhyxiuh* evidently had astringent properties, and *tecpatl* provided a thick and sticky juice. In *The Florentine Codex*, Sahagún (Book 11, p. 147) wrote of *toloa* (*Datura*) thus:

> It is also a fever medicine; it is drunk in a weak infusion. And where there is gout, there it is spread on, there one is anointed. It relieves, drives away, banishes [the pain]. It is not inhaled, neither is it breathed in.

Figure 2. *Tolohuaxihuitl* (*Datura innoxia* Miller), an ingredient in a cure for struma or scrofula, as depicted in Plate 41 of the de la Cruz–Badianus manuscript.

In de la Cruz's herbal, the illustration of the plant called *tonatiuhyxiuh ahhuachcho*, literally "Tonatiuh's dew plant" or "the sun god's dew plant," shows red flowers with either dew or rain drops attached to the leaves and stem. Emmart noted that this plant turns up several times in the herbal—as a constituent of a plaster to cure scrofula and abscessed neck glands; as an antidote against snakebite; as an antiparasitic to kill helminths; as a potion for fever, constipation, and indigestion; and in a soothing potion for the dying.

Azcapanyxhua tlahçolpahtli refers to another *Datura* species (Figure 3). Emmart (1940, p. 224) translated the Nahuatl name as "ill-smelling medicine that comes up out of an ant hill," from *azca*, *azcatl* (ant), *pen* (place, hill), *yxhua* (to come out of), *tlahçol tlazollo* (ill smelling), and *pahtli* (medicine). Locating this plant on an ant hill, pictorially reinforced by the ants on the roots and lower stem, reflects the fact that *Datura* species and related solanaceous plants usually are found growing in disturbed soil.

Tlapatl is another name that at first seems to present problems of identification but soon reveals itself as yet another term for *Datura*. Plate 21 of the de Cruz herbal reads thus: "Chapter three: Of purulence of the ears, of deafness and obstruction." It is followed by Plate 22 showing three illustrated plants and the following prescription:

When instilled into discharging ears, the root of the *maçayelli*, the seed of the herb *xoxouhquipahtli*, and some leaves of *tlaquilin* with a pinch of salt in hot water are very

Figure 3. *Azcapanyxhua tlahçolpahtli*, a flowering shrub
clearly identified as belonging to the genus *Datura*, as
depicted in Plate 20 of the de la Cruz–Badianus herbal.

helpful. And the ground leaves of two shrubs are to be smeared beneath the ears.
The shrubs are called *tololua* and *tlapatl*. The gems *tetlahuitl, tlahcalhuatzin, eztetl, xoxouhqui chalchihuitl*, pulverized and instilled together with leaves of the *tlatlanquaye*
tree ground in hot water, open obstructed ears.

Here Emmart noted that *maçayelli*, from *maça, maçatl* (deer) and *yelli, elli* (liver), is neither pictured in the herbal nor has it been identified. *Xoxouhquipahtli* (blue medicine) is
also mentioned by Sahagún (Book 11, p. 146) as being somewhat similar to *topozan* (identified as *Buddleia americana* by Martínez 1959, p. 309) but with greenish leaves that are
pulverized and ground up: "He who has festered flesh or the scabies places it there. Thus
he recovers. But it can also be drunk, he who has a fever drinks it." The plant remains
unidentified. Anderson and Dibble, in a footnote, suggested that it might be either a species of *Datura* or *Caesalpinia crista L.*, whose seeds Martínez (1959, p. 159) reported as being
used as a tonic and against menstrual pain. *Datura*, of course, would make sense, inasmuch as it figures in so many pain-alleviating medicines.

In a note on page 226, Emmart wrote of *tlapatl* var. *tlapatl*, a name that in an early edition of Sahagún was applied to the castor oil plant, *Ricinus communis* L. Inasmuch as the
plant was not native to Mexico, this must have been a later attribution. Emmart also noted
that Hernández's illustration of *tlapatl* was clearly a *Datura*.

In Book 11, Sahagún mentioned *tlapatl* not only as one of the "many different herbs which perturb one, madden one" (p. 129) but as a medicinal herb (p. 147). There is little doubt that *tlapatl* is probably *Datura stramonium*, as Anderson and Dibble identified it in a footnote (p. 129). It may also be that some of the several names applied to *Datura* pertain not to the plant itself but to medicinal potions or infusions made from it. This is how Sahagún (Book 11, p. 129) described it:

> It is small and round, blue, green-skinned, broad-leafed, white-blossomed. Its fruit is smooth, its seed black, stinking. It harms one, takes away one's appetite, maddens one, makes one besotted.
>
> He who eats it will no longer desire food until he shall die. And if he eats it moderately, he will forever be disturbed, maddened; he will always be possessed, no longer tranquil.
>
> And when there is gout, it is spread thin as an ointment in order to cure. Nor is to be sniffed, for it harms one, deranges one, it takes away one's appetite.
>
> I take *tlapatl*; I eat, I go about eating *tlapatl*.
>
> So it is said of him who goes about belittling, who goes about haughtily, presumptuously, (that) he goes about eating the *mixitl* and *tlapatl* herbs; he goes about taking *mixitl* [and] *tlapatl* to himself.

This description requires several comments. One concerns the efficacy of plasters or poultices containing *Datura* and other herbal preparations, such as tobacco, which is in the same family as *Datura*, namely, the Solanaceae. The other comment has to do with the capacity of *Datura* when administered internally in measured doses to cause insanity, as claimed by Sahagún's informant. The principal alkaloids in the fifteen to twenty species of *Datura* and its subgroups are hyoscyamine, norhyoscyamine, and scopolamine. All belong to the tropane series. Atropine is also present, but in small amounts; no visionary, or "hallucinogenic," effects have ever been reported for this alkaloid. Rather, these effects are attributable to scopolamine (Schultes and Hofmann 1979, p. 86). The tropane alkaloids are all very poisonous, however, which accounts not only for the fact that, as Sahagún reported, *Datura* can kill, but also for reliable statements that an experienced person can administer potions made from the roots and other parts in such amounts and in such ways as to cause temporary derangement and even permanent insanity (see P. T. Furst 1976, p. 140).

As for *Datura* as a pain killer, the plant has proven analgesic properties. During her stay with the Zuñi in 1879, Matilda Coxe Stevenson observed how effective it could be even in the most extreme circumstances. She observed that *Datura* was the property of the rain priests and the Little Fire and Cimex fraternities and that the medicine prepared from the powdered root was administered only by their directors. Each director had to collect his own medicine and prepare and deposit prayer-plumes to the sacred plant to assure success:

> The writer [Stevenson] observed the late Nai'uchi, the most renowned medicine man of his time among the Zuñi, give this medicine before operating on a woman's breast. As soon as the patient became unconscious he cut deep into the breast with an agate lance, and, inserting his finger, removed all the pus; an antiseptic was then sprinkled over the wound, which was bandaged with a soiled cloth. (The writer obtained samples of the antiseptic, but each time the quantity proved too small for analysis.) When the woman regained consciousness she declared that she had a peaceful sleep and beautiful dreams. There was no evidence of any ill effect from the use of the drug (Stevenson 1915, p. 46).

There is no reason to doubt that the analgesic properties can be absorbed through the skin as readily as are nicotine and other tobacco alkaloids (Wilbert 1987). Parenthetically, it is interesting to note that skin patches are replacing injections increasingly in modern medicine. Thus it has been found that small skin patches with a potent pain-killing drug called fentanyl are more effective and last longer than pain-killing drugs injected with a needle, and that transdermal delivery systems, meaning drugs delivered through the skin rather than by injection, are becoming ever more popular (*New York Times*, 30 October 1991). Judging from the de la Cruz herbal, Sahagún's Book 11, and other early colonial sources, the Aztecs knew this centuries ago.

In the section on medicinal plants in Book 11, Sahagún described *tlapatl* as follows:

> There are two kinds; the name of still another is *toloatzin*. And *tlapatl* is somewhat tall; its blossoms, its foliage, reach upward. One who has gout or whose flesh is swollen rubs himself with it. Lampblack is mixed in. It is not potable.

Just what is *mixitl*, or *mixtli*, which Sahagún mentioned among the plants that intoxicate and "madden" but that also figure in the treatment of illness? Sahagún's description seems to refer to the zombie phenomenon reported from Haiti:

> *Mixtli* is of average size, round, green-leafed. It has seeds. Where there is gout, there [the ground seeds] are spread on. It is not edible, not drinkable. It paralyzes one, closes one's eyes, tightens the throat, stops off the voice, makes one thirsty, deadens the testicles, splits the tongue.
>
> It is not noticeable that it has been drunk, when it is drunk. He whom it paralyzes, if his eyes are closed, remains forever with closed eyes. That which he is looking at, he looks at forever. One becomes rigid, mute. It is alleviated a little with wine.
>
> I take *mixitl*. I give one *mixitl*.

Martínez (1959, p. 539) included *mixtli* among his medicinal plants, but only quoted the above from Sahagún without attempting to identify it botanically. The de la Cruz herbal does not mention it at all. Most likely it, too, is *Datura*, or a closely related genus (or else it is the name of a medicine prepared from *Datura*), a judgment with which Ortíz de Montellano (1990, p. 156), in his book on Aztec medicine, agreed.

MORNING GLORY (*Turbina* [formerly *Rivea*] *corymbosa, Ipomoea violacea*)

Two species of morning glory, one with white flowers and the other with blue or purple, or, more correctly, their seeds, were among the most important ritual hallucinogens of ancient Mesoamerica, and remain so to this day. They have been the subject of a great many studies, including Schultes's 1941 monograph, which once and for all established the identity of the potent hallucinogenic seed the Aztecs knew as *ololiuqui* with the white-blossomed morning glory *Turbina corymbosa*, and Hofmann's discovery of the active principles in these seeds as ergot alkaloids closely related to *d*-lysergic acid diethylamide (LSD) (Hofmann and Cerletti 1961; Hofmann 1967). Schultes's study was especially significant because it showed Safford (1920) to have been wrong when he insisted that *ololiuqui* had to be the seeds of *Datura meteloides* (now *innoxia*), or *toloatzin*.

The identification of *ololiuqui* as the seeds not of *Datura* but of a morning glory had been well known to the early chroniclers, including Francisco Hernández, but Safford evidently had little faith in them or in the botanical expertise of the Aztecs. As early as 1919, Blas Pablo Reko (1934), an Austrian-born Mexican physician, collected seeds the

Mexicans called *ololuc* and identified them as belonging to *Rivea corymbosa* (renamed *Turbina corymbosa*). Safford agreed with the botanical determination but not with Reko's idea that *ololuc* had to be the *ololiuqui* described by Sahagún in *The Florentine Codex*, or in such other colonial sources as Ruiz de Alarcón's manuscript of 1629, which described reverence for *ololiuqui* and visionary intoxication with the seeds among pre-Hispanic "superstitions" surviving among the Nahuatl-speaking inhabitants of Guerrero (Ruiz de Alarcón 1982).

In 1934, Reko published the first historical review of the use of *ololiuqui* in the journal *El México Antiguo*. Writing in his native German, he again identified the seeds with *Rivea corymbosa*. Tests by Santesson in 1937 at least dispelled the notion that nothing in the Convolvulaceae could possibly induce visions, although the precise nature of the alkaloids was to remain a mystery until Hofmann solved the problem in the early 1960s.[4]

Although, strictly speaking, *ololiuqui* was the Nahuatl word for the seeds of *Turbina corymbosa*, it could also refer to the plant, although the latter also bore its own name, *coatl xoxouhqui* (green snake plant). Sahagún (Book 11) described *ololiuqui* both with respect to its hallucinogenic effects and its medicinal uses. In the section entitled, "The different herbs which perturb one, madden one," he wrote thus:

> Its leaves are slender, cordate, small. Its name is ololiuqui. It makes one besotted; it deranges one, troubles one, maddens one, makes one possessed. He who eats it, who drinks it, sees many things which greatly terrify him. He is really frightened [by the] poisonous serpent which he sees for that reason.
>
> He who hates people causes one to swallow it in drink [and] food to madden him. However, it smells sour, it burns the throat a little. For gout, it is only spread on the surface.

This description includes at least a brief reference to the therapeutic use of *ololiuqui* as a poultice. Another description (Sahagún, Book 11, p. 165), however, does not seem to fit the morning glory at all. Apparently, as in the case of *peyotl*, *ololiuqui* does not necessarily always refer to the same plant.

There is no such question about Martínez's (1959, p. 463) description of the visionary-divinatory use of ground-up *ololiuqui* seeds in Aztec and contemporary Oaxacan ritual. He also cited Francisco Ximénez's *Los Quatro Libros de la Naturaleza y Virtudes de las Plantas* (1615) on some of the medicinal applications of *ololiuqui* preparations. According to this early Spanish colonial source, morning glory seeds cured gonorrhea (*mal gálico*), alleviated the aches and pains of malaria, removed flatulence, and dissolved tumors. Martínez also mentioned an early Yucatec medical treatise, attributed to *El Judío*, in which *xtabentún*, the Yucatec Maya term for the morning glory, is described as having numerous therapeutic virtues, being best known for the removal of obstructions of the urinary canal.

The de la Cruz herbal does not specifically mention *ololiuqui* or *coatl xoxouhqui*, the green (or blue) snake plant, *Turbina corymbosa*. Plate 22 in the herbal, however, illustrated a plant called *xoxouhquipahtli*, the seeds and leaves of which were used in the treatment of ulcerating and discharging infections. This name also translates as "blue (or green) medicine." Sahagún (Book 11) mentioned that the plant was used as a beverage and, together with certain purgatives, in the treatment of eye diseases. Curiously, another plant with the same name, but with a slight variation in the spelling, is depicted in Plate 63 of the de la Cruz herbal. Emmart ventured no botanical identification, but according to Martínez (1959, p. 540), both *xoxoucapahtli* and *quauxoxouqui* are "probably" the same as *ololiuqui*. Perhaps so, but it is not possible to confirm this from the illustration.

PICIÉTL (*Nicotiana rustica*)

Aztec doctors prescribed *piciétl* in both the internal and external treatment of a variety of debilities, that is, in the form of quids, potions, plasters, and enemas. Considering the widespread use of *Nicotiana rustica* as a ritual intoxicant, it is curious that Sahagún did not list it specifically among the "many different herbs which perturb one, madden one." These properties are mentioned later in the section on medicinal plants, however, in the description of the plant and its uses (Sahagún, Book 11, p. 146):

> Its [*Nicotiana rustica*] leaves are wide, somewhat long; and its blossoms are yellow. It is pounded with lime. He who suffers fatigue rubs himself with it, likewise he who has the gout. And it is chewed. In this manner is it chewed: it is only placed in the lips. It intoxicates one, makes one dizzy, possesses one, and destroys hunger and a desire to eat. He who has a swollen stomach places it on the stomach and there in the navel.

The de la Cruz herbal says a great deal more about the different uses of tobacco preparations. In the legend accompanying Plate 54, tobacco is prescribed as one of the ingredients of an enema:

> Rumbling of the abdomen: For one whose bowels are murmuring because of diarrhea, make a potion, let him take it with an ear clyster of the leaves of the herb *tlatlanquaye*, the bark of *quetzalaylin*, the leaves of *yztac ocoxochitl*, and these herbs, *tlanextixiuhtontli*, *eloçacatl*, and the tree *tlanextia quahuitl* ground in bitter-tasting water with ashes, a little honey, salt, pepper and [stone] alectorium, and finally *piçietl* [*sic*].

Martínez (1959, pp. 324–325) identified *tlatlanquaye* as *Iresine calea*, a member of the amaranth family known in Mexico also as *hierba de tabardillo* and *hierba de calentura* and claimed that in Veracruz a boiled preparation of the leaves is used against malaria. Tests reported by Martínez suggested favorable results with typhus and typhoid fever. The tree called *quetzalaylin* may be a Mexican species of *Alnus*, possibly *A. arguta*, inasmuch as the compound Aztec term can be rendered as *ili verde*, the common name for *Alnus* in Veracruz (Miranda and Valdés 1964, p. 276). The other plant ingredients remain unidentified.

Piciétl appears again with Plate 82 as an ingredient for enemas to be administered for "recurrent disease":

> Recurrent disease: One who falls again into disease is to drink, before lunch, some of the sap, very like milk, which is to be pressed out of the *teohamatl*, so that he will vomit. On the third or fourth day he is to drink a potion of *tonatiuhyxiuh* stalk, also of the stalk of *tlatlanquaye* and *tlanextixiuh*, ground in hot water. On the third he is to drink of *cuecuetzpahtli* stalk ground in native wine. Which potion he is to drink before he enters the bath; upon coming out of this, he is to be anointed with the juice of ground *teohamatl* bark. His abdomen is to be purged twice with a clyster, first with the juice of the *ohuaxocoyolin* root ground in hot water; and this before he takes any food. This juice drives the pus thoroughly from the abdomen. The second time, after a lapse of a few days, with juice made from the ground herb which has the power of inebriating, which we call *piciyetl* [*sic*], salt, black native pepper and pale-colored pepper.

Of these ingredients, *teohamatl* remains unidentified, as does *tonatiuhyxiuh*, the sun god's plant; *tlatlanquaye*, according to Martínez (1959, p. 325), is *Iresine calea*, still widely employed in the Mexican countryside as a diuretic and diaphoretic and against typhus, typhoid fever, and malaria. *Tlanextixiuh*, which recurs frequently throughout the herbal in remedies for widely differing diseases, remains without secure identification, but according to Reko may be similar to *eringo* (*Eryngium* species) (Miranda and Valdés 1964, p. 250). *Ohuaxocoyolin*, illustrated in Plate 17, can be identified as a begonia. Emmart (1940, p. 222), in the plate's legend, referred to the plants only as "native bitter herbs," the tuberous root of which, dried and ground, is applied to the eye after the removal of fleshy growths. In addition, the root is used to relieve intestinal pain, the stalk in the preparation of an enema to alleviate stoppage of the urinary meatus, and, as noted above, as an ingredient of enemas in cases of recurrent disease, for which the tobacco enema constituted the final treatment.

Emmart (1940, pp. 294–295) commented on the Aztec treatment of recurrent disease as follows:

> The above reference to treatment for recurrent disease is indicative of the close observation on the part of the native doctors of recurrent fevers, similar in symptoms to tertian fever which they called *viptlatica*. Not only did they recognize the malady by its periodicity, but they knew also that treatment must be adapted to the respective phases of rise and decline of fever. The first part of the treatment consisted of an emetic. On the third or fourth day the patient was given a potion of certain herb extracts which they believed were antipyretic in effect. This was followed by a stimulant and a steam bath, which, by dilating the cutaneous blood vessels, dissipated the heat. After the bath, the body of the patient was anointed with the juice of tobacco mixed with salt and pepper.

This last passage Emmart misunderstood: the text makes it clear that treatment was not concluded with the "anointing" of the patient with tobacco juice but with its rectal administration, with the addition of pepper and salt, in a second enema that concluded the treatment.

CACAO (*Theobroma cacao* L., *T. augustifolium* DC)

A beverage made of the ground beans of the cacao tree, *Theobroma cacao* L., was highly prized in pre-Conquest Mexico, reserved, it appears, for the priesthood and nobility and for special ritual occasions. Rich in the stimulant caffeine, it clearly had overtones of the divine. A deity represented on large late Classic and early post-Classic Maya urns decorated with cacao pods presumably represents the divine patron of this tree.

Among the Mixtecs of Oaxaca and adjacent regions, who were contemporaries of the Aztecs, the chocolate drink seems to have been a virtual must in the marriage rite uniting deities and nobles, for a foaming cup of this sacred beverage is repeatedly shown standing on the marriage mat in numerous representations of this ceremony in *Codex Vienna* and other pictorial manuscripts. Also, chocolate, like pulque, tobacco, sacred mushrooms, and other precious substances was in the charge of Seven Flower, a male deity who was the Mixtec equivalent of Xochipilli, the Aztec Lord of Flowers, god of song, poetry, and spring, and divine patron of stimulants and hallucinogens (J. L. Furst 1978, pp. 139, 242–243, 272, 204).

Cacao beans were widely traded throughout Mesoamerica. Their economic importance can be seen in that they were used as currency in Aztec times and possibly also earlier. The de la Cruz herbal also prescribed *cacauaxochitl*, identified by Emmart (1940, p. 311) as *Theobroma augustifolium*, one of the major cacao trees in Mexico, as part of a medicine against "mental stupor" (Figure 4).

There is some confusion over the Nahuatl name of the chocolate tree, *Theobroma cacao*. Plate 68 of the de la Cruz herbal referred to the tree as *tlapalcacauatl* (colored cacao), from

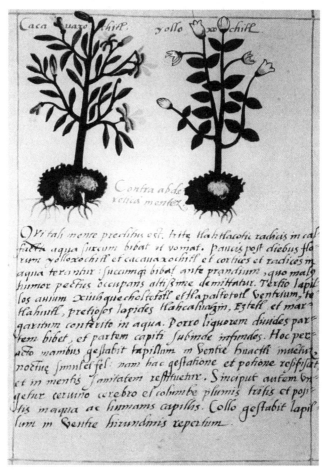

Figure 4. *Cacauaxochitl*, the cacao flower tree, as depicted in
Plate 98 of the de la Cruz–Badianus herbal, where it was
prescribed as part of the medicine against "mental stupor."
Emmart (1940, p. 311) identified the tree as *Theobroma
augustifolium*, one of the major cacao trees of the Mexican
tropics.

tlapal, tlapalli (colored) and *cacauatl* (cacao) (Figure 3). This is the earliest illustration of the
cacao tree, the caffeine-rich beans of which were ground into a stimulating ceremonial
beverage. According to Hernández (1946, pp. 908–921), however, the name applies to
only one of four varieties recognized by the ancient Mexicans, and the smallest one, at
that. The largest variety, which also bore the largest fruits, was called *cacahoaquiahuit*
(cacao tree). A second variety, of medium height, with fruits second in size to the former,
was called *mecacacahoatl*. A third kind, smaller still and with smaller fruits containing red-
dish seeds, was called *xochicacahoatl*, and a fourth subspecies, smaller than all the others
and with smaller fruit, was known as *tlalcacahoatl* (the little one). All four were similar in
character and served the same purposes, although the seeds of *tlalcacahoatl* made the best
drink, while seeds of the other three plants were preferred as currency.

    To these four varieties Hernández (1946, p. 912) added yet another cacao tree called
quahpatachtli, which he described as considerably larger in size, with bigger leaves, and
with pods containing larger and naturally sweeter-tasting seeds. Whereas the de la Cruz
herbal merely illustrated the tree without any details of its medical uses, Hernández had

a lot to say about it: the beverage made from the *cacahoatl* seeds is administered without sweeteners or other additives to bring down high temperatures and ameliorate high fevers accompanying serious illness, and also to patients suffering inflammations of the liver or other organs. Four cacao beans toasted to extract the oil were mixed with 28 grams (1 ounce) of the gum called *holly* (*ule*); the mixture, which was rendered sticky and glutinous over the fire, was said to be effective against dysentery. In excessive amounts, the chocolate drink caused constipation and triggered certain incurable illnesses.

Several classes of the chocolate beverage were known, depending on whether the beverage was made only with the beans, or other ingredients were added, among them certain kinds of flowers or other parts of plants from which flour can be made, especially maize. The first kind of chocolate drink consisted of about 100 grams (3 to 4 ounces) of well-ground raw or toasted chocolate beans mixed with ground maize. The chocolate beverage was also made by mixing ground-up fruits of *mecaxochitl*, *xochinacaztli*, and *tlilxochitl* with the chocolate beans.

Emmart (1940, p. 315) identified *mecaxochitl* as a species of *Piper*. Hernández's illustration of *tlilxochitl* clearly defined it as the vanilla plant, *Vanilla fragrans*. *Xochinacaztli* (from *xochi* [flower] and *nacaztli* [ear]) is not included among the medicinal plants in the de la Cruz–Badianus herbal, but Martínez (1959, p. 507) identified it as the *orejuela* (from *oreja* [ear]), *Cymbopetalum penduliflorum*, a medium-sized fruit-bearing tree so named because of the resemblance of its fragrant, pale yellow to greenish flowers to the human ear.

These additives to the chocolate beverage are extremely interesting. Hernández referred to the spicy fruit of the pepper plant as an analgesic that was used in combination with vanilla to promote sleep. Emmart pointed out that not only are the pepper and vanilla plants depicted next to one another in Plate 104 of the de la Cruz–Badianus herbal, but they are prescribed in combination as a charm to safeguard the traveler. More important in the present context is the fact that they are also combined with other herbs in a lotion administered against fatigue suffered by persons holding public office.

This also deserves some discussion. That the Aztecs clearly regarded lack of vigor or fatigue, especially in public officials, as a disease, is "an indication of the degree of advancement of medicine in Mexico" (Emmart 1940, pp. 277–280):

> The particular emphasis on fatigue of those administering the government and holding public office sheds light on both the attitude of mind toward fatigue and lack of vigor as well as the role of the public official of the Aztec empire before the Conquest.

The lotion against lassitude consisted of various parts of several flowering trees, shrubs, and herbs, macerated and steeped for a day and a night in spring water, "each by itself in a new pot or vessel." These juices, fortified with certain minerals, including bezoar stones found in the stomachs of three kinds of water birds, were mixed with the red sap of the brasil tree, and the blood, brains, bile, and bladders of predatory animals with a reputation for cunning, strength, and bravery, such as the jaguar, ocelot, mountain lion, wolf, and coyote. The prescription explained that

> indeed, these medicaments bestow the bodily strength of the gladiator, drive weariness far away, and, finally, drive out fear and fortify the human heart.
> In addition, a leading man or any one else, who wishes to obtain this rebuilding of the body, should eat the flesh of a white rabbit or of a white fox whelp, either roasted or boiled.

Emmart made the interesting point that the remedy, which used certain rare and precious flowers and stones, including the sweet-scented blossoms of *Plumeria* (which were used not only medicinally but as personal ornaments and as sacrifices to the gods, and whose high esteem in pre-Hispanic times is echoed in their present-day use to decorate churches), was very complex. This complexity

> suggests first that it [the remedy] was intended for the nobility and, secondly, that it might have been derived from some ritualistic ceremony used in the initiation of those holding public office. Initiation into religious, military, and secular office was usually associated with bathing, fumigating and painting the body, and fasting, together with subjecting the novice to various forms of physical pain (Emmart 1940, p. 278).

Emmart also observed that the idea of restoring vigor by anointing the body with parts of animals and eating the flesh of such rarities in nature as a white rabbit or a white fox may bear some symbolic relationship to religious cannibalism, which rested on the belief that the body of the sacrificed was holy and hence could "impart qualities desirable to the living."

European Influence on the Badianus Manuscript

How much of the de la Cruz herbal is really, as Emmart believed and as noted historian of medicine Henry E. Sigerist asserted in his Foreword to her translation, "purely Aztec," and how much is it the product of syncretism between native and European concepts? If the manuscript is a synthesis of two traditions, in which direction is it weighted—Aztec or European? Sigerist (1940, p. ix) proposed that in contrast to such early writers as Juan de Cárdenas and Francisco Hernández, who "had the tendency to project European views" on the *materia medica* of the Aztecs, the Badianus herbal "is a purely Mexican product," one that, "as far as we can see . . . shows no traces of European influence."

That is hardly accurate, but neither is the opposite view in which the Aztec contribution pales in comparison to the European contribution. In his well-known work titled *Mexican Manuscript Painting of the Early Colonial Period*, Donald Robertson (1959, p. 157) described the herbal as a "meeting of European science and native pictographic traditions." Hassig (1989, p. 32) also stressed that the illustrations fell well within the native style of stereotypical representation rather than within the European style of naturalistic representation:

> The plants are rendered in opaque paints on plain, uncolored backgrounds. They are presented as frontal, basically two-dimensional, isolated entities that are partially outlined and are of fairly uniform size. The individual parts of the plants are depicted, including the roots, which are sometimes surrounded by soil, rocks, or water. In spite of their initially naturalistic appearance, informed botanical judgment describes them as highly inaccurate, and only a very few of the painted plants have been botanically identified.

In fact, except in a very few cases, botanical identification has depended far more on the context than on the pictures. So, for example, the roots are all virtually identical; no distinction is made among the leaves, the fruits do not always match the flowers, and "the colors, while beautiful, do not always reflect patterns that are found in nature." This is true of plant representations in the pre-Hispanic codices as well.

Hassig agreed that while such conventions as snakes slithering up the sides of a plant have their parallels in European herbals (as in depictions of the adderwort), they are also well within the native pictorial tradition. For example, in the case of the snakes depicted with the plant called *coaxocotl* (serpent fruit, from *coatl* [serpent] and *xocotl* [fruit]), she pointed out that they form its name glyph (Figure 5). It seems to me it would be fairer to the Aztecs to turn this around and say that the herbal presents us with Aztec science and native conventions of Central Mexican codex-style illustration, but with some European contamination, and in a context that is, as Hassig (1989, p. 32) observed, at least in a broad sense "a direct reflection of the European herbal tradition."

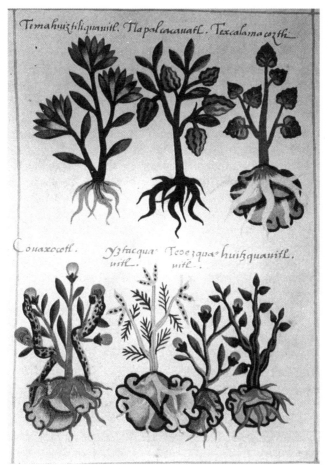

Figure 5. *Tlapalcacauatl* (*Theobroma cacao*), the principal source of chocolate, as depicted in Plate 68 (top row, center) of the de la Cruz–Badianus manuscript. As Emmart noted, this is the earliest depiction of the cacao tree. Note also the typical pre-Conquest use of snakes (lower left) as a name glyph for the plant called *coaxocotl* (serpent fruit).

It goes without saying that even the first decades of European domination and indoctrination were bound to leave their mark on the empirical and nonempirical components of indigenous medicine in Mexico, especially in areas where the priests were able to exercise tight control. Yet Ruiz de Alarcón's manuscript of 1629 shows that an extraordinary amount of the old beliefs and shamanic practices survived practically intact even a full century after the Spanish Conquest. In this herbal, too, not everything that at first glance looks suspiciously European actually is. An example is the Doctrine of Signatures, as the belief in like-curing-like was called in Europe. Although some writers have cited examples of similar concepts in mid-sixteenth-century Aztec medicine as proof of European contamination, there is good reason to believe that the Doctrine of Signatures has its purely indigenous counterpart not only in Mexico but elsewhere in the Indian Americas.

Likewise, for example, Hassig (1989, p. 36) mentioned Badianus's prescription of the crushed leaves of the wart herb, *tzotzocaxihuitl* in Nahuatl (from *tzotzoca* [wart] and *xihuitl* [plant]), as a cure for warts. Both Hassig and Emmart (1940, p. 307) saw the connection the Aztec author made between the warty appearance and rough feel of a certain plant and the therapeutic effect attributed to it as a borrowing from European herbal medicine, which most probably it was not. Emmart noted that Badianus translated the Aztec term into Latin as *verrucaria*, Pliny's name for the European wartwort or turnsole, *Euphorbia helioscopa* L., and that species of this genus are still employed in Mexico against warts. Martínez (1959) also listed several *Euphorbia* species in the treatment of skin problems.

Hassig also thought references to hot and cold were of Old World origin, but it is equally likely that the pairing of complementary opposites such as hot and cold or dry and moist was not foreign to indigenous thought.[5] True, the concept of hot and cold medicines forms a cornerstone of the humoral theory that originated in antiquity. Developed out of the writings of Hippocrates and Aristotle, and later refined by Galen, this theory of disease was still the "dominant medical ideology throughout the Renaissance," adhered to by virtually all physicians in Europe (Hassig 1989, pp. 34–36). In this system, a person's temperament is determined by a particular balance of the four humors of black bile, yellow bile, blood, and phlegm. The individual becomes ill when this balance is upset, for example, by the influence of the seasons. To ward off the illness, the afflicted person must eat foods that counteract the season (e.g., moist and hot foods in autumn, which is thought to be cold and dry), undergo treatment to remove excess phlegm, or be treated with herbs to which, in this system of classification, different degrees of heat, dryness, coldness, or moistness are attributed. As Hassig noted, the humoral theory crops up in the Aztec herbal in such terms as *noxious humors*, *evil humor*, and *black blood*, which Badianus associated with melancholia.

Such obviously European terminology is very rare even in the Latin text. The humoral theory was undoubtedly part of the intellectual baggage Sahagún and other learned friars carried with them to New Spain, but that it reflects itself in a few prescriptions of a sixteenth-century Aztec physician may have been less because of European adulteration than because it fit neatly into preexisting indigenous ideology. The Hippocratic system the Europeans brought with them "was intimately compatible with the Aztec concept of the Universe ordered on a system of balancing opposites" (Madsen 1955, p. 138). Furthermore, there is always the question whether Martín de la Cruz had adopted the humoral theory of disease or was merely borrowing its language to render his own traditional concepts more comprehensible to his Spanish readers.

Clearly, as noted at the outset, the idea of an illustrated herbal was not indigenous. Nor was the arrangement of illustrations and text within vermilion rulings that, as Hassig noted in her analysis of this and two other sixteenth-century herbals (namely, that of Sahagún in *The Florentine Codex* and that of the royal physician Francisco Hernández), conform to "standard European marginal delineations." In fact, the juxtaposition of illustration and text in this Mexican manuscript generally follows the convention set by such early Classical works as that of Dioscorides, whose herbal was initially translated into Latin from Greek and over the centuries copied, recopied, and disseminated throughout the known world in almost all European languages in painted and printed versions (Hassig 1989, pp. 36–37).

Did these early herbals serve as a model for Martín de la Cruz? The Latin version of Dioscorides was published in Siena as early as 1478 and again in 1512, years before the

Spanish invasion of Mesoamerica; a Segovian naturalist, Don Andrés de Laguna, published a Spanish translation in 1552, revised and republished in several editions that were
widely known to Spanish physicians (Hassig 1989, p. 36). Whatever might have been the
model, the two Aztecs, physician and translator, must have known what was expected of
them by their European audience, for the herbal, after all, was intended less for Nahuatl-
speakers than for people schooled in Latin and used to this kind of book.

In judging the degree of European influence on the good Aztec doctor and his Latin
translator, it may perhaps be more significant that the more than 100 maladies named in
the herbal seem to be indigenous. Not one of the great epidemic scourges introduced by
the invaders, the Old World diseases—smallpox, measles, scarlet fever, diphtheria,
whooping cough, typhus, and so forth—that ultimately killed off an estimated nine-tenths
of the indigenous population, is even mentioned. We know, for example, that half the
population of Mexico-Tenochtitlán, the Aztec capital, succumbed to smallpox between
the first entry of Cortés and his conquistadors in 1519 and his return in 1521, when, aided
by Indian allies, he conquered the demoralized city, took the last Aztec ruler, Cuauhte-
moc, prisoner and brought an end to indigenous civilization. The absence of smallpox
and other foreign diseases from the list of maladies and their treatment argues in favor of
at least relative "purity" of the manuscript, though certainly not to the degree asserted by
Sigerist and Emmart.

The *materia medica* of pre-Conquest and early colonial Indian Mexico seems to have relied far less on magic and more on the observed properties of healing plants than that of
Europe. This is not to say that supernatural or divine causality did not play a major part
in Aztec disease etiology. Of course it did. Sorcerers, the dead, and certain animals could
cause illness by magical means, and so could the gods. In indigenous medical systems it
was up to the shaman to identify the cause and prescribe what needed to be done to re-
store the patient to a state of health and balance, beyond treatment and removal of the
symptoms alone. As Ortíz de Montellano (1990) has shown, in the Aztec case, involving
as it does religion no less than practical health measures, keen observation, and detailed
knowledge of a vast array of medicinal plants, this is an extremely complex problem, one
that certainly goes far beyond the intent of this chapter.

What interests us here is whether or not the Aztecs were, by and large, capable of
recognizing and treating illness effectively with their botanical *materia medica*. By all accounts they were. As Ortíz de Montellano (1990, p. 158) has observed about one class of
Aztec remedies: "An analysis of all the botanical fever remedies in *The Florentine Codex* has
shown that 70 percent contained chemicals known to do what Aztec etiology required;
that is, they were emically effective."

It follows that when we read prescriptions of a certain plant, or combination of plants,
with or without magical or supernatural additives, however fanciful the claims might
appear at first glance, dismissing them as "primitive superstition" may be more than
doing an injustice to the old Aztec doctors. We may be harming ourselves. Here the debate over the value of the Pacific yew tree, when measured against a human life, comes to
mind. Perhaps, if pharmaceutical science had paid more and earlier attention to the numerous medical uses of this tree by Native Americans (see Moerman 1986), the efficacy of
taxol against ovarian and other deadly cancers might have been discovered years ago, and
the tree whose bark yields this miracle drug would not now be in danger of extinction.

The Roles of Author and Translator

Finally, a word about the medical qualifications of the herbal's author and the role of Sahagún. As del Pozo (1964, pp. 330, 333) observed, except for his own statement that he had

learned by experience and not by formal instruction, and that he was a physician at the College of Santa Cruz, we know nothing of the extent of Martín de la Cruz's expertise, experience, or prestige. How old was he? Did he learn his craft before the Conquest? Probably. But he was clearly not among the learned Aztec physicians who later collaborated with Sahagún to complete the massively illustrated book of "Earthly Things" (Book 11 of *The Florentine Codex*).

Sahagún returned to Tlatelolco as dean of the Colegio in 1560, eight years after the de la Cruz herbal was composed, but he remained there for only three years before moving to the main Franciscan convent of San Francisco in Mexico City. It was here that the greater part of *The Florentine Codex* was completed and copied. In 1570, he was deprived of his scribes and had to complete the work by himself. Book X was still in progress in 1576 and not finished until 1578. The final copy, in Nahuatl and Spanish, was not completed until 1585. It is logical to assume, then, as del Pozo (1964) suggested, that by the time Sahagún began working on Volume 11 with the help of Aztec physicians and herbalists, de la Cruz was no longer among the living.

Sahagún passed his last years at Tlatelolco, where, in the first days of February 1590, he suffered his final illness. Realizing that he did not have much time left, he called his beloved Aztec students and collaborators to his bedside to bid them farewell. Afterwards they carried him back to the hospital of the Convent of San Francisco in the city of Mexico, where he died on 5 February 1590 at the age of 91 (Emmart 1940, p. 25).

Emmart called Sahagún the "father of the history of Mexican medicine," a judgment with which we can heartily agree. In a wider sense, whatever the religious motivations that impelled the Aztec physician to compose his great work, by experimenting with astonishingly modern methods of ethnography, above all, by letting native informants and consultants speak in their own words rather than interposing and imposing his own interpretations, he surely merits recognition as the father not only of Mesoamerican ethnology and ethnobotany but as the pioneer of modern ethnography as such. To quote from a twentieth-century assessment of Sahagún as ethnographer of Aztec Central Mexico:

> Not until the late nineteenth and early twentieth centuries, with authors like Franz Boas . . . and Paul Radin, would fieldwork and the native voice become so important again. Thus, through their reliance on systematic fieldwork and objective description, these experiments gave birth in the sixteenth century to the first modern account of a native culture, genuinely earning for Sahagún the title of "father of modern ethnography" (Klor de Alva 1988, p. 52).[6]

NOTES

[1]A measure of the opposition among certain circles of the Spanish colonial regime to what the Franciscan scholars, especially Sahagún, had been trying to accomplish at Tlatelolco is this complaint, dated 1541, by Don Gerónimo López, Sahagún's principal adversary and a strong opponent of Indian education (quoted in Emmart 1940, 28):

> It is not enough that they teach the Indians to read and write, to bind books, to play flutes, to play cherimias, trumpets and key instruments, to be musicians and learn grammar. . . . They give them so much instruction and with so much care that there were boys, and there are more every day, who speak Latin as elegantly as Cicero.

[2]I am indebted to Jeanette Peterson for drawing my attention to this edition, for lending me her copy, and for referring me to a study of early colonial herbals by Debra Hassig (1989, pp. 30–53).

[3]The plate numbers throughout this chapter refer to the 1940 English-language edition.

[4]For a fuller account of this story, see P. T. Furst (1976, pp. 57–72).

[5]Ortíz de Montellano (1990, pp. 193–235) provides an excellent summary of the hot-cold problem and syncretism in Mexican folk medicine. See also Foster (1987, pp. 355–393).

[6]For evaluations of Sahagún's role in the birth of modern ethnography, see Klor de Alva et al. (1988) and López Austin (1974a, 1974b).

LITERATURE CITED

de la Cruz, M., ed. 1964. *Libellus de Medicinalibus Indorum Herbis*. Mexico, D.F.: Instituto Mexicano del Seguro Social.

Del Pozo, E. C. 1964. Valor Médico y Documental del Manuscrito. In *Libellus de Medicinalibus Indorum Herbis*. Mexico, D.F.: Instituto Mexicano del Seguro Social. 329–343.

Emmart, E. W. 1940. *The Badianus Manuscript (Codex Barberini Latin 241): An Aztec Herbal of 1552*. Baltimore: The John Hopkins Press.

Foster, G. 1987. On the origin of humoral medicine in Latin America. *Medical Anthropology Quarterly* 1: 355–393.

Furst, J. L. 1978. *Codex Vindobonensis Mexicanus I: A Commentary*. Institute for Mesoamerican Studies, Publication no. 4. Albany: State University of New York.

Furst, P. T. 1976. *Hallucinogens and Culture*. San Francisco: Chandler and Sharp.

Garibay Kintana, A. M. 1964. Introducción. In *Libellus de Medicinalibus Indorum Herbis*. Ed. M. de la Cruz. Mexico, D.F.: Instituto Mexicano del Seguro Social. 3–8.

Hassig, D. 1989. Transplanted medicine: Colonial Mexican herbals of the sixteenth century. *Res.* 17/18: 30–52.

Hernández, F. 1946. *Historia de las Plantas de Nueva España*, 4 vols. México D.F.: Instituto de Biología de la Universidad Nacional Autónoma de México.

Hofmann, A. 1967. The active principles of the seeds of *Rivea corymbosa* (L.) Hall f. (ololiuqui, badoh) and *Ipomoea tricolor* Cav. (badoh negro). In *Summa Antropológica en homenaje á Roberto J. Weitlaner*. Mexico, D.F.: Instituto Nacional de Antropología e Historia. 349–357.

Hofmann, A., and A. Cerletti. 1961. Die Wirkstoffe der dritten aztekischen Zauberdroge, oder die Lösung des "Ololiuqui" Rätsel. *Deutsche Medizinische Wochenschrift* 86(19): 885–887.

Klor de Alva, J. J. 1988. Sahagún and the birth of modern ethnography: Representing, confessing, and inscribing the native other. In *The Work of Bernadino de Sahagún: Pioneer Ethnographer of Sixteenth Century Aztec Mexico*. Eds. J. J. Klor de Alva, H. B. Nicholson, and E. Q. Keber. Albany, NY: Institute for Mesoamerican Studies, State University of New York. 31–52.

Klor de Alva, J. J., H. B. Nicholson, and E. Q. Keber, eds. 1988. *The Work of Bernadino de Sahagún: Pioneer Ethnographer of Sixteenth Century Aztec Mexico*. Albany, NY: Institute for Mesoamerican Studies, State University of New York.

López Austin, A. 1974a. The research method of Fray Bernadino de Sahagún: Questionnaires. In *Sixteenth Century Mexico: The Work of Sahagún*. Ed. M. S. Edmondson. Albuquerque: The University of New Mexico Press. 205–224.

———. 1974b. Sahagún's work on the medicine of the ancient Nahuas: Possibilities for study. In *Sixteenth Century Mexico: The Work of Sahagún*. Ed. M. S. Edmondson. Albuquerque: The University of New Mexico Press. 205–224.

Madsen, W. 1955. Hot and cold in the universe of San Francisco Tecospa. *Journal of American Folklore* 68: 123–138.

Martínez, M. 1959. *Plantas Medicinales de México*. Mexico, D.F.: Ediciones Bota.

McLeary, J. A., P. S. Sypherd, and D. L. Walkington. 1960. Antibiotic activity of an extract of Peyote *Lophophora williamsii* (Lemaire) Coulter. *Economic Botany* 14: 247–249.

Miranda, F., and J. Valdés. 1964. Comentarios botánicos. In *Libellus de Medicinalibus Indorum Herbis*. Ed. M. de la Cruz. Mexico, D.F.: Instituto Mexicano del Seguro Social. 243–284.

Moerman. 1986. *Medicinal Plants of Native America*. 2 vols.

Ortíz de Montellano, B. R. 1990. *Aztec Medicine, Health and Nutrition*. New Brunswick: Rutgers University Press.

Reko, B. P. 1934. Das Mexikanische Rauschgift Ololiuqui. *El México Antiguo* 3 (3/4): 1–7.

Robertson, D. 1959. *Mexican Manuscript Painting of the Early Colonial Period*. New Haven, CT: Yale University Press.

Ruiz de Alarcón, H. 1982. *Aztec Sorcerers in Seventeenth Century Mexico: The Treatise on Super-stitions by Hernando Ruiz de Alarcón*. Trans. and eds. M. D. Coe and G. Whittaker. Institute for Mesoamerican Studies, Publication no. 7. Albany, NY: State University of New York.

Safford, W. E. 1920. Daturas of the Old World and New. Smithsonian Institution Annual Report for 1916. Washington, DC: Government Printing Office. 537–567.

Sahagún, F. B. de. 1950–1963. *The Florentine Codex: General History of the Things of New Spain*. Trans. A. J. O. Anderson and C. E. Dibble. Santa Fe, NM: The School for American Research and the University of Utah.

Schultes, R. E. 1941. A contribution to our knowledge of *Rivea corymbosa*, the narcotic ololiuqui of the Aztecs. *Botanical Museum Leaflets* (Harvard University).

———. 1990. The virgin field in psychoactive plant research. *Boletim do Museu Paraense Emílio Goeldi, Botânica*, pp. 7–82.

Schultes, R. E., and A. Hofmann. 1979. *Plants of the Gods: Origins of Hallucinogenic Use*. New York: McGraw Hill Book Company.

Sigerist, H. E. 1940. Foreword to Emmart's *The Badianus Manuscript*. Baltimore: The Johns Hopkins Press. ix–xi.

Standley, P. C. 1920–1926. *Trees and Shrubs of Mexico*. Contributions of the U.S. National Herbarium, vol. 23. Washington, DC: Government Printing Office.

Stevenson, M. C. 1915. Ethnobotany of the Zuñi Indians. Bureau of American Ethnology 30th Annual Report (1908–1909). Washington, DC. 35–102.

Vogel, V. J. 1970. *American Indian Medicine*. Norman: University of Oklahoma Press.

Wilbert, J. 1987. *Tobacco and Shamanism in South America*. New Haven, CT: Yale University Press.

Ximénez, F. 1615. *Quatro libros de la naturaleza y virtudes de las plantas*. Mexico.

Gods and Plants in the Classical World

CARL A. P. RUCK

In cult and myth, many of the gods of the classical world are associated with particular plants. Athena, it was said, discovered the olive as her special tree. The laurel was sacred to Apollo, supposedly as the metamorphosis of the nymph Daphne, who had fled to avoid his amorous embrace. Demeter's chosen gift to the human race was the cultivated sheaf of grain; of equal value, it was thought, was Dionysus's gift of the vine.

Behind these common assignations lies a long tradition of cultural evolution from earlier times when the plants that were the gods' botanical attributes originally had chemical properties that made the plants more than symbolic entheogens; these properties made the plants function psychoactively in rites of shamanism. Thus, even in the classical age, Demeter was still assigned the narcotic poppy, in addition to her sheaf of grain; and her secret barley drink at the Eleusinian Mystery actually induced a visionary experience for her initiates. So, too, Apollo's chemically innocuous laurel was still responsible, if only symbolically, for precipitating the ecstatic seizures of his clairvoyant priestess at the sanctuary of Delphi. And the wine produced from the grape vines of Dionysus was not only an alcoholic inebriant, but it was treated, in social and viticultural rites, as the very embodiment of the god's possessing spirit.

At the dawn of consciousness, we may surmise that humans, apart from the other beasts, recognized the inevitable death awaiting each one and the fearful dependence of the living upon the dead for nourishment and for the continuance of generations. Thus the first science was founded. Edible matter was distinguished from inedible matter, whether poisonous or not useful or taboo, and the earliest perceptions of religion were sensed through those medicinal and magical substances that seemed to mediate between this world and the worlds of the gods and ancestors. These entheogens varied with environment and cultural traditions: availability alone would not suffice to determine a plant's sanctity; meaningful connotations in mythology and religion were also necessary.

For those peoples who took up a settled way of life—tending herds, sowing crops, and founding towns—the dark entombment of the earth became their sacred place where the spirits dwelt. This was the womb into which the seed was entrusted. Earth was the great Mother, the goddess who was the end and the next beginning of all that lives. The darkness of this chthonic realm was mirrored by the recurrent cycles of the night, lighted by the lunar phases that uncannily seemed to correlate with womankind's rhythms. Opiates, the plants that induced an irrationality and a loss of conscious control, were the pathway to Mother Earth's other world. Her special art was the discovery of how to manage the wastes produced by the continuous living of people and herds in the same place, without becoming poisoned by pollutants or famished by depletion of the land's fertility. Pollution had to be transmuted, both magically and in actual fact, into the fertilizing power that would renew Mother Earth for future generations.

In contrast to these settled peoples, nomadic peoples moved on, abandoning one location for another, and often not even burying their dead but hanging them from trees to free their spirits to the winds. Hunting and gathering required that nomadic peoples roam, migrating with the annual journey of the solar disc and seeking totemic kinship with the animals whose wild nature they would have to anticipate if they were to succeed in the hunt. Masculine strength determined precedence in the tribe, for whom god was masculine and the father. Appropriate to such a deity were wild plants, instead of cultivated ones, and his visionary realm was one of celestial enlightenment.

Such generic scenarios about earliest prehistory are misleading simplifications, although they are basically true, read back from mythological traditions and historical indications several millennia later. In what were to become the Greek lands, the Great Mother had already taken up residence, honored by the Minoans and other similar cultures, when the Indo-Europeans with their Great Father Zeus began arriving about the beginning of the second millennium B.C. By the middle of that millennium, the newcomers had taken over many of the previous settlements and assimilated their traditions with those of the earlier inhabitants.

At Mycenae, where modern archaeologists first uncovered traces of Indo-European civilization in Greece, the newcomers had imposed a kingship of their own style and they reinterpreted the name of the place to suit their own traditions. The word *Mycenae* is a feminine plural, like many names of Minoan settlements, which were named for the sisterhoods of the goddess's worshippers. In this case, the settlement was named after the Mycene girls, just as Thebes was named for the Thebe girls and Athens for the Athene girls. Even today the names of these towns are plurals grammatically. The Mycenaean Greeks, however, gave their town a false etymology, associating it with the tradition of the entheogen from their Asiatic homeland, the *mykes* or mushroom.

Perseus, the city's refounder, had dynastic ties to another branch of the Indo-European migration, the Persians. Deposing the fearful Queen of the Gorgon sisterhood on the site of what was to become the new city, Perseus was said to have picked the sacred wild mushroom. By this act, Medusa, whose name means queen, lost her power to stupefy and was changed to celestial inspiration in the form of Zeus's daughter Athena and the flying horse Pegasus, who was responsible for numerous fountains, the drink of which liberated the soul for higher visions.

In the same manner, Perseus's father, Zeus, would wield the lightning bolt as his weapon of enlightenment against the chthonic forces of darkness, planting, as it was supposed, the fungal entheogen wherever it fell to earth, as he took possession of the new Greek lands. The ensuing reconciliation of Minoans and Greek Mycenaeans would end up with the males in uneasy dominance over the females, but not without due allowance for some role for the traditions of the pre-Indo-Europeans.

The sacred mushroom of Zeus's people, as R. Gordon Wasson has shown, was the *Amanita muscaria*. The Indo-Europeans brought a remembrance of it wherever they migrated from their original home in the central Asian highlands. The Persians, for example, remembered it as *haoma*. Among the Hindus, it was *soma*.

With the passage of time, knowledge of the deity's original botanic identity was forgotten or restricted and substitutes or surrogates were employed, probably because the original was no longer easily obtainable in the new environments. The surrogates at first perpetuated certain attributes of the original, although they often were only symbolically entheogenic. In some ways this perhaps even made these surrogates more appropriate, since the Indo-Europeans were prejudiced against ultimately admitting any corporeal or material component in the experience of spiritual enlightenment.

In India, the earliest surrogate was a mushroom lacking the chemical properties of the original mushroom but symbolically appropriate nonetheless; it functioned ritually as the transmutation of corporeal putrefaction into fragrant spiritual essence through the purifying agency of fire in the making of the Mahâvîra vessel for the Pravargya sacrifice.

Fungal surrogates are also found in Greek traditions. More often, the color of *Amanita muscaria* and its relatives is remembered, or the warty scabs from its ruptured membrane, or its wild unpredictable manner of growth, or its mycorrhizal associations with certain trees, or even the intoxication of its psychoactive urine constituent. Thus, the color of *Viola odorata* (or *ion* in Greek) made it sacred to the homonymously named Iamid dynasty of clairvoyant priests who replaced earlier priestesses at Olympia when the Indo-Europeans took over the goddess's sanctuary and rededicated it to their god Zeus. Probably, however, no chemically active entheogen was used in the divination practiced by these priests, who employed rational scientific methods of prognosis based on carefully observed omens, instead of the former irrational possession that had characterized the procedure previously.

The same botanic surrogate for the fungal entheogen occurs in the traditions about Apollo's secret son Ion, who was named for the violet and was begotten in a cave at Athens. No plants other than molds can be expected to grow in such a subterranean environment, but the Queen supposedly conceived the child there as she gathered the saffron-hued *Crocus sativa* in the company of her Athena sisterhood. This child of Apollo was instrumental in shifting Athens from its previous traditions of matriliny to patriliny, as well as for purifying Apollo from the taint of female subservience, his role prior to assuming a new manifestation as a son of Zeus among the family of Olympians. It is Ion who lent his name, by a false etymology, to the "moving" electrical particle, a meaning that belies his former botanic fixity to Earth. This reinterpretation of his name was ancient and involved the myth of how he found a father, as well as his mother.

The luminous radiance of the sacred plant's color not only determined the violet and crocus as suitable surrogates for *Amanita*, but it appears to have been responsible also for the tawny hair that characterized Mycenaean princes such as Menelaus, Odysseus, and Achilles, as well as that of Apollo.

Animal surrogates also recall attributes of *Amanita*. The leopard that is sacred to Dionysus bears the warty scabs of the mushroom's ruptured membrane in the markings of its tawny pelt, and the antlered hind, an animal not found in Greece but from the Indo-European homeland, was sacred to Apollo's sister Artemis. The hind's fondness for *Amanita* and its constituents associates this animal with the entheogen in Siberian shamanism, and the golden antlers of the particular magical beast that belonged to Artemis suggest a botanical treelike surrogate of the appropriate color. Just as Perseus picked a mushroom in supplanting the religion of the goddess, Heracles numbered the plucking of these antlers among the labors he performed in claiming Greece for the religion of his father Zeus. In a similar manner, both Heracles and Perseus plucked golden apples from a sacred tree, just as another hero, Iason (or Jason), plucked the golden fleece. "Fleece" and "apple" are homonymous in Greek, and some traditions remembered that the original for what the heroes harvested from the trees was a mushroom.

The magical properties of the constituent are also recalled in the myth of the hunter Orion, one of several males who once were consorts of the goddess in the persona of Artemis, before she was assimilated to the Olympian family as a daughter of Zeus and twin sister of Apollo, who formerly also had been a version of her consort. Orion was killed for trying to rape a maiden like Artemis from the Indo-European homeland, but in dying was transmuted into a celestial configuration as the constellation. In this newer identity, it was claimed that Orion was a son of Zeus, who had inseminated Mother Earth with urine.

The Indo-European migrants, however, could not have failed to note the superior civilization of the Minoan peoples among whom they settled. That awareness, together with a tendency to equate the forward linear course of historical time with evolutionary progress, gave rise to the idea that the past is more primitive than the present and future. Hence, as new surrogates developed, the attributes of *Amanita* were also displaced upon the traditions of the previous inhabitants, equating all that was old as somehow inferior to newer manifestations in the Hellenic age that developed after the reconciliation of the two cultures. The people of the olden times, therefore, were sometimes themselves mushrooms or bore attributes of the entheogen from the original Indo-European homeland, even if in reality they must have been Minoans. At Corinth, for example, a town that also was resettled by the Mycenaean Greeks, the aboriginal populace was said to have been mushroom people before they were transformed into the new human inhabitants.

Settlements such as Mycenae and Athens were said to have been built by the Cyclopes, which were one of several versions of partial, maimed, or half figures in Greek mythology, all of them probably derived, like the one-legged man, from metaphors for *Amanita*. Each Cyclops had a single eye, suggesting the special vision afforded by the entheogen. The Cyclopes were associated with both chthonic and celestial shamanism. In the former, they tended the forge of the limping or one-legged Hephaestus in the heart of volcanoes, which were seen as a pathway to the underworld. In the later shamanic orientation, they were pressed into service of the new religion making Zeus's thunderbolts in that volcanic forge.

The same shift in orientation is represented in the encounter of the Cyclops named Polyphemus and the hero Odysseus. In the cave where Polyphemus is holding him captive, Odysseus introduces the monster to a new experience of intoxication with a powerful wine from Apollo. In escaping from Polyphemus, as in his other adventures, Odysseus is liberated from the chthonic realm to return to his homeland on the island Ithaca and to establish patriliny with his son Telemachus and father Laertes, in a place that during his absence had been in danger of becoming a queendom.

Other notorious half-men are the lame Oedipus, whose myth, like that of Ion, involves the discovery of paternity, and the one-shoed Iason, as well as Theseus, whose father's sandal was a clue that led to the hero's discovery of patriliny, and Achilles, whose heel was his only vulnerability. These lame figures probably derive originally from phallic symbolism of the Earth consort, reinterpreted through resemblance of *Amanita* to an erect phallus, for which common metaphors included the "single eye," the "lame third leg," the "little man," and so on. Sometimes the goddess even becomes involved with surrogates for the mushroom. Such is the case of the maiden (V)iole, who was responsible for the chthonic intoxication of Heracles and who bears the feminine version of Ion's name, with its reference to *Viola odorata*.

More often than through simple equation with primitivism, the Indo-European entheogen was thought to have undergone an essential hybridization from the wild plant of the Asiatic homeland (remembered as the realm of the Hyperboreans) into some cultivated substitute upon its importation and transplantation into the Mediterranean region. Thus the olive, which was supposed to have been discovered by Heracles in the Indo-European homeland, where it is not native, was transplanted to Greece and became the sacred emblem of Athena as daughter of Zeus, replacing her Minoan entheogen. It even became the symbol of Zeus at the Olympian games, after the Mycenaean Greeks took control of the sanctuary.

Among the earlier plants of Athena replaced by the olive was one the Greeks called "horse-mad" or *hippomanes*, *Datura stramonium*, the thorn-apple or jimson weed. This chemically psychoactive entheogen was associated with Athena's primordial, pre-Olympian manifestation as a maddening Gorgon Medusa and it also is characteristic of the goddess in tantric traditions. Just as Athena came to symbolize the higher inspiration of the civilized arts of the Olympian Age, the olive was thought to be superior to its botanical avatar, for it is a cultivated tree, requiring constant pruning to keep it from reverting to the useless wild olive. At Athena's Panathenaic games in Athens, instead of a wreath of olive leaves, which was the prize at Olympia, the victorious athlete received an amphora of oil pressed from the sacred olive trees. This pressing of the surrogate fruit recalls the tradition of the original Indo-European entheogen, called in Vedic lore by the metaphor of the "pressed one," which is the meaning of the name *soma*. The olive was superior not only because it was cultivated whereas *Amanita* was wild, but also because it required the further intervention of scientific procedures of manufacture to release its food.

Athens claimed to have the first olive tree that ever grew, but the same claims was made elsewhere. On the island of Delos, where Apollo and Artemis were reborn into their Olympian identities, the aboriginal olive tree retained ritual connotations of the psychoactive original. In addition to the mock flagellation of pubescent dancers who chewed on its bark in commemoration of earlier times when they would have been sacrificial victims to the goddess, the identity of the Indo-European entheogen that supplanted the Minoan religion was maintained as restricted knowledge. Each year a secret offering of *Amanita* was supposed to have been transmitted through intermediaries from the Hyperborean homeland and presented among the offerings of first fruits sent to Delos from the various Greek cities. These first fruits were symbolic of primitivism, harvested early, before the full crop had ripened to maturity. Among these gifts, the secret offering from the Hyperboreans was the most primitive avatar of the agricultural arts.

Like the olive, Apollo's bay or laurel tree, *Laurus nobilis*, was similarly considered a sacred import or transplantation from the olden times in the traditions of the god's sanctuary at Delphi. It was used for the wreaths to crown victors in the Pythian games commemorating Apollo's triumph over his atavistic former identity at the site, where in pre-Indo-European times (i.e., before Apollo was reborn as a son of Zeus) the god had functioned in chthonic shamanism as a consort of Earth. The games at Delphi included musical and athletic competitions, celebrating contests of male physical superiority and, just as the games of Athena, the harmony of the higher artistic inspiration that dispels the discord of irrational, feminine possessing spirits.

Although the laurel retained the tradition of its psychoactive original in the shamanism of the Pythian priestess, it replaced more sinister plants formerly associated with the pre-Olympian manifestations of Apollo. One of these was aconite (*Aconitum*) or wolfsbane, a metaphoric name that goes back to the Greek nomenclature. Aconite is chemically psychoactive and its flowers, like those of *Viola*, mimic the sacred color of *Amanita*. This fortuitous resemblance facilitated the merging of the Indo-European god with his indigenous chthonic precedent. Wolfsbane or *lykoktonos* originated in the prophet-deity's cults, among the northern Hyperboreans and in what was known as his other homeland among the so-called wolf-people, the matrilineal Lycians of Asia Minor. This wolf persona became characteristic not only of Apollo's darker nature, but also in general of the recidivous other self of all the heroes who were sons of Zeus. This lupine metaphor is a classical version of the werewolf mythologem and coincides with Indo-European versions of the same phenomenon. In Greek, the "wolf-madness" is rabies, the power of the she-wolf to cause the domestic dog to revert to its wild primordial ancestor. The Olympian Apollo was so dangerously unstable in his new identity that even dogs were excluded from his Delian sanctuary.

Another of Apollo's botanical surrogates was *hyacinthos*, a plant name from the pre-Indo-European language. The Greeks identified it with larkspur, *Delphinium ajacis*, perhaps since the plant's medical efficacy against ectoparasites made it a fitting analogue to their own entheogen, *Amanita*, which has the property of making flies insensate and comatose, hence its common name, fly-agaric. The annual sacrificial victim offered to Apollo at the cliffs on the island of Leukas was similarly thought to rid the populace of an infestation of flies. The Minoan *hyacinthos* may have been a different plant, probably with psychoactive properties. It bears the name of a former version of Apollo, Hyacinthos, one of the many lamented males who were mourned as dying consorts of the goddess. Apollo, in the common mythological pattern of replacement, accidentally killed his own former persona in an incident of misdirected "wind" or inspiration. The plant's flowers were said to resemble the Greek letters for the cry of lament.

The bay tree or *daphne* replaced both the *hyacinthos* and the wolfsbane. The Pythian priestess prepared for her fit of shamanic possession by commemorating the maiden Daphne who was metamorphosed into the tree to avoid the god's courtship. She chewed the leaves of laurel and became possessed by the old, darker version of Apollo, but her shrieks of frenzied rapture were transformed into enigmatic Greek verses by a male priesthood, whose masculine role symbolically was to mediate with the female past traditions and reinterpret the senseless response coherently, as befitted the newer son of Zeus. The type of questions most often answered by the Delphic oracle was consistent with this general theme of reconciliation between female and male mentalities, just as the sanctuary itself mediated between a commemoration of the original chthonic religion and the traditions of the nomadic immigrants. In addition to common problems of marital infertility (and in mythological instances, of patriliny over matriliny), the oracle was often instrumental in advising Greek cities about where new colonizations of male-dominant Hellenic civilization might be settled upon the inhospitable Earth of Mother Nature, just as the wandering Apollo had no place to call his own until he took control of Delphi.

Like the *daphne* and the olive, Poseidon's sacred plant, the celery (or what is commonly called "parsley" by classicists), *Apium graveolens*, apparently was also a chemically innocuous surrogate for a plant that originally functioned in the god's pre-Olympian religion. It, too, symbolized the triumph over the chthonic forces of primitivism. Celery was used to crown the victors in Poseidon's games at Isthmia and Nemea, sanctuaries like Delphi and Olympia that evolved from shamanic rites practiced before the coming of the Indo-Europeans. As at the other sites, the victors probably were once the sacrificial victims offered to the goddess and her consort. The celery, as a surrogate, retained its funeral connotations from those earlier times. Thus, wreaths of celery were used to adorn tombs in the classical age, and it was a homily to say that someone close to death was in need of such a chaplet.

The original entheogen may well have been the poisonous hemlock, *Conium maculatum*, which celery resembles, since hemlock was the drug employed in the classical age as a lethal potion to put criminals to death; because criminals were originally appropriate candidates for human sacrifice, the mode of execution betrays its ritual precedents. As "consort of Earth" (which is the meaning of Poseidon's name), the god in his pre-Olympian persona was a deity of death, linked with the Gorgon identity of Athena as his goddess. Thus, he, like Athena, had equine manifestations, in which form he united sexually with the Medusa.

It is, however, in the paired figures of Demeter and Dionysus that one can see most clearly the full complexity of the pattern involved in the reconciliation of the botanical and religious traditions of the Indo-Europeans and their indigenous predecessors in the Greek lands. These two deities represented the totality of human foodstuffs—Demeter, the dry, and Dionysus, the liquid. Both incorporate commemorations of their avatars in chthonic shamanism, as well as of the fungal entheogen of the Indo-European tradition. For both deities, the evolutionary perspective placed a higher value on their cultivated manifestations in the Hellenic age, as compared to their wilder, more primitive antecedents. As in other cultures, the mushroom proved to be the perfect archetypal mediating symbol. Its wildness could be tamed into cultivated hybrids, and its obvious phallic configuration could also be viewed as feminine, when the cap becomes concave upon further opening in ripeness. It grows from what looks like an egg within the earth, thus suggesting the idea of resurrection from the nether world, and it thereby in Greek lore had surrogates in various analogous bulb plants, like the crocus, the narcissus, and asphodel, the last being the flower that traditionally grew in the Elysian fields. The chthonic mushroom's sudden appearance after rainfall suggested some causal relationship with the bolt of lightning's point of impact from the celestial realm, hence the union of sky and earth.

Because of the rational bias of classical scholarship, there has been a reluctance to consider the role of entheogens in Greek religion, as though the few researchers who do were somehow imposing their own distorted ideas upon ancient society and "gods in a flowerpot." The fact that the Greeks worshipped Dionysus as a god of intoxication should alone refute any doubt that they recognized something numinous in the experience of chemically induced madness. It is not I, after all, who found a god in my wine cup.

As the Greeks saw it, there were two aspects to this state of altered consciousness caused by the drinking of wine. One was primarily effeminate, regressive, and irrational; the other, virile and inspired, with connotations of the higher arts and of the political and social institutions of their male-dominant culture. The former was the maenadism of the women who tended the god's chthonic avatars in the mountain wildernesses during the nonagricultural season; the latter, the *symposia* or "drinking parties" of the men in the city, where poetry about their mythological heritage was recited and the friendships and alliances, often basically homoerotic (and hence, excluding female) were formed that sustained the male-dominated culture. Whatever women were present at *symposia* belonged to the *hetaera* or prostitute class. They were trained in dancing, poetry, and intellectual arts, and, unlike females of the citizenry, they were adept at the lascivious sexual arousal of their male patrons—talents that would have been deemed threatening in wives and daughters. To this masculine aspect of intoxication belongs also the god's role in renewing cultural identity through the paedeutic function of the theater. In modern terms, the contrast is between the drunken brawl and a cocktail party.

Wine was recognized for what it is, basically a fungal surrogate of the "pressed one," the *soma* of the Vedic tradition. The fungal nature of fermentation was clearly observable and seemed to the ancients to be the same kind of process that occurs in cooking, whereby the raw and primitive is transformed into civilized cuisine, a process, moreover, that was thought to be a sort of putrefaction and, hence, like a resurrection from moldering matter. In fermentation, the wild, unpredictable growth of *Amanita* yielded to the civilizing arts to produce a superior inebriant.

Wine, as the drink of the new age, deposed the god's avatars, all of them from both the Indo-European and Minoan traditions, but as always in Greek religion, the deposed personae must not be dishonored. They were commemorated as part of the deity's total identity. Thus, his previous names were still maintained—like Bacchus for Dionysus—or earlier iconography was perpetuated—like the Gorgon's head that Athena wears as a breastplate, a trophy of her former persona as goddess.

So, too, wine as an intoxicant was not solely the product of the grape's fermentation, but various herbal precedents were part of its "bouquet." Among these was resin, commemorating earlier ferments from the sap of trees that were host to *Amanita*; the pine tree became sacred to Dionysus, and modern Greek wine perpetuates this association as retsina. Many of the entheogens sacred to the goddess also found their way into this

ancient drink that was a symbolic recapitulatory synthesis of the two culture's reconciliation, as well as an inebriant. Some of these additives were chemically psychoactive and
so intensified the wine's toxicity that it could be drunk safely and properly only when
greatly diluted with water. It was all of these that gave wine its "spirit," the ghosts of its
constituent gods.

Alcohol itself was a substance unknown to the Greeks, who had no name for it. Our
modern term comes from Arabic, where it first was described as the distillates of minerals for cosmetics, thence applied to the liquid distillate of ferments when it was first
discovered much later by the alchemists as *aqua vitae*. The Greek word for wine itself,
(w)oinos, appears to be Indo-European, and since viticulture was not native to these
people in the Asiatic homeland, it could not originally have described the vinous ferment
but rather their own sacred drink. They applied its name to the newer drink when they
encountered it in the course of their southern migrations. Etymologically, it appears to be
a metaphor for *Amanita* as a "circular rimmed wheel," which is a typical pictograph for
the sacred mushroom in other cultures. Cognates for *woinos* in Latin and modern languages (*vinum, vin, vino, wein,* etc.) are, therefore, derived from the Indo-European verbal
root and not assimilated, as one might expect, from whatever linguistic culture originated
viticulture.

Symbolically, the vine plant was seen as a botanical evolution of a more primitive,
related plant. This avatar was the ivy, *Hedera helix*. Without fermentation, the leaves and
berries were reputed to derange the mind. Ivy had not yet succumbed to the hybridization
that would culminate in its civilized descendent that, like the olive, through constant
pruning and tending would yield its harvest of the succulent grape, instead of the supposedly poisonous tiny berries of its ancestor.

The two aspects of Dionysus, wine and its precedents, had to be commemorated in
ritual as well. Thus, in addition to the *symposia* of the men, the female citizenry was periodically released from the strictures of their secluded and protected lives within the
innermost quarters of their houses in the city and they took off for their mountain revels
as maenads or "madwomen." They formed again into the ancient Minoan sisterhoods of
the triform goddess and reverted to wild, uncivilized behavior, and they laid claim again
to the dominant role that once was theirs and that now was denied them in their lives
within the city.

On the mountains, these women hunted the pre-viticultural manifestations of the
god's possessing spirit. Symbolic of this was the ivy. As emblem of their recidivous quest,
they bore the Minoan symbol of the *thyrsos*. This was the herbalist's staff, the implement
of those who gathered wild plants. It was composed of a fennel stalk stuffed with the
leaves of ivy that supposedly they had found. Other plants from olden times also figured
in their ritual hunt. Prominent among these was the symbolism of the opium poppy, a
plant from the Minoan religion, although it could not be expected to be found actually
growing on the mountain and in the wintertime of the revel.

Indo-European precedents also were involved in the symbolism, as is clear from the
mythical traditions about the *thyrsos*. Prometheus was said to have first brought the fiery
spirit of the celestial enlightenment from the heavens by stealing fire from the Olympians
and hiding it in a fennel stalk. Celestial fire as a bolt of lightning, especially when concealed in the herbalist's *thyrsos*, recalls the supposed involvement of lightning in the generation of mushrooms. Prometheus was the creator of the human race and his role was the
essential mediation between the chthonic realm of his own origins and the newer realm
of the Olympians, whom he tricked into accepting the right of his human creatures to
exist. It was he who taught humankind the ritual of the sacrificial meal, whereby the past
undergoes transmutation to feed the ongoing evolutionary process. The myths about
Prometheus and his brother's son Deucalion, moreover, portray the inception of a new

age of humans with the inauguration of the Olympian family of deities with the father Zeus at its head.

Animate surrogates of the primitive god also were objects of the symbolic hunt of the maenads. Such beasts were the leopard, with its suggestive spotted pelt, and burrowing animals, who made their lair in the womb of Mother Earth, like the prolific hare or the phallic serpent, the latter being a reptile with herbalist significance since it was supposed to amass the toxins for its poisonous bite from the plants it lived among. It was claimed that the maenads conducted this hunt without implements, like true primitives, and that like Mother Nature, what they found were their babies. When they caught their prey, they tore it to pieces with their bare hands and ate it raw, without the benefit of the civilizing culinary arts. Again, we are dealing with symbolic events, since it is hard to see how women untrained in hunting could have managed to capture animals barehanded or slaughter their own babies, when apparently the babies were not brought to the revel.

Like the herbalist witches of later times, who were accused of consorting with the devil, the maenads—these respectable ladies from the city—were said to engage in a wiled sexual romp with goatlike men, the ithyphallic satyrs. The brotherhood of satyrs represented the possessing spirit of the primordial Dionysus, since prime among the animal surrogates of the god was the goat. The goat was seen as the natural enemy of the plants tended in the vineyard, and the sacrifice of that animal was the appropriate offering to free Dionysus from his own former identity. The sacrificial meal of goat meat fed the new god and his worshippers in the city upon the demise of his primitive nature, but in the mountain revel, the god reverted to his role of caprine consort of the goddess from the time before he had been remade into a son of Zeus.

This god of the maenadic revel was also sometimes a bull, remembering the symbolism of the goddess as bovine in Minoan religion. The Indo-European entheogen assimilated the taurine persona as well, as in the Vedic tradition, and the maenads beat upon tympani in their mountain rites to waken the bellowing of *Amanita* as it burst suddenly into fruit with the thunderous sound of an earthquake.

Although the cultivated gardens were dormant in this winter season, the mountains would bloom with the wild flowers that were the god's bulbous surrogates. The Dionysus of the vine had departed, acquiescing to his own demise at the time of the harvest, which was a sacrifice of himself offered for human salvation. The ritual slaughter of the grape had been accomplished like a funeral, accompanied by the lamenting music of flutes; and the harvesters, disguised as satyrs, had sought to blame the murder upon the resurgent atavistic powers that were about to seize control of the world upon the death of the civilized god. The masked harvesters had trod upon the grapes and pressed the bloodlike juice, channeling it into subterranean vats, where it would be entombed, like the god's corpse, and left to molder. As his body lay fermenting through the winter months, the whole world would enter upon its regressive phase. Even the ivy would leave off its trailing, prostrate manner of summertime growth and begin to exert its regained supremacy, growing upright now in sinister mimicry of its usurped hybrid. The version of Dionysus that now took over was his primordial role as the goddess's inseminator; the erect phallus alone was this god's sign. It was borne defiantly in rural carnival-like processions, and the irrepressible lusting that it represented defied the accepted norms of civilized urban life.

This was the time when comedies originally were performed. The actors and choral dancers, costumed extravagantly as fantastic metaphors for the ithyphallus that was their prominent emblem, would hold the finger, as it were, up to the leaders of society. They would, in effect, overthrow the city as it was and remake it to the liking of their own baser instincts. Typical of the comic plot, the lower elements of society—or even women— would take control, and all that was sacred, including the Olympian gods, would have to yield. The unrestrained libido ruled the world.

When the fermentation, however, was completed, at the threshold of spring, the wilder spirit of the god's avatars would have, in turn, to give way as the cultivated god triumphed over death and returned from the grave in the guise of a divinely newborn infant, repeating the age-old miracle of the rebirth of the goddess's former consort as her son. This was not the same child as in olden times. This new Dionysus of the Hellenic age was destined eventually to resurrect even his mother and elevate her, like the Assumption later of the Blessed Virgin, to the celestial realm.

Dionysus's triumph opened, as well, the gates of the nether world, like an earlier example of Christ, and with him from the grave returned the spirits of all the dear departed. This moment, when the new wine was first breached, was celebrated as a communal banquet attended by both the living and the dead. Special table manners were in effect for this feast to ensure the proper separation of ghostly corruption and human life. These included chewing buckthorn (*Rhamnus*) as a laxative to purge the body of its own pollution, and eating and drinking from separate facilities to keep the ghosts at a respectable distance. The myths that traced the etiology of this festival recalled the coming of a new age of humans after the great flood, the redemption from madness caused by female chthonic powers, and the shift from matriliny to patriliny.

The young children of the citizenry were seen as manifestations of the infant god's miracle. At the age of three or four, the children would be indoctrinated into the metaphysical meaning of the wine as they drank it for their first experience of inebriation. We see these drunken children depicted on vase paintings as they play among the gravestones or impersonate their elders in performing various Dionysian rituals, such as the pole dance and the sacred marriage.

This symbolic renewal of the world was the context originally for the god's other type of drama, his tragedies, which would be performed at the contests held later in the spring, although the popularity of these festivals was so great that the distinction was soon blurred, so that both comedy and tragedy eventually were produced at each. Tragedy etymologically is the "goat song"; it was sung for the goat who was the sacrificial victim, the honored primitive persona of the god, who had to fall before the ascendancy of his own better self. Typically, the plot of a tragedy presents a hero whose victory would endanger the fundamental stability of the Olympian order, and hence, the hero's failure is of greater value than his personal success. The choice, for example, is between an Oedipus or an Apollo, and the worlds that each represents.

In these festivals of drama, in which all the roles were enacted by men, we see a different kind of madness. Instead of the maenadic derangement of the women in the regressive mountain revel, the male actors channeled the experience of ghostly possession into a form that furthered the evolution of the norms of Hellenic culture.

The goddess Demeter underwent a similar reconciliation of her past botanic identities with her newer Hellenic and Olympian manifestation. As "Mother Deo," which is the meaning of her name, she was recognized by the arriving Indo-Europeans as the mother goddess, with one of her pre-Greek names, and she was assimilated into the Olympian family as a sister and mate of Zeus.

Another of Demeter's names from the Minoan tradition was Persephone. Since the latter, like Deo, is not Indo-European, it has no known etymology, but the Greeks could see in it the false meaning of "deadly." This version of the goddess was assimilated as Zeus's daughter by Demeter, and her myth depicts how, unlike her mother, she was denied Olympian status and relegated to the chthonic realm as a goddess of death and resurgent life from the nether house of her consort Hades.

Many were the plants sacred to the Demeter-Persephone duo: opiates, like the poppy, for which the wild rose later became a symbolic surrogate, by virtue of its similar flower and capsulelike hips, as did also the pomegranate, which it resembles; and deranging herbs, like *Datura* and henbane (*Hyocyamus niger*), which was named in Greek for her sacred animal, the sow, a carnivorous beast that responds to the male scent of humans. By the classical age, henbane had become an abused drug by the younger generation.

As an Olympian, the goddess and her daughter were symbolized by the cultivated staff of barley. This was Demeter's antithesis to the wine of Dionysus. Hers was the dry stuff with which she nourished humankind, but like the vine, barley, too, had its atavistic precedent. This was seen as the wild and inedible weedy grass, *Lolium temulentum*, called "drunken lolium" in Latin botanic nomenclature because of the poisonous fungus (*Claviceps purpurea*) with which it is commonly infested. This fungus or ergot was called "rust" in Greek, as in English, because its reddening corruption overtakes the host kernels of grain in much the same way that the oxide of iron destroys the serviceable metal and seems to pull it back to the useless ore from which it had been manufactured. This same corruption seemed to spread from the weed to the cultivated grains, making them inedible like their ancestor. Barley, it was thought, would actually revert to lolium if it were not correctly tended to reinforce its evolutionary hybridization.

As with Dionysus, the fungus again was the ideal mediator. Grain, too, ferments, and the apparent putrefaction yields the leavening for the cooked loaf. The same triumph over atavism was seen in the transmutation of offal and dead matter into the renewed fertility of the plowland, a miracle that was commemorated by the ritual slaughter of the sow, whose decomposed remains were spread upon the fields as manure. As the Indo-European migrants traveled through the grain-growing lands to the north and east on their way toward Greece, they apparently found an early surrogate for their sacred plant in the chemical properties of *Claviceps*, the color of which perpetuated the sanctity of the original entheogen, and from which, by simple water solution, a form of LSD (lysergic acid diethylamide) can be easily separated from the other poisonous alkaloids of the ergot.

In the celebration of the Eleusinian Mysteries, which derive from Minoan precedents but which, like the other great religions of the Hellenic age, reconcile the two traditions of chthonic and celestial shamanism, ergot was employed in a drink that induced a mystical vision. The worshippers gathered at a place sacred to the goddess, beside the entrance to a subterranean tunnel considered to be one of the gateways to the nether world. There, at the village of Eleusis, near Athens, in the cavernous great Hall of Initiation, the initiates drank the potion and experienced a spiritual journey together through that passageway into the chthonic other world and then returned resurrected with the goddess, who had borne a matrilineal son during her underworld sojourn. Because of this communal rebirth, the worshippers came to feel that death, as Paul was later to preach of the Christian mystery, had lost its sting. Instead of some demonic horror, they saw that the Lord of Death, who was Persephone's son, was bound to them by ties of friendship and reciprocal hospitality in his and their own homesteads.

The initiates were sworn to secrecy under pain of death, but the myth that told of the founding of the religion was profane knowledge, including the part that listed the ingredients for the sacred potion. According to the myth, Persephone had been picking wild flowers on the frontier of this world, as queen among a maiden sisterhood, when she happened upon a particular plant, the *narkissos*. The plant's name, as we should expect, is pre-Greek, and hence its etymology is unknown, but the word was assimilated into Greek and its properties as a drug are responsible for its meaning as a "narcotic." This Minoan entheogen induced the spiritual possession that abducted Persephone to the nether world as a maenadic mate of Hades. At Eleusis, this abduction without the mother's consent was rectified by the elevation of the lost maiden to the rank of wedded wife and mother. This evolution from illicitly abducted maiden to legitimate wife culminated in the institution of the civilizing rites of agriculture. It is Persephone's mysterious son, under the name of Triptolemus, who teaches humankind the art of tending barley.

The Eleusinian potion was a symbolic drink, like wine, tracing the transition from primitivism to culture and mediating the Indo-European and Minoan religious traditions. The identity of three ingredients was not restricted knowledge. These were pennyroyal (*Mentha pulegium*), water, and barley—none of which could have been chemically responsible for the mystery experience. The water is obviously the inert medium that binds the two plants, which represent the polarity that is reconciled through the vision provided by the mystery.

Pennyroyal is a pungent aromatic mint, a wild plant that in Greek botanical lore is reputed to be an aphrodisiac; the plant's fragrance, like perfume, had connotations for the Greeks of lascivious illicit sexuality rather than of matrimonial duty and fidelity. Pennyroyal is emblematic of Persephone's abduction and the ensuing wrath of her deserted mother Demeter, who could neither accept that her daughter be a concubine nor countenance losing her to a male's control.

Barley represents the antithesis. It is the cultivated plant, symbolic of Demeter's acceptance of the periodic separation from her daughter, just as the seed is entrusted to earth only as a temporary prelude to the renewal of life and the return of Persephone with her son. Demeter becomes reconciled to her own role in the celestial realm as an Olympian, while her former self, in the persona of her daughter Persephone, resides in the chthonic realm, bound to her by family ties and cyclical visitation.

The secret ingredient of the potion—the one that made it serviceable as an abused substance—was ergot. It mediates between the polarities of wild and cultivated and of the Minoan and Indo-European traditions of shamanism. *Amanita* has no seed and defies attempts to control its unpredictable growth, but the ergot that spreads from lolium to the kernels of barley, threatening to pull the cultivated foodstuff back into primitivism, produces what appears to be an enlarged purplish seed, as the fungal mycelia permeate its host. Under appropriate conditions, the ergot-infested kernel falls to earth and enters its fruiting stage, with mushroom bodies recognizable to the naked eye. Claviceps itself is poisonous, but through the intervention of civilizing technique, the entheogenic component is separated into the solution of the potion. Nor can we doubt the association of ergot with the goddess, since "Rust" was one of Demeter's names.

There were two levels to the mystery, hence the plural Eleusinian Mysteries. The Lesser Mystery took place in the maenadic winter and involved the tradition of Persephone's abduction. It ritualized the hunt for *Amanita* or its surrogates. Part of this ceremony was the Sacred Marriage, when the woman who portrayed the role of the Queen, from the old days when Athens was a queendom, performed some secret rite in which she was possessed spiritually by Dionysus in some "taurine" form.

The Greater Mystery occurred in the fall and was experienced by the whole body of the initiates. Instead of the narcosis of the past Minoan tradition of the *narkissos*, a brilliant light of visionary illumination is described as the experience in the darkened Hall of Initiation on the Mystery night, as LSD supplanted sleep with the enlightened sight it induced in the wakeful worshippers huddled within. Although still a chthonic religion of the two goddesses, the Indo-European tradition had accommodated it to its own celestial orientation.

The Eleusinian Mystery was the most prominent initiatory religion in the classical world, but there were others that struck a different balance between the claims of earth and heaven. Some were more chthonic; others, more celestial. In the mysteries of the Kabeiroi, for example, the sacred drink enrolled the initiates into a nether world brotherhood of primordial men. In contrast, the Orphics overemphasized the Indo-European aspiration to liberate the soul from its symbiotic dependence upon the body. Their supposed founder Orpheus inadvertently abandoned his Persephone-like bride permanently in the underworld, and taught his tribesmen to shun all sexual contact with women. His followers sought to purify their bodies to attain eventually a totally spiritual existence through inhaling special herbal fumigations, like Olympians themselves, and through vegetarianism and dietary prohibitions.

LITERATURE CITED

Detienne, M. 1979. *Dionysus Slain*. Johns Hopkins. Trans. from 1977 French ed.

Ruck, C. A. P. 1976. On the sacred names of Iamos and Ion: Ethnobotanical referents in the hero's parentage. *Classical Journal* 71(3): 235–52.

Rouner, L. S., ed. 1984. The wild and the cultivated in Greek religion. In *On Nature*. University of Notre Dame. 79–95.

Toporov, N. V. 1985. On the semiotics of mythological conceptions about mushrooms. *Semiotica* 53/54. Trans. from Russian by S. Rudy.

Wasson, R. G. 1968. *Soma: Divine Mushroom of Immortality*. New York: Harcourt, Brace & World.

Wasson, R. G., A. Hofmann, and C. A. P. Ruck. 1978. *The Road to Eleusis: Unveiling the Secret of the Mysteries*. With a new translation of the *Homeric Hymn to Demeter* by D. Staples. New York: Harcourt Brace Jovanovich.

Wasson, R. G., S. Kramrisch, J. Ott, and C. A. P. Ruck. 1986. *Persephone's Quest: Entheogens and the Origins of Religion*. New Haven, CT: Yale University Press.

PART 4

Ethnobotanical Conservation

The urgency of preserving ethnobotanical knowledge still extant in indigenous societies and its importance to various fields of science, industries, and the arts is now widely recognized. With 500,000 higher plants in the world's flora, not including as many less well-known lower plants (the fungi and algae), there exists an incredibly unexplored frontier for new chemical compounds. It has been estimated that of the 80,000 species of higher plants in the Amazon Valley, fewer than one percent have been even superficially analyzed for their chemical constituents.

Indigenous peoples, particularly in the rain forest areas of the world, usually have an incredible knowledge of the properties and uses of their ambient vegetation. Amassed over many centuries or even millennia, this knowledge is one of the first aspects of their culture to disappear with the arrival and acceptance of Westernization. If only for the welfare of the human race, it is urgent that this knowledge be preserved before it is entirely lost. This urgency goes beyond the possibility of new medicinally valuable discoveries. There are in these forests many plants, still unrecognized or unused in advanced societies. The number of species now employed in the agricultural and industrial societies, almost all received from primitive societies, attests to the value of ethnobotanical studies. The increasing importance and urgency of these studies has given rise to recognition of the extreme value of what is now known as "ethnobotanical conservation."

One of the most valuable contributions that ethnobotanical conservation may give to modern science is the intricate familiarity that people in most primitive societies possess of local variants or ecotypes of the plants of their surroundings—variants of the economically valuable species as well as of those for which no use is yet known. From this knowledge and familiarity, ethnobotanical studies may be of extraordinary value in the fast-developing practical aspects of biodiversity studies.

Ethnobotanical conservation has gained wide recognition as a major element of environmental conservation. It is certainly destined to play an even greater part in numerous phases of conservation of our natural resources.

Three contributions are included in this section. Mark J. Plotkin argues for conserving biodiversity while utilizing tropical biota, a win-win situation in which economic benefits accrue by protecting and rationally utilizing species. In an article reprinted from *Conservation Biology* (vol. 1, 1987, pp. 296–310), G. T. Prance, W. Balée, B. M. Boom, and R. L. Carneiro identify the tree species and families useful to four indigenous Amazonian groups and then recommend measures to protect these species and their associated habitats. Finally, the late C. Earle Smith wrote "A Near and Distant Star" prior to his death in 1987, urging a thorough understanding of the world's plant resources before they and their culture, which depends on these resources, are completely annihilated.

The Importance of Ethnobotany for Tropical Forest Conservation

MARK J. PLOTKIN

The Need for Conservation

Biologists agree that the rate of species extinction is increasing at an alarming rate. Although many temperate life forms such as the California condor (*Gymnogyps californicus* Shaw) and the black-footed ferret (*Mustela nigripes* Audubon and Bachman) are on the verge of extinction and may disappear before the twenty-first century begins, the majority of the world's threatened species inhabit tropical forests. Covering less than 10 percent of the earth's surface, these tropical forests are believed to contain over 50 percent of the world's species (Wilson 1985).

The staggering amount of species diversity concentrated in tropical regions is best demonstrated with several comparative examples. The Rio Negro in Amazonian Brazil contains more species of fish than are found in all the rivers of the United States combined (M. Goulding, pers. com.). A hectare of forest in northeastern United States typically contains about 20 tree species, whereas a similar plot in western Amazonia may contain more than 300 (A. Gentry, pers. com.). Manú National Park in southeastern Peru is home to more bird species than are found in the entire United States (J. O'Neill, pers. com.). A single tree in the western Amazon may harbor as many species of ants as are found in the entire British Isles (E. O. Wilson, pers. com.).

One problem faced by biologists working to prevent the diminution of biological diversity is the lack of basic knowledge about most of the world's species. Carl Linnaeus inaugurated the Latin binomial system in 1753; approximately 1.7 million species of plants and animals have been described since then (Wilson 1985). In a 1964 study, it was estimated that the number of insect species alone was 3 million (Williams 1964); and more recent work has caused scientists to revise estimates sharply upwards. In the early 1980s, Terry Erwin of the Smithsonian Institution developed a technique for sampling the invertebrates of the rain forest canopy, a fauna poorly known due to the difficulties of access and the prevalence of stinging insects. The results of these studies led scientists to reconsider their ideas on the magnitude of the diversity of life on earth. When nineteen specimens of a single tree species were investigated using Erwin's technique, more than 1200 species of beetles were collected (Eckholm 1986). Erwin estimates there are more than 30 million insect species in the world (T. Erwin, pers. com.).

Compared to the insects, the angiosperms (flowering plants) are much better known. It is postulated that at least 250,000 species exist, of which 90 percent are already known to science (Eckholm 1986). Nevertheless, major expeditions to the tropics, particularly to the Amazon region, continue to bring back new species.

Yet the flora of the tropics, like the fauna, is faced with serious threats to its very survival. Consider three examples: the Brazilian Atlantic forest, the Chocó, and Madagascar.

Brazil's Atlantic forest. The Atlantic coastal forest of eastern Brazil once stretched from the state of Rio Grande do Norte to Rio Grande do Sul, an area longer than the U.S. eastern seaboard. Although this region once supported some of the most species-rich forests in the world, it was the first part of Brazil to be colonized (more than 100 years before the Pilgrims landed at Plymouth Rock) and today contains two of the three largest South American cities—Rio de Janeiro and Sào Paulo. Current estimates indicate that 95 to 98 percent of the original forest cover has been destroyed (Mittermeier 1982).

The Chocó. Lying along the northern Pacific coast of South America from southern Panama to northern Ecuador, the Chocó is the richest and least-collected region in the New World tropics. It receives over 10 meters (30 feet) of rainfall each year, making it one of the wettest regions on earth, a characteristic associated with high species diversity. Timber companies are now clear-cutting large areas of the Chocó, reducing precipitation and undoubtedly dooming many species of both plants and animals to certain extinction (Kirkbride 1986).

Madagascar. Many biologists consider Madagascar the single highest conservation priority on earth because of its exceedingly high rates of endemism and species diversity. Separated from the African mainland for over 40 million years (if in fact it was ever connected), its biota is unique due to evolution in isolation. The rates of endemism are perhaps unequaled in the world for a country its size: 40 percent of its birds, 90 percent of its primates, 98 percent of its frogs, and 100 percent of its rodents are found nowhere else in the world. As many as 80 percent of its angiosperms are found only in Madagascar and rates are even high in particular forest formations such as the southern spiny desert, where 48 percent of the genera and 95 percent of the species are endemic (Guillaumet 1984). Like those of eastern Brazil, the flora and fauna of Madagascar have been seriously affected by human inhabitants, who arrived on the island only 1000 years ago. From subfossils, we know that one-forth of the lemur fauna has become extinct in this relatively short period. As much as 80 percent of the original vegetative cover has been removed or seriously disturbed, though it is presently impossible to document exactly how many plant species may already have been lost.

Extinction *is* a natural process: since the origin of life around 10 million years ago, many species have disappeared. Yet, to view the extinctions in the above examples as natural is to misinterpret the geologic record. It has been postulated that the present rate of global species extinctions is 400 times faster than the rate in the recent geologic past, and that this rate is rapidly accelerating (Wilson 1985). The only similar situations in the history of this planet were the massive species extinctions at the end of the Paleozoic and Mesozoic eras.

A striking feature of these historic natural disasters, however, is that the extinctions were primarily faunal. Indeed, there is little evidence in the fossil record of mass extinctions of plants (Knoll 1984). In the past, plants were presumably more resistant to extinction than the dominant animal life forms, like the dinosaurs. Consequently, plant diversity has basically increased through time, with especially high rates of speciation occurring during the Carboniferous and Cretaceous periods (Knoll 1984).

The unpleasant conclusion is that the human race is causing one of the first major reductions of global, vascular plant diversity since the origin of life. That the extinction of plants may have a more serious impact on human welfare than the disappearance of animals will be discussed later.

REASONS FOR CONSERVATION

Why should anyone care whether or not species become extinct? To the layperson, our way of life seems wholly dependent on a few domesticated species that appear to be thriving—corn, wheat, cattle, and so on. Of what possible consequence could the extinction of some poorly known tropical species be for the progress of Western civilization? Several different arguments have been put forward as to why species should be protected.

Esthetic. Basically stated, the esthetic argument claims that all nature has an inherent beauty and must be protected as part of our global heritage. This argument can be put into economic terms in some cases. For example, bird-watching is a multimillion-dollar-a-year industry in the United States, while wildlife tourism is one of the major sources of foreign exchange in Kenya (Myers 1979). The nursery, foliage plant, and florist industries in the Western world are multimillion-dollar enterprises and concentrate primarily on introduced species (T. Dudley, pers. com.). Nature has certainly served as an inspiration to artists through the ages—from the cave paintings of bison in prehistoric France to the modern-day stylized pandas of Andy Warhol.

Ethical. The ethical argument claims that no one species has the right to wipe out another. Even aboriginal societies have taboos against the overhunting of certain species (e.g., Reichel-Dolmatoff 1975), although some may argue that this point of view is purely utilitarian. Nonetheless, the live-and-let-live philosophy has also been explained as a sort of mental well-being stemming from our recognition that we share the planet with many other species and that we should be willing to make sacrifices to ensure that no other life form disappears as a result of human ignorance or greed. Restated in yet another way, the ethical argument believes that civilized people sleep better knowing jungles still exist where tigers roam at night.

Diversity and stability. The argument for diversity and stability claims that ecosystems are complex mechanisms, the continued existence of which is dependent on the presence of certain key species. This argument is particularly important in the tropics, where plant-animal interactions play a much greater role in ecosystem function than in the temperate zones. A good example of this interaction is the Brazil nut tree (*Bertholletia excelsa* L.), a major source of income for many Amazonian Indians and peasants. Almost all the Brazil nut crop comes from the wild, because, with very few exceptions, plantations of the trees have not produced fruit, since the bee that pollinates the flower occurs only in natural forest (S. Mori, pers. com.). Thus, the economic value of the Brazil nut tree is eliminated if the surrounding forest does not remain intact.

Scientific value. The argument for scientific value claims that species must be protected and studied for their scientific value. Even those species which seem to be of little value commercially can teach us a great deal. A good example is the howler monkey (*Alouatta* species) from Central and South America. Living primates are divided into two basic infra-orders—Catarrhina (Old World) and Platyrrhina (New World). Because *Homo sapiens* L. is a catarrhine, many of the first primate field studies focused on African monkeys (primarily baboons and chimpanzees) in the belief that this was the best way to study the origin of human behavior. New World monkeys were regarded as an evolutionary dead-end, the study of which could teach us relatively little about our origins compared to research on Old World apes and monkeys.

In the second half of the twentieth century, paleontologists conducting excavations in the Egyptian desert unearthed the remains of *Aegyptopithecus*, the so-called Dawn Apes, which represent the most recent common ancestor of both apes and humans. Further study revealed that in terms of ecological adaptations, the closest living model for *Aegyptopithecus* is the neotropical howler monkey (Fleagle 1982). Thus, this extant non-human primate could possibly hold the key to our understanding of human evolution.

Utilitarian. The utilitarian argument claims that species must be preserved because of their utility to humankind. We simply could not survive without the plants and animals on which our civilization depends for everything, from the food we eat to the oxygen we breathe. To allow dwindling species to become extinct would be short-sighted, as there are numerous cases in which such organisms have turned out to be very important. The once-extensive herds of saiga antelope (*Saiga tatarica* L.) in Asia were reduced to fewer than 1000 individuals by the beginning of the twentieth century. Strict bans on hunting were implemented in 1919, and the herds began rapidly to replenish themselves. By the 1970s approximately 250,000 animals were harvested annually for meat and skins with little or no deleterious effects on the total population (Grzimek 1972).

ECONOMIC BENEFITS OF CONSERVATION

It is important to bear in mind that we know little or nothing about the potential utility of the flora and fauna of tropical regions. There can be little doubt that many "super species" are waiting to be discovered. When Aublet described the rubber (*Hevea*) tree in 1775, who could have predicted that its latex would be a major factor in bringing about the Industrial Revolution (Schultes 1977)?

Conservationists are beginning to understand that conservation is going to have to pay for itself. As Hugh Synge (pers. com.) put it:

> Plant conservation, with its basis in plant systematics and taxonomy, has tended to give equal weight to each threatened plant rather than emphasize the useful ones.... A change in emphasis is needed with far more effort on wild plants used by people, whether or not these plants enter the monetary economy.

Increased emphasis must be placed on the economic benefits that accrue if species are protected and rationally utilized. This approach has particular appeal in developing countries where economic conditions are often difficult. Brazil and Mexico, for example, represent two of the most important countries of the world in terms of species diversity. While there has been a tremendous growth in public awareness regarding the importance of conserving natural resources during the 1970s and 1980s, both countries nonetheless labor under the burden of enormous foreign debt, with a concomitant temptation to convert their resources into financial assets. Ever since Theodore Roosevelt set up a national park system in the United States at the beginning of the twentieth century, much of the private sector in both the industrialized and industrializing world has considered conservation to be antidevelopment (which was often the case).

How, then, can both industrialized and industrializing worlds conserve *and* utilize tropical biota? An approach is being carried out successfully on the savannas of East Africa, where wildlife tourism is a major source of revenue. Yet the situation in East Africa is very different from that in South America—the Amazon does not have the large concentrations of a relict Pleistocene mega-fauna, which has drawn so many big game hunters and camera enthusiasts to Kenya. Yet, if much of the biological diversity found in tropical forests is to be preserved for future generations, these forests will have to be managed for some sort of economic return.

Animals, which comprise only a small fraction of total tropical forest biomass, are forest products of paramount importance to local peoples who utilize wild game as sources of food and revenue. In rural Africa, up to 75 percent of the animal protein consumed by the locals comes from wild animals, and this figure may be as high as 85 percent in Amazonian Peru (de Vos 1977; UNESCO 1978). In Thailand, 52,000 birds of 40 species are harvested as food every year, and another 375,000 birds of 350 species are collected for nonfood uses (Furtado 1978). The value of skins and hides exported from the Peruvian Amazon prior to 1972 exceeded that of timber (Dourojeanni 1982). In 1985, the United States imported more than $600 million worth of wildlife and wildlife products, much of it from tropical countries (G. Hemley, pers. com.).

With adequate management, sustainable productivity of tropical wildlife could become an ecologically sound economic alternative in tropical countries. Indeed, Paucar and Gardner (1981) have shown that 500 square kilometers (200 square miles) in Amazonian Ecuador theoretically could yield over $6 million annually in wildlife products on a sustainable basis.

Two ongoing pilot projects merit special mention. Funded by the Pan American Health Organization, the Peruvian Primate Program was initiated in 1975 following the declaration of bans on the export of primates from Peru and Colombia. The program attempts to meet the demand for a number of biomedically important South American primate species by combining culling of wild populations with local captive breeding. Sufficient data have been gathered to permit safe small-scale harvesting of wild populations. Given the rising demand for animals by the biomedical community, the average price of about $300 per animal is expected to increase (R. A. Mittermeier, pers. com.). Once the captive breeding effort has been expanded and more data accrue from long-term field studies, the Peruvian Primate Program may serve as strong economic evidence of the value of implementing programs of sustained-yield management of local natural resource.

The second project is a successful crocodile management program in New Guinea. Sponsored by the Food and Agriculture Organization (FAO) and the local government, this program is already providing rural people with meat and cash income. The two local species of crocodiles, *Crocodylus porosus* Schneider and *C. novaeguineae* Schmidt, are well suited to management programs, since they produce large numbers of eggs and are extremely food efficient—1.5 kilograms (3 pounds) of food can be converted to 1 kilogram (2 pounds) of body weight versus a ratio of 6 kilograms (13 pounds) of cattle for 1 kilogram (2 pounds) of body weight (N. Vietmeyer, pers. com.). The government purchases the skins of medium-sized crocodiles but restricts the harvest of both very small and very large animals, thus ensuring that good populations of both juveniles and breeding adults will continue to flourish. By attaching an economic value to these species, the government has encouraged the local people to harvest and protect the crocodiles for their own economic well-being.

Many other tropical animals possess underexploited economic potential. South American river turtles of the genus *Podocnemis* are a relevant example. Because of their delicious meat and eggs, these turtles have been overexploited for decades, yet they are still relatively abundant in several regions. Given their tremendous reproductive capacity (a female *P. expansa* Schweiger can produce up to 150 eggs a year), it should not prove difficult to increase the size of wild populations by controlling predation of eggs and juveniles (Mittermeier 1974).

Other tropical wildlife species that have been studied include capybara (*Hydrochaerus* L.), which already have been ranched successfully in Venezuela (Williamson and Payne 1977); iguanas (*Iguana iguana* L.), which are being bred commercially in Panama (D. Werner, pers. com.); and pouched rats (*Cricetomys gambianus* Waterhouse), which are being managed successfully in Nigeria (UNESCO 1978).

PROTECTING TROPICAL FORESTS

In most tropical countries, "forest utilization" has simply meant "timber removal." Tropical forests have been utilized as nonrenewable resources, usually being turned into crop or pasture land that is only good for two or three years in many areas (Guppy 1984). The problem posed by conventional forestry practices in tropical, developing countries is that temperate technology is often inappropriate. Modern forestry originated in Europe several hundred years ago and was developed for the purpose of managing an ecosystem with relatively few species in a temperate zone and on a resilient mineral soil.

Today, the process of timber extraction demands economies of scale which usually require the use of heavy machinery. Skidding and road-building can cause tremendous damage both to fragile tropical soils (with scarification of up to 30 percent of the soil surface) and to the residual stand (Ewel 1981). Felling can result in the elimination of the established seedling crop, which is required to form the new stand. Furthermore, road damage associated with logging in tropical areas can lead to increased sediment load in streams, accelerated soil erosion, decline in water quality and uniformity of flow, as well as impoverishment of essential soil microorganisms and nutrients (Gilmour 1971; Auchter 1978).

The other characteristic of temperate-zone forestry that can conflict with conservationists' goals is the reduction of the diversity of the stand. Increasing timber production of desired species necessitates removal and/or replacement of nondesirable species (Ashton and Plotkin 1982). Furthermore, development planners in tropical countries must take note of the fact that economic policies change repeatedly within the relatively long period needed to grow tropical hardwoods.

Since the late 1800s, demand in Malaysia has changed from heavy to light hardwoods; felling, extraction, preservation, and processing techniques have been revolutionized; firewood demand has grown; labor availability and costs have fluctuated; land has progressively become more expensive, and land available for forestry has declined in quality (Wyatt-Smith 1963; Tang 1980). Maintenance of some degree of species diversity in the forest stand has preserved a degree of flexibility to meet shifting demands that otherwise would have been impossible to satisfy (Ashton and Plotkin 1982).

How can resources be utilized to replace or reduce the impact of timber extraction—which has proven environmentally so disastrous in so many tropical countries—while generating some sort of economic return? Ashton and Plotkin (1982) suggested that more attention be paid to nontimber products, such as medicines, fibers, oils, waxes, latexes, and tannins. Many of these so-called secondary forest products (which are anything but secondary to people living in or near tropical forests) already participate in international commerce, including rattan (*Calamus* species) from several tropical Asian countries ($1.2 billion per year); essential oils, resins, and other products from Indonesia ($27 million per year); illipe nuts (*Shorea* species) from Sarawak ($4 million per year); and Brazil nuts (*Bertholletia excelsa* L.) from the Amazon ($16 million per year) (Shane 1977; USDA 1978; Dransfield 1979; Myers 1979). In India, latex, forage, pesticides, tannins, dyes, resins, spices, oils, medicines, and other nonwood products extracted from the forests are estimated to be worth over $1 billion a year (UNESCO 1978).

To protect tropical ecosystems for future generations, we must demonstrate the potential value of nontimber plant products that have not yet entered the market economy. The most cost-effective method for accomplishing this is through the science of ethnobotany.

Ethnobotany and Conservation

The term *ethnobotany* was first used by Harshberger (1896) who defined it as the study of "plants used by primitive and aboriginal people." Its meaning was broadened by Robbins et al. (1916), who suggested that the science of ethnobotany should include the investigation and evaluation of the knowledge of all phases of plant life among primitive societies and of the effects of the vegetal environment upon the life, customs, beliefs, and history of these tribal peoples. Twenty-five years later, Jones (1941) offered a more concise definition: "the study of the interrelationships of primitive man and plants." Schultes (1967) expanded this to include "the relationships between man and his ambient vegetation." Ford (1987) regarded ethnobotany as "the totality of the place of plants in a culture and the *direct* interaction by the people with the plants."

Summarizing these definitions, ethnobotany is considered now to be a subbranch of the science of economic botany which "emphasizes the uses of plants, their potential for incorporation into another (usually Western) culture, and [suggests that these people] have *indirect* contact with the plants through their by-products" (Ford 1987). For the purposes of this chapter, I prefer the concept of ethnobotany promulgated by Jones. Nevertheless, I would like to restate his definition of ethnobotany as "the study of tribal peoples and their utilization of tropical plants."

Some of the earliest evidence of plant use by Native Americans are archaeological remains of peyote (*Lophophora williamsii* [Lem.] Coulter) excavated from caves in Texas and dated at about 7000 B.C. In lowland South America, artifacts from coastal Ecuador indicate that coca (*Erythroxylum* species) was in use about 2100 B.C. (Plowman 1984). Manioc (*Manihot esculenta* Crantz), achiote (*Bixa orellana* L.), pineapple (*Ananas comosus* [L.] Merrill), and peanuts (*Arachis hypogaea* L.) were domesticated by lowland South American peoples prior to the arrival of Europeans (Ford 1984). The Sinu culture (1200–1600 A.D.) of northwestern Colombia produced numerous gold pectorals with mushroomlike representations; Schultes and Bright (1979) concluded that these represented hallucinogenic mushrooms of the genus *Psilocybe*. That no South American tribe is known to consume hallucinogenic mushrooms today makes one wonder about the amount of ethnobotanical information that already has been lost.

Compared with records of Mexico and Mesoamerica, written ethnobotanical records from lowland South America at the time of the Conquest are relatively scant. Schultes (pers. com.), who has conducted extensive research on the data recorded by the Spaniards both in Mexico and in South America, has concluded that the ecclesiastical writers who helped conquer and settle Mexico were generally better educated and less exploitative than the group that subjugated the native peoples of western South America. While Mexico was still being "pacified," the king of Spain sent his personal physician Francisco Hernández to the New World for five years to study the medical plants of the Aztecs (Schultes and Hofmann 1979). Furthermore, the codices of both the Mayans and the Aztecs have been deciphered and have yielded a wealth of information on the useful plants of both tribes. Although some of the pottery from highland cultures like the Mochica of central Peru depict ritual uses of hallucinogenic fungi (*Psilocybe*) and hallucinogenic cacti (*Trichocereus*), our knowledge of ethnobotany is much less detailed for South America than for Mexico.

Ethnobotany offers a very effective approach to tropical forest conservation, since it may provide a wealth of data on nontimber products, which can often be collected in a nondestructive manner. Commodities might thus be extracted from the forest with minimal ecological and/or environmental damage, yet provide some incentive for the conservation and rational utilization of the forest.

A major barrier to developing this type of labor-intensive, capital-extensive forest management strategy is the haphazard way in which most of the early ethnobotanical data were collected. Virtually all the major figures who collected important ethnobotanical data until the twentieth century originally went to the New World for reasons other than the documentation of ethnobotanical lore. Amerigo Vespucci (1451–1512) was carrying out geographic exploration when he made the first discovery of coca (*Erythroxylum* species) chewing on the Guajira Peninsula of Colombia. Charles-Marie de la Condamine (1701–1774), after working in Ecuador to determine the shape of the earth, made the first major scientific trip down the Amazon and stumbled onto such ethnobotanical treasures as rubber (*Hevea*), ipecac (*Psychotria*), quinine (*Cinchona*), and curare (possibly *Strychnos*). The great German explorer Alexander von Humboldt (1769–1859), together with his French colleague Aimé Bonpland (1773–1858), traveled extensively in the new World tropics. They were making general collections of the biota when they observed the manufacture of both curare (from *Strychnos*?) and the hallucinogenic snuff yopo (*Anadenanthera peregrina* [L.] Speg.). While traveling in the Amazon making general collections of the flora, the British botanist Richard Spruce (1817–1893) observed the indigenous use of other hallucinogenic plants and of rubber.

A turning point in the history of South American ethnobotany occurred with the arrival of R. E. Schultes in the Colombian Amazon in 1941. After completing his Ph.D. dissertation on the ethnobotany of the Indians of Oaxaca, Mexico, Schultes decided to initiate a study of plants employed in the manufacture of arrow poisons. He remained in the northwestern Amazon until 1954, living with the Indians, participating in their rituals, and conducting ethnobotanical research. Although he eventually went to work on the USDA project to harvest natural rubber from the Amazon during the Second World War, he continued collecting plants, eventually sending home over 24,000 plant collections. More than 1500 of these species were employed medicinally or as poisons by Amazonian Indians, while others were used for purposes from clothing to contraceptives.

Most of the ethnobotanists working in South America today are following in the footsteps of Schultes, who demonstrated the importance of focusing on ethnobotany itself rather than on the mere collecting of data about useful plants as an adjunct to other studies. Schultes has always taken an interdisciplinary approach to his ethnobotanical research, incorporating botanical and anthropological aspects as well as chemical and pharmacological considerations whenever these were available. He began writing about the importance of conserving ethnobotanical lore in the tropics long before most biologists were aware that conservation in these regions was even an issue. In 1963, he wrote:

> Civilization is on the march in many, if not most, primitive regions. It has long been on the advance, but its pace is now accelerated as the result of world wars, extended commercial interests, increased missionary activity, widened tourism. The rapid divorcement of primitive peoples from dependence upon their immediate environment for the necessities and amenities of life has been set in motion, and nothing will check it now. One of the first aspects of primitive culture to fall before the onslaught of civilization is knowledge and use of plants for medicines. The rapidity of this disintegration is frightening. Our challenge is to salvage some of the native medico-botanical lore before it becomes forever entombed with the cultures that gave it birth.

Tropical forest peoples represent the key to understanding, utilizing and protecting tropical plant diversity. The degree to which they understand and are able sustainably to use this diversity is astounding. The Barasana Indians of Amazonian Colombia can identify all the tree species in their territory without having to refer to the fruit or flowers (S. H. Jones, pers. com. to E. W. Davis), a feat that no university-trained botanist is able to accomplish. And a single Amazonian tribe of Indians may use more than 100 species of plants for medicinal purposes alone. Nevertheless, to this day, very few tribes have been subjected to a complete ethnobotanical analysis. Robert Goodland of the World Bank wrote (1981):

> Indigenous knowledge is essential for the use, identification and cataloguing of the [tropical] biota. [When] tribal groups disappear, their knowledge vanishes with them. . . . The preservation of these groups is a significant economic opportunity for the [developing] nation, not a luxury.

Since Amazonian Indians are often the only ones who know both the properties of these plants and how they can best be utilized, their knowledge must be considered an essential component of all efforts to conserve and develop the Amazon. Failure to document this ethnobotanical lore would constitute a tremendous economic and scientific loss to the human race.

LITERATURE CITED

Ashton, P., and M. Plotkin. 1982. *Technologies to Increase Production from Primary Tropical Forests and Woodlands*. Report to Office of Technology Assessment, U.S. Congress, Washington, DC.

Auchter, R. S. 1978. Summation of conference, *Proceedings of a Conference on Improved Utilization of Tropical Forests*. USDA, Madison, WI.

Dourojeanni, M. 1982. *Renewable Natural Resources of Latin America and the Caribbean: Situation and Trends*. World Wildlife Fund U.S., Washington, DC.

Dransfield, J. 1979. The biology of Asiatic rattans in relation to the rattan trade and conservation. In *The Biological Aspects of Rare Plant Conservation*. Ed. H. Synge. Chichester: John Wiley and Sons. 179–186.

Eckholm, E. 1986. Species are lost before they are found. *New York Times* (14 September): 1–3.

Ewel, J. 1981. Environmental implications of utilization. *Proceedings of an International Symposium on Tropical Forest Utilization and Conservation*. New Haven, CT: Yale University, School of Forestry.

Fleagle, J. 1982. Living primates as a key to human evolution. In *Primates and the Tropical Forest*. Eds. R. Mittermeier and M. Plotkin. Pasadena: Leakey Foundation. 37–44.

Ford, R.I. 1984. Prehistoric phytogeography of economic plants in Latin America. In *Pre-Columbian Plant Migration*. Ed. D. Stone. Cambridge: Peabody Museum. 175–183.

––––––. 1987. Ethnobotany: Historical diversity and synthesis. In *The Nature and Status of Ethnobotany*. Ed. R. I. Ford. Anthropological Papers, no. 67. Ann Arbor, MI: University of Michigan Museum of Anthropology. 33–49.

Furtado, J. I. 1978. The status and future of tropical moist forest in Southeast Asia. In *Developing Economies in Southeast Asia and the Environment*. Eds. C. McAndrews and L. S. Chia. Singapore: McGraw-Hill. 73–120.

Gilmour, D. 1971. Effects of logging on streamflow and sedimentation in a North Queensland Rain Forest catchment. *Commonwealth Forestry Review* SO(1) 143: 39–48.

Goodland, R. 1981. *Economic Development and Tribal Peoples*. Washington, DC: World Bank.

Grzimek, B. 1972. *Animal Life Encyclopedia*. New York: Van Nostrand and Reinhold Company.

Guillaumet, J. L. 1984. The vegetation: An extraordinary diversity. In *Key Environments—Madagascar*. Eds. A. Jolly, P. Ogberle, and R. Albignac. Oxford: Pergamon Press. 27–54.

Guppy, N. 1984. Tropical deforestation: A global view. *Foreign Affairs*. 928–965.

Harshberger, J. W. 1896. Purposes of ethnobotany. *Botanical Gazette* 21(3): 146–154.

Jones, V. 1941. The nature and scope of ethnobotany. *Chronica Botanica* 6(10): 219–221.

Kirkbride, C. 1986. *Biological Evaluation of the Chocó Phytogeographic Region in Colombia*. Unpublished report to World Wildlife Fund–U.S. Washington, DC.

Knoll, A. 1984. Patterns of extinction in the fossil record of vascular plants. In *Extinctions*. Ed. M. Nitecki. Chicago: University of Chicago Press. 21–68.

Mittermeier, R. A. 1974. South America's river turtles: Saving them by use. *Oryx* 14(3): 222–230.

———. 1982. The world's endangered primates. In *Primates and the Tropical Forest*. Eds. R. Mittermeier and M. Plotkin. Pasadena: Leakey Foundation. 11–22.

Myers, N. 1979. *The Sinking Ark*. Oxford: Pergamon Press.

Paucar, A., and A. Gardner. 1981. Establishment of a scientific research station in the Yasuni National Park of the Republic of Ecuador. Unpublished report.

Plowman, T. 1984. The origin, evolution and diffusion of coca, *Erythroxylum* spp., in South and Central America. In *Pre-Columbian Plant Migration*. Ed. D. Stone. Cambridge: Peabody Museum. 125–163.

Reichel-Dolmatoff, G. 1975. *The Shaman and the Jaguar*. Philadelphia: Temple University Press.

Robbins, W. W., J. P. Harrington, and B. Freire-Marreco. 1916. *The Ethnobotany of the Tewa Indians*. Bureau of American Ethnology Bulletin no. 55. Washington, DC. 1–124.

Schultes, R. E. 1963. The widening panorama in medical botany. *Rhodora* 65(762): 97–120.

———. 1967. The place of ethnobotany in the ethnopharmacologic search for psychotomimetic drugs. In *Ethnopharmacologic Search for Psychoactive Drugs*. Eds. D. H. Efron, B. Holmstedt, and N. S. Kline. U.S. Department of Health, Education and Welfare Publication no. 1645. Washington, DC: Government Printing Office. 33–57.

———. 1977. The odyssey of the cultivated rubber tree. *Endeavor*, n.s., 1, nos. 3–4 (1977) 133–138.

Schultes, R. E., and A. Bright. 1979. Ancient gold pectorals from Colombia: Mushroom effigies? *Botanical Museum Leaflets* (Harvard University) 27: 113–141.

Schultes, R. E., and A. Hofmann. 1979. *Plants of the Gods: Origins of Hallucinogenic Use*. New York: McGraw Hill Book Company.

Shane, M. 1977. The economics of a Sabah rattan industry. In *A Sabah Rattan Industry*. Kuala Lumpur: Markiras Corp. Skn. Bhd. (Cited in Dransfield 1979.)

Tang, H. 1980. Factors affecting regeneration methods for tropical high forests. *Malaysian Forester* 43(4): 469–480.

UNESCO. 1978. *Tropical Forest Ecosystems: A State-of-Knowledge Report Prepared by UNESCO/ UNEP/FAO*. Paris: UNESCO.

USDA. 1978. *Agricultural Statistics, 1977*. Washington, DC: Government Printing Office.

de Vos, A. 1977. Game as food. *Unasyva* 29: 2–12.

Williams, C. 1964. *Patterns in the Balance of Nature*. New York: Academic Press.

Williamson, G., and W. Payne. 1977. *Animal Husbandry in the Tropics*. London: Longman.

Wilson, E. O. 1985. The biological diversity crisis. *Bioscience* 55(11): 700–706.

Wyatt-Smith, J. 1963. Manual of Malayan silviculture for inland forests. *Malayan Forestry Records* 23: 1–402.

Quantitative Ethnobotany and the Case for Conservation in Amazonia

G. T. PRANCE, W. BALÉE, B. M. BOOM, AND R. L. CARNEIRO

The usefulness of Amazonian forests as demonstrated by indigenous peoples who depend on them has often been cited as one reason, among others, for conservation of these forests (e.g., Myers 1982, Fearnside 1985). Upon searching the literature, however, we found few quantitative data to support this claim. We designed, therefore, a quantitative study to show how useful Amazonian forests are to indigenous Amazonian peoples in terms of the number and proportion of useful species and families therein. The purpose of this article is to present our quantitative data on the use of trees in four 1-hectare plots of terra firme dense forest by four indigenous Amazonian groups, and to show the value of these data to the field of conservation. We have also attempted to quantify the usefulness of the principal species and plant families involved in our study. Forest inventories of 1 ha were taken in the habitats of the Ka'apor, Tembé, Chácobo, and Panare Indians. Here we report the utility of trees greater than or equal to 10 centimeters diameter at breast height (DBH) from these plots to each of the four groups. Some of these results have been elsewhere discussed (Balée 1986, 1987; Boom 1985, 1986a, 1986b, 1987), emphasizing extraordinarily high percentages of tree species and individual trees from the plots that are useful in one or more ways to the Indians concerned. In these earlier studies, uses were very broadly defined, such that any species that could be used for firewood and any species that bore parts edible for game animals, if useful for nothing else (such as supplying edible parts for people, construction material, technological items, and medicine) were also considered to be useful, at least in one or both of these ways. Thus, Balée (1986, 1987) found 100 percent use for both terra firme plots by the Ka'apor and Tembé, and Boom (1987, personal communication) found 82 percent use for the Chácobo and 49 percent for the Panare, using these categories. We define "use" more narrowly here.

In this paper we do not discuss plants that supply only fuel and/or attract game animals, upon which indigenous diets depend, not because these are not a priori useful, but rather because the vast majority of trees fall into one or both of these categories anyway. Instead, we would attempt to calculate the value of tree species and families in terms of indigenously recognized uses less regularly distributed throughout the corpus of our ethnographic data and botanical collections. The objective is to evaluate the tree species and families that seem to be most useful to all four indigenous groups and to recommend measures to protect these species and their associated habitats.

Materials and Methods

The four groups on which this paper focuses speak languages of three different linguistic families: Tupi-Guarani (Ka'apor and Tembé), Panoan (Chácobo), and Cariban (Panare). They reside within the frontiers of three different Amazonian countries: Brazil (Ka'apor and Tembé), Bolivia (Chácobo), and Venezuela (Panare). Their contact histories and populations are all different. The Ka'apor were pacified in 1928 (Ribeiro 1970), and the Tembé in the 1850s (Wagley & Galvão 1949). The Summer Institute of Linguistics first effectively contacted the Chácobo in 1955 (Prost 1970). The Panare have been more or less in contact with non-Indians since the first Spanish explorers entered the middle Orinoco region in the 1600s (Henley 1982).

Group populations are: Ka'apor (ca 500), Tembé (ca 156), Chácobo (ca 400), and Panare (ca 2000). With the exception of the Ka'apor and Tembé, who reside in adjacent territories, these groups occupy different types of forest in terms of species composition and dominance. Despite these differences, we believe that there are broad similarities in terms of the ways these people use the forest and the families, if not genera and species, of trees that are most useful to them.

The method common to all these studies was the hectare forest inventory (cf Boom 1986b). Kroeber (1920), in a critique of ethnobotanical studies of his time, suggested that such studies become more quantitative. Carneiro (1978) was the first to estimate the percentage of useful trees per plot of land to the Cariban-speaking Kuikuru of the upper Xingu River basin, in Brazilian Amazonia. One can infer that the percentage of useful trees named by the Kuikuru from this plot was 76 percent, according to Carneiro's data (Carneiro 1978). Since Carneiro did not, however, obtain herbarium specimens, one cannot determine the actual percentage of useful *species* to the Kuikuru. The current project originated as an attempt to combine ethnological data on plant utility with the botanical documentation of herbarium vouchers.

Categories of Plant Use

The data on plant utility reflect declarations made by indigenous informants combined with our own observations of the cultural deployment of plants in each ethnographic case. In no case, incidentally, do informants' accounts disagree with such observations.

To compare the results of these studies, we divide uses of trees from the inventory plots into these categories: a) "edible" (including parts, such as fruits, seeds, and latex, which people consume), b) "construction material" (such as wood used in post-and-beam construction, canoes, and bridges, and leaves used for roofing thatch), c) "technology" (a very broad category, which includes lashing material, glue, pottery temper, dye, soap, pipe stem, arrow point), d) "remedy" (for sinusitis, congestion, diarrhea, headache, vomiting, fever, unwanted pregnancy, bleeding wounds, snakebite, cradle-cap, canker sores, insect repellent), e) "commerce" (boat caulking, rubber, souvenirs), and f) "other" (magic, toys, dog-fatteners, fermentation aids, and perfume). These categories do not necessarily reflect categories of use in the indigenous classification of uses, if such a classification would be reflected in the lexicons of these languages, as we believe it is (cf Berlin et al. 1974).

In fact, only one category may be considered to be readily recognized by the Indians themselves, as demonstrated in the indigenous lexicons. This category is "edible." The Ka'apor and Tembé, for example, readily identify this category as *ma'e u'u awa* ("what people eat") or *awa mi'u* ("people's food") and distinguish it from, for example, *so'o mi'u* ("game animal food" in Ka'apor) and *miyar mi'u* ("game animal food" in Tembé). The other categories may be considered to be artificial constructs based on our own collapsing of indigenous use categories, since these categories are not readily named in any of the four languages except by circumlocutions (*see* Berlin et al. 1974).

No single or short set of terms covers the semantic range of "construction material," as here defined, in any of these languages. Rather, house beams, posts, ridgepoles, thatching material, canoe-building material, and the like, are all individually named. No indigenous term in any of the four languages approximates what we term "technology." Rather, arrow points, lashing material from the bark of trees, glues, dyes, soaps, and so on, are individually named. One of the most difficult categories to define is that of "remedy." The most common American English sense of the term "remedy" is "something such as medication or therapy, that relieves pain, cures disease, or corrects a disorder" (Morris 1973). Such is not the case with the range of meaning of terms such as *puhan* (Ka'apor) and *pohang* (Tembé), which cover not only the most common American English sense of "remedy," but much more. The best gloss for these terms is, in fact, "catalyst" (i.e., that which induces change). Thus, for example, the wood of *Tetragastris altissima* is believed by the Ka'apor to be a *Kawĩipuhan* ("beer catalyst"), insofar as when it is added to the brewing ceremonial beer (*kawĩ*), it is believed to effect a stronger, more potent brew. This use of the term *puhan* seems to go beyond the American English sense of the noun "remedy." At the same time, however, *puhan* refers to "remedy" in a conventional sense. Drinking a decoction of the bark of *Fusaea longifolia*, for example, is believed by the Ka'apor to be a "diarrhea remedy" (*marikahipuhan*). We include under the category "remedy" only those species for which informants state a testable application to a human illness; that is, the culturally prescribed application, processual treatment, and effects of the plant on a given illness are stated by informants in such a way as to be falsifiable. Plant uses that are untestable in terms of the indigenous formulation of their application, given limitations in the present tools of ethnobotanical science, are placed in the category "magic" or "other."

The category "commerce" is not lexically distinguished in these languages from economic reciprocity. In Ka'apor, for example, the word for "giving" (*me'e*) is the same as that for "selling."

Having defined the range of meaning of these use categories, both in terms of indigenous perceptions (Alcorn 1981) and our own observations, we would distinguish quantitatively the relative utility of specific plants in given situations in the same terms. Some plants are clearly more useful than others for specific purposes; that is, they are explicitly preferred by the Indians themselves for a given purpose over other plants that can also fulfill that purpose, but to a less desirable degree. For example, the Ka'apor distinguish between "quite edible" (*u'u-ate-awa*) and "less edible" (*u'uwe-awa*). Thus, it seems illogical to weigh equally, for example, the small, insignificant yet sweet fruits of *Protium* spp. (Burseraceae), which in Ka'apor society are generally eaten only by children and are never the objects of intensive gathering and economic exchange and the substantial fruits of *Theobroma grandiflorum* (Sterculiaceae), which are much sought after items of food distribution among the Ka'apor. In other words, *Theobroma grandiflorum* fruits are more *important* to the Ka'apor than *Protium* spp. fruits in an indigenous and objective sense, that is, in terms of their edibility. The same distinction—more important versus less important—can be applied to plants in each of the other categories of use, according to indigenous accounts and our own observations of the extent to which people seek out certain plants vis-à-vis others that are nevertheless all useful for approximately the same ends.

Therefore, in evaluating and comparing the utility of species, each major use of a plant is counted as 1.0 and each minor use as 0.5. We would define the use value of a species as the sum of the values corresponding to its major and/or minor use(s) in each culture. Thus, if a species (i.e., some of its parts) can be used as a major technological item and at the same time serves as a minor remedy, possessing no other known uses to the culture in question, then the use value of the species is 1.5 (1 + 0.5). To calculate the familial use value per hectare plot, we sum the use values of each species in a family and divide by the number of all species in that family, whether useful or not, occurring on the plot (although species with no use value, as here defined, are excluded from our species list otherwise). This is divided to counter the high familial use values that would accrue to families that contain many species useful only in minor ways (as with Burseraceae to the Ka'apor).

Results and Discussion

Table 1 shows the results of the hectare inventories among the Ka'apor, Tembé, Chácobo, and Panare, respectively. Collection numbers only for useful species occurring on each of the plots are given. In the column "uses," codes for the major and minor uses are indicated. Major uses, which are assigned a use value of 1.0, are indicated by a capital letter; minor uses, which are assigned a use value of 0.5, are indicated by a small letter.

Most species of trees occurring on the Ka'apor and Tembé plots were useful, as here defined, in at least one way. Of the 99 species occurring on the Ka'apor plot, 76 (76.8%) were useful in some way; of the 119 species on the Tembé plot, 73 (61.3%) were useful in some way; of the 94 species on the Chácobo plot, 74 (78.7%) were useful in some way; of the 70 species on the Panare plot, 34 (48.6%) were useful in some way. The specific uses of species from the four plots are now discussed, details of which are all summarized in Table 1.

FOOD

Of the 99 species on the Ka'apor plot, 34 (34.3%) are major or minor food plants; the more important species include *Euterpe oleracea* (Palmae) and *Theobroma grandiflorum* (Sterculiaceae). Of the 199 species occurring on the Tembé plot, 26 (21.8%) are major or minor food plants; the more important species include *Oenocarpus distichus* (Palmae) and *Pourouma guianensis* (Moraceae). Of the 94 species on the Chácobo plot, 38 (40.4%) are major or minor food plants; the more important species are five palms, *Astrocaryum aculeatum*, *Jessenia bataua*, *Maximiliana maripa*, *Oenocarpus mapora*, and *Scheelea princeps*, and four Moraceae, *Pourouma cecropiifolia*, *P. guianensis*, *Pseudolmedia laevis*, and *P. macrophylla*. Of the 70 species occurring on the Panare plot, 24 (34.3%) are major or minor food plants; the more important are *Mauritia flexuosa* (Palmae), *Parinari excelsa* (Chrysobalanaceae), and an as-yet-unidentified species in the Sapotaceae.

CONSTRUCTION MATERIAL

Species useful in major and minor ways for construction account for 20.2 percent (20/99) of the species from the Ka'apor plot; the more important species include *Fusaea longifolia* (Annonaceae) and *Licania* spp. (Chrysobalanaceae), which are used as rafters and tie-beams. The reason for use of *Licania* spp. is clear: it is rot-resistant partly because of the abundance of silica found in the rays of its wood (Prance 1972, ter Welle 1976), which discourages termite infestations. In the Tembé plot, 30.3 percent (36/119) species are useful for construction; the more important species are *Minquartia guianensis* (Olacaceae), which is one of only two species used for house posts and *Xylopia nitida* (Annonaceae), commonly used for ridgepoles. The Tembé so value *Minquartia guianensis* for house posts that there is a taboo on its use as firewood: if burned, it is believed that numerous village deaths would ensue. On the Chácobo plot, 17.0 percent (16/94) of the species are sources of construction materials. The more important Chácobo species for house posts are *Lindackeria paludosa* (Flacourtiaceae), *Amaioua guianensis* (Rubiaceae), *Mezilaurus itauba* (Lauraceae), and three species of *Sclerolobium* (Leguminosae); *Vochysia vismiifolia* (Vochysiaceae) and *Diplotropis purpurea* (Leguminosae) furnish durable wood for the construction of simple bridges over small streams. In the Panare plot, only 2.9 percent (2/70) of the species are used in construction; *Mauritia flexuosa* (Palmae) furnishes leaves for roof thatch on houses, and *Amaioua corymbosa* (Rubiaceae) is a preferred species for house framework construction because of its durable wood.

TECHNOLOGY

Species used in major or minor ways for technology account for 19.2 percent (19/99) of the species in Ka'apor plot; the more important species include *Licania membranacea* (Chrysobalanaceae), the ashes of which are used in making pottery temper, and *Lecythis idatimon* (Lecythidaceae), from the bark of which is made high quality lashing material. On the Tempé plot, 21.0 percent (25/119) of the species are used in technology; the more important species are *Inga alba* (Leguminosae), from the bark of which is produced a black dye for painting the shaman's gourd rattler, and *Lacmellea aculeata* (Apocynaceae), the wood of which is used for making spoons and ladles. Species used for technology in the Chácobo plot account for 18.1 percent (17/94) of the species. The more important technological species of the Chácobo include *Astrocaryum aculeatum* (Palmae), the "wood" of which is carved into hunting bows and arrow points; *Brosimum utile* (Moraceae), the inner bark of which is made into barkcloth; and several species that supply a fibrous inner bark for lashing material: *Guatteria discolor, G. hyposericea,* and *Xylopia polyantha* (Annonaceae), and *Cecropia ficifolia* and *C. sciadophylla* (Moraceae). In the Panare plot, 4.3 percent (3/70) of the species are useful for technology; *Cochlospermum orinocense* (Bixaceae) and *Lecythis corrugata* (Lecythidaceae) provide a fibrous inner bark that is used to make tumplines for burden baskets, and the leaves of *Maximiliana maripa* (Palmae) are woven into the burden baskets themselves.

REMEDY

Species useful in major or minor ways in the preparation of remedies account for 21.2 percent (21/99) of the species on the Ka'apor plot; important remedies come from species such as *Parahancornia amapa* (Apocynaceae), the latex of which is taken orally to treat stomach ailments, and *Virola michelii* (Myristicaceae), the sap of which is used to treat canker sores. On the Tembé plot, 10.9 percent (13/119) of the species have medicinal application; important species include *Dipteryx odorata* (Leguminosae), the oil of which is used to alleviate earache, and *Carapa guianensis* (Meliaceae), the oil of which is rubbed on the body to repel blackflies (*Simulium* spp.). The Chácobo indicate that 35.1 percent (33/94) of the species on the plot are useful as remedies; examples include two species of Rubiaceae, *Calycophyllum acreanum* and *Capirona decorticans*, the barks of which are dried, powdered, mixed with water to form a paste, and applied to skin wounds to prevent or cure infections. In the Panare plot, 7.1 percent (5/70) of the species are used medicinally; important species in this category include *Tabebuia serratifolia* (Bignoniaceae), the bark of which is used as a remedy for stomach ailments, and *Simarouba amara* (Simaroubaceae), which is employed to treat snakebite.

Table 1. Ka'apor, Tembé, Chácobo, and Panare useful tree species in 1-ha forest plots. Voucher numbers are for collections of W. Balée (WB) and B. Boom (BB); first set of voucher specimens deposited at the New York Botanical Garden. Uses are as follows: A = major food, a = minor food; B = major item of construction, b = minor item of construction; C = major item of technology, c = minor item of technology; D = major remedy, d = minor remedy; E = major commerce, e = minor commerce; F = major "other use," f = minor "other use." Species without use substitutes are marked with an asterisk (*), or, if also confined to terra firme forest, with a dagger (†).

Taxon	Voucher(s)	Ka'apor Use	Ka'apor Value	Tembé Use	Tembé Value	Chácobo Use	Chácobo Value	Panare Use	Panare Value
Anacardiaceae									
Anacardium giganteum	WB1122			A	1.0				
A. microsepalum	WB1472			b	0.5				
A. parvifolium	WB225, WB1175	a	0.5	b	0.5				
Astronium sp.	BB6687							a	0.5
Tapirira guianensis	BB6629							a	0.5
Thyrsodium spruceanum	WB437	d	0.5						
	Use value subtotal		1.0		2.0				1.0
	No. species on plot		3		6				3
	Familial use value		0.33		0.33				0.33
Annonaceae									
Anaxagorea brevipes†	WP1349			a,B,d	2.0				
Fusaea longifolia	WB155, WB1319	a,B,D	2.5	a,B,d	2.0				
Guatteria discolor	BB4486					c,d	1.0		
G. hyposericea	BB4444					c	0.5		
Xylopia nitida	WB344, WB1103	B	1.0	B	1.0				
X. polyantha	BB4954					C	1.0		
X. sericea	BB6624							a	0.5
	Use value subtotal		3.5		5		2.5		0.5
	No. species on plot		5		4		4		1
	Familial use value		0.70		1.25		0.63		0.50
Apocynaceae									
Aspidosperma desmanthum	WB1150			c	0.5				
A. megalocarpon	BB4202					d	0.5		
Lacmellea aculeata	WB277, WB1091†	a,C,D	2.5	a,C	1.5				
Parahancornia amapa	WB332, WB1193	D	1.0	D	1.0				
Genus indetermined 1	BB4398					d	0.5		
Genus indetermined 2	BB6611							a	0.5
	Use value subtotal		3.5		3.0		1.0		0.5
	No. species on plot		2		4		2		2
	Familial use value		1.75		0.75		0.50		0.25
Araliaceae									
Didymopanax morototoni	BB4468					d	0.5		
	Use value subtotal						0.5		
	No. species on plot						1		
	Familial use value						0.50		
Bignoniaceae									
Jacaranda copaia	BB4094					d	0.5		
Tabebuia serratifolia	BB6704							D	1.0
	Use value subtotal						0.5		1.0
	No. species on plot						1		1
	Familial use value						0.50		1.00
Bixaceae									
Cochlospermum orinocense	BB6654							C	1.0
	Use value subtotal								1.0
	No. species on plot								1
	Familial use value								1.00

Taxon	Voucher(s)	Ka'apor		Tembé		Chácobo		Panare	
		Use	Value	Use	Value	Use	Value	Use	Value
Bombacaceae									
Eriotheca globosa	BB4275					c	0.5		
	Use value subtotal						0.5		
	No. species on plot						1		
	Familial use value						0.50		
Boraginaceae									
Cordia bicolor	WB190, BB6625	c	0.5					a	0.5
	Use value subtotal		0.5						0.5
	No. species on plot		1						1
	Familial use value		0.50						0.50
Burseracae									
Crepidospermum goudotianum	BB4133					d	0.5		
Protium altsonii	WB184, WB1141	d	0.5	d	0.5				
P. aracouchini	WB1214			a,b	1.0				
P. decandrum	WB122, WB1089	a,d	1.0	a,b	1.0				
P. heptaphyllum	BB6563							a	0.5
P. insigne	WB1216			b,e	1.0				
P. niloi	WB1391			f	0.5				
P. pallidum	WB5, WB1114	a,d	1.0	a,b,e	1.5				
P. polybotryum									
var. *polybotryum*	WB98, WB1257	f	0.5	f	0.5				
P. sagotianum	WB855, BB6655	a,d	1.0					a	0.5
P. spruceanum	WB1192			E	1.0				
P. tenuifolium	WB157, WB1276	d	0.5	b	0.5				
P. trifoliolatum	WB116, WB1184	a,b	1.0	a,b	1.0				
Tetragastris altissima†	WB8, WB1376 BB6691	a,b,F	2.0	a,b,F	2.0			a	0.5
T. panamensis	WB451	a	0.5						
Trattinnickia burserifolia	WB1408			e,f	1.0				
T. peruviana	BB4236					C	1.0		
	Use value subtotal		8.0		11.5		1.5		1.5
	No. species on plot		9		13		2		3
	Familial use value		0.88		0.88		0.75		0.50
Caryocaraceae									
Caryocar glabrum	WB114	a	0.5						
	Use value subtotal		0.5						
	No. species on plot		1						
	Familial use value		0.50						
Chrysobalanaceae									
Couepia guianensis									
subsp. *guianensis*	WB17, WB1156†	a,B	1.5	a	0.5				
C. guianensis	WB1088			a,B	3.0				
subsp. indetermined				C,e					
Exellodendrom barbatum	WB326	a	0.5						
Hirtella lightioides	BB4167					a	0.5		
Licania canescens	WB45, WB1439	a,B	1.5	a,B	1.5				
L. heteromorpha†	WB175	B,C	3.0						

(continued)

Table 1. Continued.

Taxon	Voucher(s)	Ka'apor		Tembé		Chácobo		Panare	
		Use	Value	Use	Value	Use	Value	Use	Value
var. *heteromorpha*		d,e							
L. hypoleuca	BB6677							a	0.5
L. kunthiana	WB174	a,B	1.5						
L. macrophylla	WB1116			B,d	1.5				
*L. membranacea**	WB46	C	1.0						
L. octandra subsp. *pallida*	BB4335					a,C,d	2.0		
L. sp. 1	WB1275			B	1.0				
L. sp. 2	WB1107			B	1.0				
L. sp. 3†	WB1451			a,B,C,e	3.0				
Parinari excelsa	BB6587							A	1.0
P. sp. 1	WB1272			B	1.0				
	Use value subtotal		9.0		12.5		2.5		1.5
	No. species on plot		7		9		3		5
	Familial use value		1.29		1.39		0.83		0.33
Combretaceae									
Buchenavia capitata	WB1435			f	0.5				
Combretum laxum	BB4366					d	0.5		
	Use value subtotal				0.5		0.5		
	No. species on plot				1		1		
	Familial use value				0.50		0.50		
Elaeocarpaceae									
Sloanea eichleri	BB4522					d	0.5	d	0.5
	Use value subtotal						0.5		
	No. species on plot						2		
	Familial use value						0.25		
Euphorbiaceae									
Aparisthmium cordatum	BB4049					d	0.5		
*Hevea brasiliensis**	BB4166					c,E	1.5		
Mabea caudata	WB41, WB1326	c	0.5	c	0.5				
Sagotia racemosa	WB31, WB1151	f	0.5	f	0.5				
Sapium sp.	BB6673							a	0.5
Genus indetermined 1	BB4459					a	0.5		
	Use value subtotal		1.0		1.0		2.5		0.5
	No. species on plot		5		8		3		4
	Familial use value		0.20		0.13		0.83		0.13
Flacourtiaceae									
Banara guianensis	BB6712							a	0.5
Casearia combaymensis	BB4258					A,d	1.5		
C. javitensis	BB4295					d	0.5		
C. mariquitensis	BB4016					d	0.5		
C. sylvestris	BB6570							e	0.5
Laetia procera	WB382	d	0.5						
Lindackeria paludosa	BB4332					B	1.0		
	Use value subtotal		0.5				3.5		1.0
	No. species on plot		1				4		2
	Familial use value		0.5				0.88		0.50
Guttiferae									
Caraipa grandiflora	WB1253			d	0.5				
Oedematopus sp.	BB6690							a	0.5
Rheedia brasiliensis	WB233	A	1.0						
*Symphonia globulifera**	WB102	a,D,e	2.0						
Tovomita schomburgkii	BB4350					a	0.5		
	Use value subtotal		3.0		0.5		0.5		0.5

Taxon	Voucher(s)	Ka'apor Use	Value	Tembé Use	Value	Chácobo Use	Value	Panare Use	Value
	No. species on plot		2		3		1		1
	Familial use value		1.50		0.17		0.50		0.50
Humiriaceae									
Sacoglottis sp. 2	WB369,* WB1530	a	0.5	b	0.5				
	Use value subtotal		0.5		0.5				
	No. species on plot		1		1				
	Familial use value		0.50		0.50				
Lauraceae									
Endlicheria macrophylla	BB4499					d	0.5		
Mezilaurus itauba	BB4529					B	1.0		
Nectandra acutifolia	WB1154			a,b,c	1.5				
N. cuspidata	WB1108			b,c	1.0				
N. purusensis	WB1118			b,c	1.0				
Nectandra sp.	BB4558					d	0.5		
Ocotea abbreviata	WB1076			b,c	1.0				
O. canaliculata	WB329	c	0.5						
O. caudata	WB1095			b,c	1.0				
O. rubra	WB1236			b,c	1.0				
	Use value subtotal		0.5		6.5		2.0		
	No. species on plot		3		6		3		
	Familial use value		0.17		1.08		0.67		
Lecythidaceae									
Eschweilera coriacea	WB10†, WB1077	b,c,D	2.0	b,c	1.0				
E. parvifolia	BB4311					a	0.5		
Gustavia augusta	WB193	D	1.0						
Lecythis chartacea	WB19	b,C	1.5						
L. corrugata	BB6647							C	1.0
L. idatimon	WB76, WB1278	b,C	1.5	C,d	1.5				
L. lurida	WB1383			c	0.5				
	Use value subtotal		6.0		3.0		0.5		1.0
	No. species on plot		6		4		1		1
	Familial use value		1.00		0.75		0.50		1.00
Leguminosae									
Dialium guianense	WB1146			a	0.5				
Dinizia excelsa	WB s.n.			B	1.0				
Diplotropis purpurea	WB111, BB4242	b,c	1.0			B	1.0		
Dipteryx odorata*	WB1113			B,D,F	3.0				
D. punctata*	BB6674							E	1.0
Inga alba	SB1307			A,C,e	2.5				
I. capitata	WB297, WB1102	A	1.0	a	0.5				
I. ingoides	BB6697							a	0.5
I. marginata	WB443	a	0.5						
I. cf. ruziana	BB4317					a	0.5		
I. splendens	WB505	a	0.5						
I. stipularis	WB260	a	0.5						
I. thibaudiana	WB1513			a	0.5				
I. cf. umbellata	BB6634							a	0.5
I. sp. 1	WB153	a	0.5						
I. sp. 2	BB4401					a	0.5		
I. sp. 3	BB6670							a	0.5
Parkia paraensis	WB379	C	1.0						
Poecilanthe effusa	WB178	D	1.0						
Sclerolobium chrysophyllum	BB4237					B	1.0		

(continued)

Table 1. Continued.

Taxon	Voucher(s)	Ka'apor		Tembé		Chácobo		Panare	
		Use	Value	Use	Value	Use	Value	Use	Value
Sclerolobium sp. 1	WB606	D,f	1.5						
Sclerolobium sp. 2	BB4356					B	1.0		
Sclerolobium sp. 3	BB4368					B	1.0		
Sclerolobium sp. 4	BB6678							d	0.5
Senna silvestris	BB6588							d	0.5
Swartzia laevicarpa	BB6554							d,e	1.0
Tachigalia macrostachya	WB1456			D	1.0				
T. myrmecophila	WB39, WB1508	D,f	1.5	D	1.0				
T. paniculata	WB188	D,f	1.5						
	Use value subtotal		10.5		10.0		5.0		4.5
	No. species on plot		16		18		9		12
	Familial use value		0.66		0.56		0.56		0.38
Malphigiaceae									
Byrsonima aerugo	WB1110			A	1.0				
B. laevigata	WB337	A	1.0						
B. spicata	BB6707							a	0.5
	Use value subtotal		1.0		1.0				0.5
	No. species on plot		1		1				1
	Familial use value		1.00		1.00				0.25
Melastomataceae									
Bellucia aequiloba	BB4490					a	0.5		
B. grossularioides	BB4285					a	0.5		
Miconia affinis	BB4008					a,d	1.0		
M. holosericea	BB4282					a	0.5		
M. klugii	BB4294					a	0.5		
M. longispicata	BB4445					a	0.5		
M. poeppigii	BB4535					a	0.5		
M. punctata	BB4425					a	0.5		
M. splendens	BB4283					a,d	1.0		
	Use value subtotal						5.5		
	No. species on plot						12.5		
	Familial use value						0.46		
Meliaceae									
*Carapa guianensis**	WB279, WB1244	C,D	2.0	B,C,D	3.0				
Trichilia lecointei	WB1096			c	0.5				
T. micrantha	WB1288			c	0.5				
T. cf. *pleeana*	WB1210			c	0.5				
T. schomburgkii									
subsp. *schomburgkii*	WB1305			c	0.5				
	Use value subtotal		2.0		5.0				
	No. species on plot		3		5				
	Familial use value		0.66		1.00				
Moraceae									
Bagassa guianensis	WB523	a	0.5						
Brosimum acutifolium	BB4276					a	0.5		
B. lactescens	BB4308					a	0.5		
B. utile subsp. *ovatifolium*	BB4334					B,C	2.0		
Cecropia ficifolia	BB4264					C	1.0		
C. sciadophylla	BB4257					C	1.0		
Ficus nymphaeifolia	BB4322					B,C	2.0		
Helicostylis pedunculata	WB333	b	0.5						
H. tomentosa	BB4313					a	0.5		
Perebea mollis	BB4505					a	0.5		
Pourouma cecropiifolia	BB4203					A	1.0		

Taxon	Voucher(s)	Ka'apor		Tembé		Chácobo		Panare	
		Use	Value	Use	Value	Use	Value	Use	Value
P. guianensis	WB1331, BB4501			A	1.0	A	1.0		
Pseudolmedia laevis	BB4363					A	1.0		
P. macrophylla	BB4364					A,d	1.5		
Use value subtotal			1.0		1.0		12.5		
No. species on plot			5		2		14		
Familial use value			0.20		0.50		0.89		
Myristicaceae									
Composoneura sp.	BB4437					A,d	1.5		
Iryanthera juruensis	WB1245, BB4138			d	0.5	d	0.5		
I. tessmannii	BB4291					d	0.5		
*Virola michelii**	WB255	D	1.0						
Use value subtotal			1.0		0.5		2.5		
No. species on plot			2		2		6		
Familial use value			0.50		0.25		0.42		
Myrtaceae									
Eugenia brachypoda	WB1135			b,c,e	1.5				
Eugenia sp.	BB4432					a,d	1.0		
Genus indetermined 1	BB4449					a	0.5		
Use value subtotal					1.5		1.5		
No. species on plot					2		3		
Familial use value					0.75		0.50		
Olacaceae									
Minquartia guianensis	WB			B	1.0				
Genus indetermined	BB6656							a	0.5
Use value subtotal					1.0				0.5
No. species on plot					1				2
Familial use value					1.00				0.25
Opiliaceae									
Agonandra brasiliensis	WB1363			c	0.5				
Use value subtotal					0.5				
No. species on plot					1				
Familial use value					0.50				
Palmae									
Astrocaryum aculeatum	BB4159					A,C	2.0		
A. mumbaca	WB1406			a	0.5				
*Euterpe oleracea**	WB531	A,B,C,f	3.5						
E. precatoria	BB4151					A,B,c,d	3.0		
Jessenia bataua	BB4538					A,b	1.5		
*Mauritia flexuosa**	BB s.n.							A,B	2.0
*Maximiliana maripa**	BB4573, BB s.n.					A,b,f	2.0	A,C	2.0
Oenocarpus distichus	WB530, WB1101	A,b,C	2.5	A,b,C	2.5				
O. mapora	BB4152					A,b,d	2.0		
Scheelea princeps	BB4145					A,b,C,d	3.0		
Socratea exorrhiza	BB4155					B,C,d	2.5		
Syagrus orinocensis	BB6616							A	1.0
Use value subtotal			6.0		3.0		16.0		5.0
No. species on plot			2		2		7		4
Familial use value			3.00		1.50		2.29		1.25
Quiinaceae									
Lacunaria jemani	WB220	b	0.5						
Use value subtotal			0.5						

(continued)

Table 1. Continued.

Taxon	Voucher(s)	Ka'apor Use	Ka'apor Value	Tembé Use	Tembé Value	Chácobo Use	Chácobo Value	Panare Use	Panare Value
	No. species on plot		1						
	Familial use value		0.50						
Rubiaceae									
Amaioua corymbosa	BB6603							B	1.0
A. guianensis	BB4300					B,d	1.5		
Calycophyllum acreanum	BB4419					d	0.5		
Capirona decorticans	BB4288					d	0.5		
Guettarda spruceana	BB6652							a	0.5
Palicourea grandifolia	BB4106					d	0.5		
	Use value subtotal						3.0		1.5
	No. species on plot						4		3
	Familial use value						0.75		0.50
Rutaceae									
Fagara tenuifolia	WB1298			b,c	1.0				
	Use value subtotal				1.0				
	No. species on plot				2				
	Familial use value				0.50				
Sapindaceae									
Cupania scrobiculata	WB101	a,c	1.0						
Talisia retusa	WB506	a	0.5						
	Use value subtotal		1.5						
	No. species on plot		2						
	Familial use value		0.75						
Sapotaceae									
Achrouteria sp. 1	WB1	a	0.5						
Achrouteria sp. 2	WB522	a	0.5						
Franchetella anibifolia	WB250, WB1093	a,B	1.5	a,B	1.5				
F. gongrijpii	WB215, WB1302	a,B	1.5	a,B	1.5				
F. sp. 1	WB8	a,B	1.5						
Manilkara huberi	WB1157			a,B	1.5				
*Micropholis guyanensis**	BB4526					a,C,d	2.0		
Neoxythece cladantha	WB30	a,c,f	1.5						
N. elegans	WB130	a,c	1.0						
Planchonella oblanceolata	WB14	A,c	1.5						
Pouteria caimito	WB349	A,c	1.5						
P. laurifolia	WB267	b	0.5						
P. sp. 1	WB412	b	0.5						
P. sp. 2	WB49	a	0.5						
Radlkoferella macrocarpa	WB1461			A,B	2.0				
Sprucella aerana	WB144	a	0.5						
S. guianensis	WB238	a	0.5						
Genus indetermined 1	WB519	a	0.5						
Genus indetermined 2	BB6705							A	1.0
Genus indetermined 3	BB6681							A	1.0
	Use value subtotal		14.0		6.5		2.0		2.0
	No. species on plot		17		18		1		4
	Familial use value		0.82		0.36		2.00		0.50
Simaroubaceae									
Simaba cedron	WB1094			D	1.0				
Simarouba amara	BB6630							d	0.5
	Use value subtotal				1.0				0.5
	No. species on plot				2				1
	Familial use value				0.50				0.50

Taxon	Voucher(s)	Ka'apor		Tembé		Chácobo		Panare	
		Use	Value	Use	Value	Use	Value	Use	Value
Sterculiaceae									
Sterculia pruriens	WB2, WB1260	c	0.5	c	0.5				
Theobroma grandiflorum	WB478,* WB1117	A	1.0	A	1.0				
T. speciosum	BB4275					A,d	1.5		
Use value subtotal			1.5		1.5		1.5		
No. species on plot			2		2		1		
Familial use value			0.75		0.75		1.50		
Tiliaceae									
Apeiba burchellii	WB1536			c	0.5				
A. echinata	WB491, BB4259	c	0.5			c	0.5		
Use value subtotal			0.5		0.5		0.5		
No. species on plot			1		2		2		
Familial use value			0.50		0.25		0.25		
Ulmaceae									
Ampelocera edentula	WB77	d	0.5						
Use value subtotal			0.5						
No. species on plot			1						
Familial use value			0.50						
Violaceae									
Rinorea flavescens	WB1172			c	0.5				
Use value subtotal					0.5				
No. species on plot					1				
Familial use value					0.50				
Vochysiaceae									
Qualea paraensis	BB4298					a	0.5		
Vochysia vismiifolia	BB4198					a,B,d	2.0		
Use value subtotal							2.5		
No. species on plot							2		
Familial use value							1.25		

COMMERCE

Relatively few species of plants are commercialized from the forest plots of any of the four indigenous groups studied, because of their relatively isolated situations. In the cases of the Ka'apor and Tembé, these species are used in minor commercial ways, while for the Chácobo and Panare, the commercialization of some species is a major part of their present culture. From the Ka'apor plot, 2.0 percent (2/99) of the species are in this category. These are *Licania heteromorpha* (Chrysobalanaceae), the dye obtained is used to paint *Crescentia cujete* (Bignoniaceae) bowls, which in turn are sold as souvenirs on a very small scale, and *Symphonia globulifera* (Guttiferae), the latex of which is used to glue and blacken parts of arrows, which are also sold on a very small scale as souvenirs. In the Tembé plot, 5.0 percent (6/119) of the species are commercialized; examples include *Protium* spp. (Burseraceae), the resins of which are sold on a small scale for boat caulking, and *Licania* sp. 3 (Chrysobalanaceae), the dye obtained is used to decorate *Crescentia cujete* bowls sold as souvenirs. In no case does commercialization of these tree species involve destruction of the tree itself. Collection of latex from *Symphonia globulifera* involves nonlethal scoring of the tree; resin from *Protium* spp. is collected from the ground after it is naturally exuded from the tree. Obtaining bark from *Licania* spp. to be used in the preparation of dye does not involve the girdling of the tree; the Indians affirm that their practice of stripping off

Table 2. Most important species to the Ka'apor (with use values of 2 or more). Nonsubstitutable species (*); species exclusively of terra firme dense forest (†).

Species (Family)	Use value
Euterpe oleracea (Palmae)*	3.5
Licania heteromorpha (Chrysobalanaceae)†	3.0
Oenocarpus distichus (Palmae)	2.5
Fusaea longifolia (Annonaceae)†	2.5
Lacmellea aculeata (Apocynaceae)†	2.5
Tetragastris altissima (Burseraceae)†	2.0
Symphonia globulifera (Guttiferae)*	2.0
Eschweilera coriacea (Lecythidaceae)†	2.0
Carapa guianensis (Meliaceae)*	2.0

bark pieces on one side of the tree does not kill it. In the case of the Chácobo, only 1.1 percent (1/94) of the species are commercialized; the latex from *Hevea brasiliensis* (Euphorbiaceae) is collected and, after curing and coagulating the liquid rubber over fire and forming it into large, oblong balls, it is taken to market for sale. This rubber is the Chácobo's principal source of cash; Brazil nuts (*Bertholletia excelsa*) are also collected for sale but because no trees of this species occurred in the inventory plot, these are not considered here. From the Panare plot, 4.3 percent (3/70) of the species are in the commercial category; the most important of these is *Dipteryx punctata* (Leguminosae), the seeds of which are collected and sold for the extraction of coumarin. The other two Panare commercial tree species are less important: the red exudate from *Swartzia laevicarpa* (Leguminosae) is used to paint decorative baskets that are sold as souvenirs, while the bark of *Casearia sylvestris* (Flacourtiaceae) is burned to produce a black paint that is likewise applied to decorative baskets.

OTHER

Species in the "other" category of use account for 8.5 percent (9/99) of the species on the Ka'apor plot; an example is *Tetragastris altissima* (Burseraceae), the wood of which is used as a fermentation catalyst (see above). On the Tembé plot 4.2 percent (5/119) of the species are in this category; an example is *Sagotia racemosa* (Euphorbiaceae), the fragrant roots of which the hunter rubs on his body to become lucky in the hunt. On the Chácobo plot 1.1 percent (1/94) of the species are in the "other" category; the spathe of *Maximiliana maripa* (Palmae) is used as a toy by children. In the case of the Panare there were no use-

Table 3. Most important species to the Tembé (with use values of 2 or more). Nonsubstitutable species (*); species exclusively of terra firme dense forest (†).

Species (Family)	Use value
Dipteryx odorata (Leguminosae)*	3.0
Carapa guianensis (Meliaceae)*	3.0
Couepia guianensis (Chrysobalanaceae)†	3.0
Licania sp. 3 (Chrysobalanaceae)†	3.0
Inga alba (Leguminosae)	2.5
Oenocarpus distichus (Palmae)	2.5
Anaxagorea brevipes (Annonaceae)†	2.0
Fusaea longifolia (Annonaceae)†	2.0
Tetragastris altissima (Burseraceae)†	2.0
Radlkoferella macrocarpa (Sapotaceae)	2.0

Table 4. Most important species to the Chácobo (with use values of 2 or more). Nonsubstitutable species (*).

Species (Family)	Use value
Euterpe precatoria (Palmae)	3.0
Scheelea princeps (Palmae)	3.0
Socratea exorrhiza (Palmae)	2.5
Astrocaryum aculeatum (Palmae)	2.0
Maximiliana maripa (Palmae)*	2.0
Oenocarpus mapora (Palmae)	2.0
Licania octandra (Chrysobalanaceae)	2.0
Brosimum utile (Moraceae)	2.0
Ficus nymphaeifolia (Moraceae)	2.0
Micropholis guyanensis (Sapotaceae)*	2.0
Vochysia vismiifolia (Vochysiaceae)	2.0

ful tree species in the plot that were not included in the categories already discussed above.

From Table 1 it is possible to summarize the most useful species and families to each group from the plots studied. Tables 2–5 list the most useful species (with values of 2 or more) for each group. Those species that are not substitutable, insofar as no other plant species can be used in place of them for a given purpose in a given culture, are indicated, as are those among the non-substitutable found only in terra firme dense forest. Of the 9 species listed in Table 2, which are those with a use value of 2.0 or more to the Ka'apor, 5 (55.5 percent) seem to be exclusively of terra firme dense forest (i.e., they are not encountered in old swiddens, swidden fallows, or the swamp forest). Of the 10 species listed in Table 3, which are those with a use value of 2.0 or more to the Tembé, 5 (50 percent) seem to be exclusively of terra firme dense forest. Tables 6–9 list the families with the highest familial use values (with values of 1 or more).

Examples of nonsubstitutable species include *Euterpe oleracea* (Palmae), used for bench making by the Ka'apor; *Symphonia globulifera* (Guttiferae), which serves as a contraceptive; and *Carapa guianensis* (Meliaceae), which is used as an insect repellent (Table 2). To the Tembé, only *Dipteryx odorata* (Leguminosae) is used to relieve earache, and only *Carapa guianensis* is useful as an insect repellent. Of the most important species to the Ka'apor, 8 (88.8 percent) are exclusively of terra firme dense forest and/or nonsubstitutable; of the most important species to the Tembé, 7 (70 percent) are exclusively of terra firme dense forest and/or non-substitutable. The total number of nonsubstitutable tree species from the Ka'apor plot is 6 (6 percent of all species), and the total number of nonsubstitutable species from the Tembé plot is 2 (1.7 percent of all species) (*see* Table 1).

The overall results of our analysis are summarized in Tables 10 and 11. Table 10 shows the percentages of useful tree species, arranged by category of use, in each of the four inventory plots. Table 11 shows the percentage of total useful tree species, irrespective of category of use, in each of the four inventory plots.

Table 5. Most important species to the Panare (with use values of 2 or more). Nonsubstitutable species (*).

Species (Family)	Use value
Mauritia flexuosa (Palmae)*	2.0
Maximiliana maripa (Palmae)*	2.0

Table 6. Most important families to the Ka'apor (with familial use values of 1 or more).

Family	Familial use value
Palmae	3.00
Apocynaceae	1.75
Guttiferae	1.50
Chrysobalanaceae	1.29
Malpighiaceae	1.00
Lecythidaceae	1.00

Table 7. Most important families to the Tembé (with familial use values of 1 or more).

Family	Familial use value
Palmae	1.50
Chrysobalanaceae	1.39
Annonaceae	1.25
Lauraceae	1.08
Malpighiaceae	1.00
Meliaceae	1.00

Table 8. Most important families to the Chácobo (with familial use values of 1 or more).

Family	Familial use value
Palmae	2.29
Sapotaceae	2.00
Sterculiaceae	1.50
Vochysiaceae	1.25

Table 9. Most important families to the Panare (with familial use values of 1 or more).

Family	Familial use value
Palmae	1.25
Bignoniaceae	1.00
Bixaceae	1.00
Lecythidaceae	1.00

Conclusions

Our data definitely confirm the assertion that the terra firme rainforests of Amazonia contain an exceptionally large number of useful species. The majority of tree species from four terra firme plots of 1 ha each are useful to four respective indigenous groups, at least in the use categories we have defined here. Our data further show that some of the tree species of terra firme forest are useful to the Indians in nonsubstitutable ways. Some species, in other words, are irreplaceable insofar as they are the sole species employed to achieve culturally desirable ends.

Of course, the notion of "useful" species varies from culture to culture: for example, numerous species in our data are also high-quality timber species (such as *Mezilaurus itauba*, *Carapa guianensis*, and *Tabebuia serratifolia*), which outside of indigenous contexts

Table 10. Percentage of useful tree species of all species on plots for each indigenous group by use categories.

Use category	Ka'apor	Tembé	Chácobo	Panare
Food	34.3	21.8	40.4	34.3
Construction	20.2	30.3	17.0	2.9
Technology	19.2	21.0	18.1	4.3
Remedy	21.2	10.9	35.1	7.1
Commerce	2.0	5.0	1.1	4.3
Other	8.1	4.2	1.1	0.0

are used on an industrial basis. Our data deal only with the uses nonindustrial, indigenous Amazonians have for these species, which are far more diverse than the uses to which these species are put in Western society. Although we have not discussed trees for firewood and trees that attract game, which account for the vast majority of the species on the Ka'apor and Tembé plots (Balée 1986, 1987), it is nevertheless interesting the extent to which all four plots of forest are useful to all groups in ways that are less commonly represented.

Species useful for identical purposes among different groups certainly vary, partly as a function of phytogeographic differences. For example, the Ka'apor use species of *Tabebuia* and *Brosimum* in making their bows, while the Chácobo use *Astrocaryum aculeatum* for this purpose (see above). *Astrocaryum aculeatum* has not yet been collected in the Ka'apor region, if indeed it exists there. In other words, different suites of species (i.e., endemism) within given Amazonian forests do not affect the utility of these forests per se to comparable indigenous cultures. Rather, endemism combined with the high indigenous utility of all forests in this survey suggests that the conservation of only one or a few blocs of forest would preserve, in fact, many useful Amazonian species of trees. The implications for conservation policy are inescapable: many reserves are needed throughout Amazonia.

Finally, our results suggest that certain plant families and the forest type terra firme dense forest should be of high priority for conservation. It is quite interesting that the palm family ranks consistently among those families with the highest use values for all four groups. Palms have often been described as the "grasses of the tropics," so useful are they to those who depend on them. We have offered here a positive quantitative test of this intuitive statement. Other families that rank among the highest in terms of familial use values for at least two of the four groups include Lecythidaceae (Ka'apor and Panare), Chrysobalanaceae (Ka'apor and Tembé), and Malpighiaceae (Ka'apor and Tembé). Some of the most important plant families are distributed mostly in terra firme forest, meaning that this forest type should be a priority for conservation. These important families contain many outstanding useful species which, in addition to other useful species with high use values and nonsubstitutable species not of these families, should be carefully studied, with the aim of managing and perhaps domesticating them.

Table 11. Percentage of useful species (in all categories specified) per hectare plot to the indigenous groups studied.

Indigenous group	Percentage of useful tree species from inventory sites
Ka'apor	76.8
Tembé	61.3
Chácobo	78.7
Panare	48.6

ACKNOWLEDGMENT

Republished from *Conservation Biology* 1(1987): 296–310, by permission of the Society for Conservation Biology and Blackwell Scientific Publications. Publication 54 of the Institute of Economic Botany, New York Botanical Garden.

LITERATURE CITED

Alcorn, J. B. 1981. Some factors influencing botanical resources perception among the Haustec. *Journal of Ethnobiology* 1: 221–230.

Balée, W. 1986. Análise preliminar de inventário florestal e a etnobotânica Ka'apor (Maranhão). *Boletim do Museu Paraense Emílio Goeldi, Botânica*, 2(2): 141–167.

———. 1987. A etnobotânica quantitativa does indioe Tembé (Rio Gurupi, Pará). *Boletim do Museu Paraense Emílio Goeldi, Botânica*, 3(1): 29–50.

Berlin, B., D. E. Breedlove, and P. H. Raven. 1974. *Principles of Tzeltal Plant Classification: An Introduction to the Botanical Ethnography of a Mayan-Speaking Community in Highland Chiapas*. New York: Academic Press.

Boom, B. M. 1985. Amazonian Indians and the forest environment. *Nature* 314: 324.

———. 1986a. The Chácobo Indians and their palms. *Principes* 30: 63–70.

———. 1986b. A forest inventory in Amazoniana Bolivia. *Biotropica* 18: 287–294.

———. 1987. *Ethnobotany of the Chácobo Indians, Beni, Bolivia*. Advances in Economic Botany Series, vol. 4. Bronx, NY: New York Botanical Garden. 1–68.

Carneiro, R. L. 1978. The knowledge and use of rain forest trees by the Kuikurú Indians of Central Brazil. In *The Nature and Status of Ethnobotany*. Ed. R. I. Ford. Anthropological Papers, no. 67. Ann Arbor, MI: University of Michigan Museum of Anthropology. 201–216.

Fearnside, P. M. 1985. Environmental change and deforestation in the Brazilian Amazon. In *Change in the Amazon Basin*. Vol. 1, *Man's impact on Forests and Rivers*. Manchester, England: Manchester University Press. 70–89.

Henley, P. 1982. *The Panaré: Tradition and Change on the Amazonian Frontier*. New Haven, CT: Yale University Press.

Kroeber, A. L. 1920. Review of uses of plants by the Indians of the Missouri River region, by Melvin Randolph Gilmore. *American Anthropologist* 22: 384–385.

Morris, W., ed. 1973. *The American Heritage Dictionary of the English Language*. Boston, MA: Houghton Mifflin Co.

Myers, N. 1982. Deforestation in the tropics: who wins, who loses. In *Where Have All the Flowers Gone? Deforestation in the Third World*. Eds. V. H. Sutlive et al. Studies in Third World Societies, no. 13. Williamsburg, VA: College of William and Mary. 1–24.

Prance, G. T. 1972. An ethnobotanical comparison of four tribes of Amazonian Indians. *Acta Amazonica* 2(2): 7–27.

Prost, M. D. 1970. *Costumbres, habilidades, y cuadro de la vida humana entre los Chácobos*. Riberalta, Bolivia: Instituto Lingüístico del Verano.

Ribeiro, D. 1970. *Os indios e a civilização: a integração das populações indigenas no Brasil moderno*. Rio de Janeiro: Editora Civilização.

ter Welle, B. J. H. 1976. On the occurrence of silica grains in the secondary xylem of Chrysobalanaceae. *IAWA Bulletin* 2: 19–29.

Wagley, C., and E. Glavão. 1949. *The Tenetehara Indians of Brazil: A Culture in Transition*. New York: Columbia University Press.

A Near and Distant Star

C. EARLE SMITH

Ethnobotany is the sum total of human subsistence knowledge. Without a thorough understanding of the world's plant resources, the human race would cease to exist. For example, within the last half of the twentieth century, some of us who obtain our food prepackaged and ready to eat have failed to learn those facts that enabled our ancestors to survive. Unfortunately, that same lack of knowledge is threatening the human race with disaster as we ignore the dangers of destroying the world's vegetation.

I have often speculated over the process of discovery that enabled humans to know the value of each plant species. In the far distant past, someone had to try each unknown plant to know its properties. Granted that human senses were much keener in the days before furnaces and internal combustion engines, hearing, seeing, feeling, and smelling could go only so far. Eventually, a plant species had to be tasted and tested on human physiology. Whenever the testing proved fatal, it must have been repeated many times until levels of dosage became known for the bioactive species.

Another cause for speculation has been the establishment of nutritionally sound diets. While we are assured by pediatricians that young children eventually will establish for themselves a nutritionally sound diet, how does the human mechanism signal the information that leads a primitive society to establish an adequate diet? Today, the parent pre-selects for the child the items to be tested, but who did the selection at the beginning of human time? How did the people of southeastern Asia select rice, those of southwestern Asia select wheat and barley, and the peoples of Middle America select maize? Along with the carbohydrates must be assorted protein resources (e.g., soybeans, peas, lentils, common beans) and sources of minerals and vitamins (e.g., peaches, apples, figs, eggplant, cabbage, chilies, avocados, and many more).

How, in so few generations, have we forgotten the lessons learned by our ancestors? The field of ethnobotany consists of many subareas of study, and I suspect that specialization will continue into the future. Two large subareas are classical ethnobotany and archaeological botany (or paleoethnobotany). The former refers primarily to the collecting of ethnobotanical information from local, living populations whatever their background, to preserve some knowledge of plant use in the face of massive acculturation of the world's peoples. The latter refers to the identification and interpretation of plant remains from a no-longer-extant population. These remains are often recovered with archaeological techniques and may be dry plant fragments or carbonized remains.

Other areas into which ethnobotanists are delving are economic botany, folk medicine, and herbal dietary supplements. Economic botany long has been considered a field separate from ethnobotany, but it really is ethnobotany with a financial incentive. Folk medicine and herbal dietary supplements are undoubtedly the oldest subareas of ethnobotany (see, for example, Theophrastus), but they are currently displaying a marked vigor, whipped by an enthusiasm that began with the flower-child movements of the 1960s. A visit to a local health food emporium will convince anyone that there is here also a financial incentive.

Except for economic botany, where the employment of botanists is directly concerned with the financial well-being of a commercial venture, large or small, ethnobotany exists as a supplement to other areas of botany. As a result, there are few endowed chairs in ethnobotany, and few are likely to be established in the foreseeable future. This seems somewhat of an anachronism, inasmuch as the greatest part of human existence has been directly dependent on ethnobotanical knowledge. In fact, the future of human beings may be inextricably tied to ethnobotanical knowledge. To date, the most successful chemicals for controlling human populations, for example, have come from plants, and the food supply for the current world's overpopulation is directly dependent on knowledge of how far modern commercial crops can be manipulated or their yields improved.

The field of ethnobotany is due for a change. For too long, it has been apologetic and retiring. Archaeological and anthropological colleagues have treated ethnobotany as a service due them. Ethnobotanists have forgotten the pivotal position they occupy with their special knowledge of the world's vegetation and its uses. Most of the industrial world has long since concluded that their fields have only to turn to mechanical and electronic engineering to advance to higher planes. Many of the governments of industrialized nations budget far more funds for technological development than they do for agricultural development. Unfortunately, unless more emphasis is placed on the natural products of the earth, the future of humans may be limited. We cannot continue with impunity to destroy the base upon which we live.

Warnings have been expressed by experts in many countries, and they have been constantly ignored by political and industrial leaders of the earth's strongest nations. While the United States has been at the forefront in recognizing endangered species, it consistently has found reasons for exceptions to the rules that protect those same species. Through commercial and other interests in Latin America, the United States has contributed to the further destruction of tropical forests, without emphasizing the kinds of population control that would have curbed persistent colonization. It also has continued to export pesticides and herbicides that have been banned in the parent country because of harm to the environment.

Ethnobotany is now in its beginning, just when the need for the field is most felt. Only since the late 1800s has the discipline been formalized. It might be said that the early years of the field as a formal area of study were the years of A. P. de Candolle, whose career included ethnobotany as an area of interest and the publication of *Origin of Cultivated Plants* (1886). De Candolle was one of the first botanists to recognize the value of archaeological plant remains, which he used in some of his interpretations of areas of origin. Unfortunately, in the nineteenth century, the world was still incompletely known, the full effects of population expansion had not yet been realized, and botanists of the era never knew of the destruction of natural resources that would occur in the twentieth century.

In the beginning, economic botany and ethnobotany arrived without much form and with a minimum of substance. Many plant specialists were uninterested in the fields, and some even sneered at those who had an interest in ethnobotany. In part, this attitude was justified, because some of the early ethnobotany consisted of the collection of lists of local plant names by anthropologists who then proceeded to make identifications on the basis of dictionary definitions of those names. For the most part, anthropologists were much more concerned with kinship and lineages and were willing to do their "ethnobotany" if it took a minimum of effort. At the beginning of the twentieth century, much archaeological effort was expended on the recovery of objects with value as art and with the descriptions of buildings as examples of architecture. The recovery of plant remains was not an aim of most excavators, and only the most spectacular plant preservations excited any interest.

Even at the beginning of my training in the early 1940s, the effects of population growth were not discussed, nor did there seem to be much urgency in collecting ethnobotanical information. One of my first field trips for the collection of ethnobotanical information with Richard Evans Schultes left me with the mistaken impression that there was more forest in Colombia than could ever be cut by human labor. It appeared that local customs would survive for millennia and that ethnobotanical information always would be available for collection. Return visits to Colombia have proven amply that humans with a will to survive and axes to aid them will clear-cut enormous acreages, including some of the most forbidding rain forest. I traveled a road in 1975 that I had first traveled in 1940, and the scenery was completely unrecognizable. The trees, supporting thousands of epiphytes, were all gone, and the soil surface was covered largely by herbaceous turf, except along fence rows and waterways. The forest lore had disintegrated with the destruction of the forest. Today, the human inhabitants work for wages and grow small plots primarily of maize and beans, which they supplement with canned foods and potato chips.

The Second World War was a particularly disturbing time for the inhabitants of combat zones and accessory areas where support operations were maintained. Where U.S. troops were involved, the most acculturating influences were candy bars, cigarettes, and chewing gum. The war also introduced foreign areas to enterprising young North Americans who, after the war, proceeded to spread transistor radios and plastic wares. The opening of worldwide communications after the war has probably hastened the demise of much ethnobotanical information. Only the Americas remain relatively unmarked by the passage of war, although local customs certainly succumbed to the influx of the outer world.

After the Second World War, both ethnobotany and paleoethnobotany began to change as it finally became clear to the scholarly community that a list of local names did not comprise the total botanical lore of a culture. At the same time, it became evident that plant remains which could be recovered from archaeological sites would furnish valuable information on prehistoric diets and commerce unobtainable from any other sources. It might be said that both ethnobotany and paleoethnobotany were maturing during the 1950s and 1960s. The process of maturation still proceeds, but possibly at a somewhat slower pace.

During the post–Second World War period, palynology became an additional field of interest in archaeology. Once the pollen analysts learned that pollen samples did not have to be extracted from peat bogs, the extraction of pollen from house floors, old fields, and other areas of archaeological interest opened a field of evidence to complement and support macroremains. Unfortunately, this area of study has two drawbacks that are difficult to surmount: the cost of pollen analysis is great, and the results may or may not be useful. With carbonized or dry macroremains, archaeologists can, at least, see that they have material to send to the analyst.

The latest tool to be added to the array of ethnobotanical studies is the analysis of phytoliths. Unlike palynology before it, phytolith analysis is being forwarded as a boon to archaeology before the critiques are in. The initial applications were in the detection of Old World grains (in Europe and Japan) and in unravelling the vegetational history of tropical grasslands (such as in the llanos of Venezuela). Until a substantial comparison collection of phytoliths from the world's vegetation is assembled, the method should be used with caution. It is an indication of the vitality still present in paleoethnobotany, but it may not develop with sufficient rapidity to save much archaeological information. On the other hand, the duplication of sizes and shapes of phytoliths in different species may render the evidence from phytoliths unusable.

With the maturation of ethnobotany has come the realization that people around the world have developed myriad ways to look at their portion of the world's vegetation. One of the first important studies to utilize the local knowledge of botanical lore, *Principles of Tzeltal Plant Classification* (Berlin et al. 1974) is an outstanding monograph that attempts to interpret the Tzeltal Maya understanding of the plants of their environment from their own viewpoint. Other studies have appeared, including *Huastec Mayan Ethnobotany* (Alcorn 1984), which provides much deeper insights into local understanding than ever was furnished by the old local name lists, but a true insight has yet to be achieved. From my own experience, I suggest that it will take several years of observation, which may have to be done by a member of a family in the community. As yet, none of the botanists or anthropologists has lived closely enough with a people to achieve a true understanding.

All the Native Americans with whom I have become well acquainted have impressed me with their feelings on conservation of natural resources. On the other hand, they are driven by economic reality and are forced into subsistence agriculture under less-than-optimum circumstances, considering environmental conservation. Many of the highland people in both North and South America are facing population pressures unlike any they ever have faced in historical times, and they are becoming changed by the culture with which they are surrounded. Just as Native American attitudes toward conservation have been modified, the basic knowledge of plant use has been modified. Today, as never before, the rate of change and the rate of loss of information is accelerating. Ethnobotanists are probably going to be too few in numbers and field work support to save much important information.

All this is occurring when the conservation of the world's germ plasm is becoming even more important. For the first time in the history of the human race, it is becoming possible to reintroduce ancient genetic material into modern, living tissue. One of the values of having as exact a knowledge of ancient plant use as possible is the importance of this ancient material in pinpointing old areas of variability as well as areas of cultivated plant origins. This capability must be focused more acutely in the immediate future to save as much archaeobotanical evidence as possible. Along with the potential for locating areas in which the present-day descendants of ancient cultivated plants maintain, in their genomes, quality resistance to many diseases and insects, the potential is being developed to recover part of the genetic code from dried, unburned plant tissue. This has already been explored in archaeozoology, where ancient genetic material has been transferred into laboratory organisms.

This later innovation will do little to resurrect species and information already lost. In the post–Second World War period, when new attitudes toward recording an understanding of the useful plants of other cultures was developed, little progress was made toward saving the world's plant resources. I am not convinced that massive cutting of forests is going to result in poisoning the world's atmosphere, because of the rapidity with which vegetation reclothes bare ground. I am convinced, however, that loss of the species of the original and/or old secondary forest is an irreparable loss. It would not be so threatening if only modest clearings were made in large forests stands. Propagules then would fill in the open spaces and, with time, once again would become reproductively functional individuals able to colonize new clearings and to evolve new forms.

Unfortunately, many clearing efforts have been done with large machines that eliminate vegetation from hundreds of hectares. Both distance and time act to prevent the establishment of seedlings of the original vegetational association. In the process, the local people are separated from their ethnobotanical base, which sets the stage for massive loss of ethnobotanical information. It is true that the bare surface is sometimes quickly reclothed in vegetation, but it is often alien vegetation with no ethnobotanical uses for the area's people. The magnitude of this loss is being repeated throughout the American tropics and elsewhere.

The resurrection of ancient plant variability through archaeological recovery is still in its infancy. For a few areas of the earth's surface, the recovery of information has proven to be complete enough to change previous theories of plant crop origin and distribution. For instance, for a long while Western archaeologists assumed that plant cultivation arose in the Nile Valley and that the principal Old World grains, wheat and barley, originated in the same area. Recovery of wheat and barley remains from southwestern Asia pointed to that region as the early area of cultivation, and collection of the putative ancestral forms from the natural vegetation of the uplands of the region confirmed this impression. From this reservoir of early germ plasm has come much valuable breeding stock to improve the quality and disease resistance of cultivated modern wheats and barleys and, consequently, to increase yield.

The only other part of the world where resurrection of ancient plant variability through archaeological recovery is becoming feasible is Mexico, where maize seems to have been domesticated. Unlike southwestern Asia, however, many critical areas of Mexico remain archaeologically unexplored. If they can be explored before "progress" has disturbed the sites, the evidence collected may well settle the arguments over the origin of maize and, at the same time, indicate fertile areas for the recovery of valuable genetic resources of maize from wild and residual early cultivated populations.

Evidence for many of the world's crops, such as eggplant, is not yet archaeologically known. Most of Africa and Asia remain to be surveyed for archaeological sites so that the most important can be excavated and the plant remains analyzed. In the Americas, much of Central America and lowland South America is insufficiently explored, and the archaeological evidence for the plants of those areas is being destroyed by road building, grading for construction, and even subsistence farming. Both Europe and temperate North America have a fair record of plant remains recovery, but neither of these continents is known to have been a major area for the origins of crops and thus potential sources for important germ plasm resources. Australia and New Zealand are unlikely to be pivotally important in holding critical germ plasm resources, because few of the world's crops have originated there.

Altogether, the prognosis for recovery of invaluable genetic resources from the world's vegetation is poor, and it is growing worse every day. Long before all the world's species of plants are recognized and scientifically described, they will have vanished in the face of expanding populations. The cultures of local people with tremendous reservoirs of ethnobotanical information will also disappear. Even the potentially valuable archaeological plant remains will lose their contexts and be destroyed before their potential has been realized. It seems improbable, but we are destroying our resource base so rapidly that we may find ourselves beyond the limits of recovery without recognizing those limits as we pass them—all for want mainly of adequate human population control.

A complete understanding of the world's ethnobotanical wealth was within grasp with our current capacity for storing and manipulating information. In a short while, it even would have been possible to store images as well as the language symbols which described them. Now the wealth is vanishing so rapidly that it is becoming a distant star on the horizon, perhaps never again to be approached.

180   C. E. SMITH

LITERATURE CITED

Alcorn, J. B. 1984. *Huastec Mayan Ethnobotany*. Austin: University of Texas Press.
Berlin, B., D. E. Breedlove, and P. H. Raven. 1974. *Principles of Tzeltal Plant Classification: An Introduction to the Botanical Ethnography of a Mayan-Speaking Community in Highland Chiapas*. New York: Academic Press.
de Candolle, A. 1886. *Origin of Cultivated Plants*. Trans. from French. London.
Theophrastus. 1916. *Enquiry into Plants, and Minor Works on Odours and Weather Signs*. Trans. A. Hort. London: William Heinemann.

PART 5

Ethnobotany in Education

Ethnobotanical research is entering modern educational avenues in many ways. The number of courses in schools and universities has been proliferating, and some universities have established fields of major studies or concentrations in ethnobotany. In South America, for example, particular interest is being given to ethnobotany by the several Colombian universities that teach the natural sciences and by government agencies, such as Inderena. Young Colombians are encouraged to study ethnobotany. In Ecuador, a Danish botanical group has organized an ethnobotanical section in the Botanical Department of the Universidad Católica, training ethnobotanists in this nation which still has many extant and unacculturated Indian tribes.

In North America, many university courses in ethnobotany are now being offered, and the number of doctoral dissertations based on ethnobotanical field studies has multiplied. In the United States, ethnobotanical studies have been carried out in sundry educational institutions, botanical gardens, and museums: the Botanical Museum of Harvard University, the Ethnobotanical Laboratory of the University of Michigan, the University of New Mexico, The New York Botanical Garden's Institute of Economic Botany, the Smithsonian Institution, and many others. In Canada, the University of Victoria and the Provincial Museum of British Columbia have been very actively pursuing ethnobotanical research.

The growing interest of the media, whether newspapers, magazines, or radio, and especially television and popular lectures, has been especially notable and increasing in the 1990s. Numerous popular books have been written, introducing ethnobotanical concerns to the general public. One excellent example is *Bark: The Formation, Characteristics, and Uses of Bark Around the World*, by Ghillean Prance, director of The Royal Botanic Gardens, Kew, England, and his wife, Dr. Anne Prance. By means of magnificent illustrations and an easy-to-read text, the Prances describe the diverse and creative ways in which bark has been put to use by many cultures.

In this section, two contributions address the methods and importance of ethnobotany in modern education. Edward F. Anderson, who has taught biology at the college level for more than 30 years, suggests that liberal arts institutions are the most promising source of well-trained ethnobotanists, who will not be bound by rigid, disciplinary restrictions. Judith G. Schmidt describes a technique she has used successfully for 15 years to present ethnobotanical material to students and the general public in a way that reaches the audience almost as though the viewers were in the field with the users of the plants.

Ethnobotany and the Liberal Arts

EDWARD F. ANDERSON

The liberal arts curriculum, central to the educational philosophy of many of the smaller, private colleges of the United States, continually has been criticized as irrelevant and inappropriate, especially for the sciences. In the 1980s and 1990s even stronger criticism has attacked the liberal arts as "too general" and as not professionally oriented. It is claimed that "generalists," or scientists broadly trained in a particular discipline and in other fields, are not prepared to function adequately in the highly technical scientific world of the late twentieth century.

The field of ethnobotany is a remarkable paradox in this continuing debate over the merits and disadvantages of the liberal arts curriculum in an academic world which so strongly emphasizes specialization. As a rapidly growing discipline in botany, ethnobotany is dependent on attracting broadly trained scientists with interests and expertise outside the natural sciences at a time when most disciplines of science are becoming more and more specialized. Not only do the narrower scientists fail to function well outside their own disciplines, but often they do not even recognize the value of seeing the whole picture as it really exists. Most people do not see things in the world in the very restricted, narrow way in which scientists see them. Ethnobotanists, however, must deal with the "whole picture," because they study plants within the context of human cultures. Thus, possessing expertise in only plant systematics is not enough, nor is extensive training or experience in anthropology or ethnography enough.

The discipline of ethnobotany demands breadth of knowledge in both the social and natural sciences. Such training is unusual in the academic world, especially in the large universities where students tend to study only in the narrow disciplines. Moreover, interaction of people from different disciplines, especially between the natural and social sciences, is rare; indeed, it is often frowned upon or even ridiculed. Yet such interactions are essential to ethnobotany. Good ethnobotanists are trained in one discipline (such as botany) but also have knowledge of, and training in, other disciplines (e.g., anthropology, linguistics, history). Ethnobotany is thus both relevant and timely when most scientific disciplines are becoming narrower and more restrictive in scientific training.

Another unfortunate phenomenon that exists within academic institutions is the existence of small, isolated intellectual "empires" in both the natural and social sciences. These empires are jealously protected from "outsiders," who, strictly speaking, are not trained in one of the departments or programs belonging to that empire. Some anthropologists, for example, criticize botanists who record and interpret the uses of plants in cultural activities, and some botanists, in turn, complain that anthropologists should not attempt to study the roles of plants in culture because frequently they do not understand these roles well enough to talk about them. Clearly, what is needed is a broader approach than one confined by disciplinary walls. The field of ethnobotany is a subject that cannot be dealt with from the narrow specialist's viewpoint so common in academic circles.

From where can we get broadly educated individuals who have been challenged to think about the "whole picture" and who can become well-trained ethnobotanists? Liberal arts institutions are a major source of this type of person. In actuality, ethnobotany is the liberal arts education put into practice, because it escapes the narrow confines of the disciplines so strongly advocated in many of our larger, profession-oriented universities. Even the concept of ethnobotany is difficult for some of these people to visualize—or to accept—since they have been trained in highly specialized fields, often to the exclusion of other disciplines. On the other hand, many colleges and universities with liberal arts programs offer, as part of the degree program, the exciting opportunity to combine majors and minors built upon required course work in a variety of disciplines. Two examples of interesting and relevant combinations of majors and minors would be botany (or biology)-anthropology or linguistics-botany. The liberal arts thus can educate individuals who will not be bound by rigid, disciplinary restrictions.

Ethnobotany has not yet become solidified into a "hard" discipline; thus, the ethnobotanist is not a prisoner of a particular discipline's jargon. Highly specialized scientists often are unable to communicate with people outside their discipline because they do not use the "right words." I have been frustrated and amused by anthropologists who have been critical of some of the ways I have described cultural activities in relation to plants because I did not use the "right words" (i.e., the appropriate jargon). Interestingly, when I did use the "right words," other critics then found fault with how I used them. Apparently, I had become inadvertently involved in disputes over jargon within the discipline.

Academicians often use so much jargon that they tend to talk only to people in their own discipline because it is "easier" to think and communicate in only these terms. Ethnobotanists, who have not yet become such prisoners of jargon, may help scientists communicate between disciplines by working outside the jargons involved. Furthermore, the ethnobotanist's mind can be exempted from the limitations often imposed by hard, narrow disciplines. Hopefully, ethnobotanists will not be forced to create ethnobotanical jargon in the process!

Ethnobotany is a field of activity that can, and must, go beyond simply forming bridges between specialists of different disciplines. After all, disciplines are based on artificial divisions of knowledge and experience for the convenience of study. Ethnobotany (and other so-called interdisciplinary approaches to knowledge) should remind us that knowledge, to be relevant at all, has to be set firmly into a "whole." At a time when science disciplines are proliferating and solidifying, ethnobotany has an increasingly significant role to play. Now, as validity is increasingly tested by focused narrowness and esotericism, ethnobotany can increasingly help to pull us back to a better balance in the scientific endeavor. Science in its specialization has become less and less consumable by the public and, therefore, perhaps less relevant or responsive to human needs. Ethnobotany can help alleviate that problem.

Cooperative projects using the expertise of several researchers can produce more comprehensive results in a much shorter period of time—and time is becoming a critical factor is many regions. My ethnobotanical research deals presently with the tribal people living in the highlands of mainland Southeast Asia, particularly in northern Thailand (see *Plants and People of the Golden Triangle*, Dioscorides Press, 1993). Historically, these people have practiced shifting cultivation, simply moving their villages to new sites as land is abandoned and new fields are cleared. Now, however, immigration from neighboring regions because of political unrest, high birth rates, and government reforestation programs has resulted in heavy pressures on available land, forcing these tribal people to cease moving periodically from one place to another. The government is expelling some illegal immigrants and relocating others from high-elevation watershed areas to less desirable lowland regions.

Illegal logging operations by the lowland Thai have exacerbated an already critical situation for the tribal people, who, for generations, have depended on the forests for food, medicinal plants, building materials, and ceremonial paraphernalia. The rapid disappearance of these forests with their plant and animal products is destroying significant portions of tribal cultures or is making the survival of these distinctive cultures ever more difficult.

Tribal people are forbidden to cut trees for timber to construct new houses, and suitable bamboo now often grows only at great distances from villages. Roads are being constructed, so that the former isolation of the tribes is gone, resulting in the extinction, or great modification, of traditional ways of life. Acculturation with the lowland Thai and Burmese cultures is a major threat to the tribal people; in fact, some groups are simply disappearing into the dominant cultures.

The time factor is so critical for some of these tribal groups that it is impossible for individual researchers to spend the necessary years carrying out research in their narrow, restricted fields. Anthropologists, for example, cannot wait for botanists to collect and identify the plants of a region before describing the cultural environment. Rather, botanists and anthropologists must work together, sharing information and expertise. Paul Lewis, an anthropologist and linguist who has long worked with the hill tribes, provided invaluable linguistic and anthropological data for my ethnobotanical studies of the medicinal plants of the Akha and Lahu. I, in turn, was able to provide him with scientific names for plants that he wanted to include in his Lahu-English and Akha-English dictionaries. As an ethnobotanist, I have been asked to work cooperatively with public health workers and tribal people in a refugee camp near the Laos-Thailand border. Herbal medicines are still widely used within the camp, and I have helped to identify and describe the plants used. Much can be learned through cooperative efforts such as these.

Bamboo has long played an integral role in the lives of Thailand's tribal people. A team of researchers is needed to study the types of bamboos available, their uses, and methods of propagation. This is an immediate need as bamboo is disappearing through shifting agriculture and reforestation projects, both of which exclude the perpetuation of bamboo thickets.

Another immediate need among Thailand's tribal people is for a joint research project involving anthropologists, ethnobotanists, zoologists, and agricultural specialists to study the traditional foods. The diet of these people is rapidly changing on account of the destruction of forests or the exclusion of the people from newly reforested zones, through the changing of their agricultural techniques, and with the introduction of new crops and animals by government, international aid, and missionary programs. Several studies suggest that there may be a deterioration of nutrition because of these programs.

A similar research project needs to be carried out by anthropologists, pharmacologists, and ethnobotanists on medicinal plants used by the tribal people. This knowledge cannot be permitted to disappear.

A challenge presently exists to research the effects of government reforestation programs. Ethnobotanists, foresters, ecologists, agriculture specialists, anthropologists, and economists must work with both Thai and tribal groups to avoid making political or economic decisions at the expense of the people living in the areas. Any land use program should involve a thorough consideration of both the culture involved and other realities. Can we recognize and identify native plants of cultural significance that could be either conserved or reintroduced? This is the type of question that broadly trained ethnobotanists can consider intelligently because of their understanding of the problems, their sensitivity to the people involved, as well as their knowledge of the plants and habitat under consideration.

My experiences in Thailand have demonstrated clearly that individuals with broad training, such as is provided through the liberal arts curriculum, can deal with many of the complex relationships simply because they know where to look for information. Once they find information, such individuals can understand it, owing to their prior educational experience in disciplines outside a narrow field of specialization. Even so, the most broadly trained ethnobotanists still need help from specialists. For example, although I can deal readily with the scientific names of plants and can describe accurately how I see them used, I am unable to work with the vernacular names and therefore have called upon trained linguists to commit to writing the tribal plant names that have been recorded orally on tape. I have also worked with historians, especially those dealing with oral histories and traditions; the information they have provided has helped me to interpret the changes that have occurred in tribal cultures due to migration, political activities, and interactions with outsiders. Cultural anthropologists have been helpful when I have had to deal with cultural events, such as a funeral ceremony or an ancestor offering, which involve the use of certain plants. My interactions with researchers trained in fields other than my own have been positive and exciting. All of us have benefitted from these consultations and mutually shared experiences, and clearly the advance of knowledge has been more rapid.

Ethnobotany also fulfills another important function in the liberal arts curriculum by demonstrating in courses, both for the major and the nonmajor, the interrelationships of various disciplines. Nonscientists within the liberal arts often appreciate the work of ethnobotanists more than do other scientists. For many years courses in economic botany and cultivated plants have been taught, but more recently the emphasis has often been ethnobotanical. For more than 10 years I taught an ethnobotany course on plants in human societies. This course included field methodology, the importance of preserving germ plasm pools, the origins of agriculture, regions of the world from which food plants came, and actual studies of plants used by various cultural groups. Considerable emphasis was placed on medicinal plants, including various human views of disease and the causes of illness. Student response was very favorable because it was easy for them to see the relevance and the interrelatedness of the field to other disciplines in the liberal arts.

Ethnobotany is a highly relevant field, especially in regions where cultures are changing rapidly. Researchers no longer can afford to proceed independently of one another, for time has become critical. Cooperative efforts from broadly trained researchers and from those with certain specialties are essential. Such teamwork may be a new idea to some, but to others it already has demonstrated that productive and rapid research, both in ethnobotany and in other disciplines, is possible. Furthermore, this teamwork also shows clearly that the liberal arts education can be an important vehicle by which broadly trained, thinking individuals can make significant contributions to science.

A Unique Visual Method of Sharing Ethnobotany with General Audiences

JUDITH GRACE SCHMIDT

Ethnobotanists combine, in varying degrees, the expertise of at least two disciplines, usually anthropology and botany. To these two, I have added photography, calling the triple combination "visual ethnobotany." This novel technique has enabled me to present ethnobotanical material to general audiences.

Some scientists may not be sure of the validity of my ethnobotanical focus. They may hear, for instance, of my use of common names for plants or other language that can be understood readily by laypeople. Does this make my work unprofessional or insufficiently scientific? I contend that because ethnobotanists obtain knowledge not possessed by most of the world's population, all of whom are increasingly affected by the rapidly deteriorating natural environment, ethnobotanists have a responsibility to communicate that knowledge to a wider audience.

The second half of the twentieth century has provided ethnobotanists with a remarkable opportunity for sharing their knowledge with nonscientists. The climate of thought is opening around the world. With this consciousness stirring, laypeople are glimpsing the scope of the environmental crisis. It is essential that they also understand details of its causes and solutions. Ethnobotanists have an unprecedented opportunity to forward this understanding by communicating with a wider audience.

My conviction of the responsibility and necessity of communicating ethnobotanical information with the general public is based on fifteen years of experimentation with this communication goal. During this time, my ethnobotanical studies have been characterized by four unique aspects.

The first unique aspect is the addition of photography to the usual scientific disciplines. Although this may not sound uncommon since many ethnobotanists take photographs of people and plants to illustrate aspects of their ethnobotanical research, the use of photography as a principal recording medium over many consecutive years of field work *is* uncommon. My reason for approaching ethnobotanical studies thus began with the motive for my studies: to share with a wide range of people the distilled results of ethnobotanical research.

The most suitable medium I have found for sharing what I have learned as an ethnobotanist is the slide presentation, and one of the kinds of ethnobotanical information that can best be communicated via this medium is the use of plants in other cultures. Not only are these uses interesting and important to nontechnical audiences, more so than I at first dared to anticipate, but they have proved an effective entrée to my underlying concern: to encourage a better appreciation and understanding of another culture in our midst, and to encourage the practice of a better land ethic, sorely needed today.

The second unique aspect of my ethnobotanical research is my choice of location, the so-called woodland Indians of northeastern North America. Although the words *woodland* and *Indian* are both misnomers of European designation, I use them in my presentations because they are generally recognized by my audiences. Frankly, using them offers me an opportunity to point out a misrepresentation of the indigenous people through the very words by which they are commonly called! *Woodland* describes the naturally predominant plant cover in the region of my study. My contacts are among persons of several distinct nations, including Ojibwa, Passamaquoddy, Maliseet, Mohawk, and Montagnals. For botanical purposes, these people are bound by a similar woodland flora.

What is unique about concentrating on northeastern North Americans is that today, as historically, the preponderance of ethnobotanical research is done outside the United States and Canada. When research is carried out among the indigenous people of this continent, it continues to be restricted mainly to the Southwest or Northwest or even to the romanticized horse riders of the Plains. Native Americans of the Northeast were decimated more than 300 years ago due to diseases introduced by European settlers and by warfare with these invaders. Almost all the remnant indigenous people were forcibly transplanted west of the Mississippi River by the mid 1800s. Most people living in the United States in the twentieth century do not even realize that any Native Americans are left in the Northeast!

The third unique aspect of my ethnobotanical research is the botanical component, which shows two differences from usual work in this interdisciplinary field. First, the geographic region in which I work supports a comparatively limited flora—some 3500 species of higher plants compared with some 80,000 species in the Amazon, for example. Second, my botanical focus is specifically limited to the uses of wild plants for indigenous technology, food, and ceremony.

Although many studies of indigenous uses of wild plants have focused exclusively, or at least primarily, on medicinal and hallucinogenic practices, it should be borne in mind that the definition of *medicine* in these societies is far different from the definition in so-called Western societies. Because the parts of indigenous societies are interrelated, in contrast to the parts of Western societies, Native Americans, for example, do not understand how I can try to separate medicine from food. They claim that good food is medicine, and I understand what they mean.

I have purposely excluded from my research cultigens and medicinal or hallucinogenic plants so that I could concentrate on less known uses of wild plants. This approach has satisfied my own interest more closely and, as a bonus, has made the resulting material particularly appropriate for nontechnical audiences, including schools, museums, universities, and clubs, because it does not assume the wide and specific, and often highly technical, body of knowledge required in discussions of medicinal healing. Furthermore, because medicinal plants often include species with potentially poisonous substances that are safe only if they are correctly prepared and administered, by avoiding explicit reference to these plants I have also avoided opening the door to possible misunderstanding and misuse of this knowledge by uninitiated listeners. While excluding specifics of medicines or hallucinogens in my presentation, I am at the same time able to share the general style of indigenous healing practice, showing a respect for this knowledge. Members of my audiences often are surprised to learn of the continuance of healing practices by some Native Americans in the Northeast today.

The fourth unique aspect of my ethnobotanical research is the slide show, which, perhaps, has been the most significant departure from the established norm. It is my chosen implement, a bridge of understanding designed to minimize the wide gap commonly existing between scientists and laypeople. Through color photography, the rather neglected and much acculturated woodland Indians have been introduced to an audience unaware of their presence in our society! Furthermore, some traditional and contemporary uses of wild plants by Native Americans have been brought to life before audiences unfamiliar with the abundance of traditional knowledge and indeed the still-vital importance of certain economic plants in our wild flora to these groups of Native Americans.

The slides chosen for my initial presentation represent a topical and visual balance between cultural features involving people and many of the plant species utilized. Particularly selected are many examples of so-called weed species, which are familiar to all gardeners. These species illustrate the difference between destroying plants as a nuisance or using them as an asset. Photographs of such species noticeably impact people's ideas regarding the potential usefulness of the wild species around us.

Sequences of slides show the gathering and preparing of plants for food, such as salads, desserts, and beverages. Others illustrate species used during ceremonies that have persisted among northeastern Native Americans. Two sequences on technology follow steps for making ash splint baskets and birch bark storage containers.

To further increase the audience's receptivity to the message and to capture and hold its interest, four projectors are employed for the show, together with a dissolve unit and a programmer. Two projectors are connected to the dissolve unit, permitting smooth dissolves in some slide changes, as a relief from sharp cuts. There are also lap dissolves and a variety of other special fade effects in the show. The other two projectors are used to display vertical shots on the screen side by side. This allows seasonal parallelism of plant species. It also increases the number of photographs that can be shown in a given time period—up to 450 slides in an hour without apparent haste. This set up provides a range of images more akin to motion pictures than that offered by most slide shows. These techniques have successfully kept audiences alert and attentive.

The carefully verified information in the slide show has helped dispel popular misconceptions about Native Americans, some of which have been engendered and reinforced by centuries of misinformation. Two brief examples illustrate this point. The first occurred at an anthropological conference, where I sat with an Ojibwa elder who is a practicing healer and a member of the tribe's medicine society, the Midewiwin. We listened to a well-delivered paper which incorporated several previously written accounts of the Midewiwin. The anthropologist repeated, once more, a number of incorrect concepts recorded years earlier, which evidently had not been verified with the people whose culture was being "explained."

A second and similar example of a misconception that has been reinforced by centuries of misinformation concerns Native American religions, which have been dismissed as "of the devil," mere pagan pantheism. My study reveals that their spirituality can be deep and insightful, and can lead to a life practice that is sensitive to all aspects of the natural environment and its continued preservation.

To avoid perpetuating misinformation, my primary presentation, called *Respect For Life*, was shown to the Native Americans whose cultures are illustrated before it was shared publicly. By visual documentation the presentation replaces some of the recorded inaccuracies about this continent's original people with a more correct and sensitive appreciation of the part they played and of their dynamic and evolving cultures.

During my years of travel with this presentation, it has been evident that audiences were indeed seeing both Native Americans and the plants around them in a new light. Typical remarks from the audience included the following: "You have opened my thought. I will never think of Indians the same way again!" and "Just think of all the useful plants I have destroyed in my garden!" A high school student once claimed: "It has made me realize I must not waste." At the Oregon Archaeological Society: "Fine oral history . . . scientific depth . . . superb photography." A Passamaquoddy leader said: "Your obvious and sincere interest is impressive. You have presented a positive view of the people." These comments reflect the success of the slide presentation technique in communicating data from my scientific research in an accessible context and in readily understandable terms. In this, I believe, lies the particular social merit of the program.

In general, any public awareness of North America's original people has tended to leave them in the past, as if they no longer existed. On learning of my studies of Native Americans, a highly intelligent analyst for a leading international newspaper said: "Well, they're all assimilated now, aren't they?" My aim has been to work with Native Americans in a context that would help lift our view of them to that of a living cultural tradition. Certainly, it is important to relate contemporary ethnobotany to the historic past; many if not most modern practices have come down from customs of the past. Nonetheless, we must admit that Native Americans and their cultures are not merely anachronisms from the past but vibrant elements of the present.

By becoming personally familiar with some Native Americans during fifteen seasons of field studies, I have been able to bring to my presentation an authority, a reality, which would not be present in the same degree were my research based solely on existing historical writings. During extensive interaction with these different cultures, my experiences have included sharing in practices not easily found in the late twentieth century, including morning and evening prayer, naming ceremonies, sweat lodge ceremonies, an Ojibwa wedding, and sun dances on the Plains.

Sharing ceremonies and everyday life has enhanced my own understanding and appreciation of the very concepts that I wish to communicate about Native Americans—not only their social practices but also their way of thinking. This brings into focus some of the ways in which present-day Native Americans differ from people of the European-based mainstream of our present population. It can benefit us all to incorporate some Native American values into our own societal views and practices, particularly in rescuing the endangered natural environment.

Because of its focus on a geographically less researched region, and on less well catalogued wild plant uses, my study has proved important in adding to the collected data on Native American cultures and their utilization of the flora. As these are cultures from which so much has already been lost, or at least submerged by hostile laws forbidding traditional practices, they deserve this type of publicity. The photographic record will maintain a specific, long-term scientific value. It is ethnobotanical conservation, stored against the time of further acculturation of the indigenous cultures. Because this acculturation may mean the erosion of knowledge acquired over many hundreds of years, it is vital to save what is known of Native American cultures, not only for a wider audience but for future generations of Native Americans.

For nine years *Respect For Life* has been presented as described above for the full hour as well as in shortened versions for children as young as first graders. I have also developed other slide presentations, including *The Lakota Sioux Today*, resulting from visits over ten years, to Rosebud, South Dakota (beyond the woodland region). This show has been timely as a sequel to the film *Dances With Wolves*, which left many audiences in the 1990s wondering if the Lakota even existed. To my original show, I have added *China: A Village Experience* based on three weeks of travel by bicycle in ethnically varied southern villages of Yunnan Province (following an International Congress of Ethnobiology in Kunming). This presentation gives a rare view of a corner of a vast country and reinforces indigenous knowledge of the vital importance and utilization of varied plants—principally bamboo species and upland rice. Although my primary research remains among northeastern Native Americans, I have found that "visual ethnobotany" works equally well for sharing other ethnobotanical learning.

As a freelance visual ethnobotanist I have written this chapter from the experience of successfully crossing the bridge from scientific research to a general audience. From first-hand experience with many audiences, I believe my work fulfills my responsibility and opportunity of reaching out with a wide sharing of vital knowledge. Additionally, in sharing science with general audiences through the medium of photography, my work has increased the recognition of the value of the science of ethnobotany itself, a value that relates to the continued well-being of humankind.

The way is available for ethnobotanists to cross the communications bridge to the layperson, each in his or her own unique way, to open a multitude of doors to cultural insights and ideas. These experiences, in turn, engender respect, sensitivity, and an opening of thought to the importance of plants and the values and validity of other cultures in today's telescoping and fast-changing world. Indeed, similar communication is a necessity in the full practice of ethnobotany today.

PART 6

Ethnobotanical Contributions to General Botany, Crop Improvement, and Ecology

---

Almost all domesticated plants originally were employed in primitive societies, usually in prehistoric times and in the wild state. Ethnobotanical studies, particularly archaeoethnobotanical information, can have a significant bearing on the origin and history of plant domestication. In fact, ethnobotanical knowledge among primitive people on biodiversity can still be of great value in searching out ecotypes or close relatives of cultivated plants as well as of wild species.

In many regions, this knowledge is rapidly being lost as a result of acculturation, just as literally hundreds of ecotypes, often occurring endemically, are being annihilated with extreme and uncontrolled deforestation, especially in the wet tropics. The future success of this valuable aspect of biodiversity lies in environmental and ethnobotanical conservation.

The origins of agriculture, about 10,000 years ago in the Old World and 7000 years in the New World, began with the discovery that certain wild plants encroached upon the dwelling places of humans and grew spontaneously. Undoubtedly, some human being, more observant than most of the tribe, discovered, while eating raw seeds and nuts, that if dropped, those seeds and nuts took root and grew. Although some origins of agriculture might have come about from different circumstances, it seems probable that this kind of ancient ethnobotanical observation was basic to most instances leading to early discoveries of agriculture.

Today, there is an extremely important aspect tangentially concerned with modern agriculture and the improvement of crop plants: biodiversity. Many species, if not most, have developed ecotypes which often are extremely local strains. Their great value often lies in characteristics, such as resistance to cold and disease, tolerance of heavy rainfall or of drought, early or late maturation, and a host of other features, which, through hybridization with the developed modern agricultural crop ecotype might result in descendants with desirable new characteristics. Even wild species related to the crop species may be sought for the same purpose. This utilization of wild ecotypes of the same or related species has been very successful in numerous crop plants, including certain cereals and potatoes.

It is often, particularly in the exuberant rain forest vegetation, difficult to locate these ecotypical variants. Consultation with indigenous people, who usually have deep and intimate familiarity with their ambient floras, can be of extraordinary value to botanists seeking hidden ecotypes. Knowledge of indigenous ethnobotanical information may provide leads for research or even solutions to some botanical problems.

Many investigators of the origin, history, and development of crop plants have availed themselves of ethnobotanical knowledge. The three contributors to this section are active researchers in the improvement of cultivated plants. Michael J. Balick addresses an important activity for modern ethnobotanists, namely, the collection of germ plasm of potentially useful plants. Charles B. Heiser reviews the changes in and the origins of domesticated plants from the time of Charles Darwin up to the present, emphasizing the benefits of such studies to the welfare of the human race, especially as they relate to feeding a growing population. Garrison Wilkes explores the ethnobotany of artificial selection, long an ignored aspect of plant domestication but one that is more human-centered than environmentally influenced.

Ethnobotany and Plant Germ Plasm

MICHAEL J. BALICK

The reduction in the diversity of life on this planet is increasingly being viewed as one of the major threats to long-term human existence. As natural habitats are radically modified or destroyed, the rich diversity of plant and animal life supported by these habitats disappears. In the case of the tropical rain forest habitat, much of the life that it contains has never been discovered, described, inventoried, or evaluated as to its utility.

Several strategies have been suggested to help maintain some of the biological richness found in natural areas. The most logical way is to preserve representative areas of the habitats, setting aside the acreage necessary to sustain a balanced ecosystem capable of supporting life. This strategy can be implemented through the creation of parks and biological reserves. From a realistic viewpoint, however, it would appear a most difficult task to create and protect sites sufficient to preserve all the known biological diversity. Often, socio-economic or political considerations are an obstacle to this effort; for example, as land suitable for agriculture or hunting/gathering disappears, people in search of sustenance often invade and destroy areas that are set aside for protection.

One way to ensure the survival of selected species is to "salvage" them from endangered habitats, preserving them in living collections—zoos, botanical gardens, germ plasm banks, and so forth. It is clear that both *in situ* preservation and the collection of individuals for *ex situ* preservation are important ways of helping to preserve the shrinking biological base. This chapter addresses an important activity of today's ethnobotanist: the collection of germ plasm of potentially useful plants. Ethnobotanical studies yield a wealth of information about a variety of plants and their usefulness to people. Plants are used as medicine, food, fuel, fiber, construction material, ornament, religious objects, and for many other purposes. Any germ plasm obtained during ethnobotanical studies can be put in one of three categories.

The first category of germ plasm includes the native varieties of cultivated crop plants used elsewhere in commercial agricultural production. At present, many of the major crop plants have a limited genetic base, as these have been developed through a series of selections that emphasize yield often at the expense of insect- or disease-resistance, environmental tolerance, multiple use, or other factors. This limitation has led to a narrowing of the genetic pedigree of crop plants in general. For example, 69 percent of the acreage devoted to sweet potato (*Ipomoea batatas*) production in the United States is dedicated to the cultivation of a single variety, 'Centennial'. This variety is susceptible to a number of serious diseases, including root knot, soil rot, stem rot, black rot, scruf, and sclerotial blight (Committee on Genetic Vulnerability of Major Crops, Agricultural Board 1972). Many other varieties of sweet potato are cultivated by farmers around the world, some of which are resistant to a number of these diseases.

In the recent past, the danger of depending on crop plants with such a limited genetic foundation has become apparent to many people. Organizations such as the International Board for Plant Genetic Resources (IBPGR) have taken global responsibility for the coordination of germ plasm collection and repository efforts, with the aim of helping to resolve some of the loss of genetic diversity of the major economic plants. A great number of accessions of germ plasm of the more traditional crops have been collected on a priority basis and are being curated in germ plasm banks around the world.

Ethnobotanists often have the opportunity to collect valuable genetic material, because many work in remote areas of the tropics and subtropics among people whose cultures are quickly eroding. Thus, the ethnobotanist has a chance to document and collect germ plasm of a variety of crop plants that otherwise would be given up as people shift towards a genetically more homogeneous crop base. The professional germ plasm collector, often a trained agronomist, has little chance of getting to know cultures in the way that the ethnobotanist can, and as such, the ethnobotanist is in a position to make an important contribution to gene preservation efforts.

There is an urgent need for this level of activity. As indigenous cultures come in contact with "civilization," they often give up their own varieties of plants that have been cultivated for hundreds of years in exchange for the opportunity to enter the market economy of the larger society that is engulfing their own. For example, the Guajajará and Apinayé Indians of northeastern Brazil have become much more dependent on the cultivation of commercial varieties of *Manihot esculenta* (cassava) for their livelihood. At the same time, they no longer cultivate certain more traditional crops to the degree that they did in the past.

Relative to the effort devoted to cassava, cultivation of certain species of beans, tomatoes, peppers, peanuts, and other more traditional crops by these Indians is minimal, as is extraction of babassú (*Orbignya phalerata*) kernels. Species of *Crescentia*, once widely grown to make drinking vessels, have become a mere curiosity now that a cash income affords these people the "luxury" of purchasing plastic or tin cups and vessels. This type of experience is being repeated in many thousands of places around the world, resulting in irreplaceable genetic losses.

The second category of germ plasm material includes the identification and collection of wild relatives of the more commonly cultivated plants. Traditional breeding practices have not always given due consideration to stress tolerance, disease resistance, and the production of multiple-use crop plants. Wild relatives of these crops have been able to survive without cultivation because they possess some of the desirable traits that are now considered important in a reduced-input agricultural scenario.

Ethnobotanists working in the great centers of diversity of crop plants have the chance to collect crop plant relatives as a part of their research activities. For example, South America is the center of diversity for crops such as rubber, cassava, cocoa, potato, tomato, and many other commercially important species. Wild relatives of these species are valuable as tools for crop improvement and increased adaptability in a variety of environments. There are many concrete examples of this value. A species of wild tomato, *Lycopersicon chmielewskii* from the Andes of Peru, has provided plant breeders with the genes necessary for increasing soluble solids of tomatoes used in the processing industry, a contribution worth millions of dollars annually (Prescott-Allen and Prescott-Allen 1986). Four hundred species belong to the genus *Ipomoea*; some of them might possess genes resistant to the previously mentioned diseases of the sweet potato, *I. batatas*.

The third category of germ plasm material includes plants not yet in the economic system. Ethnobotanists can make extensive contributions to the identification of these new plants. At each study site, often there are many useful plants that are little known outside the particular country, geographic region, or sometimes even the local village. Once the local people are sensitive to the interest of the ethnobotanist in these kinds of plants, they are often eager to display and discuss local varieties.

Ethnobotanists have documented hundreds of useful plants in many studies published to date, and some of these are potential candidates for broader use elsewhere. Perhaps from the agronomist's perspective, many of these species would not be of great commercial value, but one also must consider their potential role in the subsistence agricultural lifestyles prevalent in much of the world today. It is, therefore, almost impossible to assess the economic potential of many of the plants in this third category. Nevertheless, the recognition, study, collection, and distribution of these less known species may be one of the most valuable contributions of the ethnobotanist over the long term. Unfortunately, the study of underutilized species is not often of great interest to the major germ plasm banks. As a consequence, plants in this third category face an extremely high risk of genetic erosion and extinction.

Germ plasm collection is of no value without an organized infrastructure to receive, evaluate, and properly curate collected material. Some of the early collecting programs in which I have participated were unsuccessful because an adequate infrastructure for handling the material had not been set up prior to the initiation of the collection phase. The collectors were expected to develop this infrastructure upon their return to home bases. It is important, however, to know in advance of research and collection what material is of interest and where it is to be sent for germ plasm banking. A number of institutions in both developed and developing countries now are equipped properly to handle germ plasm, and they usually welcome selected material from collectors.

An additional comment is warranted at this point. If a good local facility exists for handling germ plasm, there is no need for the collector to take material outside national boundaries to accomplish the goal of genetic conservation. Germ plasm collection efforts of the Institute of Economic Botany (IEB) of The New York Botanical Garden, for example, have collaborated with many institutions in various countries. The IEB aims to support local gene banks by providing them whenever possible with material.

Indeed, in this way, local controls on transport and export of genetic material are respected while achieving the desired end—preservation. Figure 1 shows some of the areas in which the IEB staff has collected germ plasm since 1981, and the repositories to which this material has been sent. These repositories are always at other institutions, as the IEB has no germ plasm banking facilities of its own, neither in New York nor elsewhere.

Ethnobotanists have many opportunities to make important practical contributions towards improving the utilization of plants. This work is especially urgent in view of the massive on-going destruction of the world's natural habitats and the cultures that depend on these environments for their existence. This chapter is not the vehicle for providing the technical details for making proper germ plasm collections, as techniques vary among plant families. Information on collection, storage, shipping procedures, and locations of germ plasm banks is available from groups such as the Food and Agriculture Organization of the United Nations, the International Board for Plant Genetic Resources, the United States Department of Agriculture, as well as local agricultural research institutions in many countries.

In this chapter we have considered the need for the ethnobotanist to become acquainted with the importance of germ plasm collecting. Ethnobotanists could view collection of germ plasm material as a hindrance to their main goal of documenting plant use and taxonomic studies; indeed, the proper collection, documentation, and shipment of living material in compliance with local laws takes time and effort. On the other hand, the opportunity to work with genetic resources adds another dimension of intellectual and

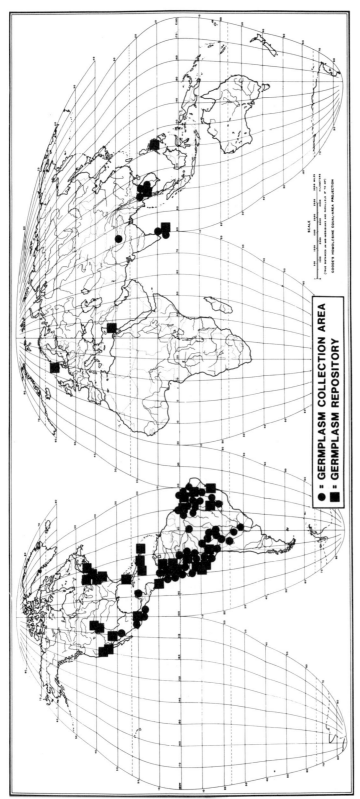

Figure 1. Selected germ plasm collections by the staff of the Institute of Economic Botany, 1982–1985.

practical interest to ethnobotanical work; it can also be a new source of funds for supporting field work and is certainly a wonderful opportunity to rescue and preserve a portion of the cultural wealth of some of the indigenous groups with which we work.

ACKNOWLEDGMENTS

I would like to thank Brian Boom, Steven King, and Christine Padoch for their valuable comments on an earlier version of this manuscript.

LITERATURE CITED

Committee on Genetic Vulnerability of Major Crops, Agricultural Board. 1972. *Genetic Vulnerability of Major Crops*. Washington, DC: National Academy of Sciences.
Prescott-Allen, C., and R. Prescott-Allen. 1986. *The First Resource*. New Haven, CT: Yale University Press.

The Ethnobotany of Domesticated Plants

CHARLES B. HEISER

Ethnobotany may be defined as the study of plants in relation to people, and, as such, deals with the study of both wild and domesticated plants. Although perhaps not ordinarily thought of as part of ethnobotany, the changes in and the origins of domesticated plants fit under a broad definition of the subject. In this chapter, certain aspects of domesticated plants will be emphasized.

Although he was not particularly concerned with the origin of domesticated plants, Charles Darwin contributed to the field. In *The Variation of Animals and Plants Under Domestication* (1868), he wanted to show "the amount and nature of the changes which animals and plants have undergone whilst under man's dominion." Darwin did not do this to show the relation between plants and people, but rather to support his theory of natural selection. In so doing, however, he brought one area of ethnobotany into focus. The human race, in its selection, Darwin informed us, "may have been trying an experiment on a gigantic scale; and it follows that the principles of domestication are important for us." Elsewhere we learn that

> although man did not cause variability and cannot even prevent it, he can select, preserve, and accumulate the variations given to him by the hand of nature almost in any way which he chooses; and thus can certainly produce a great result.... Selection [by man] may be followed either methodically and intentionally, or unconsciously and unintentionally.... We can further understand how it is that domestic races of plants often exhibit an abnormal character, as compared with natural species; for they have been modified not for their own benefit, but for that of man.

More than 150 pages of Darwin's volume are devoted to domesticated plants, including both food plants and ornamentals, and in group after group we learn that those parts of the plants most valuable to people present the greatest amount of variation. Darwin, of course, did not understand the cause of variation (see Müntzing 1959 for a consideration of Darwin's contributions in the light of modern genetics). The changes of plants under human influence have continued to be a subject of interest since Darwin's time (see Harlan 1975; Pickersgill and Heiser 1976; Schwanitz 1966).

It was Alphonse de Candolle who, first in a chapter in his *Géographie Botanique Raisonnée* in 1855, and then more fully in his *Origine des Plantes Cultivées* in 1883, developed the subject of domesticated plant origins. In his preface to the latter work, he pointed out that "the knowledge of the origin of cultivated plants is interesting to agriculturists, to botanists and even to historians and philosophers concerned with the dawnings of civilization." He treated 249 species and was able to point to a progenitor for nearly all of them. We now know, of course, that he was not always correct. Darlington (1963) found de Candolle's interpretations "over-comforting by our present standards" because of his "Linnean simplicity," but others agreed that, considering the knowledge of the day, it was an excellent assessment (Harlan 1961; Heiser 1975; Smith 1968, 1969).

In the twentieth century, the study of origins was taken up by N. I. Vavilov, who dedicated one of his principal works (1926) to the memory of de Candolle. Although Vavilov's proposed centers of origin and methods of determining them have not withstood the test of time (Harlan 1975), his contributions to the field were nevertheless enormous. Today, studies of the origins of domesticated plants are still a very active part of ethnobotany. No one person has attempted works of the magnitude of those of de Candolle and Vavilov. Probably the nearest approach is the multiauthored work edited by Simmonds (1976).

De Candolle devoted a chapter of his book to his methods of discovering or proving the origin of species. These methods, which came from botany, archeology, paleontology, history, and philology, are still applicable, although today, by far the greatest contributions come from botany and archaeology. Systematics, which furnished virtually all de Candolle's "botany," has continued to be the logical starting point in such studies. In the 1940s, cytogenetics began making many significant contributions and continues to do so. More recently, molecular studies have begun to provide important evidence (Doebley 1989; Bretting 1990).

One might think that in a hundred years of study the progenitors of all our domesticated plants would have been identified, but some of them have yet to be studied in any detail, and ancestors of others, including some of our major crops, have proved elusive. For example, there is still no agreement as to the donor of the B genome to bread wheat. Although many researchers now maintain that teosinte is the progenitor of maize (Iltis 1983), some still hold other views (Mangelsdorf 1983). Then, too, progenitors of some minor crops, such as rocoto (*Capsicum pubescens*) and naranjilla (*Solanum quitoense*), which have been the subject of my own studies, have yet to be revealed.

The reasons for the failure to identify progenitors were well known to Darwin and de Candolle. Darwin (1868) wrote that "not a few botanists believe that several of our anciently cultivated plants have been so profoundly modified that it is not possible now to recognize their aboriginal parent forms," and "equally perplexing are the doubts that some of them are descended from one species or from several inextricably commingled by crossing and variation." If, indeed, teosinte is the progenitor of maize, one of the reasons it took so long to be recognized as such is the profound morphological differences between the two plants. We also now know that many of our polyploid crop plants—bread wheat, for example—resulted from hybridization of diploid species. De Candolle (1883) pointed out the difficulties of distinguishing adventive plants, descended from the cultivated plant, from truly wild progenitors. He also raised the possibility that the ancestor of the domesticated plant might be extinct.

In view of the difficulties of tracing progenitors, why then are such studies still popular today? First of all, if the progenitor is not extinct, its identification still may be possible with more careful study and with the aid of newer tools. Second, the innate fascination of studies of plant origins appeals to many investigators.

As de Candolle indicated, such studies are of interest to many people, but other benefits may accrue from them as well. Gray (1868) pointed out that Darwin's work was a "perfect treasury . . . which governs the production, improvement, and preservation of breeds and races," so that his work will be particularly valuable to the intelligent agriculturists. Vavilov (1935, 1940), who was well aware of the value of his own studies on the systematics and origin of cultivated plants, wrote that such studies give the "plant breeder all of his necessary construction materials." This is still true; new germ plasm of potential importance is likely to turn up, particularly when field work is undertaken in connection with such studies. A modern example is provided by the discovery of *Zea diploperennis* (Iltis et al. 1979), a plant which, had Iltis not been interested in the origin of maize, might not have been discovered.

The importance of ethnobotanic studies to the welfare of the human race is obvious, and no more so than with the study of domesticated food plants. There is a growing concern as to how the rapidly increasing world population is to be fed in the future. Making our present food plants more productive, as well as finding new ones, never has been more important.

LITERATURE CITED

Bretting, P. D., ed. 1990. New perspectives on the origin and evolution of domesticated plants. *Economic Botany* 44 (supplement 3): 1–116.

Darlington, C. D. 1963. *Chromosome Botany and the Origin of Cultivated Plants.* New York: Hafner.

Darwin, C. 1868. *The Variation of Animals and Plants Under Domestication.* London. 2nd ed., 1875.

de Candolle, A. 1855. *Géographie botanique raisonnée.* Paris.

———. 1883. *Origine des Plantes Cultivées.* Paris.

———. 1959. *Origin of Cultivated Plants.* Rpt. of 1st American ed. New York: Hafner.

Doebley, J. F. 1989. Isozymic evidence and the evolution of crop plants. In *Isozymes in Plant Biology.* Eds. D. Soltis and P. Soltis. Portland, OR: Dioscorides Press. 165–191.

Gray, A. 1868. Preface to the American edition of Darwin's *The Variation of Animals and Plants Under Domestication.* New York: Orange Judd.

Harlan, J. R. 1961. Geographic origin of plants useful to agriculture. In *Germ Plasm Resources.* Ed. R. E. Hodgson. Washington, DC: American Association for the Advancement of Science. 3–19.

———. 1975. *Crops and Man.* Madison, WI: American Society of Agronomy.

Heiser, C. B. 1975. The application of biochemical systematics and genetics to the study of the origin of cultivated plants—some comments. *Torrey Botanical Club Bulletin* 102: 301–306.

Iltis, H. 1983. From teosinte to maize: The catastrophic sexual transmutation. *Science* 222: 886–894.

Iltis, H., J. F. Doebley, R. Gusman, M. Pazy, and B. Pazy. 1979. *Zea diploperennis* (Gramineae): A new teosinte from Mexico. *Science* 203: 186–188.

Mangelsdorf, P. C. 1983. The mystery of corn: New perspectives. *Proc. Amer. Phil. Soc.* 127: 215–247.

Müntzing, A. 1959. Darwin's view on variation under domestication in the light of present-day knowledge. *Proc. Amer. Phil. Soc.* 103: 190–220.

Pickersgill, B., and C. B. Heiser. 1976. Cytogenetics and evolutionary change under domestication. *Phil. Trans. Roy. Soc.* London B. 275: 55–69.

Schwanitz, F. 1966. *The Origin of Cultivated Plants.* Cambridge, MA: Harvard University Press.

Simmonds, N. W., ed. 1976. *Evolution of Crop Plants.* London: Longman.

Smith, C. E. 1968. The New World centers of origin of cultivated plants and the archaeological evidence. *Economic Botany* 22: 253–266.

———. 1969. From Vavilov to the present—a review. *Economic Botany* 23: 2–19.

Vavilov, N. I. 1926. Studies on the origin of cultivated plants. *Bul. Appl. Bot., Genet. and Pl. Breed.* 16: 139–248.

———. 1935. The origin, variation, and immunity of cultivated plants. *Chronica Botanica* 13: 1–364.

———. 1940. The new systematics of cultivated plants. In *The New Systematics.* Ed. J. Huxley. Oxford: Clarendon. 549–566.

The Ethnobotany of Artificial
Selection in Seed Plant Domestication

GARRISON WILKES

Domesticated plants represent the ultimate ethnobotanical relationship. Without the food and other useful products plants regularly supply, we would not be free to reach our potential in such activities as the arts and learning and/or to live at high densities in large metropolitan centers. The plants, in turn, are dependent on us, because, without our care and protection, most would not be very productive; in fact, many would not even survive. Because of the gene changes brought about by artificial selection, dependence now exists whereby humans and useful plants are locked into a relationship of mutual domestication.

Most cultivated food plants have lost the ability to exist in the wild; in other words, they are fully domesticated. These plants have been selected to produce gigas plant parts, soft edible tissue, thick flesh with intense color, and fruits attached by tough stems. These plants have been so altered genetically that they are dependent on us to sow them in the proper season, to protect them from competition and predation, to supply them with water and nutrients, and to harvest their seed so the cycle can be repeated. Burkill (1953) called the process "ennoblement," and Ames (1939) compared it to making plants "into wards." This unique dependence of selected plants on humans is a comparatively recent development of the last 10,000 years.

The change from being a wild plant to a plant dependent on humans has not been uniformly the same among the useful plants. Instead of one origin for all cultivated plants, each distinct crop has its own origin. Some plants are ancient domesticates; others are recent, of the twentieth century (Harlan 1971, 1975). Most of the basic food crops were established as domesticates long before written history; theories of their origins have been based on both conjecture and on the reconstruction of evolutionary pathways (Simmonds 1976). Clearly, the evolution of a domesticate is more likely the result of a sequence of genetic changes over time than it is the fixing of a particular trait. The change from being wild to being a domesticate is characterized more as a process than as an event, and this process can be separated into stages. Ironically, we know more about the changes that occurred in plant morphology through these stages than we know about the human-instigated selection forces that established the resultant traits (i.e., we know more about these plants than about the human activities that created them).

The ethnobotany of artificial selection has remained an ignored aspect of plant domestication. Most analyses have equated the domestication with the crop origin and focused more on the evolution of the crop than on the forces of human-instigated artificial selection in the domestication process. There is considerable literature on the pre-domestication (Rindos 1980), the pseudo-domestication, and environmental determinism (Ucko and Dimbleby 1969) or unrealized artificial selection (Hahn 1909; Bohrer 1972; Harlan et al. 1973) that preceded the development of full-blown agriculture. These are foraging patterns, the dung heaps, the use of fire to drive animals (Thomas 1956)—all factors that influenced gene frequencies and predisposed the plant genome to cross the threshold of domestication (Harlan 1975).

There is an equally large literature equating the origin of agriculture with the fixing of important crop-specific traits such as the nonbrittle rachis on cereals, and the increase in the human population that occurs prior to the development of urban centers (Flannery 1972). Theories of both incipient domestication and domestication itself have focused more on physical conditions than on the activities of humans, probably because the artificial selection forces in prehistory are difficult to know with certainty (Reed 1977).

Our reconstruction of prehistoric events, such as the domestication of crops, often is based on three lines of evidence. Certainly, the most direct lines of evidence are archaeological sites yielding dated artifacts in an ordered sequence. From these have come a large amount of physical evidence about events that took place in the most intensely studied regions, notably the Near East and central Mexico.

The second line of evidence consists of analysis that assumes our physical measurements and necessities for life as humans have not changed in the last 400 generations. When a small flint or sickle with a grass-stem polished surface is held in the human hand and the thumb presses a head of grain against the knife edge, we assume that our action is a reconstruction of a mode of harvest used 8000 years ago. Since humans have not changed physically, our dietary requirements must have been basically the same then as they are now. The inquisitive nature of humans probably also was the same then as now. Certainly humans had the ability to interpret the environment when in quest of food and shelter, and the ability to make abstractions and to anticipate the future.

The third line of evidence comes from looking at primitive societies and folk agriculture and from observing artificial selection in action. The assumption here is that the selective forces generated by a hunting-gathering !Kung bushman or even by a weekend mushroom/berry collector have some relevance to incipient domestication. As for post-domestication practices, there is a large number of anecdotal observations but few experimental analyses of the genetic effectiveness of folk agricultural practices. Obviously, they nevertheless have been effective because local landraces of our basic crops are very productive when compared to their progenitors. The rest of this chapter will explore this third line of evidence, the ethnobotany of artificial selection.

Through thousands of years of experiences, various cultures have selected from the available vegetation a relatively meager collection of plants upon which the world's food production is now based. The number of plant species that historically have fed the human population is only about 5000. This small number is less than a fraction of one percent of the flora of the world. Ten thousand years ago, and before the advent of agriculture, the world's human population is estimated to have been about 5 million. We were hunter-gatherers, there being about 25 square kilometers (10 square miles) per person. Today our numbers exceed 5 billion with a density of more than 25 persons per square kilometer (less than half a square mile) and a portfolio of less than 150 significant food plants that enter world commerce. (This excludes spices, medicinals, and industrial plants.) Many of the plants that once existed wild in our forage territory and from which we gathered our subsistence food, now exist only in carefully tended fields and gardens. In fact, only about fifty very productive food plants really meet our caloric or basic energy requirements (Prescott-Allen and Prescott-Allen 1990). Said another way, artificial selection has been most successful with a very few plants. A variety of fruits and vegetables is important for the vitamins, minerals, and fruity acids that are necessary for adequate nutrition, but these plants in general are not major caloric sources in our diet.

In addition to providing mere calories, usually in the form of a cereal/root carbohydrate, our symbiotic relationship with cultivated food plants has benefited us in a second way, through the natural selection for a balanced nutritional intake that tends to evolve in any system where the cultivator eats and depends on what he/she grows. Classically, over millennia, if the nutrition was balanced, the cultivator was healthy and had the energy to tend his/her plants and maximize yields. This natural selection promoted the cultivation of complementary protein plants such as corn (deficient in the amino acids lysine and methionine) and beans (deficient in cystine) in Mexico, or wheat (deficient in lysine and tryptophan) and curded milk in the Near East; cooking styles that maximized the amount of digestible protein in the final product (e.g., baked beans, quick-fried vegetables); the development of fermented or partially digested plant products such as beers from sorghum in Africa and soybean curds in the Far East; and the supplementation of cultivated carbohydrate-rich crops with trace nutrients from wild, collected pot-herbs for gruels and soups.

Many of these food systems rich in folk knowledge are passing out of existence with the worldwide trend of shifting from growing crops for household use to growing crops for sale. While the subsistence level of these food producers has not increased with the shift, the intensified use of land/water/fertilizer has made possible the occupation of more people per unit area. The current condition constitutes a rapid reversal, in one or two generations, of eight millennia of *in situ* evolutions where the cultivators ate the crops they cultivated (Wilkes 1977b).

The ethnobotany of the domestication of plants relative to their nutritional mix in the diet never has been seriously investigated. It is a fact that agricultures with seed plants (e.g., India, Near East, Mexico, Peru, China) have originated only where there existed a natural, high-quality, and concentrated plant protein in the local native flora. Each of the major regions has developed both cereal and legume plant protein sources that have been the nutritional package for the civilization that developed and flourished *in situ*. Clearly, the Indian center would not have been able to develop a national vegetarian diet were it not for the wealth of legumes under cultivation. In China, high protein vegetables still supplement soybeans in the diet, and a quick-fry cooking style has evolved to minimize the role of more expensive and less available animal proteins. In Aztec Mexico, where the only domesticated animals were the turkey and the duck, beans were a major component in the diet.

Even in the Middle Ages in Europe, legumes were important in the diet wherever turnips (prior to the introduction of potatoes from the South American Andes) and peas were the staples. Europe, in fact, was helped through the food-for-subsistence to food-for-sale transition of the industrial revolution by the sudden plant wealth from new crops, notably the common bean, introduced with the discovery of the New World. The replacement of peas and turnips by potatoes in Europe is a good example of the replacement of one crop by another that is more productive but maintains the nutritional balance. In Mexico, grain amaranths have been replaced by maize; in the Andean Highlands, wheat, barley, and potatoes are displacing quinoa (Wilkes 1977b). These examples should give meaning to the ethnobotanical rule of food production, which states that in indigenous agriculture where the crops are consumed and not sold, there evolves and is maintained a reasonable level of nutritional adequacy.

A second rule of food production states that in indigenous agriculture where the crops are grown only for sale, there develops an expanding surplus. The overall objective of agricultural systems under the second rule is to replace a pre-existing (natural) community with a cultivator-made community. If the potentially unstable increase from an artificial community is to be maintained, it must be consistent with three aims:

1. To operate at a maximal profit (labor/yield).

2. To minimize year-to-year instability in production.

3. To operate so as to prevent long-term degradation of the production capacity of the agricultural system.

It is under the unspoken assumptions of this "common sense" that cultivators make the decisions that set in motion the artificial selection which guides crop evolution.

The first point is self-evident; no subsistence, peasant agriculture can be long sustained that does not supply an adequate diet. Currently, many peasant agricultures do not sustain the cultivators, and they are forced into seasonal off-land employment or into the cultivation of cash crops such as coffee or opium poppies.

The second criterion of stability is exemplified in the following account told by the late Professor E. Hernández-X, Post-Graduate College and National School of Agriculture in Mexico (Hernández 1970). In Tlaxcala, Mexico, Hernández encountered an old Indian working in his cornfield and asked him what kind of corn he planted. The old man responded that he grew yellow-corn, cream-corn, and white-corn. When asked which was the earliest maturing corn, the Indian replied that the yellow took five months, the cream six months, and the white seven months to mature. When asked which yielded the most, he informed Hernández that the yellow-corn gave a little, the cream more, and the white-corn was the best. Hernández then asked him why, if the white-corn was the most productive, he didn't plant the best corn. The old man smiled and said, "That is the question my son who works at the factory asks. Tell me, Mr. Agriculturist, exactly how much and when will it rain next year?" At this point, Hernández responded that he could not divine the future. With a knowing look, the old man said "Exactly! Therefore, I plant all three, so if there is a little rain, I always have some yellow-corn to eat. If there is more rain, I'll have enough to eat with the cream-corn, and if it's a good year with plenty of rain, I will have white-corn to sell." He added drolly, "Usually it isn't a good year."

Admittedly an anecdotal account of biological stability, this story is also an example of effective artificial selection, since the seeds with differing growth potentials were color coded. Because of the color-coding system (seed color under genetic control of primarily recessive genes which would show off-types due to outcrossing), the frequency of differing growth potentials and water requirements is maintainable over the years. If there were not this selection, either one of the three genotypes would dominate after a few years, or the three interfertile genotypes would homogenize into a single type of corn. This artificial selection for seed color was linked genetically to growth potentials, which becomes an example of the third point.

Charles Darwin (1868) considered artificial selection to be human-speeded "telescoped evolution"; and indeed it is because recessive genes can be fixed in a population within two generations and dominant genes with close inbreeding almost as quickly. I am not aware of any ethnobotanical study that has tested by progeny trials the genetic effectiveness of folk agricultural selection practices. Obviously, such practices are effective because of the racial diversity present in cultivated crops, but the point here is that the rigor of genetic analysis has not been applied as a measure for folk agriculture selection practices. Much of the literature in this area is anecdotal notes about color preferences for special occasions (blue corn) or for bits of cloth tied around certain mature opium poppy capsules (did they yield more opium?). Presumably they were seed capsules for the next year's crop. Worldwide, the opportunity to collect folk wisdom about crops *in situ* is disappearing with the spread of elite germ plasm; the time to be about these studies is now.

I shall end with some anecdotal observations for maize in the Americas, a crop with which I have personally worked. To the best of my knowledge, the rigors of genetic analysis have never been applied to these often-observed, artificial selection practices. I am sure that specialists with other sexually reproducing cross-breeding seed crops have made comparable observations, but I am limiting the examples here to maize.

Throughout the U.S. Southwest and Mexico, blue kernels are esteemed for making tortillas on special occasions. These colored corns are controlled by interaction of the alleles of the A locus and R locus with genes controlling anthocyanin and related pigments in the seed aleurone, pericarp, and vegetative parts of the plant. The blue (really purple) aleurone dyes the corn meal, and the resulting tortillas are blue. These color genes could be easily swamped by pollen from the more abundant white or yellow maize varieties, but special planting sites within the field (usually uphill and in a corner of the field) insure the degree of genetic isolation needed to maintain true-breeding seed stock. The color form does not have to be blue; it could be red, striped, or marbled, and often these are called "Sangre de Cristo" or "Rayo de Dios" and are maintained at a low frequency in the field. I am aware of no genetic analysis of the effectiveness for the selection practices that maintain the genetic isolation of these morphs from the larger population.

Another example was given by Edgar Anderson when he was measuring racial variation in maize in Jalisco, Mexico. He would select twenty-five ears at random from the corn crib to establish a racial measurement for the cornfield of a specific cultivator. Much to his displeasure, the farmer kept bringing him ideal ears to measure; to please the fellow, Anderson measured them also, but put a check by these data so he could ignore them in the final analysis (Anderson 1946, 1954). This is a beautiful example of an opportunity lost because here were two measurements, the variation of the field and the ideal goal of the cultivator. A comparison of the two could have established a measurement of the effectiveness of selection over time for this particular cultivator.

In my own investigations, I have found a difference between the ears selected for seed ears and the harvest ideal (Wilkes 1977a). I did some preliminary genetic analysis of the maize race Conejo from Guerrero, Mexico, but never went on to measure the effectiveness of the selection process. Obviously, I, too, missed an opportunity, proving that the rigors of genetic analysis have not been used to the extent to which they could to explore the effectiveness of indigenous selection practices.

In the future, when ethnobotanical observations of both cultivator practices and genetic progeny tests are made, we will be able to expect, I think, truly new insights and new models for seed plant domestication (Heiser 1969; Harris 1972; Libby 1973).[1] Maybe we should call this process anthro-selection instead of artificial selection, because it is much more human-centered than environmentally influenced.

NOTES

[1]Since this chapter was written, several significant articles and books have appeared, of which the reader should be aware. Nonetheless, the central conclusion of this chapter remains, namely, that we know little about the dynamics of gene frequencies under anthro-selection/domestication. The sources are as follows:

Altieri, M. A., M. K. Anderson, and L. C. Merrick. 1987. Peasant agriculture and the conservation of crop and wild plant resources. *Conservation Biology* 1: 49–58.

Altieri, M. A., and L. Merrick. 1987. *In situ* conservation of crop genetic resources through maintenance of traditional farming systems. *Economic Botany* 41: 86–92.

Harris, D. R., and G. C. Hillman, eds. 1989. *Foraging and Farming: The Evolution of Plant Exploitation*. London:: Unwin Hyman.

Oldfield, M. L., and J. B. Alcorn. 1987. Conservation of traditional agroecosystems. *BioScience* 37: 199–208.

Rindos, D. 1984. *The Origin of Agriculture: An Evolutionary Perspective*. New York: Academic Press.

LITERATURE CITED

Ames, O. 1939. *Economic Annuals and Huan Cultures*. Cambridge, MA: Botanical Museum of Harvard University.

Anderson, E. G. 1946. Maize in Mexico—A Preliminary Survey. *Ann. Missouri Botanical Garden* 33: 147–247.

———. 1954. *Plants, Man and Life*. London: Andrew Melrose.

Bohrer, V. L. 1972. On the relation of harvest methods to early agriculture in the Near East. *Economic Botany* 26: 145–155.

Burkill, I. H. 1953. Habitat of man and the origin of the cultivated plants of the Old World. Proc. Linn. Soc. 164: 12–42.

Darwin, C. 1868. *The Variation of Animals and Plants Under Domestication*. 2 vols. London.

Fannery, K. V. 1972. The cultural evolution of civilization. *Annual Review of Ecology and Systematics* 3: 399–426.

Hahn, E. 1909. *Die Entstehung der Pflugkultur*. C. Winter

Harlan, J. R. 1971. Agricultural origins: Centers and noncenters. *Science* 174: 468–474.

———. 1975. *Crops and Man*. Madison, WI: American Society of Agronomy.

Harlan, J. R., J. M. J. de Wet, and G. Price. 1973. Comparative evolution of cereals. *Evolution* 27: 311–325.

Harris, D. R. 1972. The origin of agriculture in the tropics. *American Scientist* 60: 180–193.

Heiser, C. B., Jr. 1969. Some considerations of early plant domestication. *BioScience* 19: 228–231.

Hernández-X, E. 1970. *Exploración Etnobotánica y su Metodología*. Chapingo, Mexico.

Libby, W. J. 1973. Domestication strategies for forest trees. *Canadian Journal of Forestry Research* 3: 256–276.

Prescott-Allen, R., and C. Prescott-Allen. 1990. How many plants feed the world? *Conservation Biology* 4: 365–374.

Reed, C. A., ed. 1977. *Origins of Agriculture*. The Hague: Mouton.

Rindos, D. 1980. Symbiosis, instability and the origin and spread of agriculture: a new model. *Current Anthropology* 21: 751–772.

Simmonds, N. W., ed. 1976. *Evolution of Crop Plants*. London: Longman.

Thomas, W. L., Jr., ed. 1956. *Man's Role in Changing the Face of the Earth*. Chicago: University of Chicago Press.

Ucko, P. J., and G. W. Dimbleby, eds. 1969. *The Domestication and Exploitation of Plants and Animals*. Chicago: Aldine.

Wilkes, H. G. 1977a. Hybridization of maize and teosinte in Mexico and Guatemala and the improvement of maize. *Economic Botany* 31: 254–293.

———. 1977b. Native crops and wild food plants. *The Ecologist* 7: 312–317.

PART 7

Ethnobotany and Geography

One of the most frequent ways in which ethnobotanical studies have been carried out and published is by geographic areas: countries, provinces, states, and even smaller areas. Among the many significant modern contributions of this kind are S. K. Jain's *Dictionary of Indian Folk Medicine and Ethnobotany* (Deep Publishers, New Delhi, India, 1991), García Barriga's *Flora Medicinal de Colombia*, Barrau's *The Oceanians and Their Food Plants* (Universidad Nacional de Colombia, Bogotá, 1975), Taylor's *Plants Used as Curatives by Certain Southeastern Tribes* (Harvard Botanical Museum), J. M. Watt and M. G. Breyer-Brandwijk's *The Medicinal and Poisonous Plants of Southern and Eastern Africa*, and Towle's *The Ethnobotany of Pre-Colombian Peru* (2nd ed., E. & S. Livingstone, Edinburgh, 1962). An outstanding article because of the paucity of material available is "Neglected Aspects of North American Ethnobotany" by David H. French (*Canadian Journal of Botany* 59 (1981): 2326–2330). Also noteworthy are *Islands, Plants, and Polynesia: An Introduction to Polynesian Ethnobotany* (Eds. Paul Cox and Sandra Banack, Dioscorides Press, 1991) and *Plants and People of the Golden Triangle: Ethnobotany of the Hill Tribes of Northern Thailand* (Edward Anderson, Dioscorides Press, 1993). Focusing on two completely different geographical areas, both works point out the urgent need to study the interaction between culture and environment in indigenous societies and to preserve the knowledge of plants and their uses in these cultures before it vanishes. Although delimiting ethnobotanical studies and publications by geographical boundaries may not be acceptable to some anthropological specialists, it is, nevertheless, an extremely efficient and practical approach to field work in this period of both rapid development of research and the threat of acculturation of primitive societies which in many regions are in danger of extinction.

The seven contributions in this section examine various studies that contribute both to ethnobotany and geographic understanding. Eduino Carbonó shows how ethnobotanical research in Colombia, with its commitment to the dignity and productive capacity of Indian cultures, permits the integration of these people into the nation's mainstream without any trace of discrimination. John O. Kokwaro presents the case for "development without destruction" in Africa. Weston La Barre provides an overview of the various studies in North America by tribal group and by plant species. J. K. Maheshwari covers current ethnobotanical research among tribal groups in the hot, arid zones of India. The late George R. Morgan analyzed some of the important aspects of geography in relation to the ethnobotanically very important hallucinogenic cactus, peyote, which is the basis of ancient Mexican sacred ceremonies and of the Native American Church, a modern religious organization with 300,000 members in the United States. Ong Hean Chooi describes the many uses of plants native to the tropical rain forest of northern Malaysia. Finally, Nancy J. Turner outlines a much-needed historical development of ethnobotany in northwestern North America from its origins to the present day.

Current Outlook for Ethnobotany in Colombia

EDUINO CARBONÓ

Colombia, with its high floristic diversity and a variety of indigenous cultures within its territory, is a country with excellent conditions for ethnobotanical research. Experienced investigators estimate that there are between 35,000 and 50,000 species in the flora of Colombia (Schultes 1951; Patiño 1980) at a time when Indian settlements still exist throughout the country, except in the departments of Atlántico, Bolívar, Quindío, Santander, and the Intendencia of Santander (Ministerio de Gobierno 1982). Furthermore, the marginal cultures of segments of the mestizo population enrich the endless sources of ethnobotanical information.

Nonetheless, this very positive scenario that served as the foundation for many pre-Hispanic cultures was affected by colonization. The role of the original players was changed, and the stage was set for the present outlook, namely, mass destruction of natural vegetation and accelerated extinction of the last aboriginal strongholds. Because Indian communities provide multiple manifestations of direct dependence on the environment, as indicated by the knowledge generated from the interaction between humans and the surrounding flora, it is therefore necessary to carefully explore and rescue all that creative capacity for the purpose of incorporating it into the nation's productive potential.

Past and Present Ethnobotanical Studies

Colombian and foreign botanists, anthropologists, sociologists, and specialists in other areas have ventured into the ethnobotany of Colombia. One of the pioneer works on the use of the most common plants is credited to Enrique Pérez Arbeláez (1947), an outstanding botanist dedicated to the diligent observation of the use of the country's natural resources. In the field of medical botany, Hernando García Barriga has collected voluminous information on popular medical practices, which is summarized in the three volumes of his *Flora Medicinal*. Victor Manuel Patiño, who is credited with the laudable documentation of the history of the natural resources of the country, has also been a driving force in advancing our knowledge of the use of plants (Patiño 1957, 1959). Other investigators have studied the use of plants and vegetable products in Indian communities, establishing basic similarities and differences between the customs of different groups (Schultes 1942, 1970; Uscátegui 1954, 1956, 1961, 1963).

One of the greatest advocates in this discipline in Colombia is Richard Evans Schultes, who travelled throughout the Amazon jungle for more than thirty years making ethnobotanical observations. One of the first theoretical essays elaborated in the country that stimulated new research in this area can be credited to him (Schultes 1941). Schultes's publication was amplified by Yépez's (1953) essay that appeared some years later and attempted to establish guidelines for future research.

In the last half of the twentieth century, numerous ethnobotanical publications have appeared on work carried out in Indian communities (von Hildebrand 1975; Forero-Pinto 1980; Pineda 1982; Pradilla and Espinosa 1982; Glenboski 1983; La Rotta 1983). Even though programs of supervised research with specialties in this area do not exist, a number of researchers have oriented their efforts towards gathering the knowledge of popular botany. Their marked concern reflects an instinctive response to the need to compile the accumulated experiences of survival in marginal communities. The environment's growing deterioration is forcing us to look among cultural relicts for alternatives that impede the total obliteration of plant life for indigenous knowledge of its uses.

The known studies cover diverse topics: thus, while some offer general information about the ethnic groups they study, others gather information on useful plants that could be bases of new resources for industry. The most common tendency is to evaluate the validity of these studies based on the potential industrial use that can be obtained from the information. In this spirit, the work consists of extracting from a community the largest number of plants with potential industrial use.

If it is true that the contribution of a new resource as industrial raw material represents the justification for the research to satisfy human necessities, it is also true that ignoring the cultural context from which that resource is extracted means isolating an element without evaluating the possible repercussions of this action on the whole. In the culture of Indian communities, each element is connected to the life system in a network of relationships, and the apparent superficiality represents, most of the time, a strategy in which common and practical sense are veiled by magic rituals through which the rules of behavior towards the environment for the management and use of each resource are established.

The majority of works that have appeared to date are the result of individual efforts; much of the information gathered continues to be incidental and therefore disperses. Although all this information constitutes a valuable tool for future work, it is necessary to give a multidisciplinary focus to new programs. Future investigations can broaden their scope if interdisciplinary work teams are incorporated into their execution to articulate aspects of different scientific areas in the search for more faithful interpretation of themes in the relationship between humans and their environment.

Besides obtaining the advantages of a holistic vision and the agility that would be given to the execution of the studies, it would be possible to integrate experiences of different regions and to experiment in the use of techniques and resources between them. The failure of many development plans in Indian reservations can be explained by the difficulties in establishing a balanced cultural interchange. Subjected to the judgment of technological superiority, the plans do not take advantage of adaptive local experiences, but impose unidirectional assimilation that in most cases leads to the adoption of false solutions.

Modern studies in Colombia tend to emphasize the human character in ethnobotanical research and to expose the interesting relationships between marginal cultures and their environments. The results obtained by these investigations show that the ever-growing concern for this discipline cannot be a simple nostalgia for a pleasant past. Rather, in light of the negative results obtained in the attempts to manage the resources of the tropics, the concern is that the design of appropriate strategies to use tropical forests would have more solid foundation if based on the experiences and knowledge of the Indian groups and farmers that inhabit these regions.

It is clear that the lack of knowledge in either over- or underestimating indigenous practices of resource management has been the largest source of undesirable consequences for improvement programs. In the process of modernization or civilization (as the application of modern techniques in the development of certain areas is called), organizational schemes, norms, and management techniques traditionally used in the intervened upon regions are unrecognized and therefore destroyed without evaluating their importance. This sudden alteration in ways of thinking and behaving generates unexpected and unpredictable changes that affect the entire life system—soil, water, air, the forest, the human population. Furthermore, forced colonization and evangelization has not only demolished indigenous cultures, but, as a corollary, the natural environment of these cultures.

It is appropriate to include a phrase from Patiño (1975–1976) that denotes the difference between the indigene and the European colonizer in the use of the environment: "The first lives in harmony with his environment and takes from nature what he needs without plundering and squandering the resources." The agricultural, silvicultural, and other techniques used for centuries of interaction with mountains and neotropical forests become ever more attractive in the conservation of these environments.

The ample diversity of the neotropical landscape is subutilized; hundreds of hectares of forest are destroyed to extract a single product, limiting any possibility of making use of the variety of resources present therein. Here is an enormous challenge for ethnobotany in Colombia: to design the most appropriate strategy to take advantage of the biological diversity found within its territory. There are several essays on forms and techniques that could be used to obtain resources without exhausting the natural landscape (Pineda 1974, 1975; Reichel-Dolmatoff 1982; La Rotta 1983).

Perspectives for Future Ethnobotanical Studies

The outlook of ethnobotany is broad. The floristic inventory of Colombia is yet to be completed, and the sampling intensity must be carried out at the speed with which the vegetation is being swept away or faster. Through ethnobotanical research, new species can be discovered as well as the elements related to the better use of natural resources. The case of *yoco* (Schultes 1942) is a good example of the existence of undescribed plant species already known and used by Indian communities for many years but identified only as a result of ethnobotanical studies.

In silvicultural management there is much to be gained, especially in the conservation of species or areas of undisturbed vegetation. Many apparently unimportant rituals and customs contain precise rules for the protection of some species from extinction. Many wild species are sources of plant germ plasm indispensable for genetic improvement, and the preserved areas are natural reserves. In exploiting the forest, the effect of extracting particular species has hardly been considered, nor has there been much advance towards multiple use of the diverse neotropical forest.

In the field of medicine, abundant information on the popular use of plants is available, and studies continue to appear showing the high number of wild plants with medicinal use in Indian communities. The intensity of this search and the complementarity of pharmacological studies offer hope of finding new, therapeutically valuable chemical compounds.

In agriculture, a high ecological price is paid for the inappropriate use of tropical soils in the Andean zones as well as in other rain forests; in these areas we have not taken advantage of traditional experiences related to the control and planning of agricultural activities. This aspect, of singular interest to the country, represents the groundwork of great expectations for experimentation in so-called tropical agriculture.

With respect to its human content, ethnobotanical research allows the gathering of many cultural elements fundamental to elaborating a strategy of national integration. The most common attitude towards Indian communities is to underestimate them and treat them with indifference. This attitude is the result of a conscience created by colonial subjugation that keeps us looking at aborigines as the inhabitants of invaded lands, peoples whose values were downtrodden, whose principles were disregarded, and whose rights were ignored.

Ethnobotanical research can bring the survivors of the first inhabitants of Colombia to their rightful place, not as a protectionist posture that eventually segregates them, but rather out of respect towards their human condition, and in recognition of their customs and productive capacity. Only the complete understanding of what is left of Indian culture will bring about integrationist policies lacking any form of discrimination.

LITERATURE CITED

Forero-Pinto, L. 1980. Etnobotánica de las comunidades Cuna y Waunana del Chocó, Colombia. *Cespedesia* 9: 115–306.

Glenboski, L. 1983. *The ethnobotany of the Tukuna Indians, Amazonas Colombia*. Biblioteca José Jerónimo Triana. Bogotá: Instituto de Ciencias Naturales. 4: 1–92.

von Hildebrand, P. 1975. Observaciones preliminares sobre utilización de tierras y fauna por los indígenas del río Mirití-Paraná. *Rev. Colom. de Antropología* 18: 191–258.

La Rotta, C. 1983. *Observaciones etnobotánicas sobre algunos especias utilizadas por la comunidad indígena Andoque (Amazonas, Colombia), Dainco Corporación Araracuara*. 1–117.

Ministerio de Gobierno. 1982. *Colombia Indígena, Litografía Arco*. Bogotá. 1–229.

Patiño, V. M. 1957. Aspectos especiales de la vegetación natural en América Equinoccial. Guaduales y manglares. *Rev. Colom. de Antropología* 6: 159–191.

———. 1975–1976. *Historia de la vegetación natural y sus componentes en la América Equinoccial*. Imprenta Departamental. Cali. 1–429.

———. 1959. Plátanos y bananos en América equinoccial. *Rev. Colom. de Antropología* 7: 295–337.

———. 1980. *Los recursos naturales de Colombia. Aproximación y retrospectiva*. Bogotá: Carlos Valencia Editores. 1–149.

Pérez Arbeláez, E. 1947. *Plantas útiles de Colombia*. Contraloría General de la República Bogotá. 1–537.

Pineda, R. 1974. *El país de la garza del centro*. Estudios antropológicos. Universidad del Cauca. Popayán. 1–174.

———. 1975. La gente del hacha. Breve historia de la tecnología según una tribu amazónica. (Mimeografiado). Bogotá. 1–35.

———. 1982. *Chagras y cacerías de la garza siringuera: Sistema hortícola Andoque, Amazonía colombiana*. Bogotá: Edit. Comité de publicaciones de Oram. 1–57.

Pradilla, H., and A. Espinosa. 1982. Los Tunebo. Adaptación al medio ambiente y salud. Bogotá: FES, Universidad Nacional, Ministerio de Salud Pública. 1–104. Unpublished.

Reichel-Dolmatoff, G. 1982. Cultural change and environmental awareness; a case study of the Sierra Nevada de Santa Marta, Colombia. *Mountain Research and Development* 2(3): 289–298.

Schultes, R. E. 1941. La etnobotánica: Su alcance y sus objetivos. *Caldasia* 1(3): 7–12

———. 1942. Yoco: A stimulant of southern Colombia. *Botanical Museum Leaflets* (Harvard University) 10: 301–324.

———. 1951. La riqueza de la flora Colombiana. *Rev. Acad. Colom. Ciénc. Exact. Fís. Nat.* 7(30): 230–242.

———. 1970. Notas etnotoxicológicas acerca de la flora amazónica de Colombia. In *Simposio y Foro de la Biología Tropical Amazónica*. Ed. J. M. Idrobo. Bogotá: Editorial Pax. 177–196.

Uscátegui Mendoza, N. 1954. Contribución al estudio de la masticación de las hojas de coca. *Rev. Colomb. de Antropología* 3: 207–289.

———. 1956. El tabaco entre las tribus indígenas de Colombia. *Rev. Colom. de Antropología* 5: 11–52.

———. 1961. Algunos colorantes vegetales usados por las tribus indígenas de Colombia. *Rev. Colom. de Antropología* 10: 331–340.

———. 1963. Notas etnobotánicas sobre el ají indígena. *Rev. Colom. de Antropología* 12: 89–96.

Yépez Agredo, S. 1953. *Introducción a la etnobotánica Colombiana*. Bogotá: Publicaciones de la Sociedad Colombiana de Etnología. 1–48.

Ethnobotany in Africa

JOHN O. KOKWARO

The term *ethnobotany* can be defined briefly as folk botany, or the description of the various methods by which local peoples utilize plants. Even though two or more groups of people share the same environment, each group will have its own traditional method of utilizing a particular plant.

The large continent of Africa has a variety of ethnic groups from north to south, and a similar variety of vegetation from bare deserts to tropical rain forests. Ethnobotanical activities reflect these variations and will, therefore, be less pronounced in the Sahara region than in the region south of it. Consequently, the ethnobotany discussed in this chapter primarily concerns Africa south of the Sahara, where there is a greater wealth of plant species, but the broad traditional use of local plants, such as in ethnomedicine, applies to the whole continent. There is no question that Africans have utilized and continue to utilize plants in many ways. This knowledge about plant uses is passed on from parents to children from generation to generation. Indeed, the ability to discriminate between edible and nonedible or medicinal and poisonous plants has been essential for the survival of these people.

The beginning of Western botany, as we can read from the works of Theophrastus (375 B.C.) and others, was primarily ethnobotanical. This was later followed by a major interest in systematic botany, when botanists became preoccupied with making inventories of the floras. In Africa, the trend was the same, although inventories were not written; instead, people memorized the names of plants useful both for humans and for their domestic livestock.

In the Western world, the science of botany moved to areas like anatomy and physiology during the nineteenth century and to genetics and palynology in the twentieth century. Whereas such modern disciplines of botany are learned and restricted to schools and research institutions, ethnobotany in its original meaning is still predominantly practiced by unlettered people in most parts of Africa.

The continent of Africa, bisected by the equator, extends northwards to the Mediterranean Sea and as far south as the Cape of Good Hope. Geologically of very ancient pre-Cambrian rocks, the continent has had a surprisingly tranquil history of steady denudation, except in the east where the rift system associated with volcanic action has caused depression and upheaval. North Africa is cut off effectively from tropical and southern Africa by the huge Sahara desert, which extends from coast to coast and continues as the Arabian desert to the east. The southern part of the continent is affected also by a combination of the Namibia and Kalahari deserts. Mild temperate climates occur at the northern and southern extremities and around the high mountains in the tropics, and these conditions tend to modify the vegetation.

The lowland equatorial Zairian basin with its West African continuation constitutes the tropical rain forest region with moderate to plentiful rainfall. To the north and south of this forested region, the climate shows a steady change from a double rainy season to a single one, which trails off to nil in the deserts. The vegetation responds to the climate by grasses progressively taking over from the trees: first woodland with some grass, then real savanna occurs where the more widely spaced trees grow among tall grass; and finally the trees become fewer and the grass shorter.

Owing to these large stretches of deserts in the continent, the number of species composing the flora is estimated at about 40,000 compared with some 80,000 in the Amazon alone. This estimate, however, may be far from the reality, since documentation of Africa's flora is not complete. Furthermore, the estimate may be low since ethnobotany is much practiced in the continent and some species may have been eliminated in certain areas even before their documentation in the flora. In this chapter we will look first at indigenous medicinal plants of Africa, then consider various plants used as fibers and flosses.

Ethnomedical Practices

One of the leading ethnobotanical practices in Africa has been, and still is, the use of herbs as a source of medicament for health care. It is the kind of ethnobotany that has given scientists some challenging background for modern drug development. Owing to its importance in the continent, we shall now discuss it in detail.

Gradually, through trial and error, primitive people learned the uses of plants, especially those that served as food and as remedies for disease. To be lasting, this kind of classification of plants required the existence of names, and the practice of giving names to useful plants marked the beginning of systematic botany. This ability to know the names of plants and to identify them in their natural habitats is what I refer to as ethnosystematics (i.e., a folk knowledge of botanical classification).

There are several reasons why ethnosystematics is particularly well developed in Africa. First, the continent has a very rich and diverse flora, which has served people for many thousands of years. Plants are by far the most widely used source of medicaments in African traditional medicines. In these systems of medicine, people are regarded not simply as isolated organisms, but their sociological environment is also taken into account. Africans have long associated plants with ritual, symbolism, and religious beliefs. More ritual symbols are drawn from the Plant Kingdom (particularly trees) than from any other part of the environment.

A second reason why ethnosystematics is well developed in Africa is that the continent is particularly rich in languages and dialects, and every ethnic group has names for the plants growing in its environment. In many cases, knowledge of plants (and their names) is shared by neighboring cultures. This is probably because some of the people who make the greatest use of herbal remedies are from pastoral tribes that come in contact with other ethnic groups while traveling great distances in search of grazing for their livestock.

A good knowledge of ethnosystematics has survived in Africa largely because traditional medicine is still widely practiced. In most African countries, the majority of people lives in rural areas where some of the common problems are lack of communication (transport and telephone) services, low income, and lack of modern medical facilities. The nearest hospital or dispensary may be many kilometers away and out of reach. Even if a hospital can be reached, the sick person may prefer traditional medicine. Well-established traditional practitioners usually have many customers in rural areas, so it is important they be able to identify and know their plants.

The last factor in the survival of a well-developed ethnosystematics in Africa is the oral tradition, which results in knowledge being widely held in the community. Knowledge of traditional medicines is handed down orally by medical practitioners, parents, elders, or priests and priestesses. This involves naming the plants to be used and the diseases to be treated and observing the preparation and application of the medicaments. This tradition of relaying medical information orally is found in all African countries. Even Ethiopia, which has had its own written language for more than 2000 years, has no written record of traditional medicine.

This oral tradition has all the advantages of a living culture, being continually enriched and reflecting attitudes, beliefs, and style of life. It also has drawbacks. Some of the knowledge is liable to be distorted or lost completely during transfer, therefore becoming both erroneous and dangerous to the recipient. In the absence of an African pharmacopeia, our knowledge of African traditional medicine and the ethnosystematics on which it relies needs to be placed on a firmer basis.

TRADITIONAL PRACTITIONERS

Traditional medicine men and women continue to occupy an important position in African societies. Sometimes referred to as traditional practitioners, healers, witch doctors, and so on, they can be grouped according to their practice:

1. Herbalists usually use plants for treating patients.

2. Diviners are also herbalists but use divinatory procedures for treatment.

3. Spiritualists hardly use plants at all for treatment.

4. Great therapists utter prayers, incantations, and invocations, which are used as weapons against powerful, supernatural or invisible forces. They give people "vitalized" charms, amulets, bracelets, belts, and pendants of all kinds as protection against diseases, accident, and bewitchment.

5. Traditional midwives, also called birth attendants, may be obstetricians, herbalists, gynecologists, or pediatricians. They provide health care before, during, and after birth, and also care for newborn infants and young children.

6. Traditional surgeons use special knives, sharpened and tempered according to esoteric procedures, for circumcisions and excisions. Cassava leaves, liquid from snails, and various other ingredients are used as agents to check hemorrhage.

7. Traditional psychiatrists deal with a patient's socioreligious antecedents, using a series of rites, which include chants, incantations, and ritual dances, and in which music is played using certain musical instruments.

It is the diviners, spiritualists, and great therapists who normally perform ritual ceremonies and prepare ritual medicines to protect people from unnaturally caused diseases, curses, or sorcery. What is interesting here is that there is no cure in the world, not even Western medicine, for a person who has been cursed or bewitched—except through his or her own traditional beliefs. The victims of mystical misfortunes caused by the sorcery of an enemy or by the curse of a family member (living or dead) can be treated only by those whom I call advanced traditional practitioners. A ritual diviner or spiritualist will determine, usually by divination with animal shells or marbles, whether a patient has been cursed or bewitched and, if so, will treat him or her with magical medicine. The treatment involves primarily a ritual purification, in which the original curse or sorcery is removed, thus removing the physical symptoms as well.

Traditional medicine as practiced in Africa embraces a wide field of medicine and pharmacology, including pharmacognosy. Though distinct professions in modern medicine, these fields are usually combined by a single traditional practitioner. A good herbalist must have an excellent memory to carry all this information in his or her head. A traditional practitioner generally provides health care by using plant, animal, and/or mineral substances as tangible objects for treating patients, with plants being by far the most widely used substances in African traditional medicine.

I have observed that in the fields of traditional pediatrics, obstetrics, and gynecology, women are excellent and authentic herbalists. They are also much more liberal in passing on their herbal knowledge to their daughters than are men. A man normally will wait until he is old before choosing one of his sons (usually the firstborn) and teaching him the medical art. Although the medicine man is usually more prominent and more popular in our society, a woman was the first physician, probably because of her close association with children, who need the most medical attention in a family.

Whereas many practitioners have mastered small lists of herbal remedies, only a few attain eminence. It must be appreciated that a whole traditional pharmacopeia known to each practitioner is carried in his/her head, and it includes the recollection of where the plant is to be found as well as the identification of the prepared decoctions, infusions, ash or other dried leaves, roots, and bark stored in the personal pharmacy. The respect the medical practitioner commands among African tribes is reflected by the very eminent names he or she is referred to in every dialect. The reputation of *Bwana Mganga* (Swahili) depends on the appreciation of the practitioner's medicine; a person is usually credited with having good medicine rather than with being a good doctor.

CONCEPT OF DISEASES

The concept of diseases among Africans reflects their culture and can be broadly grouped into two categories. The first group, naturally caused diseases, are those due to tangible material that affects the body organs. Such natural diseases are regarded as minor or normal because they can be described by the patient and treated by the doctor in strictly physical terms.

It is a common preconceived belief (fear) that as soon as a disease becomes more acute or severe, it is due to unnatural causes or intangible forces, which implies that a hostile person is using supernatural powers against the patient, or that the victim may have transgressed the moral code and incurred the wrath of ancestors. This second group of diseases is generally classified among Africans as complicated and serious. Cases are characterized by persistent illness. Bewitched or cursed persons require special types of treatment, medicine, and traditional doctors. In such complicated cases, traditional treatment not only makes use of material substances like herbs but sometimes includes resources from the immaterial world.

Whatever the source of the disease, we can define traditional medicine as "the sum total of all the knowledge and practices used in diagnosis, prevention, and elimination of physical, mental, or social imbalance, and relying exclusively on practical experience and observation handed down from generation to generation." The term *medicine* as used here refers not only to herbs or drugs but also includes a whole range of charms, amulets, spells, and incantations.

PREPARATION AND DISPENSING OF DRUGS

The part of a plant used for preparing a drug depends primarily on the structure of the species. For trees and shrubs, it is a common to use the bark and/or roots. It is from such usage that East Africans coined a colloquial Swahili name for herbal medicine: *miti shamba*, meaning "medicine from the tree." With small plants and herbs, the tendency is to use the whole plant and (by contrast with trees) the leaves also are frequently used.

Traditional pharmacognosy is limited in that an extract from one plant is normally dispensed alone. Only occasionally is an infusion with extracts from two or more plant species given to a patient.

The methods of preparing plant drugs by African healers are uniform and usually are accomplished by one or two of the following procedures.

1. Boiling is a common method, especially with roots and bark of trees or shrubs. The decoction is then taken orally or used for bathing, depending on the nature of the disease.

2. Soaking in cold water is generally used with crushed leaves or small herbaceous plants. The concoction is used as in the first method.

3. Burning is also used with leaves and small herbs after the material has been dried. The ash can be licked, rubbed directly onto the wound, or soaked in water and drunk or gargled in the mouth.

4. Chewing is a first-aid method of preparing a drug, especially for treatment of snakebite, stomach disorders, or mouth and throat ailments.

5. Heating or roasting is usually employed in preparing succulent leaves or other plant parts for a poultice (a kind of moist dressing applied on an inflamed body part).

6. Crushing or pounding normally precedes other methods such as boiling, soaking, or burning. Crushed material may alternatively be applied directly to a wound, usually after being mixed with some kind of oil.

What is interesting about all six methods of preparing herbal remedies is that each of them attempts to extract whatever active principle is contained in the plant before it is dispensed to the patient. To the herbalists, the preparation is done to obtain the "power force" of the drug. Practically all herbalists have some kind of pharmacy with ready-made decoctions, infusions, instillations, powdery ash, inhalations, plasters, fomentations, enemas, embrocations, fumigations, and bandages (usually made from dry pseudo-stem portions of banana).

Another common feature of leading herbalists is the tendency to own a private garden where special medicinal plants are grown that are not available naturally in the village. In this way, the supply of fresh drugs is made available to the customers as the demand may arise.

The method of dispensing the drug depends largely on the type of disease to be treated. Aromatic drugs for treating influenza or similar diseases are used customarily in the form of steam. Traditionally, drugs are frequently taken with various types of foodstuffs. The pastoral tribes take their drugs usually in milk; other groups use soup, porridge (especially that made from the African millet flour, *Eleusine coracana*), honey, blood, and various kinds of local beers. Most foodstuffs employed are in liquid form and are both nutritious and appetizing. Since most plant parts used for preparing herbal remedies can be sour, bitter, or with offensive smell, such drugs are generally mixed with a favorite food to make them more palatable.

HIGHER PLANTS IN TRADITIONAL MEDICINE

The total number of drug plants used in Africa is so large that it would be unrealistic to try to list all the plants in this chapter. Specific examples can be read in the already available literature (e.g., Kokwaro 1976).

Apart from indigenous species, a number of introduced or naturalized plants now have encroached upon traditional medicine. The aromatic leaves of a number of gum trees (*Eucalyptus* species) introduced from Australia often are used for treating influenza. *Citrus* species are used for the same purpose. Bark of pride of India (*Melia azedarach*) is used in Zaire and West Africa as an anthelmintic, and the castor oil plant (*Ricinus communis*), which grows widely throughout Africa, is much used as a purgative.

Lower plants are not so frequently employed by Africans in their traditional medicine as are higher plants.

RECOMMENDATIONS

Owing to the rapid change in climate in many parts of Africa, from wet to dry, and also due to rapid population growth, there is an urgent need for conservation of drug plants. While conservation of ecologically important habitats should be encouraged throughout the continent, vigorous cultivation of drug plants through organized botanic gardens must also be immediately implemented. Collection of data from authentic herbalists and compilation of national and regional pharmacopeias of medicinal plants must be done quickly. These activities can be followed by the more orthodox, modern scientific investigation of drug plants with the aim of developing modern drugs from them. Close collaboration between the traditional herbal practitioners and modern medical research institutions should be encouraged.

Ethnobotany of Vegetable Fibers and Flosses

While the amount of ethnobotanical literature on African timber trees and medicinal, food, and poisonous plants has increased tremendously, very little is known about plants that provide fibers and flosses. The following list of uses for fibers and flosses gives botanical names for some of the common indigenous and exotic plants that are of value commercially or locally. The list is not intended to be comprehensive, but rather illustrative of the less known ethnobotanical practices. Because of the rough nature of their products, their scattered occurrence, or the difficulty of their preparations, the majority of species are not likely to be of value as commercial fibers or flosses. Those few that are of high value commercially have been marked with an asterisk.

BRUSH FIBERS
  Brooms
    Cocos nucifera
    Panicum maximum
    Phragmites communis
    P. mauritianus
    Raphia species
    Sida acuta
    S. alba
    S. cordifolia
    S. rhombifolia

    Smithia speciosa
    Tacca involucrata
    Wissadula amplissima
      var. rostrata

  Brushes
    Aloe species
    Borassus aethiopium
    Elaeis guineensis
    Hibiscus micranthus
    Hyphaene species
    Raphia species

CORD AND CORDAGE
   Fishing lines
      Ananas comosus
      Calotropis procera
      Caperonia palustris
      Crotalaria striata
      Elaeis guineensis
      Hibiscus aspera
      H. esculentus
      H. micranthus
      Pergularia extensa
      Sida rhombifolia
      Urena lobata

   Ropes
      Agave sisalana
      Borassus aethiopium
      Brachystegia taxifolia
      Cannabis sativa
      Cocos nucifera
      Crotalaria retusa
      Dombeya rotundifolia
      Entada phaseoloides
      Ficus petersii
      Glyphaea laterifolia
      Grewia bicolor
      G. forbesii
      G. holstii
      Hibiscus tiliaceus
      Isoberlinia globiflora
      Lasiosiphon lampranthus
      Phoenix reclinata
      Sansevieria cylindrica
      S. ehrenbergii
      S. robusta
      Securidaca longipedunculata
      Sesbania bispinosa
      S. sesban
      Sterculia africana
      S. rhyncocarpa
      Tamarindus indica
      Uvaria species
      Vigna schimperi

   Strings
      Abutilon usambarensis
      Bidens magnifolia
      Cissampelos pariera
      Crotalaria intermedia
      Elaeis guineensis
      Grewia bicolor
      G. forbesii

      G. holstii
      Juncellus laevigatus
      Rhynchospora corymbosa
      Sansevieria robusta
      Sansevieria species
      Securidaca longipedunculata
      Tamarindus indica
      Trema guineensis
      Triumfetta bartramia
      T. macrophylla
      Vigna schimperi

   Twine
      Ananas comosus
      Borassus aethiopium
      Cannabis sativa
      Corchorus capsularis
      C. olitorius
      Elaeis guineensis
      Hexalobus monopetalus

   Other tying materials
      Abutilon mauritianum
      Barringtonia racemosa
      Brachystegia species
      Calamus asperrimus
      Callopsis volkensii
      Cissampelos pariera
      Clematis simensis
      Cordia holstii
      Culcasia scandens
      Cyperus immensus
      Dombeya cincinnata
      Entada sudanica
      Harungana madagascariensis
      Hibiscus tiliaceus
      Lannea barteri
      Metaporana densiflora
      Parkia filicoidea
      Paullinia pinnata
      Periploca linearifolia
      Piliostigma thonningii
      Raphia mombuttorum
      R. ruffa
      Stephania species
      Sterculia africana
      Stictocardia laxiflora
      Triclisia sacleuxii
      Triumfetta tomentosa
      Uvaria species

FABRICS

Bags, bagging, or sacking
Abutilon usambarensis
*Brachystegia randii
*Cannabis sativa
Corchorus capsularis
C. olitorius
Crotalaria striata
Dichrostachys glomerata
Dombeya nairobiensis
*Sansevieria robusta
*Sansevieria species

Bark cloth
Adansonia digitata
*Brachystegia boehmii
*B. edulis
*B. randii
*Ficus species
Isoberlinia globiflora

Carpets
*Cannabis sativa
Corchorus capsularis
C. olitorius

Cloth
*Ananas comosus
Bauhinia thonningii
*Cannabis sativa
Corchorus capsularis
C. olitorius
Elaeis guineensis
*Ficus species
*Gossypium species
Musa ensete
Musa species
Securidaca longipedunculata

NETS AND NETTING
Agave sisalana
*Asclepias semilunata
*Calotropis procera
*C. retusa
*C. striata
Cyperus papyrus
Entada phaseoloides
Hibiscus esculentus
H. tiliaceus
Pachyrrhizus erosus
*Poulzolzia hypoleuca
Securidaca longipedunculata
Sida rhombifolia

*Thespesia populnea
Typhonodorum lindleyanum

PLAITING AND WEAVING
Baskets and basketwork
*Borassus aethiopium
*Cocculus hirsutus
*Cocos nucifera
*Crotalaria species
Cyperus alternifolius
var. flabelliformis
C. maranguensis
Dichrostachys glomerata
Eleusine indica
Oreobambos buchwaldii
Oxytenanthera species
Pandanus species
*Phoenix reclinata
*Raphia mombuttorum
*R. ruffia
*Smilax kraussiana
Sporobolus pyramidalis

Hats
*Borassus aethiopium
*Cocos nucifera
Luffa cylindrica
Musa species
*Sansevieria species

Mats and matting
*Borassus aethiopium
Clappertonia ficifolia
*Cocos nucifera
Cyperus alternifolius
var. flabelliformis
C. articulatus
C. haspan
C. papyrus
Pandanus species
Pennisetum purpureum
Phoenix dactylifera
*P. reclinata
*Phragmites communis
*P. mauritianus
Pycreus umbrosus
*Raphia mombuttorum
*R. ruffia
*Sansevieria species
*Scirpus corymbosus
Sterculia africana
Torulinium confertum

Summary

Plant utilization in Africa includes a vast array of foods, drugs, building and other raw materials, fuels, fibers, and even a much more recent utilization of ornamental plants. I have summarized, as examples, two major areas, namely, ethnomedical practices and the ethnobotany of vegetable fibers and flosses. In both examples, the majority of employed plants are indigenous to Africa; however, a few introduced or naturalized species are also being incorporated into African ethnobotany.

The majority of Africans (and non-Africans, for that matter) may assume that there are enough plants in cultivation to supply all the continent's diverse needs and that the gene pool, consisting of hundreds of thousands of species not currently cultivated, may well be dispensed with. As modern scientists, however, we must now begin to consider plants as invaluable resources that nature has given us, and we must conserve, propagate, and utilize these plants without adversely affecting them.

One important thing that we must teach Africans is the idea of conservation of vegetation in general. Indigenous vegetation is sometimes referred to as weeds, uneconomic forest, or scrub, and is often destroyed to make room for other agricultural or domestic activities. Indeed, if various lower plants, such as molds thriving in certain habitats, had been judged useless at the beginning of the twentieth century and it had been feasible to render them extinct, any person attempting to save them would have been considered mad and the importance of modern antibiotics, which have saved so many lives, would have been unforeseen. In fact, no plant species should be considered valueless, even if it has been scientifically studied for a long time, because it may have value in a quite different and unknown field in the future.

Numerous products are still obtained from wild plants, including drugs, foods, resins, and excellent hardwoods, but most drugs normally are synthesized eventually, while woods are now obtained from plantations of fewer species or totally replaced by new synthetic materials. Although drug chemicals are synthesized, biological screening of large members of plants frequently brings structures to the notice of modern scientists that might not have been discovered in generations of blind empirical groupings in the laboratories. The number of biochemical mechanisms of plants awaiting discovery is large. Certain plants, no doubt, will be developed into new food plants, while others possessing valuable attributes can be utilized for improving known food plants of the same genus, to feed the ever-increasing populations, particularly in Africa where the food situation has continued to deteriorate.

The role of ethnobotany in development cannot be overemphasized. African countries need to develop industrially, but developers should not be permitted ruthlessly to destroy vegetation without planning how to replace it or at least to conserve parts of it. Our goal should be "development without destruction" whenever we plan any activities, especially those that affect the natural vegetation.

BIBLIOGRAPHY

Ampofo, O. E., and F. D. Johnson-Ramauld. 1978. Traditional medicine and its role in the development of health services in Africa. World Health Organization Regional Office for Africa, Brazzaville. Afro. Tech. Paper no. 12.

Ayensu, E. S. 1978. *Medicinal Plants of West Africa*. Algonac: Reference Publications.

Hedberg, I., ed. 1979. *Systematic Botany, Plant Utilization and Biosphere Conservation*. Stockholm: Almquist and Wiksell.

Hedberg, I., and O. Hedberg. 1968. Conservation of vegetation in Africa south of the Sahara. *Acta Phytogeographica Suecica* 54. Uppsala.

Kokwaro, J. O. 1976. *Medicinal Plants of East Africa*. Nairobi: East African Literature Bureau.

———. 1980. *Classification of East African Crops*. Nairobi: East African Literature Bureau.

———. 1983. An African knowledge of ethnosystematics and its application to traditional medicine. *Bothalia* (Pretoria) 14(2): 237–243.

Sofowora, A. 1982. *Medicinal Plants and Traditional Medicine in Africa*. Chichester: John Wiley & Sons.

Watt, J. M., and M. Breyer-Brandwijk. 1962. *Medicinal and Poisonous Plants of Southern and Eastern Africa*. Edinburgh/London: E. and S. Livingstone.

The Importance of Ethnobotany
in American Anthropology

WESTON LA BARRE

For anthropologists, taxonomy in their field begins with the classic study by Emile Durkheim and Marcel Mauss, *De quelques formes primitives de classification* (1903), which has been translated into English as *Primitive Classification*. Interest in primitive classification and its rationales has since broadened into the important field of cognitive anthropology, which concerns itself with all manner of indigenous folk sciences and systems. It is significant, however, that an important early paper on color nomenclature among the Malay and Sumatran Batak was written by a botanist, H. H. Bartlett (1929), and that anthropologist H. C. Conklin's study (1955) of Hanunoo color categories was prompted by the chromatic differences upon which this Philippine people's taxonomy of plants often depends. Indeed, it might be stated that indigenous biological systems are one major source of primitive taxonomy of frogs (Bulmer and Tyler 1968), of marsupials and rodents (Bulmer and Menzies 1972–1973), and of reptiles and fishes (Bulmer et al. 1975), although more complex philosophical systems in cognitive anthropology derive from African studies (e.g., for the Dogon, Griaule and Dieterlin 1954) and ethnolinguistics (Sturtevant 1964).

That lower-level categories in folk taxonomies sometimes correspond with biological species in a one-to-one relationship has often been used as evidence that our biological species reflect objective natural groupings (e.g., Diamond 1966 on Fore birds in montane New Guinea). In what is evidently the most extensive investigation of the principles of native plant classification (Berlin et al. 1974), however, the Mexican Tzeltal "groupings do not correspond to botanical categories in any meaningful fashion" (Berlin et al. 1966, p. 63). The high proportion of such correspondence, in forty of sixty-eight cases, occurred in plants introduced *after* the Spanish Conquest; twenty-four of the twenty-seven species of high cultural significance for which there is a one-to-one correspondence belong to this post-Conquest group of forty.

Furthermore, lexically labelled taxa here do not exhaust conceptually and crucially significant unnamed or unlabelled covert taxa (Berlin et al. 1968). The 1941 study by Wyman and Harris on Navaho medical ethnobotany "is the first work to our knowledge, where the logically comparable notions of scientific genera are presented in a more or less explicit fashion" (Berlin et al. 1973, p. 233), although these authors (1968, p. 298, fn 3) noted that in 1939 Whiting presented some evidence for unnamed categories of plants in Hopi ethnobotany. The physician and naturalist Dennber (1939) presented a zoological classification of mammals that was devised by the Guaraní of Argentina and is of interest to biologists. In the potato taxonomy of the hundreds of named Aymara varieties of *Solanum tuberosum* in the Titicaca plateau, morphological criteria—not even color, much less genetic traits—appear not to be botanically significant (La Barre 1948). It is safer to proceed empirically as though classificatory criteria were *sui generis* for each tribal system and were not necessarily commensurate with the logic of Linnean groupings.

Ethnobotanical studies have been made primarily by American ethnologists and, since the majority of these are Americanists by geographic area, the result is that Native American groups have been especially well covered. The pioneering study in ethnobotany was by J. Walter Fewkes (1896). For various interests, botanical and nonbotanical, it may be useful to list the principal groups, hence areas, that have been covered: Apache (Bourke 1892; Castetter and Opler 1939), Arapaho (Kroeber 1907), Aymara (La Barre 1959a), Blackfoot (Johnston 1970), California Indians (Chestnut 1902; Harrington 1932; Romero 1954), Cherokee (Mooney 1891, 1932), Cheyenne (Grinnell 1905), Chippewa (Densmore 1910, 1913, 1928, 1929a; Reagan 1921), Chiricahua (Castetter and Opler 1939), Choctaw (Bushnell 1909), Comanche (Carlson and Jones 1939; Peters 1968), Creek (Swanton 1922, 1928), Dakota Sioux (Howard 1957), Gosiute (Chamberlin 1911), Hopi (Whiting 1939), Houma (Speck 1941), Iroquois (Fenton 1940, 1942), Jamaicans (Steggerda 1929), Kiowa (Vestal and Schultes 1939; La Barre 1947), Menomini (Smith 1923), Mescalero Apache (Castetter and Opler 1939), Meskwaki (Smith 1928), northern Mexico (Hrdlicka 1908), Micmac (Wallis 1922), Mississippi, lower and Gulf (Swanton 1911), Missouri River Indians (Gilmore 1915), Mohegan (Tantaquidgeon 1928), Navaho (Wyman and Harris 1941; Elmore 1944), Ojibwa (Smith 1932), Pacific Northwest tribes (Turner 1974), Papago (Densmore 1929b; Castetter and Underhill 1935), Paiute (Wheat 1967), Potawatomi (Smith 1933), Rappahannock (Speck et al. 1942), Salish, coastal (Turner and Bell 1971), Seneca (Fenton 1940), Southwestern Indians (Hrdlicka 1908; Taylor 1940), Teton Sioux (Densmore 1918), Tewa (Robbins et al. 1916), Thompson (Teit 1930), Tzeltal (Berlin et al. 1974), Upper Rio Grande (Curtin 1947, 1949), Ute (Chamberlin 1909), and Zuni (Stevenson 1915),

Studies of ethnobotanical interest on specific plants include those on *Amanita muscaria* (Wasson 1967a, 1972a; Schultes and Hofmann 1980, pp. 44–55), Amerindian drink plants (Havard 1896; La Barre 1938), hallucinogens (Safford 1917; Schultes 1960; Schultes and Holmstedt 1969–1970, 1972; Emboden 1972a; La Barre 1972, 1975; Harner 1973; Schultes and Hofmann 1980), *Banisteriopsis caapi* (Reichel-Dolmatoff 1972; Harner 1973; Schultes and Hofmann 1980, pp. 163–183), the myristicaceous hallucinogens (Schultes and Holmstedt 1968, 1969–1970; Chagnon et al. 1971; Schultes and Swain 1976; Schultes et al. 1978), *Anadenanthera* (*Piptadenia*) species (Safford 1916; von Reis Altschul 1967, 1972); *Cannabis* (Emboden 1972b; La Barre 1977, 1980; Schultes and Hofmann 1980), *Datura* (Gayton 1928; Schultes and Hofmann 1980); *Ilex* (Hale 1891; Emboden 1972a, pp. 71, 86), *Lophophora williamsii* (Diguet 1907; Safford 1915; Rouhier 1927; Klüver 1928; Lewin 1931; Hollander 1935; Schultes 1937a, 1937b; Gusinde 1939; Aberle 1966; Anderson 1969; La Barre 1969), *Nicotiana* (Harrington 1932; Wilbert 1972), *Nymphaea caerulea* or narcotic blue water lily (Emboden 1989), *Psilocybe* species (Wasson 1962a; Schultes and Hofmann 1980), *Sophora secundiflora* (Howard 1957, 1960, 1972; Campbell 1958; Troike 1962; Adovasio and Fry 1976), *Tabernanthe iboga* (Fernandez 1972), and *Trichocereus pachanoi* (Sharon 1972).

In the first half of the twentieth century, ethnobotanical studies by anthropologists tended to be no more than folkloristic lists of native plants with their uses and botanical identification. These were conceived as no more than incidental addenda to ethnographies. But there was also early interest in native herbs as potential medicines. Well known among these folk contributions from the Americas are cascara, cinchona, cocaine, cohosh, dockmarkie, lobelia, pipsissewa, puccoon, the aspirin-related salixes (much used by the original inhabitants of the Americas), sassafras (in colonial times a cinnamon equivalent), and tobacco.

The interest in new Native American plants as potential medicines is as early as Hernández (1651), Beverly (1705), and Bartram (1791). Summaries of useful plants have been made by Millspaugh (1887), Youngken (1924, 1925), Stone (1962), and Vogel (1970), as well as of principles of traditional medicine relying on the supernatural "power" in certain plants (Nieuwenhuis 1924; Rogers 1948; La Barre 1959a, 1959b, 1970a, 1970b, 1970c). A related study on the contacts between Iroquois herbalism and colonial medicine was written by Fenton (1942), while others have written about indigenous doctoring (Dixon 1908; Clements 1932; Corlett 1935; Hallowell 1935).

Ethnobotany is not merely an odd neologism. Botany and ethnology are on occasion considerable mutual aids. Palynology, the study of pollens, is critical in some kinds of archeology, as also is Carbon-14 count on organic materials. Botanical data are highly significant in the paleoethnology of religion, including Wasson's *soma* and the ancient use of cannabis in the Hindic sphere (Wasson 1958, 1967a, 1970, 1972a; La Barre 1980), in ancient Greek religion (Wasson et al. 1978) and in the Americas (Wasson 1962a, 1962b, 1972a, 1972b; Adovasio and Fry 1976). Likewise, in the reverse direction, ethnological data at times have been useful in the understanding of botanical problems, such as the imbalance in Old and New World narcotic plants, perhaps owing to profound cultural differences (La Barre 1970b). (Botanical archeology and paleobotany are more fully treated elsewhere in the present volume.)

Shortly before the middle of the twentieth century, indigenous knowledge began to be treated by ethnologists more respectfully than as mere folklore. La Barre (1942), writing on folk medicine and folk sciences as good potential sources of useful information, discussed Kiowa folk sciences (1947), and, in a monograph on a Native American group of the Lake Titicaca plateau (1948), he presented materials on Aymara ethnometeorology, ethnohistory, ethnobotany, ethnoanatomy and ethnophysiology, ethnoanthropology, ethnopsychology, ethnoethnology, and ethnogeography. Ethnobotany was, of course, by then a subject of established interest among American anthropologists, and he took his cue from ethnobotany for his other folk sciences. The interest in folk sciences has since flourished with the rise of cognitive anthropology, and we now have studies of the ethnoichthyology of the Hong Kong boat people (E. N. Anderson 1967) and the ethnoentomology of the Navaho (Wyman and Bailey 1964).

Aymara medicine is very highly developed, as is also the tribe's largely botanical *materia medica* (La Barre 1959a). Indigenous folk medicine and psychiatry have become an important interest in modern medical anthropology (Kiev 1964). For various reasons, ethnobotany and ethnozoology have led studies in folk science, culminating in the works of Berlin (1978) and Berlin et al. (1966, 1968, 1973, 1974), and in a special issue on folk biology in *American Ethnologist* (vol. 3, no. 3, August 1976). Radin (1927) has also contributed importantly to these developments, and, somewhat independently, Reichel-Dolmatoff (1971) in sensitive field work in Amazonia.

Although he claimed to be neither professional botanist nor anthropologist, opportunity must be taken to pay homage to the late R. Gordon Wasson, whose work ranks with the best in both professions. Wasson founded ethnomycology virtually single-handedly (Wasson 1956, 1958, 1961, 1962a, 1963, 1967a, 1967b, 1970, 1972a, 1972b; Wasson and Wasson 1957), besides contributing to ethnobotany in general (Wasson 1962b, 1966, 1973). He resolved the three-and-one-half millennia mystery of the identification of the Vedic hallucinogen *soma* (Wasson 1967a, 1967b, 1970, 1972a), sufficient in itself for a sound scholarly reputation (see La Barre 1970a, 1970c, 1980). For an appreciation of some of the ramifications in the anthropology of religion resulting from Wasson's insights, see his writings on both primitive (Wasson 1956, 1958, 1961, 1962a, 1963) and early Greek religion

(Wasson et al. 1978). Ethnobotanists will relish especially Wasson's acute understanding of "flower" in Aztec iconography (Wasson 1973).

LITERATURE CITED

Aberle, D. F. 1966. *The Peyote Religion Among the Navaho*. Chicago: Aldine.

Adovasio, J. M., and G. F. Fry. 1976. Prehistoric psychotropic drug use in northeastern Mexico and Trans-Pecos Texas. *Economic Botany* 30: 94–96.

Anderson, E. F. 1969. The biogeography, ecology, and taxonomy of *Lophophora* (Cactaceae). *Brittonia* 21(4): 299–310.

Anderson, E. N. 1967. *The Ethnoichthyology of the Hong Kong Boat People*. Ph.D. thesis, University of California, Berkeley.

Bartlett, H. H. 1929. Color nomenclature in Batak and Malay. *Papers of the Michigan Academy of Sciences, Arts and Letters* 10: 1–52.

Bartram, W. 1791. *Travels through North and South Carolina, Georgia, East and West Florida, the Cherokee Country . . . with Observations on the Manners of the Indians*. Philadelphia: James and Johnson.

Berlin, B. 1978. Ethnobiological classification. In *Cognition and Categorization*. Eds. E. Rosch and B. B. Lloyd. Hillsdale, NJ: Lawrence Erlbaum Associates; distributed by John Wiley and Sons. 11–26.

Berlin, B., D. E. Breedlove, and P. H. Raven. 1966. Folk taxonomies and biological classification. *Science* 155(14) (October): 273–275.

———. 1968. Covert categories and folk taxonomies. *American Anthropologist* 70(2): 290–299.

———. 1973. General principles of classification and nomenclature in folk biology. *American Anthropologist* 75(1): 214–242.

———. 1974. *Principles of Tzeltal Plant Classification: An Introduction to the Botanical Ethnography of a Mayan-Speaking Community in Highland Chiapas*. New York: Academic Press.

Beverly, R. 1705. *History of Virginia, by a Native and Inhabitant of the Place*. London: B. and S. Tooke. Rpt. 1947, Chapel Hill: University of North Carolina. 2nd ed., London, 1722.

Bourke, J. G. 1892. The medicine men of the Apache. Bureau of American Ethnology 9th Annual Report. Washington, DC. 443–603.

Bulmer, R., and J. I. Menzies. 1972–1973. Karem classification of marsupials and rodents. *Journal of the Polynesian Society* 81: 472–499; 82: 86–107.

Bulmer, R., J. I. Menzies, and F. Parker. 1975. Karem classification of reptiles and fishes. *Journal of the Polynesian Society* 84: 267–307.

Bulmer, R., and M. Tyler. 1968. Karam classification of frogs. *Journal of the Polynesian Society* 77: 333–385.

Bushnell, D. I. 1909. *The Choctaw of Bayou Lacomb, St. Tammany Parish, Louisiana*. Bureau of American Ethnology Bulletin no. 48. Rpt. 1917.

Campbell, T. N. 1958. The origin of the Mescal bean cult. *American Anthropologist* 60: 156–160.

Carlson, G. G., and V. H. Jones. 1939. Some notes on the uses of plants by the Comanche Indians. *Papers of the Michigan Academy of Sciences, Arts and Letters* 25: 517–542.

Castetter, E. F., and M. E. Opler. 1939. *The Ethnobiology of the Chiricahua and Mescalero Apache*. University of New Mexico, Ethnobiological Series 4(5).

Castetter, E. F., and R. M. Underhill. 1935. *The Ethnobiology of the Papago Indians*. University of New Mexico Bulletin 275, biological series, vol. 4, no. 3.

Chagnon, N. A., P. Le Quesne, and J. M. Cook. 1971. Yanomamó hallucinogens: Anthropological, botanical and chemical findings. *Current Anthropology* 12: 72–74.

Chamberlin, R. V. 1909. Some plant names of the Ute Indians. *American Anthropologist* 11: 27–40.

———. 1911. Ethno-botany of the Gosiute Indians of Utah. *Memoirs of the American Anthropological Association* 2(5).

Chestnut, V. K. 1902. Plants used by the Indians of Mendocino County, California. *Contributions to the United States National Herbarium* 7(3).

Clements, E. E. 1932. Primitive concepts of disease. *University of California Publications in American Archaeology and Ethnology* 32: 185–252.

Conklin, H. C. 1955. Hanunoo color categories. *Southwestern Journal of Anthropology* 11: 339–344.

Corlett, W. T. 1935. *The Medicine-man of the American Indian*. Baltimore: Charles C. Thomas.

Curtin, L. S. M. 1947. *Healing Herbs of the Upper Rio Grande*. Sante Fe, NM: Santa Fe Laboratory of Anthropology.

———. 1949. *Ny the Prophet of the Earth*. Santa Fe, NM: San Vincente Foundation.

Dennber, J. G. 1939. Los nombres indígenas en guaraní de los mamíferes de la Argentina y países limítrofes y su importancia para la systémica. *Physis* 16: 225–244.

Densmore, F. 1910. Chippewa music. Bureau of American Ethnology, Bulletin no. 45. Washington, DC. 1–216.

———. 1913. Chippewa music II. Bureau of American Ethnology Bulletin no. 53. Washington, DC. 1–341.

———. 1918. Teton Sioux music. Bureau of American Ethnology Bulletin no. 61. Washington, DC. 1–561.

———. 1928. Uses of plants by the Chippewa Indians. Bureau of American Ethnology 44th Annual Report. Washington, DC. 275–397.

———. 1929a. Chippewa customs. Bureau of American Ethnology Bulletin no. 86. Washington, DC. 1–204.

———. 1929b. Papago music. Bureau of American Ethnology Bulletin no. 90. Washington, DC. 1–229.

Diamond, J. M. 1966. Zoological classification system of a primitive people. *Science* 151: 1102–1104.

Diguet, L. 1907. Le "peyote" et son usage chez les indiens du Nayarit. *Journal de la société des Américanistes de Paris* 4: 21–29.

Dixon, R. B. 1908. Some aspects of the American shaman. *Journal of American Folklore* 21: 1–12.

Durkheim, E., and M. Mauss. 1903. De quelques formes primitives de classification. *Année sociologique* (1901–1902) 6: 1–72. (Trans. R. Needham. 1963. *Primitive Classification*. Chicago: University of Chicago Press.)

Elmore, F. H. 1944. *Ethnobotany of the Navajo*. Albuquerque: University of New Mexico Press.

Emboden, W., Jr. 1972a. *Narcotic Plants*. 2nd ed. New York: Macmillan.

———. 1972b. Ritual use of *Cannabis sativa* L.: A historical-ethnographic survey. In *Flesh of the Gods: The Ritual Use of Hallucinogens*. Ed. P. T. Furst. New York: Praeger. 214–236.

———. 1989. The sacred journey in dynastic Egypt: Shamanistic trance in the context of the narcotic water lily and the mandrake. *Journal of Psychoactive Drugs* 21: 61–75.

Fenton, W. N. 1940. An herbarium from the Allegheny Seneca. In *Historic Annals of Southwestern New York*. Eds. Doty, Congdon, and Thornton. New York: Lewis Historical Publishing Company. 787–796.

———. 1942. Contacts between Iroquois herbalism and colonial medicine. Smithsonian Institution Annual Report for 1941, Publication 3670. 503–526.

Fernandez, J. W. 1972. *Tabernanthe Iboga*: Narcotic ecstasy and the work of the ancestors. In *Flesh of the Gods: The Ritual Use of Hallucinogens*. Ed. P. T. Furst. New York: Praeger. 237–260.

Fewkes, J. W. 1896. A contribution to ethnobotany. *American Anthropologist*, old series, 9: 14–21.

Gayton, A. H. 1928. *The Narcotic Plant Datura in Aboriginal America*. Thesis, University of California, Berkeley.

Gilmore, M. R. 1915. Uses of plants by the Indians of the Missouri River region. Bureau of American Ethnology Annual Report 1911–1912. Washington, DC. 33: 43–154.

Griaule, M., and G. Dieterlin. 1954. The Dogon. *African Worlds*. Ed. D. Forde. London: Oxford. 83–110.

Grinnell, G. B. 1905. Some Cheyenne plant medicines. *American Anthropologist* 7: 37–43.

Gusinde, M. 1939. Der Peyote-Kalt, Entstehung und Verbreitung. In *Festschrift zum 50-jahrigen Bestandsjubiläum des Missionhauses St. Gabriel*. Vienna-Mödling: St. Gabriel Studien 8. 401–499.

Hale, E. M. 1891. *Ilex cassine, the Aboriginal North American Tea*. USDA Bulletin 14.

Hallowell, A. I. 1935. Primitive concepts of diseases. *American Anthropologist* 46: 364–368.

Harrington, J. P. 1932. Tobacco among the Karuk Indians of California. Bureau of American Ethnology Bulletin no. 14. Washington, DC. 1–84.

Harner, M. J., ed. 1973. *Hallucinogens and Shamanism*. New York: Oxford University Press.

Havard, V. 1896. Drink plants of the North American Indians. *Torrey Botanical Club Bulletin* 23: 34–46.

Hernández, F. 1651. *Nova Plantarum, Animalium et Mineralium Mexicanorum Historia*. Rome: B. Deuersini et Z. Masetti.

Hollander, A. N. J. 1935. De Peyote-cultus der Noord-amerikaanise Indien. *Mensch en Maatschapeije* 11(1–2).

Howard, J. H. 1957. The Mescal bean cult of the central and southern plains: An ancestor of the peyote cult? *American Anthropologist* 59: 75–87.

———. 1960. Mescalism and peyotism again. *Plains Anthropologist* 5: 84–85. Mimeo.

———. 1972. Potawatomi mescalism and its relationship to the diffusion of the peyote cult. *Plains Anthropologist* 7: 125–135. Mimeo.

Hrdlicka, A. 1908. *Physiological and Medical Observations Among the Indians of Southwestern United States and Northern Mexico*. Bureau of American Ethnology Bulletin no. 34. Washington, DC. 1–460.

Johnston, A. 1970. Blackfoot Indian utilization of the flora of the northwestern great plains. *Economic Botany* 24(3).

Kiev, A., ed. 1964. *Magic, Faith, and Healing*. New York: Free Press.

Klüver, H. 1928. *Mescal [Peyote] The "Divine" Plant and its Psychological Effects*. London: K. Paul, Trench, Trubner & Company. Rpt. 1966 as part of *Mescal, and Mechanisms of Hallucination*. Chicago: University of Chicago Press.

Kroeber, A. L. 1907. *The Arapaho*. American Museum of Natural History, Bulletin 18.

La Barre, W. 1938. Native American beers. *American Anthropologist* 40(2): 224–234.

———. 1942. Folk medicines and folk science. *Journal of American Folklore* 55(218): 197–203.

———. 1947. Kiowa folk sciences. *Journal of American Folklore* 60: 105–114.

———. 1948. The Aymara Indians of the Lake Titicaca Plateau, Bolivia. *Memoirs of the American Anthropological Association* 68.

———. 1959a. Materia medica of the Aymara, Lake Titicaca Plateau, Bolivia. *Webbia* 15(1): 47–94.

———. 1959b. *The Peyote Cult*. Hamden, CT: Shoe String Press.

———. 1969. *The Peyote Cult*. 5th ed. Norman: University of Oklahoma Press.

———. 1970a. *The Ghost Dance; Origins of Religion*. New York: Doubleday, London: Allen & Unwin 1972; rev. 2nd ed. Delta Books 1972, 1978. Prospect Heights, IL: Waveland Press, 1990.

———. 1970b. Old and New World narcotics: A statistical question and an ethnological reply. *Economic Botany* 24(1): 73–80.

———. 1970c. Review of Wasson's *Soma*. *American Anthropologist* 72: 368.

———. 1972. Hallucinogens and the shamanic origins of religion. In *Flesh of the Gods: The Ritual Use of Hallucinogens*. Ed. P. T. Furst. New York: Praeger. 261–278.

———. 1975. Anthropological perspectives on hallucination and hallucinogens. In *Hallucination*. Eds. R. K. Siegel and L. J. West. New York: John Wiley & Sons. 9–52.

———. 1977. Anthropological views of *Cannabis*. *Reviews in Anthropology* 4: 237–250.

———. 1980. History and ethnography of *Cannabis*. In *Culture in Context: Selected Writings of Weston La Barre*. Durham, NC: Duke University Press.

Lewin, L. 1931. *Phantastica: Narcotic and Stimulating Drugs*. Trans. from 2nd German ed. by P. H. A. Wirth. New York: E. P. Dutton & Company. Rpt. 1964.

Millspaugh, C. F. 1887. *American Medicinal Plants, an Illustrated and Descriptive Guide to the American Plants used as Homeopathic Remedies*. 2 vols. New York: Boericke & Tafel.

Mooney, J. 1891. The sacred formulas of the Cherokee. Bureau of American Ethnology 7th Annual Report (1885–1886). Washington, DC. 301–397.

———. 1932. The Swimmer manuscript: Cherokee sacred formulas and medicinal prescriptions. Bureau of American Ethnology Bulletin no. 99. Washington, DC. 1–319.

Nieuwenhuis, A. W. 1924. Principles of Indian medicine in American ethnology and their psychological significance. *Janus, Archives internationales pour l'Histoire de la Médecine et le Géographie Médicale* 28: 305–356.

Peters, A. 1968. Comanche plant medicines. *Papers on Anthropology*. University of Oklahoma 9(1).

Radin, P. 1927. *Primitive Man as Philosopher*. New York: D. Appleton.

Reagan, A. B. 1921. Some Chippewa medicinal receipts. *American Anthropologist* 23: 246–249.

Reichel-Dolmatoff, G. 1971. *Amazonian Cosmos*. Chicago: University of Chicago Press.

———. 1972. The cultural context of an aboriginal hallucinogen. *Banisteriopsis caapi*. In *Flesh of the Gods: The Ritual Use of Hallucinogens.* Ed. P. T. Furst. New York: Praeger. 84–113.

von Reis Altschul, S. 1967. Vilca and its use. In *Ethnopharmacologic Search for Psychoactive Drugs.* Eds. D. H. Efron, B. Holmstedt, and N. S. Kline. U.S. Public Health Service Publication no. 1645. Washington, DC: Government Printing Office. 307–314.

——. 1972. *The Genus Anadenanthera in Amerindian Cultures.* Cambridge, Massachusetts: Botanical Museum, Harvard University.

Robbins, W. W., J. P. Harrington, and B. Freire-Marreco. 1916. *The Ethnobotany of the Tewa Indians.* Bureau of American Ethnology Bulletin no. 55. Washington, DC. 1–124.

Rogers, S. L. 1948. Disease concepts in North America. *American Anthropologist* 35: 559–564.

Romero, J. B. 1954. *The Botanical Lore of the California Indians.* New York: Vantage Press.

Rouhier, A. 1927. *La plante qui fait les yeux merveilles, Le Peyotl.* Paris: G. Doin.

Safford, W. E. 1915. Identification of Teonanacatl of the Aztecs with the narcotic cactus *Lophophora williamsii.* Address to Botanical Society of Washington, DC.

——. 1916. Identity of cohoba, the narcotic snuff of ancient Haiti. *Journal of the Washington Academy of Sciences* 6: 547–562.

——. 1917. Narcotic plants of the ancient Americans. Smithsonian Institution Annual Report for 1916. Washington, DC: Government Printing Office. 387–424.

Schultes, R. E. 1937a. Peyote and plants used in the peyote ceremony. *Botanical Museum Leaflets* (Harvard University) 4(7).

——. 1937b. Peyote and the plants confused with it. *Botanical Museum Leaflets* (Harvard University) 5(5).

——. 1960. Native narcotics of the New World. *Pharmaceutical Sciences,* 3rd lecture series, 142–185.

Schultes, R. E., and A. Hofmann. 1980. *The Botany and Chemistry of Hallucinogens.* 2nd ed. Springfield, IL: Charles C. Thomas.

Schultes, R. E., and B. Holmstedt. 1968. The vegetal ingredients of the myristicaceous snuffs of the northwest Amazon. *Rhodora* 70: 113–160.

——. 1969–1970. The New World Indians and their hallucinogenic plants. *Bulletin des Stupéfiants* 21(3): 3–16; 21(4): 17–30; 22(1): 25–53.

——. 1972. An overview of hallucinogens in the Western Hemisphere. In *Flesh of the Gods: The Ritual Use of Hallucinogens.* Ed. P. T. Furst. New York: Praeger. 3–54.

Schultes, R. E., and T. Swain. 1976. De plantis toxicariis e Mundo Novo tropicale commentationes. XIII. Further notes on *Virola* as an orally administered hallucinogen. *Journal of Psychedelic Drugs* 8: 317–324.

Schultes, R. E., T. Swain, and T. Plowman. 1978. *Virola* as an oral hallucinogen among the Boras of Peru. De plantis toxicariis e Mundo Novo tropicale commentationes. XVII. *Botanical Museum Leaflets* (Harvard University) 25: 259–272.

Sharon, D. 1972. The San Pedro Cactus in Peruvian Folk Healing. In *Flesh of the Gods: The Ritual Use of Hallucinogens.* Ed. P. T. Furst. New York: Praeger. 114–135.

Smith, H. H. 1923. *Ethnobotany of the Menomini Indians.* Museum of the City of Milwaukee, Bulletin 4: 1–174.

——. 1928. *Ethnobotany of the Meskwaki Indians.* Museum of the City of Milwaukee, Bulletin 4: 175–326.

——. 1933. *Ethnobotany of the Ojibwa Indians.* Museum of the City of Milwaukee, Bulletin 4: 327–525.

——. 1933. *Ethnobotany of the Forest Potawatomi.* Museum of the City of Milwaukee, Bulletin 7: 1–230.

Speck, F. G. 1941. A list of plant curatives obtained from the Houma Indians of Louisiana. *Primitive Man* 14(4).

Speck, F. G., R. B. Hassrick, and E. S. Carpenter. 1942. Rappahannock herbals, folklore and science of cures. *Proceedings of the Delaware County Institute of Science,* Media, PA 10(1).

Steggerda, M. 1929. Plants of Jamaica used by natives for medicinal purposes. *American Anthropologist* 31: 431–434.

Stevenson, M. C. 1915. Ethnobotany of the Zuñi Indians. Bureau of American Ethnology 30th Annual Report (1908–1909). Washington, DC. 35–102.

Stone, E. 1962. *Medicines Among the American Indians.* New York: Hafner.

Sturtevant, W. G. 1964. Studies in ethnoscience. In *Transcultural Studies in Cognition, American Anthropologist* 66(3), part 2. Eds. A. K. Romney and R. G. d'Andrade. 99–131.

Swanton, J. R. 1911. Indian tribes of the lower Mississippi valley and the adjacent coast of the Gulf of Mexico. Bureau of American Ethnology Bulletin no. 43. Washington, DC. 1–387.

———. 1922. Early history of the Creek Indians and their neighbors. Bureau of American Ethnology Bulletin no. 7. Washington, DC. 1–492.

———. 1928. Religious beliefs and medical practices of the Creek Indians. Bureau of American Ethnology Annual Report 1924–1925. Washington, DC. 473–672.

Tantaquidgeon, G. 1928. Mohegan medical practices, weather-lore, and superstition. Bureau of American Ethnology Bulletin no. 43 (1925–1926). Washington, DC.

Taylor, L. A. 1940. *Plants Used as Curatives by Certain Southeastern Tribes*. Cambridge, MA: Botanical Museum of Harvard University.

Teit, J. A. 1930. *The Ethnobotany of the Thompson Indians of British Columbia*. Bureau of American Ethnology 45th Annual Report (1927–1928). Washington, DC. 441–552.

Troike, R. C. 1962. The origin of plains mescalism. *American Anthropologist* 64: 946–963.

Turner, N. J. 1974. Plant taxonomic systems and the ethnobotany of three contemporary Indian groups of the Pacific Northwest (Haida, Bella Coola, and Lillooet). *Syesis* 7 (Supplement 1).

Turner, N. J., and M. A. M. Bell. 1971. The ethnobotany of the Coast Salish Indians of Vancouver Island. *Economic Botany* 25(1): 63–104.

Vestal, P. A., and R. E. Schultes. 1939. *The Economic Botany of the Kiowa Indians as It Relates to the History of the Tribe*. Cambridge, MA: Botanical Museum of Harvard University.

Vogel, V. J. 1970. *American Indian Medicine*. Norman: University of Oklahoma Press.

Wallis, W. D. 1922. Medicine used by the Micmac Indians. *American Anthropologist* 24: 24–30.

Wasson, R. G. 1956. Lightning bolt and mushrooms: An essay in early cultural exploration. In *Festschrift for Roman Jakobson*. Ed. M. Halle. The Hague: Mouton. 105–112.

———. 1958. The divine mushroom: Primitive religion and hallucinatory agents. *Proceedings of the American Philosophical Society* (24 June) 102(3): 221–223.

———. 1961. The hallucinogenic fungi of Mexico: An inquiry into the origin of the religious ideas among primitive peoples. *Botanical Museum Leaflets* (Harvard University) 19(7): 137–162.

———. 1962a. The hallucinogenic mushrooms of Mexico and psilocybin: A bibliography. *Botanical Museum Leaflets* (Harvard University) 20(2): 25–73.

———. 1962b. A new psychotropic drug from the mint family. *Botanical Museum Leaflets* (Harvard University) 20: 77–81.

———. 1963. The mushroom rites of Mexico. *The Harvard Review* 1(4): 7–17.

———. 1966. *Ololiuqui* and the other hallucinogens of Mexico. In *Summa Antropológica en Homenaje a Roberto J. Weitlaner*. Mexico: Instituto Nacional de Antropología e Historia. 329–348.

———. 1967a. *Soma, Divine Mushroom of Immortality*. New York: Harcourt, Brace & World.

———. 1967b. The fly-agaric and man. In *Ethnopharmacologic Search for Psychoactive Drugs*. Eds. D. H. Efron, B. Holmstedt, and N. S. Kline. U.S. Public Health Service Publication no. 1645. Washington, DC: Government Printing Office. 405–414.

———. 1970. Soma of the Aryans: An ancient hallucinogen. *Bulletin of Narcotics* 22(3): 25–30.

———. 1972a. The divine mushroom of immortality. In *Flesh of the Gods: The Ritual Use of Hallucinogens*. Ed. P. T. Furst. New York: Praeger. 185–200.

———. 1972b. Review of *A Separate Reality: Further Conversations with Don Juan*, by Carlos Castaneda. *Economic Botany* 26(1): 98–99.

———. 1973. The role of "flowers" in Nahuatl culture: A suggested interpretation. *Botanical Museum Leaflets* (Harvard University) 23(8): 305–324.

Wasson, R. G., A. Hofmann, and C. A. P. Ruck. 1978. *The Road to Eleusis: Unveiling the Secret of the Mysteries*. With a new translation of the *Homeric Hymn to Demeter* by D. Staples. New York: Harcourt Brace Jovanovich.

Wasson, V. P., and R. G. Wasson. 1957. *Mushrooms, Russia, and History*. New York: Pantheon Books.

Wheat, M. P. 1967. *Survival Arts of the Primitive Paiutes*. Reno: University of Nevada Press.

Whiting, A. F. 1939. Ethnobotany of the Hopi. *Museum of Northern Arizona Bulletin* 15.

Wilbert, J. 1972. Tobacco and shamanistic ecstasy among the Warao Indians of Venezuela. In *Flesh of the Gods: The Ritual Use of Hallucinogens*. Ed. P. T. Furst. New York: Praeger. 55–83.

Wyman, L. C., and F. L. Bailey. 1964. *Navaho Indian Ethnoentomology*. Albuquerque: University of New Mexico Press.

Wyman, L. C., and S. K. Harris. 1941. Navaho Indian medical ethnobotany. *University of New Mexico Bulletin* 366, Anthropological Series 3.5. Albuquerque.

Youngken, H. W. 1924. Drugs of the North American Indians. *American Journal of Pharmacy* 96(July): 485–502.

———. 1925. Drugs of the North American Indians. *American Journal of Pharmacy* 97(March): 158–185; (April): 257–271.

Ethnobotanical Resources of
Hot, Arid Zones of India

J. K. MAHESHWARI

The arid zone in India covers approximately 12 percent of the geographical area of the country and is spread over eight states. The hot, arid zone occupies more than 320,000 square kilometers (120,000 square miles) and is spread over the states of Rajasthan (62 percent), Gujarat (19 percent), Andhra Pradesh and Karnataka (10 percent), and Punjab and Haryana (9 percent). The cold, arid tracts are located in the northwestern Himalayas—in Ladakh region (in the state of Jammu and Kashmir) and in Lahaul and Spiti (in the state of Himachal Pradesh).

The northern portion of the arid zone is the continuation of the great Saharo-Sindhian Desert and is bounded on the east by the Aravalli Ranges. Three-fifths of Rajasthan lies west of the Aravallis and constitutes the bulk of the country's hot, arid zone (Figure 1). The region is characterized by aridity, sparseness of vegetation, and extremes of temperature. The aridity of the climate coupled with the sandy soil and anthropocentric activities produces vegetation of desertic type with low biological productivity. The extreme arid zone in a small portion of western Rajasthan does not experience any rain for more than a year at a stretch. In other cases, all twelve months of a year may be dry, or there may be sporadic showers of very low density and duration.

A study of ombrothermic diagrams shows that the hot, arid zone has a long dry season of eight to ten months, alternating with a very short rainy period. The rainfall decreases from about 500 millimeters (20 inches) at the foot of the Aravallis to as low as 75 millimeters (3 inches) in western Rajasthan and 15 millimeters (0.6 inch) at Leh (capital of Ladakh region, Kashmir). The mean annual rainfall ranges from 100 millimeters (4 inches) in the northwestern sector of Jaisalmer district to 450 millimeters (18 inches) in eastern Rajasthan, 300 to 500 millimeters (12 to 20 inches) in Gujarat, and 200 to 450 millimeters (8 to 18 inches) in Punjab and Haryana.

The most characteristic feature of the climate of the Indian desert is the great variation in temperature. May and June are the hottest months of the year with average maximum temperatures ranging between 40°C (104°F) and 42°C (108°F). January is the coldest month with average minimum temperatures ranging between 4.7°C (40°F) and 10.2°C (50°F). Cold waves during winter and heat waves during summer are common phenomena in this part of the country. The highest maximum temperature recorded was 50°C (122°F) in June 1934, and the lowest minimum temperature recorded was 5.9°C (42°F) in January 1967.

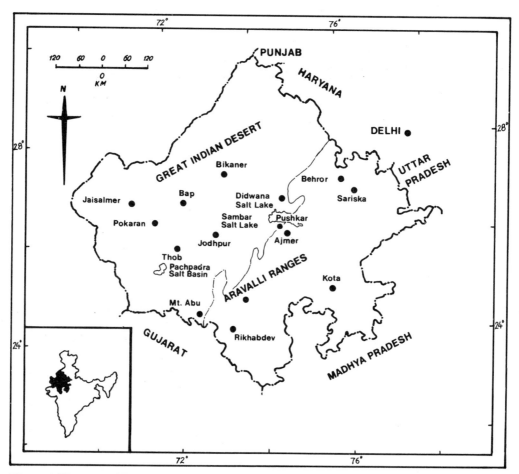

Figure 1. The state of Rajasthan, which includes the Great Indian Desert, various salt lakes, and the Aravalli Ranges.

The region experiences high annual evaporation with most of the Thar Desert having a potential evaporation rate of more than 250 millimeters (10 inches). The rate of evaporation is lowest in the month of January (3 millimeters per 0.1 inch per day) and highest (13 millimeters per 0.5 inch per day) during the months of May and June.

The frequency of dust storms is quite high in the region. The erratic rainfall results in an imbalance between the human and animal populations on the one hand, and the plant, water, and land resources on the other. This imbalance leads to progressive depletion of resources, to permanent loss of valuable plant species and of topsoil by water or wind erosion, and ultimately to the production of large areas of barren wasteland (Maheshwari 1976, 1984). The general terrain shows a conspicuous series of sand ridges, dunes, and hillocks.

The important saline lakes and depressions (called *ranns*) are located at Sambhar, Lunkaransar, Jamsar, Didwana, Bap, Pokhran, Pachpadra, Thob. Many other smaller saline lakes are also distributed within the desert. The Rann of Kutch is a dry bed of the remnant of the sea that once connected the Marmada rift with Sind and separated Kutch from the mainland. It is a vast expanse of tidal mud flats with saline efflorescence. The average annual rainfall is about 350 millimeters (14 inches) with a very long dry season of eight to ten months.

Tribal Groups of the Arid Zone

The scheduled tribes in the country have been variously termed as adivasis, hill tribes, primitive tribes, tribals, and so forth. They are presumed to form the oldest ethnological sector of the population and are also known as aboriginals. They live in their tribal ways and have their own superstitious and magico-religious beliefs and customs, including medicines for various diseases and ailments. India has the second largest concentration of scheduled tribes in the world. Predominant among its 563 scheduled tribes are the Bhil, Gond, Ho, Kachari, Kawar, Khasi, Khond, Kol, Mina, Munda, Naga, Oraon, Santal, and Saora. Tribal population has increased from approximately 20 million (1951 census) to 52 million (1981 census).

The state of Rajasthan is inhabited by a large number of tribes, including the Bhil, Damor, Garasia, Kathodia, Mina, and Sahariya (Figure 2). Among the state's nomadic tribes are the Banjara and the Gadia Lohar. The Banjara or Lambadis of Gujarat, Rajasthan, Maharashtra, Madhya Pradesh, and Andhra Pradesh originally carried merchandise, with the aid of their pack animals, from one part of the country to another. Even today they move from place to place and wear very colorful dresses and ivory ornaments. The Gadia Lohar, so named because they move about in their *gadia* or carts, are the blacksmiths of Rajasthan, Gujarat, and Madhya Pradesh. Wherever they travel, their heavy and richly decorated bullock carts go with them. The Mina or Maina, many of whom have settled down as agriculturists, are concentrated in Bundelkhand district, Uttar Pradesh; in Hoshangabad, Nima, and Sagar districts of Madhya Pradesh; and in some parts of Rajasthan.

Gujarat has been the home of several tribes since time immemorial, including the Halpati, Kathodi, Kolgha, Koli, and Kotwalia (Figure 3). The Kukna, Kunbis, and Warli live in the hilly region and forest tracts, whereas other tribes such as the Choudhary, Dhodia, Dubla, and Naikdas live in the plains (Figure 4).

Many of India's tribals depend on plants for their livelihood and collect tubers, barks, roots, rhizomes, flowers, fruits, seeds, leaves, fibers, gums, lac, honey, and wax to use as food, medicines, vegetable oils, gums, tanning materials, dyes, fuel, fiber, timber, charcoal, domestic articles, agricultural or hunting tools, and for religious worship, witchcraft, magic, and esthetic purposes. There thus exists an interrelationship between these people and plants, a relationship that has been termed *ethnobotany*.

Tribal people are the repository of accumulated experience and knowledge of indigenous vegetation that has not been properly utilized for the economic development of the region. These people are not only familiar with the thousands of plant species in their ecosystems, but they also understand the ecological interrelations of the various components of their resource base. Field work by the author and others among the tribals of the arid and semi-arid zone has brought to light new information that can be utilized in developing the indigenous economy of the tribals by organizing systematic collection of the forest produce and by locating cottage industries—especially of herbal drugs, oil seeds, foods, gums, fibers, cordage, mats and basketry, combs, brooms, dyes, tannins, incenses, water bottles, and musical instruments—in the tribal areas.

The Bhils inhabit a wide belt covering the Banswara, Chittorgarh, Dungarpur, Sirohi, and Udaipur districts of Rajasthan; the Dhar, Jhabua, Khargone, and Ratlam districts of Madhya Pradesh; the Panchmahal and Sabarkantha districts of Gujarat; and the Khandesh district of Maharashtra. They constitute the third largest tribe of India, the first two being Gonds and Santals. References to Bhils in the literature date back to the Mahabharata period, where one finds the name of Prince Eklavya. Probably, the name Bhil derived from the Dravidian word *bil* or *vil*, meaning "a bow." Formerly, the chief occupation of the tribe was hunting and free exploitation of forest produce. Now it is cultivating the

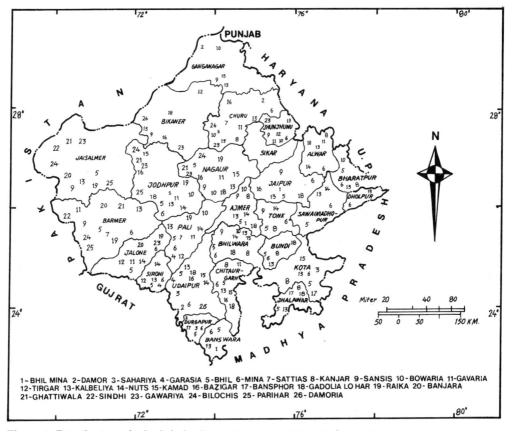

Figure 2. Distribution of scheduled tribes in the state of Rajasthan, by district.

land; some of the Bhils live as farm servants and laborers, working in the forests or in the fields.

Two Bhil medicine men, Badwa and Bhopa, would not talk with me about medicinal uses of herbs, but they revealed information about the traditional system of medicine. Some important medicinal plants used by the Bhils are listed below:

AILMENT	SPECIES	PLANT PART USED
Scorpion bite	*Argemone mexicana* L.	Latex
Asthma	*Calotropis procera* R. Br.	Rootstock
Mumps/diabetes	*Cayratia trifolia* Domin.	Whole plant
Cuts/wounds/sores	*Cocculus hirsutus* Diels	Leaves
Jaundice	*Cuscuta reflexa* Roxb.	Stem
Dysentery	*Diospyros melanoxylon* Roxb.	Seed kernels
Fever	*Enicostema hyssopifolium* Verdoorn	Leaves
Ringworm	*Euphorbia hirta* L.	Whole plant
Asthma	*Pennisetum americanum* Leeke	Grains
Urinary inflammation	*Pergularia daemia* Chiov.	Leaf
Jaw ache	*Solanum surattense* Brum. f.	Seeds
Anti-hemorrhagic	*Tridax procumbens* L.	Plant juice
Jaw ache	*Ziziphus nummularia* W. & A.	Roots

The common recipe for abortion among the Bhils comprises stem bark of *Mangifera indica* L. and *Syzygium cumini* Skeels, and root of *Ziziphus nummularia* W. & A. The gum

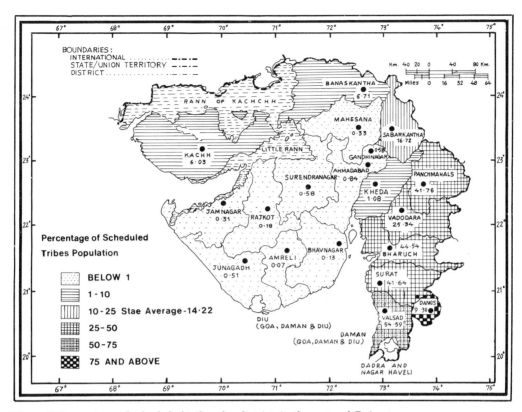

Figure 3. Percentage of scheduled tribes, by district, in the state of Gujarat.

resin of *Boswellia serrata* Colebr. is burnt as an incense, and the colorful seeds of *Abrus precatorius* L. are strung as necklaces. Flowers of *Madhuca longifolia* var. *latifolia* Chev. are fermented and used to produce *maudi*, a liquor consumed on various occasions. Tribal brooms and brushes are made from twigs of *Malvastrum coromandelianum* Garcke and grasses like *Vetiveria lawsonii* B. & M. The whole plants of *Chrozophora rottleri* Juss., *Verbascum chinense* Sant., and *Blumea* species, and the roots of *Balanites aegyptiaca* Del. are valued as fish poison. The Bhils often extract Katha from heartwood of *Acacia catechu* Willd., make charcoal, and collect herbal drugs, gums and resins, firewood, fruits, honey, beeswax, lac, bamboo, and bidi leaves.

The Dangs district in southeastern Gujarat is inhabited predominantly by the Bhil, Choudhary, Dhodia, Dhorkoli, Dubla, Gamit, Kathodi, Kokani, Kotwalia, Kunbi, Naikda, Pardhi, and Warli tribes. Because the Bhils, Kunbis, and Warlis occur in significant numbers in this district and in many respects share a common culture pattern, they are commonly called Dangi.

The staple food of the Dangi tribals is ngali (*Eleusine coracana* Gaertn.). Millets like varai (*Panicum sumatrense* R. & S.), samo (*Echinochloa colonum* Link), banti (*E. frumentacea* Link), and kodra (*Paspalum scrobiculatum* L.) are cultivated, but the Dangi also utilize several wild species for food: tubers of *Dioscorea bulbifera* L., *Paracalyx scariosa* Ali, and *Tacca leontopetaloides* O. Ktze.; tender shoots of *Dendrocalamus strictus* Nees; leaves of *Bacopa monniera* Penn. and *Schrebera swietenioides* Roxb.; fruits of *Capparis zeylanica* L. and *Gmelina arborea* Roxb.; receptacles of *Ficus racemosa* L.; and seeds of *Holoptelea integrifolia* Planch.

Several wild plants are important sources of medicines for the Dangi: *Cochlospermum religiosum* Alst. (diuretic, cough), *Embelia tsjeriam-cottam* DC. (to increase lactation), *Fla-*

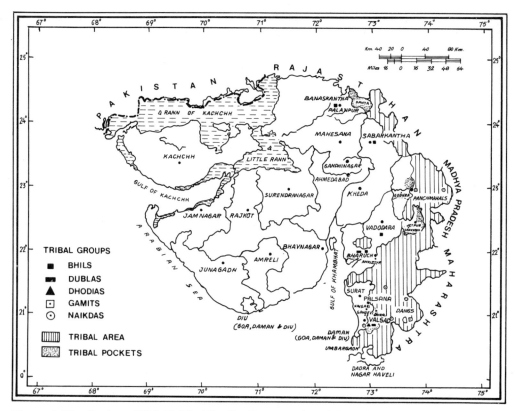

Figure 4. Distribution of Bhil, Dubla, Dhodia, Gamit, and Naikda in the state of Gujarat.

courtia indica Merr. (fever), *Soymida febrifuga* Juss. (malarial fever and blood pressure), and *Tacca leontopetaloides* O. Ktze. (leprosy). Wild plants are also important sources of other products: cordage (*Ventilago denticulata* Willd.), condiment (*Zingiber cernuum* Dalz.), fish poison (*Casearia graveolens* Dalz. and *Verbascum chinense* Sant.), and arrow poison (*Euphorbia acaulis* Roxb. and *E. tirucalli* L.).

The Maldhari tribals of Saurashtra utilize many species of the forest flora for food: fruits of *Ampelocissus latifolia* Planch., *Alangium salvifolium* Wang., *Capparis decidua* Edgew., *Grewia subinaequalis* DC., *Momordica dioica* Roxb. ex Willd.; buds of *Capparis decidua* Edgew.; shoots of *Leucas aspera* Spr.; and leaves of *Melochia corchorifolia* L. and *Tribulus terrestris* L. For fodder, this group uses *Alhagi pseudalhagi* Desv.

Other products obtained by the Maldhari from the forest include fiber from the stems of *Hibiscus vitifolius* L., *Melochia corchorifolia* L., and *Sterculia villosa* Roxb., and from the bark of *Helicteres isora* L.; toothbrushes from *Bergia suffruticosa* Fenzl, *Cassia auriculata* L., and *Indigofera caerulea* var. *monosperma* Sant.; walking sticks from *Ixora arborea* Roxb.; dyes from flower buds of *Solanum incanum* L; bidi wrappers from leaves of *Bauhinia racemosa* Lamk.; brooms and mats from leaves of *Phoenix sylvestris* Roxb.; fish poison from fruits of *Xeromphis spinosa* Keay; a soap substitute from bark of *Acacia torta* Craib; a coffee powder substitute from seeds of *Borreria articularis* F. N. Will.; and moth repellant from twigs of *Artemisia parviflora* Roxb.

The Maldhari tribe uses wild plants for a wide range of medicinal purposes: bark of *Peganum harmala* L. as a substitute for cinchona bark, and bark of *Hymenodictyon excelsum* Wall. as a substitute for quinine; leaf paste of *Biophytum sensitivum* DC. to check bleeding;

seeds of *Buchanania lanzan* Spr. as a brain tonic; seeds of *Mucuna prurita* Hook. to improve digestion; seed paste of *Psoralea corylifolia* L. to treat leukoderma; a root decoction of *Zornia gibbosa* Span. to induce sleep; and seeds of *Annona squamosa* L. to induce abortion.

Ethnobotanical Resources

As mentioned earlier, tribals utilize a large number of wild and cultivated plant species. Much useful information on plants associated with primitive societies and ethnic groups living in the arid and semiarid zone is scattered in ancient treatises, travelogues, gazetteers, ethnographies, archaeological accounts, herbals, pharmacological literature, notings on herbarium sheets, and reports of various Integrated Tribal Development Projects. From an ethnobotanical viewpoint, however, there is no comprehensive account of the plants employed by tribals of the arid zone, even though the region is inhabited by numerous tribes.

Preliminary studies have been made by King (1869), Bhandari (1974), Shah et al. (1981), Shah and Gopal (1982), Joshi (1982), and Maheshwari et al. (1984, p. 21). The National Botanical Research Institute, Lucknow, has embarked upon ethnobotanical surveys and collection, identification, and documentation of plant products used by tribals in each district. An exhaustive inventory of plants and plant products used by tribals for food, medicine, fodder, fiber, abortion, house-building, firewood, oil, narcotics, beverages, musical instruments, cordage and basketry, fish poison, and snakebite is being prepared. The plant resources of the arid zone are very limited, but, surprisingly, there is a large number of potentially economic plant species that deserve mention.

WILD EDIBLE PLANTS

Famine is often a serious challenge to the people inhabiting the Great Indian Desert, and many plant species are used as famine food. Some of these species are of great importance as emergency foods, while others are used as supplementary foods in times of scarcity. The following species are noteworthy:

SPECIES	LOCAL NAME	PLANT PART
Roots & other underground parts		
Ceropegia bulbosa Roxb.	Hedulo	Tubers
Cyperus bulbosus Vahl	Moth	Bulbs
C. rotundus L.	Motho	Roasted tubers
Portulaca tuberosa Roxb.	Safed mushali	Roots
Green vegetables		
Caralluma edulis B & H.	Pimpa	Young shoots
Chenopodium album L.	Chilaro	Leaves & young shoots
Commelina benghalensis L.	Bakhana	Leaves
Emex spinosa Campd.	Khato palak	Leaves
Euphorbia caducifolia Haines	Thor	Leaves
Portulaca oleracea L.	Kulfo	Leaves
Rivea hypocratiformis Choisy	Rotabel	Leaves & young shoots
Salvadora persica L.	Mitha jal	Leaves
Sesuvium sesuvioides Verdc.	Lunio	Leaves
Solanum nigrum L.	Makoi	Leaves & young shoots
Talinum portulacifolium Schweinf.	–	Leaves
Trianthema portulacastrum L.	Santo	Leaves

(continued)

SPECIES	LOCAL NAME	PLANT PART
Flowers & flower buds		
Calligonum polygonoides L.	Phog	Buds as salad
Fruits		
Acacia leucophloea Willd.	Arunja	Young pods
Capparis decidua Edgew.	Ker	Unripe fruits as pickles & vegetables
Cordia gharaf Ehrenb. & Aschers.	Gondi	Berries
Cucumis melo var. *momordica* Cogn.	Kachro	Unripe fruits
Ephedra foliata Boiss.	Lana, Suo-phogaro	Fruits
Glossonema varians Bth. & Hk.f.	Dodha	Follicles (*khiroli*)
Grewia tenax Fiori	Gangerun	Fruits
G. villosa Willd.	Gangeti	Fruits
Momordica balsamia L.	Barh-karelo	Fruits as substitute for balsam pear
M. dioica Willd.	Kankero	Unripe fruits as vegetable
Moringa concanensis Nimmo	Sarguro	Unripe fruits
Prosopis cineraria Druce	Khejri	Young pods as vegetable (*sangri*)
Rhus mysorensis G. Don	Dansaro	Fruits
Salvadora oleoides Decne.	Jal	Fruits
S. persica L.	Pilu	Fruits
Ziziphus mauritiana Lamk.	Bor	Fruits
Z. nummularia W. & A.	Bordi	Fruits
Seeds		
Acacia nilotica subsp. *indica* Brenan	Babul	Seeds (*hilario*)
A. senegal Willd.	Kumat	Seeds as vegetable
Achyranthes aspera L.	Andi jaro	Seeds as tonic
Cenchrus biflorus Roxb.	Bhurat	Grains eaten as bread
C. setigerus Vahl	Dhaman	Grains
Citrullus colocynthis Schrad.	Tumba	Seeds (*sogra, khankara*)
Dactyloctenium aegyptiacum Willd.	Makara	Grains (*kheech*)
D. sindicum Boiss.	Tantia ghas	Grains
Indigofera cordifolia Roth	Bekario	Seeds
I. linifolia Retz.	Lambio bekario	Seeds used as flour
Lasiurus hirsutus Boiss.	Sevan	Grains (*sogra*)
Sarcostemma acridum Voigt	Khir khimp	Seeds
Sesamum indicum L.	Til	Seed refuse
Tamarix troupii Hole	Imli	Seeds
Tribulus terrestris L.	Kanit	Seeds
Gums		
Acacia jacquemontii Benth.	Bu banvali	Gum
A. nilotica subsp. *indica* Brenan	Babul	Gum as masticatory
Barks		
Acacia leucophloea Willd.	Arunja	Bark
A. nilotica subsp. *indica* Brenan	Babul	Bark
Prosopis cineraria Druce	Khejri	Bark

MEDICINAL PLANTS

The following plants are used in folklore medicine:

SPECIES	LOCAL NAME	PLANT PART USED	AILMENT
Ammannia baccifera L.	Jal bhangro	Whole plant	Guineaworm disease
Argemone mexicana L.	Satyanasi	Yellow latex	Eye infections & rheumatic pains
Aristolochia bracteolata Lamk.	Hookah bel	Whole plant	Used in native medicines
Balanites aegyptiaca Del.	Hingot	Fruit pulp	Cough & skin diseases
Bergia suffruticosa Fenzl	Kakria	Leaves	Sores & bone fractures
Boswellia serrata Colebr.	Salaran	Gum resin	Sores & skin eruptions
Citrullus colocynthis Schrad.	Tumba	Fruits	Used as a purgative
Cleome vahliana Farsen.	Nodi	Leaves	Skin diseases
		Seeds	Scabies & leukoderma
Commiphora wightii Bhand.	Guggul	Gum resin	Ulcers, pyorrhea, & chronic bronchitis
Convolvulus microphyllus Spr.	Santari	–	Laxative, bronchitis
Corallocarpus conocarpus Cl.	Nai kandi	Root paste	Snakebite
C. epigaeus Cl.	Kadawi nai	Root paste	Swellings
Corchorus depressus Stocks	Bahuphali	Whole plant	Tonic for gonorrhea
Datura innoxia Mill.	Daturo	Seeds	Hydrophobia
Enicostema hyssopifolium Verdoorn	–	Whole plant	Snakebite & exhaustion
Euphorbia caducifolia Haines	Thor	Latex	Cough & blisters
Glinus lotoides L.	Bakda	Whole plant	Urinary troubles
Lycium barbarum L.	Morali	Leaves	Abscess
Maytenus emarginata Ding Hou	Kankero	Leaves	Sores
		Fruit	Used as a blood purifier
Pedalium murex L.	–	Leaf mucilage	Gonorrhea, dysuria, & calculi
Pulicaria crispa Bth. & H.f.	Dhola lizru	Leaves	Headache
Salvia aegyptiaca L.	–	Seeds	Diarrhea
Sarcostemma acridum Voigt	Arr thor	Roots	Snake & dog bites
Schweinfurthia papilionacea Merr.	Sanipat	Leaves & fruits	Typhoid fever
Solanum surattense Burm.f.	Bhur hingani	Berries	Cough & toothache
Tephrosia uniflora subsp. *petrosa* Gill. & Ali	Bishoni	Leaves	Syphilis
Viola cinerea var. *stocksii* Beck.	Musa karni	Plant decoction	Fever
Withania somnifera Dun	Asgandh	Roots & leaves	Lumbago & rheumatism

PLANTS USED AS REFRIGERANTS

SPECIES	LOCAL NAME	PLANT PART USED	HOW USED
Abutilon indicum Sweet	Tara kanchi	Seeds	To cool drinks
Convolvulus auricomus var. *volubilis* Bhand.	Rotabel	Whole plant	To cool drinks
Lepidagathis bandraensis Blatt.	Unt-katala	Mucilaginous seeds	To cool drinks
Ocimum americanum L.	Rantulsi	Seeds	To cool drinks
Phoenix sylvestris Roxb.	Khajur	Roots	To cool drinks
Sisymbrium irio L.	Khub khala	Seeds	To cool drinks
Tamarindus indica L.	Amli	Fruit pulp	For "looh" & sunstroke
Tamarix troupii Hole	Imli	Fruits	For "looh" & sunstroke

PLANTS USED FOR TOOTHBRUSHES AND TOOTHACHES

SPECIES	LOCAL NAME	PLANT PART USED	USE
Acacia nilotica subsp. *indica* Brenan	Baval	Twigs	Toothbrush
Azadirachta indica A. Juss.	Neem	Young twigs	Toothbrush
Cassia auriculata L.	Aval	Twigs	Toothbrush
Commiphora wightii Bhand.	Guggul	Young twigs	To strengthen teeth
Dicoma tomentosa Cass.	Vajradanti	Roots	Toothbrush
Fagonia cretica L.	Damasha	Stems	Toothbrush
Ficus benghalensis L.	Vad	Twigs	Toothbrush
Indigofera oblongifolia Forsk.	Jhil	Twigs	Toothbrush
Peganum harmala L.	Harmal	Plant smoke	For toothache
Salvadora persica L.	Pilvo	Twigs	Toothbrush
Striga gesnerioides Vatke	Missi	Plant	To strengthen teeth

PLANTS USED FOR FIBERS, MATS, BASKETS, AND STUFFING

SPECIES	LOCAL NAME	PLANT PART USED	HOW USED
Acacia jacquemontii Benth.	Bu-baonli	Shoots	To make baskets
	Bast fiber	To make rope	
A. leucophloea Willd.	Reonja	Shoots	To make baskets
	Bast fiber	To make rope	
Aerva persica Merr.	Bui	Seeds	Stuffing for beds & pillows
Alhagi pseudalhagi Desv.	Jawasa	Twigs	To make "tatties," baskets
Bauhinia racemosa Lamk.	Asundro	Bark	To make hanging baskets
Butea monosperma Taub.	Palas	Leaves	As umbrellas, dishes, & cups
		Root bark	To make hanging baskets
Calotropis procera R. Br.	Aak	Stem	To make cords & rope
		Seed floss	Stuffing for pillows & quilts
Crotalaria burhia Buch.-Ham.	Sannia	Twigs	To make ropes & baskets
Leptadenia pyrotechnica Decne.	Khimp	Twigs	To make ropes & baskets
Phoenix sylvestris Roxb.	Khajur	Pinnules	To make mats & baskets
Saccharum bengalense Retz.	Munj	Leaf sheaths	To make mats, baskets, & cordage
Sida cordifolia L.	Bal	Fiber	Excellent for ropes
Tephrosia falciformis Ramas.	–	Twigs	To make baskets & ropes
Vetiveria zizanioides Nash	Khas	Roots	To make screens, curtains, & "tatties"
Wattakaka volubilis Stapf	Padal bel	Stem	To make ropes & thread

PLANTS USED FOR FIREWOOD AND CHARCOAL

SPECIES	LOCAL NAME	SPECIES	LOCAL NAME
Acacia leucophloea Willd.	Arunja	*Butea monosperma* Taub.	Phalas
A. nilotica subsp. *indica* Brenan	Babul	*Calligonum polygonoides* L.	Phog
		Prosopis chilensis Stuntz	Vilayti-khejra
A. senegal Willd.	Kumat	*P. cineraria* Druce	Khejri, shami
Anogeissus pendula Edgew.	Dhau	*Woodfordia fruticosa* Kurz	Dhawai
Bauhinia racemosa Lamk.	Asundro	*Ziziphus* species	–

FODDER PLANTS RELISHED BY LIVESTOCK

SPECIES	LOCAL NAME	PLANT PART	LIVESTOCK
Acacia nilotica subsp. *indica* Brenan	Babul	Leaves & pods	Goats & camels
Alhagi pseudalhagi Desv.	Jawasa	Leaves	Camels
Calligonum polygonoides L.	Phog	Twigs	Camels
Clerodendrum phlomidis L.f.	Arni	–	Camels
Crotalaria medicaginea Lamk.	Gugario	–	Camels
Gisekia pharnacioides L.f.	Sareli	–	Camels
Haloxylon salicornicum Bunge	Lana	–	Camels
Indigofera cordifolia Roth	Bekar	–	Cattle & goats
I. oblongifolia Forsk.	Jhil	–	Goats & sheep
Melilotus indica All.	Gorha dal	–	Horses
Pithecellobium dulce Benth.	Vilayti imli	Leaves	Goats
Prosopis chilensis Stuntz	Vilayti-khejra		Greedily eaten by cattle & goats
P. cineraria Druce	Khejri	Leaves & pods	Valued for goats
Suaeda fruticosa Forsk.	Lunki	–	Camels
Tecomella undulata Seem.	Rohiro	Leaves	Cattle & goats
Ziziphus nummularia W. & A.	Bordi	Leaves (palla)	Valued for goats & camels

FODDER GRASSES RELISHED BY LIVESTOCK

SPECIES	LOCAL NAME	RATING	LIVESTOCK
Acrachne racemosa Ohwi	–	Good	Cattle
Bothriochloa pertusa A. Camus	Bhalka	Excellent	Cattle
Brachiaria ramosa Stapf	Murat	Much liked	Cattle
Cenchrus ciliaris L.	Dhaman	Valuable	Sheep, cattle, & horses
Cymbopogon schoenanthus Spr.	Sugani	Good	Camels
Dichanthium annulatum Stapf	Karad	Valuable	Cattle
Echinochloa colonum Link	Jirio	–	Greedily eaten by cattle & horses
Eremopogon foveolatus Stapf	Buari	Much liked	Cattle
Lasiurus ecaudatus Saty. & Shank.	Shevan	Much relished	Camels, cattle, & goats
Panicum antidotale Retz.	Garmano	–	Goats, cattle, & camels
P. turgidum Forsk.	Munt	Excellent	Camels
Sehima nervosum Stapf	Seran	Excellent	Cattle
Tetrapogon tenellus Chiov.	Chinki	–	Cattle

PLANTS WITH MISCELLANEOUS USES

SPECIES	LOCAL NAME	PLANT PART USED	HOW USED
Anogeissus pendula Edgew.	Dhau	Leaves	Dark green dye
Aristolochia bracteolata Lamk.	Hookah bei	Seeds	Hair softener
Arnebia hispidissima DC.	Ram-bui	Roots	Red dye
Balanites aegyptiaca Del.	Hingot	Fruit pulp / Drupes	Detergent for silk / In fireworks
Bauhinia racemosa Lamk.	Asundro	Leaves	Bidi wrapper
Calligonum polygonoides L.	Phog	Whole plant	To make coal used by ironsmiths
Cassia auriculata L.	Anwal	Bark	Tanning agent
Cordia gharaf Ehrenb. & Aschers.	Gondi	Bark	To redden lips (as a substitute for *pan*)
Eclipta prostrata L.	Jal bhangro	–	To blacken gray hair
Haloxylon recurvum Boiss.	Khar	Plant ash	Soap substitute
Lycium barbarum L.	Morali	Twigs	To crystalize salt (in local salt industry)
Withania coagulans Duns.	Paneer-bandh	Fruits	To coagulate milk
Wrightia tinctoria R. Br.	Kerno	Leaves	Blue dye

The Forest in Tribal Economy

The forest plays a vital role in the economy as well as in the daily needs of tribals, who depend on its flora for their livelihood and who collect wild and cultivated plants growing in and near the forest. Several cottage and rural industries derive the bulk of their raw material from wild-growing plants—herbal drugs, fibers, flosses, cordage, mats, basketry, gums, resins, lac, tannins, gugal or incense materials, toys, musical instruments, Katha extraction, agricultural implements, brooms, and brushes, to name a few. Forest resources can, therefore, form the basis for plant-based cottage industries in tribal areas.

A number of valuable medicinal plants grow wild in the region's forests and can have a good market, including *Adhatoda zeylanica* Medic., *Asparagus racemosus* Willd., *Barleria prionitis* L., *Centratherum anthelminticum* O. Ktze., *Citrullus colocynthis* Schrad., *Cressa cretica* L., *Eclipta prostrata* L., *Helicteres isora* L., *Hemidesmus indicus* Schult., *Holarrhena pubescens* G. Don, *Operculina turpethum* S. Manso, *Psoralea corylifolia* L., *Solanum surattense* Burm.f., and *Withania somnifera* Dun.

Tribals also make attractive household articles, such as umbrellas, combs, brooms, mats, fans, water bottles, musical instruments, drums, and fishing nets, which are sold in urban centers through traders and contractors. Thus, the present study has brought to light a wealth of new information that can be utilized in improving tribal economies and incorporated in the Integrated Tribal Development Projects located in the arid zone.

Strategies and Programs for Ethnobotanical Research

The rich and untapped flora that human societies have been using for their many needs must be investigated for the purpose of developing new sources of proteins, fats, starches, alkaloids, therapeutic agents, and pharmacodynamic compounds. Much of our present knowledge of plant resources has its origin in indigenous cultures. The oral cultures of the arid tropics represent an invaluable knowledge base that is largely undocumented. The rich plant lore has been passed on by word of mouth and by tradition from generation to generation in different parts of the world. Today, some aboriginal tribals still live in forests. Their knowledge of the uses of plants is often kept secret and passed on only by oral traditions. It is to uncover these hidden and secret uses of the flora that ethnobotany has become an important part of our investigation (Schultes 1962; Gorinsky 1980). Future investigations should be undertaken along the following lines:

1.   Ethnobotanical surveys and intensive field work in tribal areas.

2.   Collection, identification, and inventorization of plants and development of ethnobotanical herbaria and musea.

3.   Phytochemical, biological, and pharmacological screening.

4.   Commercial cultivation and utilization of promising species.

5.   Conservation *ex situ* of rare and endangered species in ethnobotanical gardens, with special reference to primitive and ancient cultivars, and wild relatives of crop plants.

Conclusion

There is a growing realization of vast potential still hidden among the so-called primitive cultures. The fact that tribals utilize or do not utilize certain plant species for food or medicine, for example, already provides an empirical screening of plants based on the historical process of trial and error, and based on empirical selection. It is, however, essential to gain a scientific understanding of the empirical choices through phytochemical screening and pharmacological investigations of plants, and through a nutritional analysis of tribal foods and diets. Such analyses are of a potential economic relevance as they may lead to a wealth of new information on underexploited and/or new uses of plants as sources of proteins, fats, starches, alkaloids, and tannins.

One example of a potentially useful economic plant is *Momordica dioica* Roxb. ex Wild. Tribals use the tuberous roots of this plant to make *tikhur*, a food product. Phytochemical screening of the roots revealed a total sugar content of 92.5 percent; starch content (by glucose oxidase methods, after hydrolysis), 90 percent; oliogosaccharides, 0.12 percent; and crude protein (nitrogen × 6.25), 3.4 percent. The chemical composition of other useful plants is shown in Table 1.

Table 1. Chemical composition of edible seeds and grains used by tribes of the Valsad and Dangs districts, Gujarat, India.

Species	Moisture (%)	Ash (%)	Protein (%)	Starch (%)	Fat (%)	Crude fiber (%)	Minerals (mg/100 g) K	P	Fe
Eleusine coracana Gaertn.									
Red-grained	8.23	10.23	5.70	53.74	1.32	7.92	420	600	09
White-grained	5.56	9.25	2.13	56.25	2.07	8.34	380	600	05
Hibiscus sabdariffa L.	5.24	14.82	24.41	13.19	19.11	16.80	726	400	02
H. cannabinus L.	5.00	12.07	24.85	13.79	23.26	15.98	960	1800	08
Panicum miliaceum L.	7.50	10.06	7.42	36.00	3.70	7.20	372	766	08

The importance of ethnobotanical research is now keenly felt, as it represents one of the best avenues for searching out promising new economic plants in arid lands, many of which are becoming rare, endangered, and scarce. Among these are *Acacia nilotica* subsp. *cupressiformis* Ali & Faruqui, *Caralluma edulis* Bth. & Hk.f., *Commiphora wightii* Bhand., *Indigofera caerulea* var. *monosperma* Sant., *Lepidagathis bandraensis* Blatt., *Moringa concanensis* Nimmo, *Rhus mysorensis* G. Don, *Talinum portulacifolium* Schweinf., *Tecomella undulata* Seem., and *Withania coagulans* Don.

Euphorbia acaulis Roxb., used by tribals for gout, rheumatism, fever, and dysentery, and as arrow poison, was screened at the Regional Research Laboratory, Jammu (Singh et al. 1984). It has shown promising results, and its anti-inflammatory and anti-arthritic activity was found to be superior to that of phenylbutazone. Known locally as *Scirpus kysoor* Roxb. or *bid*, this species is used as food by the Pahars of Gujarat. Phytochemical screening of this plant's tubers show the following chemical composition:

Protein	4.280%
Ash	3.000%
Moisture	6.880%
Fat	0.600%
Crude fat	8.130%
Reducing sugars	6.770%

Nonreducing sugars	7.467%
Total sugars	14.630%
Starch	51.130%

Minerals present include sodium (72 milligrams per 100 grams), potassium (280 milligrams per 100 grams), and iron (36 milligrams per 100 grams). Table 2 shows the anti-inflammatory potential of another *Euphorbia* species. Many important drugs are still extracted from medicinal plants that were originally employed as indigenous remedies.

Table 2. Anti-inflammatory activity of rootstocks of *Euphorbia fusiformis* Buch.-Ham. ex D. Don on carrageenan-induced edema in groups of five mice.

Treatment	Dose (mg/kg)	Edema (mg mean ± S.E.)	Percent inhibition	P-value
Oral				
Control	—	75 ± 5.2	—	—
n-Hexane extract	50	54 ± 12.4	28.00	<0.010
	100	24 ± 5.4	68.00	<0.001
Phenylbutazone	100	40 ± 4.7	46.66	<0.001
Intraperitoneal				
Control	—	70 ± 6.6	—	—
n-Hexane extract	25	36 ± 2.8	48.57	<0.001
	50	22 ± 2.8	71.42	<0.001
	100	20 ± 4.7	71.42	<0.001
Phenylbutazone	50	43 ± 3.2	38.57	<0.001

The impact of deforestation, urbanization, and modernization has shaken the very base of diverse tribal cultures; many of them are declining and becoming acculturated at an alarming rate (Maheshwari 1983). There are schemes for shifting tribals from their natural habitats to other areas for the purpose of providing them with alternative occupations. This process has already led to a great loss of the empirical knowledge that the forest-dwelling tribes have had, particularly with respect to herbal drugs, contraceptives, abortifacients, and birth control. Our great concern, therefore, lies in the progressive divorcement of people in traditional societies from dependence upon the immediate environment and rapid disintegration of aboriginal traits with acculturation. There is an urgent need for ethnobotanical documentation in arid lands, before information is permanently lost.

LITERATURE CITED

Bhandari, M. M. 1974. Famine foods in the Rajasthan Desert. *Economic Botany* 28: 73–81.

Gorinsky, C. 1980. Ethnobiology—the old synthesis. *Journal of Ethnopharmacology* 2: 304.

Joshi, P. 1982. An ethnobotanical study of Bhils—a preliminary survey. *Journal of Economic and Taxonomic Botany* 3: 257–266.

King, G. 1869. Famine foods of Marwar. *Proceedings of the Asiatic Society of Bengal* 38: 116–122.

Maheshwari, J. K. 1976. Phyto-amelioration problems and conservation of biota in Rajasthan Desert. First Indian Conference on Desert Technology, Jodhpur, 1976, Pts. 2 and 3. 13–15.

———. 1983. Developments in ethnobotany. *Journal of Economic and Taxonomic Botany* 4(1): I–V.

———. 1984. Deserts' menacing march. *Northern India Patrika* 13 June 1984: 9.

Maheshwari, J. K., B. S. Kalakoti, and L. Brij. 1984. Ethnomedicine of Bhil tribe of Jhabua district. M.P. All-India Seminar on Anthropology: Theory and Practice, Lucknow, 20–22 January 1984.

Schultes, R. E. 1962. The role of the ethnobotanist in the search for new medicinal plants. *Lloydia* 25: 257–266.

Shah, G. L., and G. V. Gopal. 1982. An ethnobotanical profile of the Dangies. *Journal of Economic and Taxonomic Botany* 3: 355–364.

Shah, G. L., A. R. Menon, and G. V. Gopal. 1981. An account of the ethnobotany of Saurashtra in Gujarat State (India). *Journal of Economic and Taxonomic Botany* 2: 173–182.

Singh, G. B. et al. 1984. Anti-inflammatory activity of *Euphorbia acaulis* Roxb. *Journal of Ethnopharmacology* 10: 225–233.

Geographic Dynamics
and Ethnobotany

GEORGE R. MORGAN

Kennst du das Land, wo die Zitronen blühn?
Knowest thou the land where the lemon-trees bloom?
GOETHE

Plant geography and ethnobotany share a fundamental interest in the distribution of plants. In fact, the formal beginnings of ethnobotany emerged from the science of plant geography, which began in the early nineteenth century with Alexander von Humboldt, a naturalist, geographer, and father of the discipline. Alphonse de Candolle expanded Humboldtian plant geography by emphasizing the geographic origins and dispersals of cultivated plants; his classic works *Géographie botanique raisonnée* (1855) and *Origin of Cultivated Plants* (1886) have remained important ethnobotanic references.

The acknowledged founder of ethnobotany was John W. Harshberger (1869–1929), a botanist of wide-ranging interdisciplinary interests, especially in plant geography and ecology. In 1895, Harshberger introduced the term *ethnobotany* and formulated the tenets of the discipline in a lecture paper titled "The Purposes of Ethno-botany" and addressed to the University Archaeological Association, University of Pennsylvania. In this lecture, Harshberger announced the need for ethnobotanic studies in explaining past distributions of plants utilized by tribal peoples, specifically Native American tribes of North America. Six years later he clearly interpreted ethnobotany as a subdivision of plant geography: "Ethnobotany throws light upon the past distribution of plants, and, as such, becomes a department of phytogeography" (Harshberger 1906). He further stated the need for interdisciplinary effort in the search for geographic origins of cultivated plants:

Phytogeography, or plant geography in its widest sense, is concerned not only with the distribution of wild plants, but also with laws governing the distribution of cultivated plants. In order to determine the latter, that is, the original center from which the cultivation of such plants has spread, it is necessary to examine the historic, archaeologic, philologic, ethnologic, and botanical evidence of the past uses of such plants by the aboriginal tribes of North America (Harshberger 1906).

Most plant geographers, whether botanists or geographers, have never seriously included the distribution of cultivated plants in their studies. Instead, they have described and interpreted the floristic and vegetational distribution of wild plants in relation to variables of the natural environment, including human alteration of the habitat. Such studies have sufficiently absorbed the time and interests of plant geographers. Ethnobotany has been an outgrowth of plant geography, and, as such, has interested ethnobotanists in the origins and distribution of useful plants.

In the early twentieth century, the Russian geneticist and plant geographer N. I. Vavilov published his monumental work *Origin, Variation, Immunity and Breeding of Cultivated Plants*. In a worldwide search for domesticated crop origins, he studied plant distribution, as well as genetics, cytology, and anatomy, and concluded that mountainous regions were the centers of origin of major crops, because those were the areas of greatest plant diversity.

Although centered within the domain of botanical science, the field of ethnobotany has been open and invitational to other disciplines. A few anthropologists, notably ethnologists and archaeologists, have paralleled botanists historically in their ethnobotanical interests. Archaeological techniques for analyzing and dating plant remains have been of utmost importance in deciphering prehistoric-historic origins and distributions of early useful plants. Furthermore, ethnographic studies have added much knowledge about useful plants of tribal societies.

In more recent and seemingly inevitable alliance with ethnobotany have been phytochemists and pharmacologists. Chemists have accompanied botanists in the field to search for medicinal and hallucinogenic plants, but the major arena of biochemical exploration has been the laboratory, where chemists seek to isolate and identify a compound or compounds responsible for a plant's alleged virtues. The final links between practitioners of traditional medicine and pharmacists, namely, pharmacological testing of isolated chemical compounds on living animal systems and sometimes even studies on the effects of these compounds on human beings, often are the slowest and costliest links to forge.

Harshberger (1895) heralded the practical application of ethnobotany: "Ethnobotany is useful in suggesting new lines of manufacture at the present day . . . we may learn by this study new uses of plants of which we were in ignorance." The plant search for new medicines, foods, and fibers spontaneously engendered teamwork among botanists, anthropologists, chemists, and pharmacologists.

Persons from disciplines and backgrounds other than botany, agricultural sciences, anthropology, and chemistry have also contributed to the ethnobotanical literature. Alfred W. Crosby Jr., historian, in his delightful book, *The Columbian Exchange* (1973), has aptly subtitled the work "Biological and Cultural Consequences of 1492." R. Gordon Wasson, a former New York businessman, became a world authority on psychoactive mushrooms used by Old and New World peoples; his many articles and books reveal a holistic approach, with in-depth research into diverse fields such as linguistics, art history, and religion. He pioneered a new branch of ethnobotany known as *ethnomycology*.

Aside from their biogeographic investigations, most geographers have focused their studies on the description and analysis of landscapes and the spatial interactions of people and land. Plants are an important component of landscape studies, but they are only one component of the landscape. Cultural geographers have contributed studies of ethnobotanical interest; especially important have been their studies of domesticated plant origins and dispersals and tropical agriculture. Plants entering large-scale trade arouse the interest of economic geographers whose studies are complementary to economic botany. Historical geographers incorporate useful information about plant distributions in past landscapes. For example, Douglas R. McManis (1975) noted the importance of sassafras to the British and their search for the plant:

His chief complaint . . . was that he could not find sassafras in the area. The English of the time valued the tree highly, believing that infusions made from its barks and roots had nearly miraculous curative powers. . . . But after entering Massachusetts Bay (inexplicably bypassed by Gosnold), Pring found sassafras in abundance.

In 1965, Carl O. Sauer, in his address to the Association of American Geographers entitled "The Education of a Geographer," urged geographers to merge their geographic interests with biology:

The field of biogeography requires more knowledge of biology than can be demanded of most of us. It is, however, so important to us [that] we should encourage the crossing of geography with natural history wherever the student is competent. In particular, we need to know much more of the impact of human cultures on plant cover, of man's disturbance of soil and surface, of his relation to the spread or shrinkage of individual species, of human agency in the dispersal and modification of plants. To these questions a few of us are, and more should be, addressing ourselves. . . . Especially do we need more workers who like and are able to live on frontiers, such as those of biology.

A predominant theme of geography has been the study of people and phenomena. Geographers also study and describe lands and peoples of the world, but, because of limited time in a vast and complex world, their research vistas must focus on selected areas, peoples, and phenomena. My own research interests, for example, have been on the frontiers of geography and ethnobotany. I have been interested in the distribution and environment of wild plants, especially those species used and traded by Native Americans. Furthermore, I have been curious about historic changes in plant distribution caused by human activity, particularly plants with limited ranges.

The following exposition illustrates the geographic content of ethnobotany. Plants and humans are biophysical systems in space and time. As living systems, they change organically and spatially within the dimension of time. Time and space are reciprocal: *Locus nullus nisi in tempore* (The Modisae, Latin grammarians). The relationships of plants and people are also reciprocal and dynamic interactions. As people expand the geographic ranges and abundance of their beneficial plants, human space and numbers expand, while avoidance of harmful plants may contract human living space.

Exploration initiated dramatic changes in vegetal and economic geography. The Spanish discovery of maize, for example, led to dissemination of the plant in the Old World. Although Native American inventiveness led to domestication of the plant, European resourcefulness expanded its geography, and its greater adaptation and variability. The domestication of maize was a slow process; its geographic spread was relatively rapid, almost fully accomplished during the sixteenth century. Other major economic plants have become cosmopolitan since the fifteenth century. Many transplanted species have been more productive in their new environments than in their place of origin. Ethiopian coffee has become maximally productive in Central and South America; tropical America's cacao has become most commercial along the lowlands of West Africa; Southeast Asia's sugar cane has become most successful in the West Indies and South America; Hawaii has become the major producer of South America's pineapple; and the center of the rubber industry is Malaysia, not Brazil, where the plant is native.

European botanical gardens have been of the utmost importance for the introduction of economic plants throughout the world. These institutions have continued to be centers of plant research, as well as public gardens of pleasure. Foremost among botanic gardens in the transfer of economic plants has been the Royal Botanic Gardens at Kew,

England. Plant collectors from Kew explored the world's flora and its distribution, with emphasis on plant geography. The first two directors of the Kew Gardens, Sir Joseph Dalton Hooker and his successor, W. T. Thiselton-Dyer, were plant geographers *par excellence*. In 1848, a museum of Economic Botany was established at Kew for the study of the world's useful plants. Plants from the British Empire and elsewhere were grown and studied at Kew's greenhouses. In addition, Kew Gardens transferred many New World plants, such as rubber and cinchona, to the Asian tropics for commercial production.

The worldwide distribution of economic plants increased market competition, lowered commodity prices, and fostered trade agreements. Monopolies collapsed as plants could be obtained from more than one geographic source. A foremost example is the European spice monopoly. For centuries, secrecy and intrigue surrounded the geography of spices. Scarcity of supply enabled spice traders to demand high prices in European markets.

In traditional societies, most plant products are collected, produced, and consumed locally. Subsistence farmers and gatherers engage in limited local trade. Most traditional agricultural societies cultivate a few species that were domesticated long ago (in the time scale of oral tradition) by alien cultures in far away lands. Often, the foreign domesticates became a major source of food. Other American plants, such as peanuts, manioc, and sweet potatoes, for example, became staple African foods. The cosmopolitan distribution of most major food plants occurred during the sixteenth century. Within this short span of time, the world's population reached proportions unparalleled in history. The distribution of major food plants, along with improved medicines, fed the population increases, but the shadow of Malthus now looms over civilization with chronic food shortages, malnutrition, sickness, and death.

Only a minority of the earth's inhabitants can afford the world's cornucopia of plant products. A country's wealth may be expressed by the quantity and variety of its botanical imports. A distinguishing mark of industrial-commercial societies is the supermarket, an instructive place to study the geography of foods. Commercial societies consume the plant luxuries (mostly tropical) of the world, as well as produce, export, and import its necessities. Traditional societies do well to produce necessities. Of course, wealthy peoples may view imported luxuries (e.g., coffee, bananas, even West Indian handmade cigars) as necessities. Standards of living are, indeed, socio-economic variables.

The geographic spread of many temperate economic plants occurred with the growth of overseas settlements. Europeans migrated to lands of opportunity for farming. They sought areas of mild climates and good soils, where they could grow their familiar plants. They settled in the temperate lands of Australia–New Zealand, the Americas, and South Africa. Russians migrated east to Asiatic Russia, settling the thin strip of woodland-grassland wedged between the subarctic coniferous forest and the deserts of Central Asia. The major semi-arid temperate grasslands of the world were not settled and farmed by Europeans until after 1850.

Overseas settlements have been termed *new lands*, because they were largely virgin. Indigenous populations had made little imprint on the soil: peoples were few and widely dispersed, and hunting and gathering were the predominant ways of economic life. Europeans produced farm surpluses from these rich and spacious lands of few people. The world's major exports of wheat have come from these "new lands" of European settlement, which have played a major role in feeding the world's peoples.

Manchuria was the one major temperate grassland recently settled by a non-European people until millions of North Chinese flooded its virgin grasslands at the beginning of the twentieth century. The Manchu lost their imperial grazing lands to Chinese wheat farmers.

Europeans also pushed indigenous peoples aside in their quest for lands to settle.

Many native inhabitants died of foreign diseases, many others were killed, some were absorbed, and the remainder have survived in marginal country, exiled in their own lands. Many tribes fought to retain their territories and resources, their ways of life, and their very lives. Among the valiant were the Maori of New Zealand, the Bushmen and Zulu of South Africa, the Tupinamba and Araucanian of South America, and the Sioux, Apache, Seminole, Aztec, and Iroquois of North America, but ultimately their territories were taken by people who came from across the waters and pastured strange animals and planted foreign seeds.

Tropical lands attracted few European settlers; they settled in the cooler highlands of the tropics, namely, the high plateaux of Eastern Africa, the tierra templada of the New World, and the "hill stations" of Asia. At higher altitudes they could plant temperate crops, particularly grains and fruits.

Europeans largely avoided settling the interior lowlands of hot-wet regions. The climate was uncomfortable and unhealthy; malaria and other tropical diseases were feared; tropical vegetation and swampy mazes were communication barriers. The rain forests contained surprisingly little for people to eat; agricultural clearings were quickly invaded by the jungle. It was long believed that the luxurious rain forest indicated fertile soils for agricultural potential, but eventually tropical soils were proven to be exceptionally low in nutrient bases.

Geographic perceptions of the tropical world were simultaneously attractive and repulsive: "El Dorado" and "White Man's Grave." Tropical isles were perceived as places of paradise, while the continental tropics were perceived as lands of mystery and danger. The word *interior* still connotes primitive backwardness and danger. Tropical interiors contained concealed dangers: poisoned arrows, head hunters, black magic, venomous snakes, and a myriad of vermin emanating from miasmic sludge. European literature influenced perceptions of tropical interiors as dangerous and mysterious:

> Tyger! Tyger! burning bright
> In the forests of the night,
> What immortal hand or eye
> Could frame thy fearful symmetry?
> (WILLIAM BLAKE, "The Tyger")

The late Marston Bates, zoologist and naturalist intimately knowledgeable in tropical environments, described the rain forest as a place of mystery:

> Awe and wonder come easily in the forest, sometimes exultation—sometimes for a man alone there, fear. Man is out of scale, the forest is too vast, too impersonal, too variegated, too deeply shadowed. The rain forest is perhaps more truly a silent world than the sea. The wind scarcely penetrating it is not only silent, it is still. All sound then gains a curiously enchanted mystery.

Minerals and plant resources have been the economic lures of the tropical world. The flow of tropical plants to Europe increased as the Industrial Revolution expanded; manufactured goods, in turn, flowed from the industrial machines to world markets. Colonialism guaranteed the nations of Europe control of natural resources and markets.

Initially, economic plant species were exploited in the wild. In the Amazon, rubber trees produced small amounts of latex because of improper harvesting methods, which mutilated and killed the trees and decimated whole populations of Indians. Production was limited because rubber trees were scattered in the forest. In their quest for money and power, the rubber barons abused their captive Indian labor. To coerce the Indians to collect increasing quotas of latex, rubber barons and their henchmen whipped, mutilated, and killed many Indians; latex production became associated with human blood. By 1910, the British had successful plantations of Brazilian rubber in Southeast Asia, and they had developed a more productive and protective method of tapping. Labor was plentiful and efficient; Tamils from southern India worked on the plantations, and there were many productive smallholders, mainly Chinese. The Brazilian rubber industry that had lasted through the later decades of the nineteenth century could not match Southeast Asia's controlled plantation production. Furthermore, because of exploitation, the Amazon's rubber trees and Indian labor could not have lasted much longer.

The plantation system was a European creation; its *raison d'être* was to supply the European market with tropical plant products that were not produced at home. From its beginnings in the fifteenth century, the plantation system expanded to become a most important type of agriculture in the tropics; eventually it extended into the subtropics, where tobacco, cotton, and tea were grown for export.

The plantation is a highly specialized type of agriculture; it has always been linked to market demand. It requires large tracts of land devoted to one crop, such as coffee, tea, bananas, rubber, or sugar cane. Plantations tend to be located near navigable rivers or near the sea coast, and in areas of low population density. Furthermore, they demand a large labor supply, but native populations have not been the main work force, because of low numbers and a disinclination for plantation work. Malays, Hawaiians, and Native Americans were among indigenous peoples uninterested in plantation work. Thus plantation owners have had to import their labor supply: African slaves to the Americas; Tamils from southern India to Ceylon's tea plantations and Malaysia's rubber plantations; Biharis of northern India to the tea plantations in Assam; Chinese, Japanese, Filipinos, and Portuguese to Hawaii's sugar cane plantations. Labor has been the largest cost factor in plantation operations, usually amounting to more than half the total cost, for plantation owners provided laborers with necessities, including housing, food, and medical care.

Sugar cane plantations were a major force in human geographic change. In the early sixteenth century, the Spanish and Portuguese introduced African slave labor to New World sugar cane plantations. The British and French followed in the seventeenth century, with plantations in the West Indies. Between 1500 and 1870, ten million Africans were taken to the Americas, mostly to the West Indies and Brazil. Only about 5 percent of the African slaves went to the United States to work on cotton and tobacco plantations. D. B. Grigg's (1974) fascinating historical geography work points out that slave numbers in the United States grew by natural increase; in the West Indies and Brazil the death rate exceeded the birth rate, necessitating a constant replacement of imported slaves. Infant mortality was as high as 50 percent; disease and malnutrition were common. Much of the land was planted with sugar cane, leaving little space for food production. The welfare of slaves was negligible; most British and French owners were absentees.

Britain established the profitable triangular trade with West Africa and the West Indies. Sugar cane was an integral part of that trade. Lancastrian textiles and other British goods were exchanged for West African slaves and items such as pepper. The slaves were shipped across the Atlantic (known as the "Middle Passage") to the West Indies, where they were traded for homeward-bound sugar, molasses, and rum. Profits from sugar products permitted the purchase of more British goods for another phase of the triangular trade. Noel Deerr (1950) summarized the situation thus: "It was not only the trade in sugar that was of value: the cycle of trade that developed out of the sugar industry was of greater value still."

New England also entered the triangular trade, but the items traded to West Africa were different: rum to Africa, slaves to the West Indies, homeward-bound molasses for the making of more rum. By the mid-seventeenth century, there were so many rum distilleries in New England that it had become a social problem (McManis 1975). New England also traded fish and packed meats to the food-poor West Indies, in exchange for sugar and tobacco.

In 1807, Britain declared slave trade illegal, but clandestine trade continued until the latter part of the nineteenth century. By an act of Parliament in 1837, the British were able to obtain indentured labor from India for sugar cane plantations throughout the British Empire. Between 1838 and 1917, 370,000 Indians immigrated to Trinidad and British Guiana. Many went to other British colonies of the Caribbean, such as Jamaica and Grenada. In 1879, Indian immigration to the Fiji Islands began, but, by 1919, the natural increase of the immigrants was so high that immigration ceased. In 1860, Indian indentured labor began arriving in Natal, South Africa. In all, Indian emigration to British sugar colonies numbered about 1.5 million (Klein 1979).

France also sought Indian indentured labor from French possessions. Thus, Indians from Chandernagore and Pondicherry emigrated to Martinique and Guadeloupe. On a smaller scale, Chinese indentured labor went to the West Indies. Although there were difficulties and troubles within the system of indenture, it did afford a better way of life for many, and it supplied sugar cane plantations with labor.

Imported labor brought useful plants and plant folklore to distant lands. In new areas of occupation, laborers also sought plants similar to homeland plants, reasoning that plants with similar morphologies would have similar attributes. The reintroduction of the peanut is a good example of a plant accompanying immigrant labor. Archaeologists have found the peanut, which is native to southern Brazil and Paraguay, stored in containers within tombs of pre-Columbian Peru. The Portuguese introduced it into Portugal in the sixteenth century and thence into West Africa, where it became a staple subsistence food among African societies. The seeds were eaten raw, roasted, or boiled in soup. African slavers recognized the value of the peanut as a survival food for human cargoes crossing the Atlantic, and slaves introduced it into the West Indies and the southern United States as a subsistence crop. In both hemispheres, Africans called the plant *pindar*, a word adapted from the Portuguese *pinda*.

It was not until after the Civil War, however, that the peanut became an important commercial crop. Civil War soldiers, discovering the energy food value of the peanut, helped to disseminate the seeds. The peanut became an ideal substitute crop for cotton fields ravaged by the boll weevil. Also, new uses increased the plant's popularity; George Washington Carver (1864–1943), an African-American agricultural scientist, discovered about 300 uses for the peanut. In contemporary North America, the peanut has become a commercially profitable enterprise.

This brief exposition of post-Columbian transmission of plants and animals demonstrates the relationship between geography and ethnobotany-economic botany. In conclusion, we may say that geography is unavoidably enmeshed with ethnobotany and that the relationship is pervasive to the extent that one is not normally aware of the connections. Harshberger pointed to the relationships about a century ago. When we refer to a plant as wild or domestic, are we not, in part, referring to its geography? We know much less about the geography of most wild plant species. The etymology of *domestic* (*domus*, a house) specifies proximity to people. Is not a weed a spatial rogue? And is not a *cultigen* a plant of which we are uncertain as to its geographic origin?

The relationships between people and plants are geographically transactional. Within environmental limitations, people have manipulated the global range and abundance of useful plants, while the plants, in turn, have more or less influenced where people can live and the extent to which their own populations can grow.

LITERATURE CITED

Bates, M. 1960. *The Forest and the Sea*. Rpt. 1988. New York: Lyons & Burford.

Blake, W. 1794. The tyger. *Songs of Experience*. Rpt. 1984. New York: Dover.

Crosby, A. W., Jr. 1973. *The Columbian Exchange: Biological and Cultural Consequences of 1492*. Contributions in American studies, no. 2. Westport, CT: Greenwood.

de Candolle, A. 1855. *Géographie botanique raisonnée*. Paris.

————. 1886. *Origin of Cultivated Plants*. Trans. from French. London.

Deerr, N. 1950. *The History of Sugar*. Vol. 2.

Grigg, D. B. 1974. *The Agricultural Systems of the World*. New York: Cambridge University Press.

Harshberger, J. W. 1895. The purposes of ethno-botany. Paper presented to the University Archaeological Association, University of Pennsylvania.

————. 1906. *Phytogeographic Influences in the Arts and Industries of American Aborigines*. Bulletin of the Geographical Society of Philadelphia.

Klein, R. M. 1979. *The Green World: An Introduction to Plants and People*. 2nd ed. 1986. New York: HarperCollins.

McManis, D. R. 1975. *Colonial New England: A Historical Geography*. New York: Oxford University Press.

Sauer, C. O. 1982. The education of a geographer. In *Land and Life: A Selection from the Writings of Carl Ortwin Sauer*. Ed. J. Leighly. Rpt. of 1950 ed. Berkeley: University of California Press.

Vavilov, N. I. 1987. *Origin, Variation, Immunity and Breeding of Cultivated Plants: Phytogeographic Basis of Plant Breeding*. Trans. K. Starr. Redwood Seed.

A Case for Ethnobotany in Malaysia

ONG HEAN CHOOI

The tropical rain forests of Southeast Asia are one of only three major blocks of tropical rain forests in the world (Pringle 1969; FAO 1975; Whitmore 1975). The other two are the forests of Central and South America and those of Africa (mainly in the Congo Basin). The Southeast Asian tropical rain forests cover an area of about 250 million hectares (600 million acres), extending from southwestern India, Sri Lanka, Myanmar, southeastern China, Vietnam, Laos, Cambodia, Thailand, the Philippines, Malaysia, and Indonesia to Papua New Guinea and northeastern Australia. In terms of area, the tropical rain forests of America are the largest; those of Southeast Asia are second.

The Southeast Asian tropical rain forest came into existence as early as 70 million years ago, as indicated by the presence of fossils (Muller 1968, 1973). It is quite logical not to expect that the original forests closely resembled the present-day forests, which are believed to have come into existence about 1.5 million years ago, during the Pleistocene era (Ashton 1973).

Even though Peninsular Malaysia (comprising Malaysia and Indonesia) is a comparatively small area, its tropical rain forest is not homogeneous in physiognomy or species composition. The forest, which becomes different with changes in altitude, topography, and soil types, can be divided into fourteen or more types, the major ones being lowland dipterocarp forests, hill dipterocarp forests, upper dipterocarp forests, montane oak-laurel forest, ericaceous forests, mangrove swamp forests, peat swamp forests, limestone vegetation, quartz ridge vegetation, and strand vegetation. These forests are considered by botanists, ecologists, and economic botanists alike as the largest reserve of plant species of potential economic importance and are also the center of origin and diversity of many present-day and potential future crop-plants (Vavilov 1951; Frankel and Bennett 1970; Li 1970; Schery 1972; Jung et al. 1973; Frankel and Hawkes 1975; Sastrapradja 1975; USNAS 1975; Soepadmo 1979a, 1979b).

The number of species in many of these forest types is also large. The rich Malaysian flora is estimated to contain about 10,000 species of angiosperms and 1500 species of ferns. The number of species of fungi and algae are not known but are believed to be several thousand. In comparison, the Amazon basin may support 80,000 species of higher plants.

That tropical rain forests are useful to humans is obvious, for they produce timber, rattan, food (fruits, carbohydrate from various sources, green vegetables, etc.), medicine, essential oils (perfumes and spices), resins, gums, dyes, latex, vegetable oil, tannin, and also ornamental species. The forest is also a gene bank, providing a rich and diverse source of genetic characters for cross-fertilization to produce new and better hybrids of useful plants.

The tropical rain forest, however, is surely and rapidly in danger of losing plants in both quantity and quality. For one thing, the forested area is continuously being reduced, and for another, the quality of the remaining forests is deteriorating due to overexploitation. The main reason for the shrinking of forest area is development; large tracts of forests are cleared for agriculture and human habitation. As population increases, the forest is continuously pushed back to accommodate more people and their needs for agricultural and commercial produce. The building of hydroelectric dams also results in large areas of forest being lost forever through rises in water level. Forests are also being cleared for the building of roads and railway lines and for the establishment of hundreds of miles of high-tension wires that take electricity from the dams to distant cities.

Forests further suffer from exploitation for timber. Modern methods of felling are swift, and large tracts of forest are destroyed by the use of bulldozers with caterpillar wheels for pulling logs out of the forest. Not only is the logging area destroyed by the bulldozer, but breaking the canopy cover brings about a change that affects an even larger area. Pioneer plants colonize the exposed area and may overcome the original species, especially those that are unable to tolerate sudden changes in environmental conditions. Furthermore, timber exploitation has even wider implications in contributing to flooding and thus to lowering the quality of water in streams and rivers; valuable topsoil, as well, is lost, causing erosion. The effects of forest exploitation, therefore, may be felt all the way downstream to the coast. Silting occurs, which in turn causes additional problems.

The complex web of initial human activity and of subsidiary effects in various chains of reactions results in changes that are far more severe than first envisaged. The effects are sometimes felt far away from and long after the initial action. Therefore, the connection between acts and results may not be easily recognized, especially by those who seek not to associate their actions with detrimental effects.

An appropriate phrase to apply here may be, "a small step forward for development, a big leap backwards for nature." Certainly, progress will continue as far as civilization is concerned, and it is not the intent nor can one expect a complete halt to forest exploitation. Instead, people who are entrusted with the power and license to foster development and who are pushing the last frontier (forests) ever backwards should take time to ponder the implications of their initial and seemingly small chipping at nature's rock. Progress must and will go on, but we should try to be more careful not to turn natural forests into human-made jungles which are not self-sustaining and which may cause hardships for future generations.

The Orang Asli

Having considered the shrinking forests of Malaysia, let us now consider the people who are most adept at utilizing these forests. This does not include those who exploit the forests for timber and rattan on a large commercial basis but, rather, those who are partly or fully dependent on the forests for food, medicine, tools, building materials, and so forth, and who have been dependent for many generations; to them, the forest is home. These people are the Orang Asli.

The Orang Asli are the aboriginal people of Peninsular Malaysia. The total population has been variously quoted as between 53,000 (Voon et al. 1979) and 60,000 (Carey 1979). A census carried out by the Jabatan Hal Ehwal Orang Asli (Department of Aboriginal Affairs) gave the total population of Orang Asli in 1974 as 56,900 persons. The Orang Asli are divided into three groups: Semang, Senois, and Proto-Malays, based on physical characteristics, language, cultural traits, and ecological adaptation. Generally, the Semang are foragers, the Senois are swidden farmers, and the Proto-Malays are horticulturists, but there is considerable overlap among the three groups.

The Orang Asli can be further divided into eighteen tribes (Department of Aboriginal Affairs), which occupy differing habitats in the forests or on the forest fringes. Some Orang Asli are hill dwellers, while others live in the lowlands; still others inhabit riverine or coastal settlements. The methods of utilizing natural resources and the species used are quite different for various tribes, although some overlap occurs. Even different communities from the same tribe may utilize the same plants for different purposes. Thus, the scope of the study of ethnobotany among the Orang Asli in Peninsular Malaysia is tremendous.

Time Constraints

Despite the vast potential for ethnobotanical studies in Peninsular Malaysia, time may be the limiting factor. On the one hand, with the passing of time, the forest is being continuously reduced in size, and, on the other hand, the Orang Asli are continuously exposed to and encouraged by the authorities to become part of mainstream civilization. Various resettlement policies have been implemented by the authorities since British colonial days to bring the Orang Asli out of the forests for a variety of reasons. One of the main reasons is for national security, since Orang Asli communities can serve as sources of food, information, and other materials for communist insurgents in the forest (Carey 1979).

As the forests are destroyed, some species will become extinct long before the forest itself is completely wiped out. Many plants are susceptible to the adverse effects of minor changes in environmental conditions and thus will not survive in exposed or exploited forests. Others cannot compete with the regenerating species during succession from barren ground to new forest. The number of plants is greatly reduced by deforestation, thus lowering the viability of self-sustaining populations of many of the species.

These changes in the forest will affect the Orang Asli way of living. Dependence on the forests will be reduced. Furthermore, those people who live as self-sufficient farmers or fishermen, or who are employed by the private or public sector and who have access to market produce and treatment in clinics and hospitals, or who live in houses built by the authorities eventually will lose their knowledge of forest resources. Thus, there is an urgent need for more extensive and intensive studies of ethnobotany among the Orang Asli before either the forests lose most of the species utilized by these people or the people themselves lose most of their knowledge and practice of forest resource utilization.

The Interdisciplinary Nature of Ethnobotany

The study of ethnobotany is not confined to botanists. In fact, it cannot and must not be restricted to them. It attracts people from many disciplines, including chemists, biochemists, phytochemists, pharmacologists, medical doctors, anthropologists, historians, and others. Ethnobotany is multidisciplinary in nature. As Alcorn (1981) said, "A renewed focus on the useful plant lists that traditionally defined ethnobotany may provide the important and necessary starting point for the systematic, multidisciplinary inquiry that is the unrealized potential of ethnobotany."

A primary role of the ethnobotanist is to provide updated lists of useful plants. This function is of especially great importance in Peninsular Malaysia, as most botanical and ecological information recorded previously in books and journals has become outdated.

Also, information on the Orang Asli is far from complete: "Orang Asli know and utilize a wide range of wild plant species for medicine, food and making cultural artifacts, but no complete inventory has been made for any aboriginal group" (Rambo, quoted in Rambo and Sajise 1984, p. 247).

Ecologists play an important role in this process of inventorying useful plants. When an ecological assessment of species is made in the forests where the plants are encountered, information can be provided on habitat preferences, soil types, abundance of each species, probable methods of vegetative propagation, and related data. The ecological status of selected species of particular interest may have to be ascertained through autecological studies.

Botanists and ecologists must coordinate their work with that of various botanical gardens, so that the rarer species can be grown. Information collected from field records may be used to provide proper conditions for survival of these species under cultivation. The work of the botanist, ecologist, anthropologist, and botanical gardens should be extensive and all-encompassing, since information on all useful plants, whether a species be of little or great economic importance, is vital.

Work in other disciplines will necessarily be more selective. The chemist, biochemist, and phytochemist, for example, will have to select only certain species for analysis. The magnitude of analyzing all potentially valuable chemical contents from all reputedly useful plants is so extensive a task that it may be impractical for them to try to analyze the thousands of different uses of the great number of species involved.

Once the medicinal potential of a plant has been scientifically established and medically tested, horticulturists and agriculturalists will need to assess the agricultural potential of the selected species. Plant breeders must choose the most qualified variety for optimum production of the active ingredient. The costs of producing the plant and of extracting the active ingredient will indicate roughly the final cost of producing a new drug for general consumption.

One area where coordination seems to be lacking at present is among medical doctors, botanists, ecologists, anthropologists, and sociologists. Researchers from all these disciplines have at times reported plants of reputed medicinal values. There is a need, however, for these specialists to record accurately the symptoms of patients treated by traditional methods, the plants used, and the methods of usage. Many patients who seek the help of traditional medicine may have been treated in clinics and hospitals for various lengths of time, especially those suffering from heart ailments, diabetes, cancer, and stroke, without much noticeable improvement. These are the most interesting cases, as the visible progress (if any) obtained by being treated traditionally can be compared with medical records. Patients who are seemingly cured by traditional methods should be persuaded to visit a hospital for a medical check-up to verify the cure. If a cure is verified, the particular plant or plants used should be investigated promptly. This can shorten the process of screening plants for potential use, as many plants may be reputed to cure, but few will finally prove to be useful as actual cures.

Another area of interest concerns the traditional species of foods. There are many species of underutilized food plants in Malaysia. Still, Malaysia imports millions of dollars' worth of food. Even the Orang Asli have been growing and consuming foreign food plants for a long time, a practice that is unavoidable and probably irreversible, as most of the introduced species, especially vegetables, are cultivars best suited for quick growth and early harvest. Nevertheless, there is ample opportunity for research on underutilized local food plants and on methods for increasing productivity, for improving and maintaining quality, and for testing food values, palatability, shelf life, and so on. With increased availability and improved quality, these species might be able to compete with introduced species in the market.

There is an urgent need for the compilation of a complete list of plants utilized by Malaysians and by the Orang Asli in particular, as they face evermore rapid changes in their environment and lifestyles. The traditions of the Orang Asli are succumbing to changes brought about by exposure to mainstream Malaysian civilization and by the diminution of forests. To study the Orang Asli, one must be with them to observe, follow, and enquire about practices in their natural environment. Field work, then, is of the utmost importance and necessity, and it must be coordinated with botanical gardens, so that plants in immediate danger of extinction or of becoming rare may be propagated under close supervision. If possible, of course, botanical gardens should grow all potentially useful plants.

Potential Benefits of Ethnobotany

History has taught us that the traditional practices of various ethnic groups are not to be denigrated. The human race has benefitted in many ways from ethnobotany and will continue to derive benefits from the study of so-called primitive people whose cultures and practices have evolved through many generations.

Interest in traditional medicine is on the rise; many institutions of higher learning and research laboratories are investigating indigenous cures. Traditional medicine is no longer only for the poor and uneducated in Malaysia; the rich and highly educated often seek treatment from its practices. It has become an alternative medicine for some and a last resort for others. Traditional medicine will not replace modern medicine, even in Malaysia, nor do practitioners of traditional medicine claim to be able to replace the skill and knowledge of medical doctors. Each has its own niche and its own role to play. A person may seek treatment from clinics or hospitals on one occasion but consult traditional medicine for certain complaints on another.

Medicinal plants of the Orang Asli that may be of interest include those used for treating diabetes, high blood pressure, hemorrhoids, anemia, and ulcers, and those employed in family planning and as aphrodisiacs and tonics. For example, the fungus *Microporus xanthopus* is valued for birth control by both Orang Asli men and women. Normally, it is administered to women, after which they are said to no longer to conceive. If at a later date a woman wants to become pregnant, sterility is said to be nullified by eating *Epirizanthe cylindrica*. For men who have eaten the sporocarp of *M. xanthopus*, however, sterility is reputedly irreversible, although the ability to perform sexually is not lost. I have observed three women who took the fungus for birth control and have not conceived since. One woman ate the sporocarp five years ago and has never again conceived.

In preliminary experiments with domestic cats, the fungus proved effective when fed to kittens before sexual maturity. In the case of sexually mature female cats, the fungus was effective in preventing pregnancy for a period of two years, after which the cats became pregnant again. No mature male cats were available for testing. Male kittens fed with the fungus failed to develop masculine features, such as broad skull structure; and both the penis and testes did not develop appreciably, although general physical growth was not retarded.

As we continue to seek cures for ailments and maladies, some answers to our troubles may lie in the forests waiting to be tapped. For as long as the forests remain and for as long as the unlettered indigenous peoples of the world preserve their knowledge of the properties of plants, ethnobotanical investigations will certainly be a key to progress in modern medicine.

LITERATURE CITED

Alcorn, J. B. 1981. Factors influencing botanical resource perception among the Huastec: Suggestions for future ethnobotanical inquiry. *Journal of Ethnobiology* 1(2): 221–230.

Ashton, P. S. 1973. The quaternary geomorphological history of Western Malaysia and lowland forest phytogeography. *Trans. Second Aberdeen-Hull Symp. Males. Ecol.* 35–62.

Carey, I. 1979. The resettlement of the Orang Asli from a historical perspective. *Fed. Mus. Jour.* 24: 159–174.

FAO. 1975. *Formulation of a Tropical Forest Cover Monitoring Project.* Rome: FAO/UNEP.

Frankel, O. H., and E. Bennett, eds. 1970. *Genetic Resources in Plants.* IBP Handbook No. 11. Oxford: Blackwell Scientific Publications.

Frankel, O. H., and J. G. Hawkes. 1975. *Crop Genetic Resources for Today and Tomorrow.* IBP publication no. 2. Oxford: Cambridge University Press.

Jung, K., B. C. Stone, and E. Soepadmo. 1973. Malaysian tropical forests: An under-exploited genetic reservoir of edible fruit tree species. *Proc. Symp. Biol. Res. and Nat. Dev.*, Malay. Nat. Soc., Kuala Lumpur. Eds. E. Soepadmo and K. G. Singh. 113–121.

Li, H. L. 1970. The origin of cultivated plants in S. E. Asia. *Economic Botany* 24: 3–19.

Muller, J. 1968. Palynology of the Pedawan and Plateau sandstone formations (Cretaceous-Eocene) in Sarawak, Malaysia. *Micropaleontology* 14: 1–37.

———. 1973. Palynological evidence for change in geomorphology, climate and vegetation in the Mio-Pliocene of Malaysia. *Trans. Second Aberdeen-Hull Symp. Males. Ecol.* 6–34.

Pringle, S. L. 1969. World supply and demand of hardwoods. *Proc. Conf. Trop. Hardw.* Syracuse.

Rambo, A. T., and P. E. Sajise. 1984. *An Introduction to Human Ecology Research on Agricultural Systems in Southeast Asia.* Laguna, Philippines; University of Philippines Publ.

Sastrapradja, S. 1975. Tropical fruit germ plasms of S.E. Asia. In *S.E. Asian Plant Genetic Resources.* Eds. J. T. Williams et al. 33–46.

Schery, R. W. 1972. *Plants for Man.* Prentice Hall.

Soepadmo, E. 1979a. Genetic resources of Malaysian fruit trees. *Malays. Appl. Biol.* 8: 33–42.

———. 1979b. The role of tropical botanic gardens in the conservation of threatened valuable plant genetic resources in S.E. Asia. In *Survival or Extinction.* Eds. H. Synge and H. Townsend. Kew: Royal Botanic Gardens. 63–74.

USNAS. 1975. *Underexploited Tropical Plants with Promising Economic Value.* 1–186.

Vavilov, N. I. 1951. The origin, variation, immunity and breeding of cultivated plants. *Chronica Botanica* 13: 1–366.

Voon, P. K., S. H. Khoo, and Z. H. Mahmud. 1979. Integrated surveys for socio-economic change among the Orang Asli in Peninsular Malaysia. *Fed. Mus. Jour.* 24: 145–155.

Whitmore, T. C. 1975. *Tropical Rain Forests of the Far East.* Oxford: Clarendon Press.

Ethnobotany Today in Northwestern North America

NANCY J. TURNER

The ideal definition of ethnobotany is, in my view, a broad one. Since the term is derived from *ethno-*, pertaining to "race, people, cultural group, nation," and *botany*, "the science of plants," a logical definition is "the science of people's interactions with plants." Some prefer to restrict the discipline to the study of aboriginal,[1] pre-industrial peoples and their relationships with plants, but this definition does not recognize the complex relationships and interdependence between plants and modern societies of all types. It should be acceptable to study ethnobotany among Canadian Chinese, Canadian Ukrainian, Anglo American, and Black American cultures as much as among modern Native American groups.

Most of the ethnobotanical research that has been done to date in northwestern North America has concentrated on aboriginal peoples and their interactions with indigenous plants, although some historical work was done in 1984 by Buell and his colleagues on medical ethnobotany of early Chinese American herbalists. Ethnobotanically, northwestern North America is a region of both challenge and promise. Bordered by the Rocky Mountains on the east and the Pacific Ocean on the west, the area to be discussed here extends roughly from northern California to central Alaska, covering Oregon, Washington, Idaho, western Montana, and most of the province of British Columbia, Canada. The region is geographically and vegetationally diverse, and there is exceptional linguistic and cultural variation among its indigenous peoples. Depending on where exact boundaries for the region are drawn, there are between sixty and eighty distinct language groups represented, within three to five major cultural areas.

Early Ethnobotanical History

It is inevitable, given the richness of the flora and aboriginal cultures in this region, that the recording of ethnobotanical knowledge of the indigenous peoples, particularly on the practical utility of plants, should commence immediately with the entry of Europeans. Explorers and early naturalists—such as William Ellis, assistant surgeon and artist who travelled with Captain James Cook; José Mariano Moziño, Spanish botanist and explorer; Archibald Menzies, physician and botanist for Captain George Vancouver; Meriwether Lewis and William Clark; and David Douglas—reported on the indigenous residents and plant uses they encountered (Newcombe 1923; Cutright 1969; Wilson 1970; Morwood 1973; Turner 1978a).

The information they provided is fragmentary and anecdotal because of the transience of these early explorers, their lack of familiarity with the flora, and, in most cases, their inability to speak the aboriginal languages. Nevertheless, their observations are invaluable to modern ethnobotanists because they allow some reconstruction of pre-contact cultural patterns and comparisons with more recent ethnobotanical information. Furthermore, the value of these observations is enhanced by being first-hand accounts; most of the observed activities involving plants would not be witnessed today.

In the later nineteenth and early twentieth centuries, several scientifically trained researchers—notably ethnographers Franz Boas, John Swanton, James Teit, A. F. Chamberlain, and Charles F. Newcombe, reporter James G. Swan, Roman Catholic missionary father A. G. Morice, and geologist-naturalist George M. Dawson—contributed substantially to the growing body of ethnobotanical data (see Morice 1893; Swan 1857; Dawson 1891; Chamberlain 1892; Newcombe 1895–1910; Teit 1906; Boas 1921; Cole and Lockner 1989). They were able to remain with communities of aboriginal people for longer periods of time than did their explorer predecessors, and they were able to pursue actively the detailed recording of ethnobotanical traditions of these peoples. The resultant records, both published and unpublished, serve as a foundation for modern ethnobotanical work in the region, although they were made when ethnobotany as a named discipline was largely unknown.

Fortunately, these early reporters generally were careful and methodical in recording information and were either themselves relatively knowledgeable in botany or were in contact with botanists who could confirm the identifications of the plants about which they wrote. James Teit, for example, corresponded with a well-known botanist of the day, John Davidson of Vancouver. Teit also sent specimens for determination to Dominion government botanists in Ottawa.[2] Except in the cases of Franz Boas and John Swanton, however, the linguistic knowledge of these early ethnobotanical researchers was limited, and their renderings of aboriginal terminology often were inaccurate. Further complications arose from the use of different orthographies for recording the complex sounds in the diverse aboriginal languages. Nevertheless, the works of these early reporters have provided a starting point, enabling later linguists and other researchers to check, improve, and standardize the recording of folk botanical terminology in the different languages.

From the late eighteenth century to the late nineteenth century a drastic and tragic loss of aboriginal populations occurred, mainly due to the ravages of European diseases, such as smallpox, measles, and tuberculosis, for which the indigenous peoples had no natural immunity (see Boyd 1990). This catastrophic situation certainly must have influenced the recording of ethnobotanical information in many ways. The type and application of herbal and other medicinal treatments would have been determined to a major extent by the need to treat this new range of devastating diseases. The loss of cultural and linguistic knowledge of those who died at this time is itself an irreconcilable tragedy.

Ethnobotany Comes of Age

By the first half of the twentieth century, ethnobotany was coming to be recognized and treated as a distinct discipline. In northwestern North America, several publications reflected this. One was an edited version of James Teit's meticulous ethnobotanical notes, *The Ethnobotany of the Thompson Indians of British Columbia* (Steedman 1930). Another, by ethnologist Erna Gunther, was *Ethnobotany of Western Washington* (1945), which, according to the author's description, was still a "catchall" publication. Although rich in detail, it was not intended to reflect a systematic approach to the accumulation of ethnobotanical material, but rather was a compilation of botanical information from the author's wide-ranging ethnographic work and from various literature sources (E. Gunther, pers. com. 1974). Other ethnologists and linguists also accumulated ethnobotanical information in conjunction with their mainstream work. A study of Makah herbal medicines by Frances Densmore, for example, was published as part of her book, *Nootka and Quileute Music* (1939).[3]

Archaeologist and Bureau of Indian Affairs employee Albert Reagan (1934) published ethnobotanical research on the Hoh and Quileute of Washington during this period. Archaeologist Harlan I. Smith's *Materia Medica of the Bella Coola and Neighbouring Groups* (1928) is another example of sideline research. Smith also made copious notes on other aspects of ethnobotany of the Bella Coola (now known as Nuxalk), Carrier, and neighboring groups; these unpublished notes are housed at Canada's National Museum of Civilization in Ottawa. Other researchers who have made significant contributions to ethnobotanical knowledge in northwestern North America include ethnographers Wayne Suttles, Homer Barnett, Leslie Spier, Verne Ray, William Elmendorf, and Thomas F. McIlwraith (see Suttles 1990).

Ethnobotanical Work in the Last Half of the Twentieth Century

Ethnobotanical studies of the 1970s and 1980s were severely and increasingly restricted by a loss of traditional cultural knowledge and practice among aboriginal peoples within this century. In large part this was a result of active suppression of aboriginal languages and cultures that took place in Indian residential schools and churches (Levine and Cooper 1976). Some government, religious, and educational officials, with little understanding of indigenous cultures, could see no value in them and sought to eliminate them. Thus, in the 1980s, with a few exceptions in remote, rural areas, only some members of the eldest generation of aboriginal peoples recalled days when wild plants were a major source of foods, materials, and medicines. In comparison with past use, a comparatively small number of wild plant products are still being used in traditional ways (see Turner 1991a for further discussion of historical change in plant use and plant knowledge) (see Figures 1, 2).

Despite this general and tragic loss of culture and language traditions, the discipline of ethnobotany has grown and expanded in several ways in this region. First, there has been a proliferation of basic ethnobotanical field studies for a variety of cultural and linguistic groups. Ethnobotanical reports on aboriginal groups such as the Coast Salish of Vancouver Island (Turner and Bell 1971), Southern Kwakiutl (Kwakwaka'wakw) (Turner and Bell 1973), Bella Coola (Nuxalk) (Turner 1973), Haida (Turner 1974), Upper Stó:lō (Halkomelem) (Galloway 1982), Lillooet (Turner 1974), Shuswap (Palmer 1975), Okanagan-Colville (Turner et al. 1980), Hesquiat (a Nuu-chah-nulth or Nootkan group) (Turner and Efrat 1982), Nitinaht (Ditidaht, also a Nuu-chah-nulth group) (Turner et al. 1983), Tanaina (Kari 1987), Flathead (Hart 1979), and Thompson (Nlaka'pamux) (Turner et al. 1990) have already been published, and other, similar works will be published in the future.[4]

Eugene Hunn at the University of Washington, Seattle, has been working on Sahaptin ethnobiology and published an award-winning book with his Sahaptin collaborator, James Selam (Hunn 1990; see also Hunn 1981). David French of Reed College, Portland, has been doing ethnobotanical research at the Warm Springs Indian Reservation in Oregon for many years and has collaborated with Eugene Hunn (see French 1981; Hunn and French 1981; see also Hunn et al. n.d.). Steven Gill (1983) completed a doctoral dissertation

Figure 1. Margaret Lester (left) and the late Nellie Peters, Lillooet speakers from Mount Currie, gather reed canary grass stalks (*Phalaris arundinacea*) for decoration of cedar-root baskets.

for Washington State University at Pullman on Makah ethnobotany and has undertaken other ethnobotanical studies as well. Helen Norton, a doctoral student under Eugene Hunn at the University of Washington, also did ethnobotanical research in Washington, as well as in Alaska (Norton 1979a, 1979b, 1981, 1985).

Randy Bouchard and Dorothy Kennedy of the British Columbia Indian Language Project, Victoria, have been researching Salishan ethnobotany since the early 1970s as part of an ongoing, active ethnographic and linguistic field program. Specifically, they have accumulated ethnobotanical data and produced a wide variety of reports and unpublished manuscripts on Shuswap (Secwepemc), Lillooet, Thompson (Nlaka'pamux), Okanagan-Colville, Bella Coola (Nuxalk), Sechelt, Squamish, and Mainland Comox (Sliammon) (see Kennedy and Bouchard 1983). Much of their Okanagan-Colville information is contained in Turner, Bouchard, and Kennedy (1980). Other ongoing ethnobotanical research includes that of Brian D. Compton (University of British Columbia, Vancouver) on North Wakashan and Southern Tsimshian peoples (information currently being compiled and analyzed in a doctoral dissertation); Leslie Johnson Gottesfeld (doctoral student at the University of Alberta, Edmonton) on Gitksan, Wet'suwet'en, and Tsimshian (see Gottesfeld and Anderson 1988; Gottesfeld 1992); Michele Kay (master's degree student at the University of Victoria) on Ulkatcho Carrier; and Elizabeth Ritch-Krc (completed master's degree at University of British Columbia) on the Carrier plant medicines of herbalist Sophie Thomas of Stoney Creek (Ritch-Krc 1992).

Many of these studies follow the descriptive modes set by Steedman's and Gunther's work, but with one important distinction: rather than being carried out by ethnologists as a "sideline" to other ethnographic work, with varying degrees of involvement by botanists and linguists, modern ethnobotanies have been undertaken mostly with ethnobotanical research as the primary goal, and as collaborative efforts among ethnobotanists,

Figure 2. The late Nellie Peters (right) exhibits some of her coiled cedar-root basketry, decorated with bitter cherry bark (*Prunus emarginata*) and split, cured stems of grass. Her husband, Alec Peters (left) holds some of the snowshoes he makes; the frames are of western yew (*Taxus brevifolia*) sapling wood, the webbing of deer hide strips.

linguists specializing in the study of aboriginal languages, and aboriginal plant specialists and researchers. This has led to a higher degree of accuracy in botanical identifications and linguistic transcriptions of aboriginal botanical terminology.

Greater detail in the reporting of ethnobotanical information has been an observed trend in recent years. Active participation by the researcher in the harvesting, preparation, and consumption of traditional foods, for example, is a common occurrence and allows a better understanding of the practical aspects of plant knowledge. With increasing frequency, aboriginal peoples are participating in and initiating ethnobotanical research and interpretive programs.

Attention to research methodology also has been stressed (Norton and Gill 1981). Most modern ethnobotanical researchers use tape recorder and camera in addition to preparing herbarium collections to document their findings; this further enhances the value of their work.

Special Topics in Ethnobotany

SPECIES INVESTIGATIONS

As a direct outcome of the quest for greater detail in ethnobotanical research, numerous papers have been published on aboriginal use and interaction with particular plant species or groups of species, such as black tree lichen (*Bryoria fremontii*), soapberry (*Shepherdia canadensis*), bracken fern (*Pteridium aquilinum*), springbank clover and Pacific silverweed (*Trifolium wormskioldii* and *Potentilla anserina* subsp. *pacifica*), blue camas and riceroot (*Camassia* and *Fritillaria* species), devil's-club (*Oplopanax horridus*), *Lomatium* species, cow-parsnip (*Heracleum lanatum*), cottonwood mushroom (*Tricholoma populinum*), coniferous tree species, and spiny wood fern (*Dryopteris expansa*) (Turner 1977, 1981b, 1982; Norton 1979b; Hunn and French 1981; Turner and Kuhnlein 1982, 1983; Kuhnlein and Turner 1987; Turner et al. 1987; Turner 1988a; Turner et al. 1992). These reports include details of aboriginal nomenclature, distribution of use, methods of harvesting and preparation, and specific information on chemical composition and nutritive values. Cultural adaptations to some non-native plants have also been investigated (see Suttles 1951b) and should be a continuing subject for research.

PLANT FOODS

Aside from basic, descriptive ethnobotanical work, several general topics within the discipline have been investigated. The role of traditional plant foods in aboriginal diets has become a particularly important subject in the 1980s, and it is notable that to both aboriginal people and the general public this subject seems to be highly relevant. Two British Columbia Provincial Museum (now Royal B.C. Museum) handbooks on aboriginal foods (Turner 1975, 1978b) have been distributed widely and are being used in aboriginal studies curricula in various parts of the province. The people of 'Ksan (Gitksan) compiled a fascinating account of their traditional foods (People of 'Ksan 1980). The Coast Tsimshian people of Port Simpson have also produced, in collaboration with the local school board, an illustrated handbook of their traditional food (People of Port Simpson 1983) as part of their school curriculum development. The Coqualeetza Education Training Centre (1981) and the Secwepemc Cultural Education Society (1986) have published books on ethnobotanical topics specifically for school use. Other publications on traditional plant food use include a cookbook, *Indian Foods* (Medical Services Pacific Region, 1972), an article on aboriginal foods of southeastern Alaska (Jacobs and Jacobs 1982), and a pamphlet on Interior Salish food preparation (Surtees 1974). Although outside the range of this chapter, a book on northwest Alaskan Eskimos is outstanding for its treatment of traditional plant foods and deserves mention: *Nauriat Nigiñaqtuat: Plants That We Eat*, by Anore Jones (1983).

There has been a corresponding interest in traditional foods in the academic world, particularly among nutritionists studying past dietary regimes of aboriginal peoples and the potential of traditional foods for enhancing contemporary diets (see Kuhnlein 1984). Studies on the nutrient content of aboriginal plant foods in the region have been carried out by Benson et al. (1973), Keely (1980; Keely et al. 1982), and Kuhnlein (Kuhnlein et al. 1982; Turner and Kuhnlein 1983; Kuhnlein and Turner 1987; Laforet et al. n.d.). As much as possible, these studies incorporate aboriginal knowledge of traditional foods and follow aboriginal preparation techniques (Figure 3). A summary of the nutrient content of major traditional vegetal foods of the Pacific Northwest is provided by Norton et al. (1984). This type of work is continuing, and the results show that plant foods were indeed an essential source of nutrients in the traditional aboriginal diet (see Kuhnlein and Turner 1991 for an overview of ethnonutritional plant research in Canada).

Scientists and indigenous peoples are collaborating closely to learn more about traditional foods and how they may be used for future benefit. Harriet V. Kuhnlein, for example, coordinated a major nutrition study with the people of the Nuxalk Nation, Bella Coola, British Columbia. This project involved, among other elements, nutrient assessment of traditional plant foods and availability studies to determine whether these foods could be utilized without detriment to species populations and whether they are plentiful enough for their use to be practical (Lepofsky et al. 1985). The knowledge of Nuxalk elders was a crucial component of the project, and active use of traditional foods was

Figure 3. A cooking pit, with the traditional types of heating stones and plants for lining and covering, will be used to prepare a sample of Pacific silverweed roots, springbank clover rhizomes (*Trifolium wormskioldii*), and edible blue camas bulbs (*Camassia* species) prior to nutrient analyses and taste evaluation. Samples were also analyzed of the raw "root" foods, and those prepared by modern stove-top cooking (see Kuhnlein, Turner, and Kluckner 1982; Turner and Kuhnlein 1982, 1983).

encouraged. Potential nutrient deficiencies among the Nuxalk population were determined, and traditional foods promoted to improve the quality of the Nuxalk diet. Handbooks with general nutritional information and descriptions of traditional foods and their preparation were distributed to each household in the community (Kuhnlein 1984, 1989a, 1989b, 1989c, 1990; Nuxalk Food and Nutrition Program 1984, 1985). A demonstration garden of wild food plants was established at the entrance of the reserve's health clinic to allow first-hand identification of the food plants and to show how many of them could be propagated for home use.

Similar cooperation between nutritionists and aboriginal people is taking place elsewhere in the region (see Hilty et al. 1980). The growing of traditional food and medicinal plants is seen by some indigenous peoples (e.g., Cowichan on Vancouver Island, and Nesconlith in the British Columbia interior) as a means of conserving valued and dwindling cultural resources. The Nesconliths and the British Columbia Ministry of Agriculture sponsored a symposium at Salmon Arm (October 1992) on traditional Shuswap foods to raise awareness of their past and future significance.

The study of traditional plant foods has resulted in documentation of one factor in northwestern North America that has received scant attention in the past. Aboriginal peoples in this region have been perceived as having a hunting-fishing economy, with the gathering of plant resources as only a secondary occupation. Since hunting and fishing were done almost exclusively by men, and gathering by women, the implications were that women were not as important in maintaining subsistence in traditional economies. This assumption has been contested by several ethnobotanical researchers, who presented evidence to show that women gathering plant foods were making a major contribution to subsistence and that plant foods comprised a substantial proportion of the total diet (Hunn 1981; Hunn and French 1981; Norton 1981, 1985).

Although, with the exception of tobacco (Turner and Taylor 1972), conventional agriculture was not practiced traditionally in northwestern North America, a better understanding of strategies for habitat enhancement (i.e., for increasing productivity of certain economically important species) is emerging as ethnobotanical knowledge increases. Controlled burning was widely employed to maintain lowland meadows for the production of edible camas (*Camassia* species) and other species (see Suttles 1951a; Norton 1979a; Turner and Kuhnlein 1983; Thoms 1989; Turner 1991b). Mountainsides and upland meadows were burned periodically to enhance production of berries such as thimbleberry (*Rubus parviflorus*), blackcap (*Rubus leucodermis*), and blueberries and huckleberries (*Vaccinium* species). Species with edible underground parts, such as tiger lily (*Lilium columbianum*), yellow avalanche lily (*Erythronium grandiflorum*), and spring beauty or "Indian potato" (*Claytonia lanceolata*), were also encouraged by periodic burning (Turner et al. 1980; Barrett and Arno 1982; Turner et al. 1990; Turner 1991b).

Foods such as springbank clover rhizomes (*Trifolium wormskioldii*) and Pacific silverweed roots (*Potentilla anserina* subsp. *pacifica*) on the coast, and spring beauty (*Claytonia lanceolata*), yellow avalanche lily (*Erythronium grandiflorum*) and bitterroot (*Lewisia rediviva*) in the interior were all harvested intensively from the same digging grounds over many years. In some cases, harvesting sites were owned by certain families and passed as property from generation to generation. That these sites maintained a sustained yield over many years indicates a sophisticated harvesting strategy, involving specific timing, selective gathering, soil enhancement through cultivation, and intentional or unintentional propagation from fragments of root, bulb, or rhizome, or even from seed (Edwards 1979; Turner et al. 1980; Turner and Kuhnlein 1982, 1983). This type of information is significant because it allows insights into possible proto-agricultural practices that may have been used by early humans in the tentative beginnings of selection and crop production.

PLANT MATERIALS AND ARCHAEOBOTANY

Plant materials used in aboriginal technologies have also been a subject of study (see Turner 1979). More than 100 species are now known to have been used in British Columbia alone. Microscopic identification of plant tissues in artifacts is also being carried out (see Croes 1977, 1980; Friedman 1978; Croes and Blinman 1980; Schaeffer 1981; Florian 1982) and is adding significantly to basic ethnobotanical and archaeobotanical knowledge in the area. The work of Croes and Friedman and their colleagues is particularly relevant for various archaeological sites, such as Ozette and Hoko River in Washington, where living and resource gathering areas were suddenly covered, in prehistoric times, by mud and, because of anaerobic conditions, plant remains and artifacts did not decay and have been relatively well preserved. Careful excavation, documentation, preservation, and identification of artifacts of wood and fiber from these sites have yielded considerable new information.

It is difficult to find plant remains in other types of archaeological sites, except for charred material from fire pit and hearth sites, and some material from caves and arid sites. Archaeologists have become increasingly interested in the plant materials being discovered from sites (see Pokotylo and Froese 1983; Lepofsky 1987; Hayden 1992) and are attempting to put these discoveries in perspective through ethnobotanical work with indigenous consultants, as part of a field known as ethnoarchaeology (see Alexander 1992). In short, archaeobotany is still a relatively new field in this region but is showing more promise than would have been believed a few decades ago.

Systematic identification of archaeological plant materials, such as seeds, charred woods, and cordage fibers, and of museum artifact materials, such as the wood of masks and the materials used in various types of baskets, is ongoing at the Royal British Columbia Museum and other museums. Eventually identification of the construction materials of all museum artifacts should be possible, and articles whose botanical features may have been only superficially documented or incorrectly identified will be more accurately described, and hence become more meaningful components of collections.

Also of particular significance is the work of paleobotanists correlating pollen records with archaeobotanical and ethnographic records to reconstruct the interrelationships between people and important economic plants in the distant past (see Hebda and Mathewes 1984). Another approach to plants in aboriginal technologies has been taken by Hilary Stewart, whose work with plant materials has had a practical and experimental orientation (see Stewart 1977, 1984).

MEDICINAL PLANTS

Medicinal plants of northwestern North America, though numbering in the hundreds of species, are only now starting to be investigated for their chemical composition and pharmacological properties. Relatively few have so far found use in "Western" medicine. Aside from the works of Smith (1928) and Densmore (1939) and a few as yet unpublished general surveys,[5] to date no major works centering on traditional medicines of northwestern North America are available. Many scattered references to the medicinal use of plants are given in the ethnobotanies mentioned previously, and some specific information is contained in some shorter articles (e.g., Edwards 1980; Smith 1983; Turner 1982, 1984, 1989b; Turner and Hebda 1990; Gottesfeld and Anderson 1988), but the ethnomedicinal information from all these sources needs to be collated and evaluated so that thorough pharmacological analyses of medicinal species can be carried out where indicated. Also lacking is a theoretical framework for traditional medicinal practices in northwestern North America.

Fortunately, some research in this field is ongoing. David and Kathrine French at Reed College, Portland, have been doing extensive research on traditional medicines of the people of the Warm Springs Reservation in Oregon and are preparing their work for publication. Biochemist Neil Towers and his students, Elizabeth Ritch-Krc and Alison MacCutcheon and colleagues at the University of British Columbia,[6] have maintained a deep interest in the chemical properties of medicinal plants and have done some preliminary research indicating positive results for antibiotic properties of some species.

An event that has significantly increased general awareness of and interest in traditional medicines is the discovery of a promising new anti-cancer agent, taxol, in the bark of western yew (*Taxus brevifolia*) (see McAllister and Haber 1991). Yew has been used in a variety of medicinal preparations by aboriginal peoples (Turner and Hebda 1990). Several contemporary Haisla people are currently using it as a heart medicine (Alison Davis, pers. com. 1992). Taxol is now indicated in preliminary tests to be effective against ovarian and other difficult-to-treat cancers, and it has gained tremendous prominence in the media and among medical researchers (see Borman 1991; Manes 1992). Now that the name of this hitherto relatively obscure tree is a household word, people are asking how many other medicinal plants are growing in our temperate forests. The potential utility of native plants may be an important factor in the environmentalists' battle to save remaining northwestern old-growth forests from being eliminated by logging.

With the increasing realization of the inseparable links between the psychological and physiological aspects of healing, more attention needs to be paid to the processes of healing in aboriginal cultures. The prayers to medicinal plants and medicines of the Kwakiutl (Kwakwaka'wakw), as recorded by Boas (1930), are an example of the esteem and reverence directed towards healing agents. This attitude continues to the present and may play a role in the effectiveness of medicinal plants.

FOLK PLANT CLASSIFICATION

The study of folk plant classification systems as pioneered by Brent Berlin and his colleagues (see Berlin et al. 1974; Berlin 1992), has been a topic for some research in northwestern North America (e.g., Hunn 1982; Turner 1974, 1987, 1988a, 1989a). Studies such as these have shown similarities as well as significant differences in folk taxonomies of indigenous peoples in our region as compared with societies previously studied. There is yet much more to learn in this field, however, and much that is already lost or altered because the basic fabric of indigenous classification systems has been so disrupted by European influence. Still, it is important to remember that even the changes in folk classes and nomenclature are valid subjects for study and can give significant insights into the evolution of folk taxonomic systems. Also important to consider are the means for evaluating the relative significance of plants in aboriginal cultures (see Turner 1988b).

Aboriginal plant names are being used in linguistics research to help explain the diffusion and proliferation of aboriginal languages in the region (see Hess n.d.; Kinkade 1986, 1989; Turner, Ignace, and Compton 1992). More of this type of comparative research becomes possible as more ethnobotanical knowledge concerning individual groups is analyzed.

Public Interest in Ethnobotany

As already indicated, use of ethnobotanical information within indigenous communities and among members of the general public has increased significantly in the last half of the twentieth century. Public awareness of this field has increased correspondingly. Extension courses in ethnobotany through major universities, local colleges, and museums are well attended (see Environmental Studies Program 1992). The Royal British Columbia Museum has produced a travelling exhibit, *Wild Harvest*, which centers on the harvesting and preparation of traditional indigenous plant foods in western Canada. The museum also has sponsored a series of public speaking tours on this topic throughout the province. Publications on aboriginal foods and others of a more general nature (Carrier Linguistic Committee 1973; Kari 1987) are further evidence of increasing public appreciation of ethnobotany. In 1992, the Knowledge Network produced a film on ethnobotany in British Columbia, and there have been many newspaper and magazine articles on aspects of traditional plant knowledge.

Concern for public safety has been a topic of considerable importance in the "popularizing" of ethnobotanical knowledge. Poisoning through mistaken identification or misuse of edible species or through wrongful application of herbal medicines by inexperienced persons must always be considered a potential outcome of a high public profile for ethnobotany. It is the responsibility of ethnobotanists and other educators to minimize such possibilities through careful and persistent reminders of the dangers of the improper use of plants. There is still much that we do not know about the content and effects of many plant medicines, and, even with some traditional foods, there are concerns over potential toxicity (Kuhnlein and Turner 1987, 1991; Turner and Szczawinski 1991).

The conservation of populations of edible, medicinal, or otherwise useful wild plants is another matter of concern. Many botanists and environmentalists are disturbed by the potential harm that can be done to indigenous plant populations through intensive use and overharvesting by people. With all the other pressures on native vegetation from urban agricultural encroachment, overgrazing, logging, highway construction, and invasion of habitats by aggressive introduced weeds and exotic pests, this fear is well founded, at least in the case of certain sensitive, rare, or endangered species. Propagation and cultivation of traditional native plants is in some cases the only means by which the integrity of the species can be maintained. I personally hope that experimentation with wide-scale growing of some outstanding traditional food plants will be taken as a future direction for ethnobotanical research in this region (Turner 1981a).

Ethnobotany Today and Tomorrow

In this chapter I have tried to set in perspective the status of ethnobotany in northwestern North America by tracing its origins. The basic body of recorded and published ethnobotanical knowledge has grown significantly, particularly since the 1970s. More is known about the names and traditional uses of plants in different aboriginal groups in northwestern North America, and, because of an increasing sophistication of methodology, the extent and detail of knowledge have also been enhanced. Interest in the discipline from both practical and theoretical perspectives has never been greater. Researchers in many fields—botanists, ecologists, ethnologists, archaeologists, linguists, medical doctors and health care workers, nutritionists, and historians—are exchanging information and ideas and actively collaborating in the collective goal of increasing understanding of cultural and natural systems.

Aboriginal experts and educators, such as Daisy Sewid-Smith, Kwakwaka'wakw historian of Campbell River; the late John Thomas, a speaker of Nitinaht (Ditidaht); the late Annie York of Spuzzum and Mabel Joe of Merritt, both Thompson (Nlaka'pamux) speakers (Figures 4, 5); and Mary Thomas, Shuswap (Secwepemc) elder of Enderby, are also making substantial contributions to the discipline as educators. Staffs of museums, universities, and schools likewise have been active in disseminating ethnobotanical information to the public. Aboriginal communities have taken a major role in cultural and language education, including plant knowledge, for themselves and others. In this sense, ethnobotany has been part of a recent and continuing societal trend of increasing interest in many aspects of aboriginal traditions and cultural heritage, including art, weaving, dancing, potlatching, and other ceremonial events, many of which were previously harshly suppressed by legal, religious, and educational institutions.

Aboriginal land claims, based in part on traditional economies and seasonal movement patterns, and hence, on natural resource use, have also been prominent in this trend. In this way, ethnobotanical knowledge has entered the legislative and political realm, and will probably continue to do so in the future until the many outstanding land claims and resource use conflicts in the region are settled. There is also increasing support in Canada for aboriginal self government, in which First Nations will gain more control of their own lands and resources. Ethnobotanical knowledge will certainly provide important background information on traditional land management.

The ceremonial, religious, and mythological characteristics of plants have received relatively little attention by ethnobotanists to date (see Blanchette et al. 1992). The sacredness of plants, their spiritual attributes, and the metaphors for ecological knowledge and sustainable use provided in some of the myths and traditional ceremonies and dances relating to the natural world should be carefully studied. Aboriginal perceptions of plants and the environment may well provide our dominant "Western" society with some keys to sustainable living (see Turner 1992b, 1992c).

What does the future hold for ethnobotanical work in northwestern North America? Since I began my field research in ethnobotany in the late 1960s, most of the aboriginal

Figure 4. The late John Thomas, Nitinaht (Ditidaht) speaker, language specialist, and teacher from Vancouver Island's west coast holding a root-digging stick he has carved of western yew (*Taxus brevifolia*).

elders with whom I worked have passed away. Others have passed from their seventies into their eighties and nineties and no longer can participate actively in research. Some aboriginal languages already have become extinct; other languages are spoken by only a very few elderly speakers who are familiar with the many aboriginal names and uses of plants. Even fewer original sources of traditional ethnobotanical knowledge survive. The loss of cultural knowledge, like the extinction of species, is absolute.

This does not mean that ethnobotanical research will cease. The gap left with the passing of the older generation of aboriginal people will be immense and never will be filled, but many younger aboriginal people have learned from the elders so that many facets of knowledge about indigenous plants will perpetuate into the foreseeable future. The cultural adaptations made within the historic period (i.e., 1850–1950) by aboriginal groups in relation to their foods, materials, medicines, and systems of nomenclature and classification are significant subjects for ethnobotanical study. Historical, archaeological, and particularly ecological components of ethnobotanical research will likewise continue. Furthermore, much of the ethnobotanical information gathered thus far consists largely of basic facts; these lend themselves to many forms of analysis, compilation, and comparison that will lead to significant findings, with implications for many fields of research. For example, pre-contact subsistence strategies could be reconstructed for some peoples with information that is already in existence.

The use of computers in sorting and analyzing large volumes of ethnobotanical data is still in its infancy in our region. Already computer sorting has been used in itemizing and comparing ethnobotanical information in the study of folk taxonomies (Turner 1974). Eugene Hunn has been utilizing the computer for statistical analysis and for sorting ethnobotanical information (Hunn 1975; Hunn and Norton 1984). Some data from our region already have been incorporated into the NAPRALERT Retrieval System for Natural Prod-

Figure 5. Mabel Joe, Thompson (Nlaka'pamux) speaker and herbal specialist of Merritt, in the Nicola Valley, holds cleaned samples of edible "pine mushroom" (*Tricholoma magnivelare*).

uct Data at the University of Illinois, Department of Pharmacognosy and Pharmacology, Chicago. This information will be much used by taxonomists, chemists, pharmacologists, nutritionists, and ethnobotanical researchers.

Nutrient analyses of significant plant foods and chemical analyses of medicinal species will continue to add significantly to the understanding of species of past economic importance and potential utility. Toxicological investigations into potentially harmful species also should merit priority. Horticultural experimentation with indigenous plants in northwestern North America may result in the production of new types of food and in the enhancement of existing foods for aboriginal and non-aboriginal peoples alike.

There is a continuing trend at various universities and other institutions towards integrated and multidisciplinary research in which ethnobotanical studies are combined with research on indigenous land use patterns, archaeology, botany, ecology, linguistics, ethnogeography, ethnozoology, and other disciplines. This should lead to a better understanding of the wide-ranging significance of ethnobotany from both practical and theoretical perspectives. University courses in ethnobotany and ethnobiology in northwestern North America are also on the increase.

Ethnobotanical studies on ethnic groups in northwestern North America other than aboriginal peoples should be encouraged since each community has its own attitudes, names, and uses for plants and plant products relating to other aspects of local culture. Furthermore, relationships with plants change over time, along with cultural and environmental factors, and this process of change is itself important and should be a consideration in any ethnobotanical studies.

The Declaration of Belém, drafted and passed in 1988 in Belém, Brazil, at the First International Congress of the International Society for Ethnobiology and supported unanimously by a motion of the Society of Ethnobiology in Riverside, California, in 1989 heralds a new era in the recognition and appreciation of indigenous peoples as knowledgeable stewards of most of the world's biodiversity. A newsletter on Traditional Ecological Knowledge, *TEK Talk*,[7] is another manifestation of this recognition (see Turner 1992, n.d. (b)). The imperative for fair and appropriate acknowledgement and compensation of indigenous peoples for their knowledge is as important in northwestern North America as it is elsewhere in the world. Continued collaboration, consultation, and fair exchange of knowledge and funding for research and publication are essential components of any ethnobotanical work today. Ethnobotany in northwestern North America will continue to be a living, expanding discipline. Many contributions are yet to be made; much is yet to be learned.

NOTES

[1]In Canada, the current, preferred terms for First Nations Peoples are *aboriginal* or *indigenous* rather than *Indian* or *native*.

[2]James Teit, unpublished papers, ca. 1898–1918, American Philosophical Society Library, Philadelphia, Pennsylvania.

[3]Densmore was particularly interested in ethnobotany and had previously conducted ethnobotanical research among the Chippewa (Ojibwa).

[4]For example, a major project on Shuswap (Secwepemc) ethnobotany is under way, as a collaborative venture, with a number of academic researchers (including myself, Marianne Boelscher Ignate, and Brian Compton) and many Shuswap elders and researchers. A number of publications from this research are planned, including a general, comprehensive Shuswap ethnobotany, to be published by the Secwepemc Cultural Education Society.

[5]These include unpublished observations on "Materia Medica of the Northwest Coast Indians of British Columbia" by David Rollins (1972) and "Medicinal Plants of the Native Peoples of the Coastal Pacific Northwest" by Robin Marles (1977).

[6]Neil Towers has retired from the University of British Columbia, but he and his students continue their work in this field. Elizabeth Ritch-Krc completed a master's degree in 1992 on Carrier medicines, in which she undertook analysis of antibiotic and anticancer properties of selected medicinal plants, with promising initial findings. Alison MacCutcheon and her colleagues also did preliminary antibiotic screening of nearly 100 medicinal plants of coastal British Columbia, obtaining impressive initial results. These were reported at a symposium sponsored by the Department of Botany, the University of British Columbia and Science World in a talk titled "Current Research on Antibiotics from BC plants" (9 May 1992).

[7]*TEK Talk* is published in Ottawa, Canada, by UNESCO, as part of the World Decade for Cultural Development.

ACKNOWLEDGMENTS

I am indebted to the aboriginal people of northwestern North America who have maintained their interest in plants and plant use and who have generously shared with me and other researchers their knowledge, experience, and deep appreciation of plants. Special thanks go to Alec Peters, the late Nellie Peters, Margaret Lester, the late Edith O'Donaghey, the late Annie York, Mabel Joe, and the late John Thomas for allowing their photographs to be used. Since I first drafted this article in the mid-1980s, half of these elders and many others knowledgeable about plants have passed away. They are sadly mourned, but in their teachings and in the strength of their characters, they leave behind them a lasting legacy of knowledge and wisdom (Turner 1992b). Special thanks also go to some of the younger aboriginal people who have inspired me and helped me towards a better understanding. These include, among many, Ron Ignace of Skeechestn; Daisy Sewid-Smith of Campbell River; Kim Recalma-Clutesi of Qualicum; Maggie and Jack Sedgemore of Victoria; Philip Paul of Tsartlip, Brentwood Bay; Jill Harris of Duncan; and Mary Thomas and family of Enderby.

My sincere thanks go to the following people for reading earlier drafts of this manuscript and for their helpful advice: Brian D. Compton, The University of British Columbia, Vancouver; Dr. Richard I. Ford, University of Michigan, Ann Arbor; Dr. Eugene S. Hunn, University of Washington, Seattle; Randy Bouchard and Dorothy Kennedy, British Columbia Indian Language Project, Victoria; Dr. David H. French, Reed College, Portland, Oregon; Dr. Harriet V. Kuhnlein, School of Dietetics and Human Nutrition, Macdonald College, McGill University, Ste. Anne de Bellevue, Quebec; Dr. Richard J. Hebda and Dr. Robert T. Ogilvie, Botany Unit, Royal British Columbia Museum, Victoria; Kate Seaborne, University Extension, University of Victoria; Alison Davis, Environmental Studies Program, University of Victoria; and my husband, Robert D. Turner. Sandra Peacock carefully retyped the manuscript onto computer disk so that it could be typeset directly.

REFERENCES

Alexander, D. 1992. A reconstruction of prehistoric land use in the Mid-Fraser River area based on ethnographic data. In *A Complex Culture of the British Columbia Plateau: Traditional Stl'atl'imx Resource Use.* Ed. B. Hayden. Vancouver: University of British Columbia Press.

Barrett, S. W., and S. F. Arno. 1982. Indian fires as an ecological influence in the Northern Rockies. *Journal of Forestry* 80(10): 647–651.

Benson, E. M., J. M. Peters, M. A. Edwards, and L. A. Hogan. 1973. Wild edible plants of the Pacific Northwest. *Journal of the American Dietetic Association.* 62: 142–147.

Berlin, B. 1992. *Ethnobiological Classification: Principles of Categorization of Plants and Animals.* Princeton: Princeton University Press.

Berlin, B., D. E. Breedlove, and P. H. Raven. 1974. *Principles of Tzeltal Plant Classification. An Introduction to the Botanical Ethnography of a Mayan-Speaking People of Highland Chiapas.* New York: Academic Press.

Blanchette, R. A., B. D. Compton, N. J. Turner, and R. L. Gilbertson. 1992. Nineteenth century shaman grave guardians are carved *Fomitopsis officinalis* sporophores. *Mycologia* 84(1): 119–124.

Boas, F. 1921. *Ethnology of the Kwakiutl.* Bureau of American Ethnology, 35th Annual Report 1913–1914, part 1. Washington, DC.

————. 1930. *Religion of the Kwakiutl.* Columbia University Contributions to Anthropology 10, parts 1 and 2.

Borman, S. 1991. Scientists mobilize to increase supply of anticancer drug taxol. *Chemical and Engineering News* (2 September 1991): 11–18.

Boyd, R. F. 1990. Demographic history, 1774–1874. In *Northwest Coast.* Ed. W. P. Suttles. Vol. 7 of *Handbook of North American Indians.* Gen. ed. W. C. Sturtevant. Washington, DC: Smithsonian Institution. 135–148.

Buell, P. D., D. W. Lee, J. L. MacDonald, C. Muench, and M. Willson. 1984. *Chinese Medicine on the Golden Mountain.* Ed. H. G. Schwarz. An Interpretive Guide for the Washington Commission for the Humanities travelling exhibit, "Chinese Medicine in Washington and the Northwest: Past and Present." Seattle, Washington.

Carrier Linguistic Committee. 1973. *Hanuyeh Ghun 'Utni-i (Plants of Carrier Country).* Central Carrier Language. Fort St. James, British Columbia (The first part of this book is written in Carrier, the second in English.)

Chamberlain, A. B. 1892. *Report on the Kootenay Indians of southeastern British Columbia.* 8th Report on the Northwestern Tribes of Canada. British Association for the Advancement of Science, Edinburgh Meeting, U.K.

Cole, D., and B. Lockner. 1989. *The Journals of George M. Dawson: British Columbia, 1875–1878.* Vancouver: University of British Columbia Press.

Coqualeetza Education Training Centre. 1981. *Upper Stó:lō Fraser Valley Plant Gathering*. Sardis, British Columbia.

Croes, D. R. 1977. *Basketry from the Ozette Village Archaeological Site: A Technological, Functional and Comparative Study*. Ph.D. thesis, Washington State University, Pullman.

———. 1980. *Cordage from the Ozette Village Archaeological Site: A Technological, Functional and Comparative Study*. Laboratory of Archaeology and History, Project Reports No. 9, Washington State University, Pullman.

Croes, D. R., and E. Blinman, eds. 1980. *Hoko River: A 2500 Year Old Fishing Camp on the Northwest Coast of North America*. Laboratory of Anthropology, Reports of Investigations No. 58, Washington State University, Pullman.

Cutright, P. R. 1969 *Lewis and Clark: Pioneering Naturalists*. Chicago: University of Illinois Press.

Dawson, G. M. 1891. Notes on the Shuswap People of British Columbia. *Transactions of the Royal Society of Canada*, Section II, part 1. 3–44.

Densmore, F. 1939. *Nootka and Quileute Music*. Bureau of American Ethnology Bulletin no. 24. Washington, DC.

Edwards, G. T. 1979. Indian spaghetti. *The Beaver* (Autumn): 4–11. (Nuxalk use of Springbank Clover)

———. 1980. Bella Coola. Indian and European medicines. *The Beaver* (Winter): 4–11.

Environmental Studies Program, University of Victoria. 1992. *Plants for All Reasons: Culturally Important Plants of Aboriginal Peoples of Southern Vancouver Island*. Ed. N. J. Turner. Teaching Manual, written by students of Environmental Studies 400C Class and University Extension Class, July 1992.

Florian, M. E. 1982. *Plant Materials Used in West Coast Indian Basketry: Their Identification and Implication of Ecological, Cultural and Technical Response to their Usage*. Paper presented at the Canadian Ethnology Society's 19th Annual Meeting, 7–11 May 1992, Vancouver, British Columbia.

French, D. H. 1981 Neglected aspects of North American ethnobotany. *Canadian Journal of Botany* 59 (11): 2326–2330.

Friedman, J. 1978. *Wood Identification by Microscopic Examination. A Guide for the Archaeologist on the Northwest Coast of North America*. Heritage Record No. 5, British Columbia Provincial Museum, Victoria.

Galloway, B. 1982. *Upper Stó:lō Ethnobotany*. Sardis, British Columbia: Coqualeetza Education Training Centre.

Gill, S. J. 1983. *Ethnobotany of the Makah and Ozette People, Olympic Peninsula*. Ph.D. thesis, Washington State University, Pullman.

Gorman, M. W. 1896. Economic botany of southeastern Alaska. *Pittonia III*. Part 14: 64–85.

Gottesfeld, L. M. J. 1992. The importance of bark products in the Aboriginal economies of Northwestern British Columbia, Canada. *Economic Botany* 46(2): 148–157.

Gottesfeld, L. M. J., and B. Anderson. 1988. Gitksan traditional medicine: Herbs and healing. *Journal of Ethnobiology* 8: 13–33.

Gunther, E. 1945. *Ethnobotany of Western Washington*. University of Washington Publications in Anthropology 10(1). University of Washington Press, Seattle. Rev. ed. 1973.

Hart, J. 1979. The ethnobotany of the Flathead Indians of western Montana. *Botanical Museum Leaflets* (Harvard University) 27 (10): 261–307.

Hayden, B., ed. 1992. *A Complex Culture of the British Columbia Plateau: Traditional Stl'atl'imx Resource Use*. Vancouver: University of British Columbia Press.

Hebda, R. J., and R. W. Mathewes. 1984. Holocene history of cedar and native Indian cultures of the North American Pacific Coast. *Science* 225: 711–713.

Hess, T. n.d. *Borrowed Words and B.C. Prehistory*. Sardis, British Columbia: Coqualeetza Education Training Centre.

Hilty, I. E., J. H. Peters, E. M. Benson, M. A. Edwards, and L. T. Miller. 1980. *Nutritive Values of Native Foods of Warm Springs Indians*. Oregon State University Extension Service Circular 809, Corvallis.

Hunn, E. S. 1975. A measure of the degree of correspondence of folk to scientific biological classification. *American Ethnologist* 2(2): 309–327.

———. 1981. On the relative contribution of men and women to subsistence among hunter-gatherers of the Columbia Plateau: A comparison with *Ethnographic Atlas* summaries. *Journal of Ethnobiology* 1(1): 124–134.

————. 1982. The utilitarian factor in folk biological classification. *American Anthropologist* 84(4): 830–847.

————. 1990. *Nch'i-wana, "The Big River": Mid-Columbian Indians and Their Land.* Seattle: University of Washington Press.

Hunn, E. S., and D. H. French. 1981. Lomatium: A key resource for Columbia Plateau native subsistence. *Northwest Science* 55(2): 87–94.

Hunn, E. S., and H. H. Norton. 1984. Impact of Mt. St. Helens ashfall on fruit yield of mountain huckleberry, *Vaccinium membranaceum*, important Native American fruit. *Economic Botany* 38(1): 121–127.

Hunn, E. S., N. J. Turner, and D. H. French. n.d. Ethnobiology and subsistence. In *Plateau.* Ed. D. E. Walker. Vol. 12 of *Handbook of North American Indians*, gen. ed. W. C. Sturtevant. Washington, DC: Smithsonian Institution. In press.

Jacobs, M., and M. Jacobs. 1982. Southeast Alaska native foods. In *Raven's Bones.* Ed. A. Hope. Sitka, AK.

Jones, A. 1983. *Nauriat Nigiñaqtuat. Plants That We Eat.* Kotzebue, AK: Maniilaq Association.

Kari, P. R. 1987. *Tanaina Plantlore. Den'ina K'et'una.* An Ethnobotany of the Dena'ina Indians of Southcentral Alaska. National Parks Service, Alaska Region, Anchorage. (Original version published 1977 as *Dena'ina K'et'una. Tanaina Plantlore.* Adult Literacy Laboratory, Anchorage Community College, Alaska.)

Keely, P. B. 1980. *Nutrient Composition of Selected Important Plant Foods of the Pre-Contact Diet of Northwest Native American Peoples.* M.Sc. thesis, University of Washington, Seattle.

Keely, P. B., C. S. Martinsen, E. S. Hunn, and H. H. Norton. 1982. Composition of Native American fruits in the Pacific Northwest. *Journal of the American Dietetic Association* 81: 568–572.

Kennedy, D., and R. Bouchard. 1983. *Sliammon Life, Sliammon Lands.* Vancouver, British Columbia: Talonbooks.

Kinkade, M. D. 1986. *Blackcaps and Musqueam.* Paper Presented to the 21st International Conference on Salish and Neighboring Languages, Seattle, Washington, 14–16 August 1986.

————. 1989. *Prehistory of Salishan Languages.* Paper read at the 88th Annual Meeting of the American Anthropological Association, Washington, DC.

'Ksan, People of. 1980. *Gathering What the Great Nature Provided. Food Traditions of the Gitksan.* Vancouver, British Columbia: Douglas & McIntyre; Seattle: University of Washington Press.

Kuhnlein, H. V. 1984. Traditional and contemporary Nuxalk foods. *Nutrition Research* 4: 789–809.

————. 1989a. Change in use of traditional foods by the Nuxalk native people of British Columbia. In *Perspectives in Dietary Change.* Ed. G. H. Pelto and L. A. Vargas. Cambridge, UK: Internal Nutrition Foundation.

————. 1989b. Factors influencing use of traditional foods among the Nuxalk people. *Journal of the Canadian Dietetic Association* 50: 102–108.

————. 1989c. Nutrient values in indigenous wild berries used by the Nuxalk People of Bella Coola, British Columbia. *Journal of Food Composition and Analysis* 2: 28–30.

————. 1990. Nutrient values in indigenous wild plant greens and roots used by the Nuxalk People of Bella Coola, British Columbia. *Journal of Food Composition and Analysis* 3.

Kuhnlein, H. V., N. J. Turner, and P. D. Kluckner. 1982. Nutritional significance of two important root foods (springbank clover and Pacific silverweed) used by native people on the coast of British Columbia. *Ecology of Food and Nutrition* 12: 89–95.

Kuhnlein, H. V., and N. J. Turner. 1987. Cow-parsnip (*Heracleum lanatum* Michx.): An indigenous vegetable of native people of northwestern North America. *Journal of Ethnobiology* 6(2): 309–324.

————. 1991. *Traditional Plant Foods of Canadian Indigenous Peoples. Nutrition, Botany and Use.* Vol. 8 of *Food and Nutrition in History and Anthropology.* Ed. S. Katz. Philadelphia: Gordon and Breach Science Publishers.

Laforet, A., N. J. Turner, and A. York. n.d. Traditional foods of the Fraser Canyon Nlaka'pamux. In *American Indian Linguistics and Ethnography in Honor of Lawrence C. Thompson.* Festschrift. Eds. T. Montler and A. Matina. University of Montana Press, Special Publications in Linguistics. Missoula. In press.

Lepofsky, D. 1987. *Fraser River Investigations into Corporate Group Archaeology: Report on Floral Remains.* Unpublished report to B. Hayden, Department of Archaeology, Simon Fraser University, Burnaby, British Columbia.

Lepofsky, D., N. J. Turner, and H. V. Kuhnlein. 1985. Determining the availability of traditional wild plant foods: An example of Nuxalk foods, Bella Coola, British Columbia. *Ecology of Food and Nutrition* 16: 223–241.

Levine, R. D., and F. Cooper. 1976. The suppression of B.C. languages: Filling in the gaps in the
documentary record. *Sound Heritage* 4(3/4): 43–75.

McAllister, D. E., and E. Haber. 1991. Western yew—precious medicine. *Canadian Biodiversity*
1(2): 2–4.

Manes, C. 1992. In praise of yew. *Orion. People and Nature* 11(1): 30–39.

Medical Services, Pacific Region. 1972. *Indian Food. A Cookbook of Native Foods from British Columbia*. Health and Welfare Canada, Victoria, British Columbia.

Morice, A. G. 1893. Notes archaeological, industrial and sociological on the western Déné.
Transactions of the Canadian Institute, Session 1892–1993.

Morwood, W. 1973. *Traveller in a Vanished Landscape. The Life and Times of David Douglas*. London: Gentry Books.

Newcombe, C. F. 1895–1910. Unpublished notes on native Indians of British Columbia. Provincial Archives of British Columbia, Victoria.

———. 1923. *Menzies' Journal of Vancouver's Voyage, April to October, 1792*. Victoria, British
Columbia: King's Printer.

Norton, H. H. 1979a. The association between anthropogenic prairies and important foods
plants in western Washington. *Northwest Anthropological Research Notes* 13: 175–200.

———. 1979b. Evidence for bracken fern as a food for aboriginal peoples of western Washington. *Economic Botany* 33: 384–396.

———. 1981. Plant use in Kaigani Haida culture: Correction of an ethnohistorical oversight.
Economic Botany 35: 434–449.

———. 1985. *Women and Resources of the Northwest Coast. Documentation from the 18th and Early
19th Centuries*. Ph.D. thesis, University of Washington, Seattle.

Norton, H. H., and S. J. Gill. 1981. The ethnobotanical imperative: A consideration of obliga-
tions, implications and methodology. *Northwestern Anthropological Research Notes* 15(1):
117–134.

Norton, H. H., E. S. Hunn, C. S. Martinsen, and P. B. Keely. 1984. Vegetable food products of
the foraging economies of the Pacific Northwest. *Ecology of Food and Nutrition* 14: 219–228.

Nuxalk Food and Nutrition Program. 1984 *Nuxalk Food and Nutrition Handbook*. Bella Coola,
British Columbia: Bella Coola Health Clinic.

———. 1985. *Kanusyam a Snknic: "Real Good Food."* A Nuxalk (Bella Coola) Recipe Book. Bella
Coola, British Columbia: Bella Coola Health Clinic.

Palmer, G. 1975. Shuswap Indian ethnobotany. *Syesis* 8: 29–81.

Pokotylo, D. L., and P. D. Froese. 1983. Archaeological evidence for prehistoric root gathering
on the Southern Interior Plateau of British Columbia: A case study from Upper Hat Creek
Valley. *Canadian Journal of Archaeology* 7(2): 129–157.

Port Simpson, People of. 1983. *Port Simpson Foods*. A Curriculum Development Project. The
People of Port Simpson and School District No. 52 (Prince Rupert), Prince Rupert, British
Columbia.

Reagan, A. B. 1934. Plants used by the Hoh and Quileute Indians. *Transactions of the Kansas
Academy of Science* 37: 55–70.

Ritch-Krc, E. M. 1992. *A Selection of Traditional Medicinal Remedies Important to Contemporary
Carrier People in their Treatment of Disease*. M.Sc. thesis, University of British Columbia,
Vancouver.

Schaeffer, E. 1981. *Identification of vegetable fibres of woven ethnological artifacts*. ICOM Committee for Conservation, 6th Triennial Meeting, Ottawa, Ontario.

Secwepemc Cultural Education Society (SCES). 1986. *Shuswap Foods*. Shuswap Cultural Series.
Kamloops, British Columbia: SCES.

Smith, G. W. 1983. Arctic Pharmacognosia II. Devil's-club, *Oplopanax horridus*. *Journal of
Ethnopharmacology* 7: 313–320.

Smith, H. I. 1928. *Materia Medica of the Bella Coola and Neighbouring Tribes of British Columbia*.
National Museum of Canada Bulletin No. 56, Ottawa, Ontario.

Steedman, E. V., ed. 1930. *The Ethnobotany of the Thompson Indians of British Columbia*. Bureau of
American Ethnology 45th Annual Report. Washington, DC.

Stewart, H. 1977. *Indian Fishing. Early Methods on the Northwest Coast*. Seattle: University of
Washington Press.

————. 1984. *Cedar: Tree of Life to the Northwest Coast Indians*. Vancouver, British Columbia: Douglas & McIntyre.

Surtees, U. (with M. Thomas). 1974. *Lak-la Hai-ee: Interior Salish Food Preparation*. Kelowna, British Columbia: Lamont-Surtees Publishing.

Suttles, W. P. 1951a. *Economic Life of the Coast Salish of Haro and Rosario Straits*. Ph.D. thesis, University of Washington, Seattle.

————. 1951b. The early diffusion of the potato among the Coast Salish. *Southwestern Journal of Anthropology* 7(3): 272–288.

————. 1990. *Northwest Coast*. Vol. 7 of *Handbook of North American Indians*. Gen. ed. W. C. Sturtevant. Washington, DC: Smithsonian Institution.

Swan, J. G. 1857. *The Northwest Coast; or Three Years' Residence in Washington Territory*. New York.

Teit, J. A. 1906. *The Lillooet Indians*. Memoirs of the American Museum of Natural History. The Jesup North Pacific Expedition, vol. 2, part 5. Eds. F. Boas and G. E. Stechert. New York.

Thoms, A. V. 1989. *The Northern Roots of Hunter-Gatherer Intensification: Camas and the Pacific Northwest*. Ph.D. thesis, Washington State University, Pullman.

Turner, N. J. 1973. The ethnobotany of the Bella Coola Indians of British Columbia. *Syesis* 6: 193–220.

————. 1974. Plant taxonomic systems and ethnobotany of three contemporary Indian groups of the Pacific Northwest (Haida, Bella Coola, and Lillooet). *Syesis* 7 (supplement 1)

————. 1975. *Food Plants of British Columbia Indians. Part I: Coastal Peoples*. British Columbia Provincial Museum Handbook No. 34, Victoria.

————. 1977. Economic importance of black tree lichen (*Bryoria fremontii*) to the Indians of western North America. *Economic Botany* 31: 461–470.

————. 1978a. Plants of the Nootka Sound Indians as recorded by Captain Cook. In *Nu:tka: Captain Cook and the Spanish Explorers on the Coast. Sound Heritage* 7(1): 78–87.

————. 1978b. *Food Plants of British Columbia Indians. Part II: Interior Peoples*. British Columbia Provincial Museum Handbook No. 36, Victoria.

————. 1979. *Plants in British Columbia Indian Technology*. British Columbia Provincial Museum Handbook No. 38, Victoria.

————. 1981a. A gift for the taking: The untapped potential of some food plants of North American native peoples. *Canadian Journal of Botany* 59(11): 2331–2357.

————. 1981b. Indian use of *Shepherdia canadensis*, soapberry, in western North America. *Davidsonia* 12(1): 1–14.

————. 1982. Traditional use of devil's-club (*Oplopanax horridus*; Araliaceae) by native peoples in western North America. *Journal of Ethnobiology* 2(1): 17–38.

————. 1984. Counter-irritant and other medicinal uses of plants in Ranunculaceae by native peoples in British Columbia and neighbouring areas. *Journal of Ethnopharmacology* 11: 181–201.

————. 1987. General plant categories in Thompson and Lillooet, two Interior Salish languages of British Columbia. *Journal of Ethnobiology* 7(1)55–82.

————. 1988a. Ethnobotany of coniferous trees in Thompson and Lillooet Interior Salish of British Columbia. *Economic Botany* 42(2): 177–194.

————. 1988b. The importance of a rose: Evaluating the cultural significance of plants in Thompson and Lillooet Interior Salish. *American Anthropologist* 90(2): 272–290.

————. 1989a. All berries have relations: Midlevel folk plant categories in Thompson and Lillooet Interior Salish. *Journal of Ethnobiology* 9(1): 69–110.

————. 1989b. Really strong people: Native peoples of British Columbia and traditional herbal medicine. *The Herbarist* 55: 74–82.

————. 1991a. Wild berries. In *Berries*. Ed. J. Bennett. Toronto, Ontario: Harrowsmith Books. 49–69.

————. 1991b. Burning mountain sides for better crops: Aboriginal landscape burning in British Columbia. In *Fire: Anthropological and Archeological Perspectives*. Ed. K. Canon. Archeology in Montana, Special Issue. 32(2): 57–73.

————. 1992a. Plant resources of the Stl'atl'imx (Fraser River Lillooet) people: A window into the past. In *A Complex Culture of the British Columbia Plateau: Traditional Stl'atl'imx Resource Use*. Ed. B. Hayden. Vancouver: University of British Columbia Press. 405–469.

———. 1992b. Just when the wild roses bloom: The legacy of a Lillooet basket maker. *TEK Talk, a Newsletter of Traditional Ecological Knowledge* 1(2).

———. 1992c. "The earth's blanket": Traditional aboriginal attitudes towards nature. *Canadian Biodiversity* 2(4): 5–7.

Turner, N. J., and M. A. M. Bell. 1971. The ethnobotany of the Coast Salish Indians of Vancouver Island. *Economic Botany* 25(1): 63–104.

———. 1973. The ethnobotany of the Southern Kwakiutl Indians of British Columbia. *Economic Botany* 27(3): 257–310.

Turner, N. J., R. Bouchard, and D. I. D. Kennedy. 1980. *Ethnobotany of the Okanagan-Colville Indians of British Columbia and Washington*. British Columbia Provincial Museum Occasional Paper No. 21, Victoria.

Turner, N. J., and B. S. Efrat. 1982. *Ethnobotany of the Hesquiat Indians of Vancouver Island*. British Columbia Provincial Museum Cultural Recovery Paper No. 2, Victoria.

Turner, N. J., L. M. J. Gottesfeld, H. V. Kuhnlein, and A. Ceska. 1992. Edible wood fern rootstocks of western North America: Solving an ethnobotanical puzzle. *Journal of Ethnobiology* 12(1): 1–34.

Turner, N. J., and R. J. Hebda. 1990. Contemporary use of bark for medicine by two Salishan native elders of southeast Vancouver Island. *Journal of Ethnopharmacology* 229 (1990): 59–72.

Turner, N. J., M. Boelscher Ignace, and B. D. Compton. 1992. *Secwepemc (Shuswap) Tree Names: Key to the Past?* Paper presented to the 27th International Conference on Salish and Neighbouring Languages. 6–8 August 1992, Kamloops, British Columbia.

Turner, N. J., and H. V. Kuhnlein. 1982. Two important "root" foods of the Northwest Coast Indians: Springbank clover (*Trifolium wormskioldii*) and Pacific silverweed (*Potentilla anserina* subsp. *pacifica*). *Economic Botany* 36(4): 411–432.

———. 1983. Camas (*Camassia* spp.) and riceroot (*Fritillaria* spp.): Two liliaceous "root" foods of the Northwest Coast Indians. *Ecology of Food and Nutrition* 13: 199–219.

Turner, N. J., H. V. Kuhnlein, and K. N. Egger. 1987. The cottonwood mushroom (*Tricholoma populinum* Lange): A food resource of the Interior Salish Indian peoples of British Columbia. *Canadian Journal of Botany* 65: 921–927.

Turner, N. J., and A. F. Szczawinski. 1991. *Common Poisonous Plants and Mushrooms of North America*. Portland, OR: Timber Press.

Turner, N. J., and R. L. Taylor. 1972. A review of the Northwest Coast tobacco mystery. *Syesis* 5: 249–257.

Turner, N. J., J. Thomas, B. F. Carlson, and R. T. Ogilvie. 1983. *Ethnobotany of the Nitinaht Indians of Vancouver Island*. British Columbia Provincial Museum Occasional Paper No. 24, Victoria.

Turner, N. J., L. C. Thompson, M. T. Thompson, and A. Z. York. 1990. *Thompson Ethnobotany. Knowledge and Usage of Plants by the Thompson Indians of British Columbia*. Royal British Columbia Museum, Memoir No. 3, Victoria, British Columbia.

Wilson, I. H., ed. 1970. *Noticias de Nutka: An Account of Nootka Sound in 1792 by José Mariano Moziño*. Seattle: University of Washington Press.

PART 8

Ethnopharmacology

One of the subdivisions of modern ethnobotany is the study of aboriginal knowledge of the physical properties of plants and the accompanying familiarity with those components of the vegetal environment that permit indigenous peoples to use various plants for their presumed medicinal value. This subdivision is now known as ethnopharmacology and represents one of the most active fields of ethnobotanical research.

Early in human history, familiarity with those plants capable of causing noticeable physical and psychological changes when ingested or otherwise applied to the human body became associated with certain individuals, namely, the medicine men and women. These "specialists" gradually acquired great powers, a characteristic that continues today in many primitive societies. While medicine men and women are usually repositories of vast knowledge of plant properties, the general population (often the women) may likewise be conversant with the medicinal use of their flora.

Some of the oldest pre-literate archaeological records in both the Old and New Worlds concern the medicinal use of plants. A find in Iraq (dated from before the Christian era) has plant material of a number of species, some horticultural but a number undoubtedly employed in local medicine. The Egyptian pharaohs sent scouts far and wide searching for medicinal plants. The ancient Greeks possessed an extensive vegetal pharmacopoeia. One of Theophrastus's students, Alexander the Great, sent back from his campaigns in Asia medicinals for cultivation. India had an extensive pharmacopoeia several thousand years ago. Chinese ethnopharmacological knowledge, still very much alive, goes back several millennia, and some of the written records date from the beginning of the Christian era.

Some of the oldest archaeological records in the New World indicate the importance of bioactive plants. Some of the interesting "mushroom stones" of Guatemala are dated as early as 900 B.C. The statue of Xochimilco, Aztec god of inebriating plants, and many engravings in Mexico and Guatemala depict medicinal or hallucinogenic plants. The Spanish conquerors of Mexico fortunately had a few ecclesiastics who, like Sahagún, wrote sympathetically about native economic and bioactive plants. The first herbal of the New World, the Badianus manuscript, is outstanding not only because of its thorough treatment and expert illustrations of medicinal plants but because of its first-hand information. In the same period, the Spanish king's personal physician, Dr. Francisco Hernández, produced a truly scientific ethnopharmacological volume on the medicine plants, animals, and stones of "New Spain." Much information on medicinal use of plants in Brazil was recorded by the physician Guilherme Piso in *Historia Naturalis Brasiliae* dated 1648.

It must be borne in mind that in many indigenous groups there are in reality two systems of medicine: that of the medicine man or woman and that of the general population. In the first system, psychoactive plants, considered to have a resident spirit and therefore to be sacred, are used to communicate with the spirit world via visions and other hallucinations. Failing a diagnosis through this medium, the medicine man or woman has recourse to prescribing medicinal plants. The second system of medical practice is based wholly on general familiarity with medicinal plants, knowledge amassed by experimentation over millennia and passed on orally from generation to generation.

This knowledge, of great potential value to humanity as a whole, appears to be doomed to extinction in many regions due to rapid acculturation and Westernization of indigenous peoples. The loss of this knowledge is especially serious in our efforts to seek potentially valuable chemical constituents in the great number of species of unstudied plant, a major goal of ethnobotany and particularly ethnopharmacology.

The Amazon basin is an example of a rich ethnobotanical region. Its almost 3 million square miles support the world's largest rain forest with an estimated 80,000 species of flowering plants, approximately 15 percent of the world's half million species. The northwestern sector, particularly the Colombian Amazonia, is home to 70,000 Indians in 50 ethnic groups speaking many languages in more than 12 linguistic families. The knowledge that these Indians possess of the bioactivity of their plants is deep. In this relatively small sector of the Amazon basin, ethnopharmacological research has recorded almost 1600 flowering plants in 596 genera and 145 plant families employed as medicines or poisons (*The Healing Forest: Medicinal and Toxic Plants of the Northwest Amazonia*, Richard Evans Schultes and Robert F. Raffauf, Dioscorides Press, 1990). Only a minute fraction of this assemblage has been chemically studied. According to an outstanding Brazilian phytochemist, fewer than one percent of the plants in the Amazonia of Brazil have ever been even superficially analyzed for bioactive compounds. A similar wealth of unstudied plants lies waiting for ethnopharmacological investigation in other tropical parts of the world, particularly Southeast Asia and Africa.

Many pharmacological surveys have been carried out as random samplings of a flora very costly of time and money. If phytochemists must procure sufficient material of 80,000 species from distant and often wild regions of difficult transport, the task undoubtedly could never be finished. Here is where ethnobotany can be a great help. Concentrating on those plants which aborigines, by long experimentation, have found to be physiologically or psychologically active and which they have bent to medicinal use would be a kind of "short cut." Any plant that is bioactive has a least one active chemical, and we should know what this compound is. Although we may never have any use for these new compounds in our society, they could perhaps be employed for a completely different purpose, or they might, on rare occasions (as with quinine) be valuable for the same purpose.

Two excellent examples illustrate this principle. The primary source of one of the many curares (*Chondrodendron tomentosum*) used by South American Indians to kill animals yields an alkaloid, tubocurarine, which is now a valuable adjunct of modern medicine and is used as a muscle relaxant before surgery. The other example is rotenone, a complex ketone from a leguminous vine, the bark of which is used by Indians as a fish poison but from which the active principle is now used as a rapidly biodegradable pesticide in our agricultural regions.

In some European nations, herbal medicine still is popular, as in the Culpepper Shops of Great Britain and the Paracelsus Shops in Switzerland. In the United States, however, interest in herbal medicine has almost disappeared. Part of the reason for this disinterest is due the extraordinary exaggeration of fake "cures" with herbs and herbal concoctions during the latter half of the nineteenth century. There eventually began a reaction against this abuse which more or less coincided with the advent of the development of more sophisticated chemistry. The result was the expectation that most, if not all, advances in pharmacology would come from synthetic chemicals. Chemistry did make notable strides in medical fields, but a change in outlook took over in the second half of the twentieth century.

The 1930s saw a resurgence of interest in plants as sources of new medically promising constituents. Realization of the value of plants, particularly those employed in the traditional medicine of less advanced societies, began with the discovery of tubocurarine, the first in a series of so-called Wonder Drugs, some of which revolutionized modern medicine. Tubocurarine, from the arrow poison curare, was followed by the antibacterial activity of penicillium and the many subsequent antibiotics, most of which, like penicillin, came from fungi, and some of which, like penicillium, were recorded as medical agents in ancient and medieval ethnobotanical writings.

A series of new and potent drugs followed these discoveries. Many of them were scientifically investigated as the result of ethnobotanical reports—either in literature or from field studies—of use in primitive societies: cortisone from the Mexican yam, reserpine from the Indian snake-root, the cytotoxic properties of the May apple or *Podophyllum*, the anti-leukemia use of chemicals in the Madagascan periwinkle, the hypotensive activities of alkaloids of the hellebore and sundry other examples. It would seem that the limit of the discovery of new therapeutics has just begun, especially with the extraordinary new chemical techniques now available.

The challenge of modern ethnopharmacological research is this: How many valuable therapeutic agents reside in the chemically unstudied 500,000 species of higher plants? Fortunately, phytochemical studies are now increasingly occupying the attention of specialists in numerous parts of the world. The number of excellent articles and books, the numerous symposia and congresses dedicated to ethnopharmacological topics, the growing interest evident in pharmaceutical companies—all indicate a long overdue recognition of the vast chemical reservoir lying in wait for study in the floras of the world.

Studies of the ethnopharmacology of primitive or undeveloped societies around the world probably comprise the most active aspect of ethnobotany in terms of the number of investigators—botanists, ethnobotanists, anthropologists, phytochemists, pharmacologists, medicinally trained personnel, historians, and others in many parts of the world. The listing of those involved would be voluminous. Not included in this volume, but noteworthy nonetheless are Mark J. Plotkin's "*Strychnos medeola*: A New Arrow Poison from Suriname" (*Ethnobiology: Implications and Applications*, 1990, Brazil) and Werner Wilbert's "New Concepts of Traditional Medicine Evident in Warao Ethnopharmacognosy" (*Social Science and Medicine*). Plotkin demonstrates the importance of the discovery of curare not for its potential medicinal value, but as evidence of the existence of species with biodynamic potential known locally but not to Western medicine, and Wilbert documents the existence of a pneumatic (i.e., by mystical causation through contagion) theory of illness among the Warao Indians of lowland South America, and the healing of diseases by herbal remedies. Plotkin has also written *Tales of a Shaman's Apprentice* (Penguin Books, New York, 1993).

The seven original contributions and two reprints included in this section have been selected because they combine field and laboratory work, for if any of the various aspects of ethnobotany is interdisciplinary, it certainly is ethnopharmacology. The late N. G. Bisset discusses the role of selected arrow- and dart-poison plants in the development of medicinal agents. Memory Elvin-Lewis and Walter H. Lewis outline several new approaches to medical and dental ethnobotany. Albert Hofmann reports examples of his research showing the fruitful collaboration between the ethnobotanist and the medical chemist. Two extremely significant and thought-provoking articles are reprinted here, namely, B. R. Holmstedt's inaugural lecture at the First International Congress on Ethnopharmacology, "Historical Perspective and Future of Ethnopharmacology" (*Journal of Ethnopharmacology* 32 (1991): 7–24) and "Ethnopharmacology—A Challenge" by B. R. Holmstedt and J. G. Bruhn (*Trends in Pharmacological Sciences* 3 (May 1982)). The former traces the history of selected ethnopharmacological discoveries in the nineteenth century, including strychnine, cathine, and ephedrine, while the latter argues for the study of traditional medicine as a powerful aid to biological research. Jan-Erik Lindgren reviews research concerning Amazonian psychoactive indoles. Dennis J. McKenna, L. E. Luna, and G. B. N. Towers discuss the chemical or biodynamic properties of the many plants mixed with the hallucinogenic beverage ayahuasca. Plutarco Naranjo offers six reasons for urgent ethnopharmacological research and suggests several genera worthy of additional study as new sources of secondary organic compounds. Finally, Peter A. G. M. de Smet summarizes important botanical, pharmacological, and chemical considerations in researching the use of hallucinogenic plants by indigenous peoples.

Arrow Poisons and Their Role in the Development of Medicinal Agents

NORMAN G. BISSET

In the past, ethnobotany has focused largely on the knowledge possessed by less advanced or pre-literate communities—a restriction that is no longer felt to be appropriate. Since the 1960s, ethnobotany has been called "the study of the relationships between man and his ambient vegetation" (Schultes 1967) or "the study of direct interrelations between humans and plants" (Ford 1978). These two, very similar definitions embrace the increasing interest in the spiritual and cultural dimension of human contacts with plants. Economic botany, on the other hand, to R. E. Schultes is a part of ethnobotany, while R. Ford reserves the term for indirect contact with plants or their derived products, emphasizing among other things "their potential for incorporation in another (usually Western) culture." Thus, it is not only the uses to which peoples with an intimate knowledge of their surrounding vegetation put the plants available to them that are of potential interest, but also the uses that peoples of other cultures make of such knowledge.

A group of plants of interest in this context comprises those that are incorporated into the arrow and dart poisons that have been, and still are, used by hunters in various parts of the world (Bisset 1989). Many of these plants have active principles with drastic biological effects (i.e., contain biodynamically active molecules). At the same time, they also often play a part in local medicine and thus illustrate the aphorism propounded by Paracelsus: *dosis sola facit venenum*, it is the dose alone that makes a poison (Bisset 1991). Such materials have attracted attention in the continuing search for potentially useful medicinal agents. The following sections consider briefly some of the contributions that the ethnobotany of arrow- and dart-poison plants has made to the present-day medicinal armamentarium, but it must be borne in mind that the literature has its pitfalls, including the dart poison that never was (Bisset 1990).

Strychnos L. (Loganiaceae)

Western medicine aims to cure illness as rapidly as possible, and it has not been averse to making use of heroically acting medicines. The 1644 edition of *Herbarivs oft Crvydt-boeck* by Rembert Dodoens (Dodonaeus 1644), under the heading *Nux Vomica*, recommends two drachms of the powder mixed with the same quantity of powdered Anise or Fennel and Honey and drunk with warm water; and through the vomiting it drives out much phlegm and gall from the body. But it brings death to the hind or stag; and it makes birds, as well as fish, so docile and stupefied that they can easily be caught.

Two drachms, approximately 0.25 ounce or 8 grams, of *Nux vomica* could contain more than 90 milligrams of strychnine—and the lethal dose of the alkaloid for an adult is put at 50 to 100 milligrams (Reynolds 1989)! Heroic indeed!

In the third edition of *Pharmacopoeia Universalis* (James 1764, p. 197), we find these words:

> *Nux Vomica* . . . These . . . are not, or at least ought not to be, used in Medicines, for they are extremely narcotic and virulent, exciting Inquietudes, Rigors, Convulsions, Horrors, Tremors, and an irregular Respiration. They are principally used for poisoning Dogs, Cats, Crows, and other Animals, by a Barbarity peculiar to Mankind.

During the first half of the nineteenth century, *Nux vomica* was less used in Great Britain, where it was considered to be a dangerous and uncertain remedy, than on the Continent, where its preparations were used to treat certain forms of paralysis, tremors, epilepsy, hypochondria, hysteria, neuralgia, dyspepsia and other stomach conditions, dysentery, rectal prolapsus, and impotence, and where these preparations were found occasionally to be beneficial in dealing with intermittent fevers and intestinal worms (Woodville 1794, pp. 29–32; 1832, 2: 222–226; 5: 136–142; Pereira 1850, pp. 1479–1498).

The investigation of *Strychnos* species is due the important pharmacological properties of their alkaloids. Past studies of the Asian species were determined largely by their connection with the old pharmaceutically important drugs *Lignum colubrinum* (snake wood), *Nux vomica*, and *Faba sancti Ignatii* (St. Ignatius bean); arrow poisons from Asia provided a second focus of attention. Most of the work on the South American species relates to the group of arrow and dart poisons known as curare. Research into the African species of the genus developed relatively late and followed on from chemical and pharmacological screening programs. All this work, during which more than 300 different alkaloids have been isolated and their structures determined (Bisset 1980, 1992; Ohiri et al. 1983; Massiot and Delaude 1988), reveals *Strychnos* as one of the most thoroughly studied alkaloid-containing plant genera. Nevertheless, there is a serious lacuna in the knowledge thus acquired. Although the essential reason for investigating these plants has been the pronounced activity of their alkaloidal constituents, advances in their pharmacological study have not kept pace with the phytochemical investigations.

Many Southeast Asian arrow and dart poisons have been prepared with *Strychnos Nux-vomica* L. and *St. ignatii* P. J. Bergius. These plants furnished *Nux vomica* and St. Ignatius beans, respectively, and it is from these drugs that Pelletier and Caventou (1819) first isolated strychnine (Figure 1), the archetype of the violently convulsant poisons. Subsequently, the alkaloid was shown to be the active ingredient not only in Southeast Asian arrow and dart poisons, but in some Central African arrow and ordeal poisons as well (Bisset 1989). Although strychnine has been used as a circulatory stimulant, it is applied now chiefly as a respiratory stimulant in certain cases of poisoning. Being very bitter, strychnine aids the appetite and digestion, but it has been widely misused as a so-called general tonic (W. C. Evans 1989, p. 614). The introduction into medical practice of such highly active substances which are not aimed at any specific target represents a "blunderbuss" or kill-or-cure approach.

An interesting point in investigating the botany of the *Strychnos* species incorporated into such poisons is that in Malaysia the aborigines distinguished many more entities

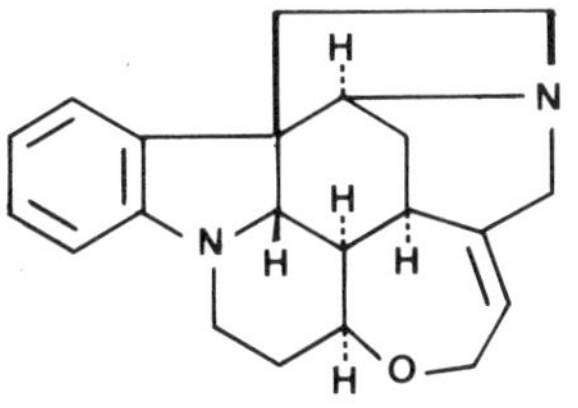

Strychnine

Figure 1. Strychnine is the violently convulsant alkaloid present in the seeds of only 6 of 200 *Strychnos* species, including *Nux vomica* and *Faba St. Ignatii* (St. Ignatius bean). It is present in some arrow and dart poisons, and it has also been used medicinally as a bitter tonic and respiratory stimulant.

than did the taxonomist who identified the specimens collected. In some cases, however, the descriptions of the plants, obtained before material was collected, did not agree with the specimens subsequently brought in (Leenhouts 1962; Bisset and Woods 1966). It has not as yet been possible to determine whether there are corresponding chemical differences (see the discussion on *Acokanthera* below).

Strychnos species are widely used for medicinal and other purposes in Asia and Africa. To learn about the basis of their supposed activity, extensive screening programs have been carried out. There are several approaches to this type of work, two of which are exemplified by published work on the genus. Concomitantly with taxonomic revisions, several hundred herbarium fragments, belonging to most of the Asian and almost all the African species, were examined for alkaloids as the putative source of activity. Although these studies enabled data to be acquired for many species that otherwise would have been difficult to obtain, their scope is clearly limited by the small size of the samples and the range of plant parts normally available from herbarium material. In spite of this, among other things, they have singled out some of the species that are rich in alkaloids and have revealed that strychnine, far from being a common alkaloid in the genus, is present in only about 6 or so of its estimated 200 species (Bisset 1980). Subsequent investigation of material from about two-thirds of the South American species has served to confirm the limited utility of such samples, at least as far as the genus *Strychnos* is concerned (Quetin-Leclercq et al. 1990).

What understandably has proved to be a greater source of information has been a program based on a full-scale pharmacological screening primarily for convulsant and muscle-relaxant activity. The findings of this and other work have been summarized by Ohiri et al. (1983). There is no doubt that where it is possible to carry out such a program, it provides a more effective and informed basis for future detailed studies, but that it is nevertheless limited by the availability or accessibility of the primary material. Among the interesting findings is that the bis-tertiary indole alkaloids present in one of the African species of *Strychnos* used as chew sticks, *S. afzelii* Gilg, have considerable antimicrobial activity (Verpoorte et al. 1978).

Chondrodendron (Menispermaceae)—*Strychnos* (Loganiaceae)

In contrast with the foregoing account, the paralyzing (muscle-relaxant) South American dart- and arrow-poison curare represents one of the best examples of the transformation of an ethnic material into an indispensable aid in modern surgery (Bisset 1992). When Roulin and Boussingault (1828) and Pelletier and Petroz (1829) investigated the poison, it was not strychnine they found but, rather, what is now known to be a highly complicated mixture of tertiary and quaternary (ammonium) bases which the then-available chemical techniques were incapable of separating. It was not until the application of chromatographic methods in the 1930s and 1940s that the composition of curare could begin to be unraveled and an insight could be gained into the chemistry of its active constituents.

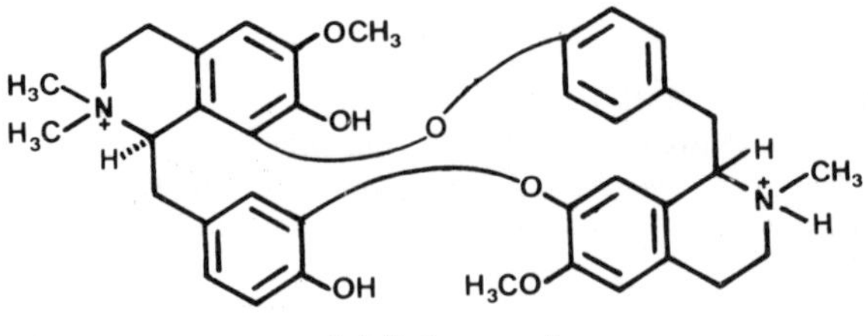

(+)-Tubocurarine

Figure 2. (+)-Tubocurarine, the first pure muscle-relaxant alkaloid to be isolated from the South American dart poison curare, has been extensively used as an adjunct in surgery.

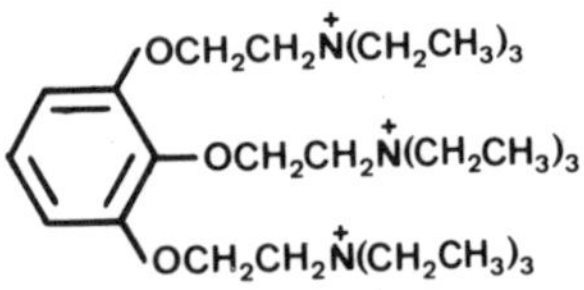

Gallamine (Flaxedil®)

$(CH_3)_3\overset{+}{N}-(CH_2)_{10}-\overset{+}{N}(CH_3)_3$

Decamethonium

$(CH_3)_3\overset{+}{N}-(CH_2)_2O\cdot CO\cdot (CH_2)_2\cdot CO\cdot O(CH_2)_2-\overset{+}{N}(CH_3)_3$

Suxamethonium

Figures 3–5. Gallamine (Flaxedil®), decamethonium, and suxamethonium (disuccinylcholine) are simplified (synthetic) molecules with structures based on that of (+)-tubocurarine (Figure 2) and with short-lasting muscle-relaxant properties. Only suxamethonium has kept its place in the medical arsenal.

(+)-Tubocurarine (Figure 2), the active principle of *Chondrodendron tomentosum* Wedd., was the first of the naturally occurring muscle-relaxant substances to be obtained in crystalline form and for which a structure (albeit much later proven to be incorrect) was proposed (King 1935). Investigation of the molecular features responsible for bringing about muscle relaxation led to the introduction into clinical practice of such drugs as gallamine (Figure 3), decamethonium (Figure 4), and suxamethonium (Figure 5), in which there was a drastic simplification of the (+)-tubocurarine molecule. Studies on their mechanism of action revealed that these synthetic substances functioned in a different way from the active principles of curare (Bovet 1959; Bovet-Nitti 1959). Improved understanding of what is required of a muscle relaxant as an adjunct to anaesthesia in surgery, as well as of the mechanisms by which it acts, has led to the production of purpose-built

molecules that are targeted to act at a very specific site in the body, namely, the acetylcholine receptors on the post-synaptic membrane of the neuromuscular junction.

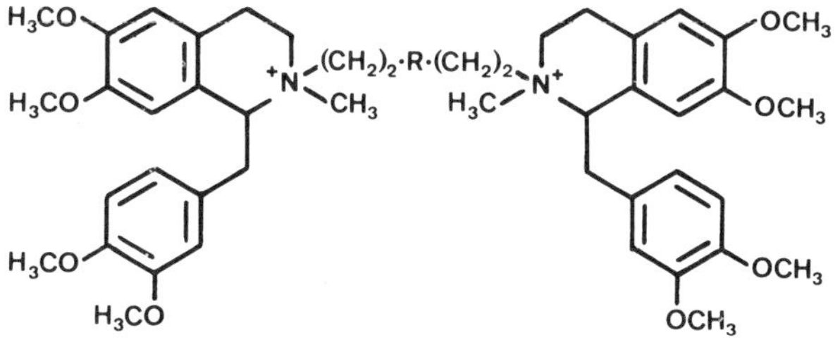

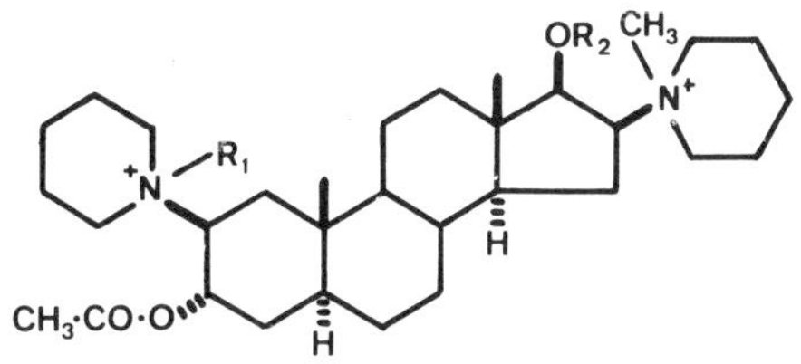

Pancuronium R$_1$ = CH$_3$, R$_2$ = CO·CH$_3$

Vecuronium R$_1$ = H, R$_2$ = CO·CH$_3$

Laudexium R = (CH$_2$)$_6$

Atracurium R = CO·O·(CH$_2$)$_5$·O·CO

Figures 6–7. Pancuronium and vecuronium are steroidal muscle relaxants with features derived
in part from the (+)-tubocurarine molecule; they have improved properties and are currently
being widely used in surgical operations. Laudexium, now withdrawn, and the newer
atracurium, with advantages that allow it to be given to patients who would have problems
with other muscle relaxants, are also modeled on (+)-tubocurarine.

Such a highly specific modification of the original naturally occurring molecule aimed
at a very specific body tissue may be termed the "laser" approach. Examples of the molecules developed are the steroid derivatives pancuronium (Figure 6a) and vecuronium
(Figure 6b) (see Bowman and Rand 1980) and the bis-benzylisoquinoline derivative
laudexium (Figure 7a), which, along with (+)-tubocurarine, is giving way to atracurium
(Figure 7b) and related compounds. Atracurium not only has fewer side effects, so that it
can be used in cases where the older drugs would be hazardous, but its molecule is so constructed that it has a built-in fail-safe mechanism. At the pH of the body it undergoes
decomposition through spontaneous hydrolysis and Hofmann degradation, thereby
ensuring a limited duration of action (Bisset 1992).

Modeled on the highly active bis-quaternary bis-indole alkaloids of the toxiferine
(Figure 8a) type, present in some of the South American species of *Strychnos* utilized in the
preparation of curare, is the semisynthetic, short-acting muscle-relaxant alcuronium (Figure 8b), which has two quaternary allyl functions (Schlittler 1971).

It was not until 1970 that the African species *Strychnos usambarensis* Gilg was found to
be the main constituent of an arrow poison made by the Banyambo, a small group of
hunters living in the Akagera National Park in eastern Rwanda along the frontier with
Tanzania. Chemical investigation showed that the active principles were bis-quaternary
bis-indole alkaloids identical with some of those found in South American species of the

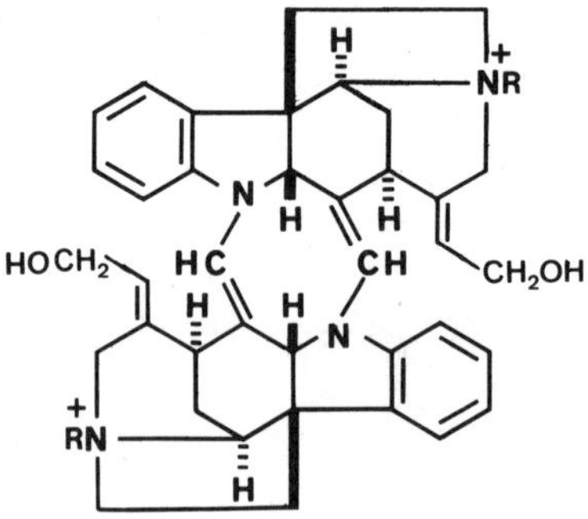

Toxiferine R = CH$_3$

Alcuronium R = CH$_2$·CH:CH$_2$

Figure 8. Toxiferine comes from a type of curare other than (+)-tubocurarine and has served as the model for the short-acting muscle relaxant alcuronium, which is made by semisynthesis from strychnine.

genus. In other words, the Banyambo arrow poison is a curare (Angenot 1971). More than seventy alkaloids have been isolated from the plant and some of them have been found to possess a degree of antimitotic activity against certain experimental animal tumors (Bassleer et al. 1982; Quetin-Leclercq 1990).

Euphorbiaceae and Thymelaeaceae

During the Spanish conquest, slow-acting arrow poisons were encountered in the Caribbean region, along the coasts of Central and northern South America, and in other parts of the continent, the victims of which died *rabiando*, "mad with pain." These war poisons were entirely different from the curares used by hunting tribes of the inland forests (Bisset 1992). Among the plants used to make war poisons were *Hippomane mancinella* L. (the notorious manchineel tree), *Hura crepitans* L. (the sandbox tree), *Colliguaja odorifera* Molina, *Euphorbia cotinifolia* L., *Sapium* species (all Euphorbiaceae), and *Schoenobiblus peruviana* Standl. (Thymelaeaceae). Elsewhere in the world, species of *Sapium* and *Croton* have been major arrow-poison constituents and the latex from species of *Euphorbia* and *Jatropha* has been added to the poisons to improve their consistency and to act as adhesives.

The active principles of these various plants are potent pro-inflammatory and co-carcinogenic ester derivatives of certain diterpenoid alcohols (F. J. Evans 1989). *Hippomane mancinella* has been the subject of several investigations. The hydrophobic fraction from the latex contains cryptic irritants of the tigliane and daphnane groups, while the hydrophilic fraction has irritant factors, also based on the tigliane and daphnane skeletons, which include mancinellin (Figure 9) and huratoxin (Figure 10a); the irritant factors of the daphnane group had an ID$_{50}^{24}$ of 0.02 microgram per mouse ear (Adolf and Hecker 1984).

Although some writers have been reluctant to accept the accounts of the Carib arrow poisons as related by the Spanish chroniclers, because of the supposedly exaggerated effects, the descriptions are not inconsistent with the consequences of a subcutaneous or intravenous injection of the pro-inflammatory diterpenes just discussed. These compounds can induce acute inflammation, while their long-term effects give rise to a chronic inflammatory reaction, which leads to tissue damage as a result of vasoconstriction and cell-membrane deterioration. This may be associated with the ability to stimulate prosta-

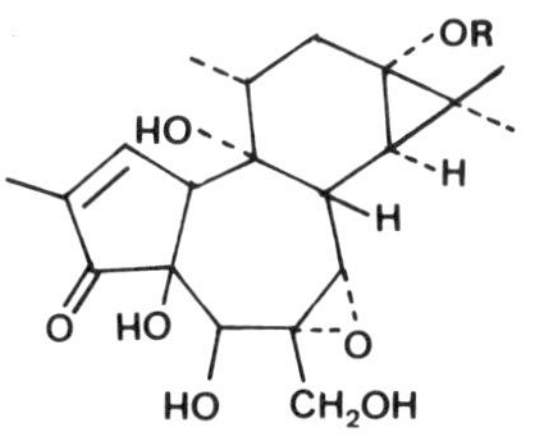

Mancinellin R = CO·(CH:CH)$_3$·(CH$_2$)$_8$·CH$_3$

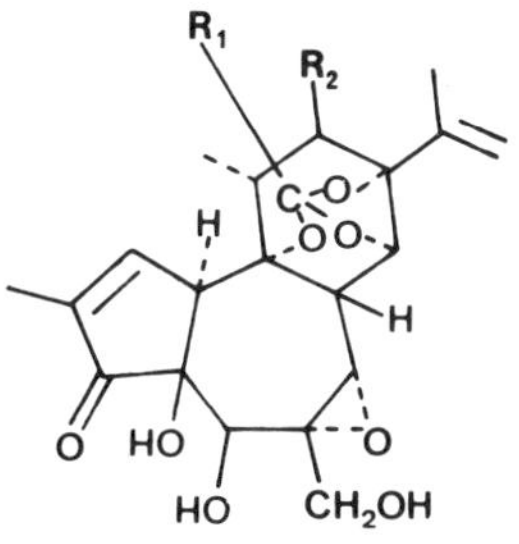

Huratoxin R$_1$ = (CH:CH)$_2$·(CH$_2$)$_8$·CH$_3$, R$_2$ = H

Mezerein R$_1$ = C$_6$H$_5$, R$_2$ = O·CO·(CH:CH)$_2$·C$_6$H$_5$

Figures 9–10. Mancinellin and huratoxin are from the notorious manchineel tree, at one time used as an arrow poison in the Caribbean region, and mezerein has been isolated from mezereum (*Daphne* species). These substances have a range of interesting physiological properties and are used to probe the workings of the body.

glandin (PGE$_2$) production. The substances are active at extremely low doses and in this respect are hormonelike (Edwards et al. 1985). The intense pain which features so prominently in the descriptions of the effects of the early poisons may result from the prostaglandin produced leading to a condition of hyperalgesia (i.e., lowering of the threshold to pain-stimulating substances such as bradykinin arising from the tissue damage), and also from the release of serotonin by the blood platelets, causing pain through stimulation of sensory nerve endings (Bowman and Rand 1980).

Hura crepitans has a translucent yellow sap rather than a latex. It also contains irritant diterpene compounds, including the daphnane derivative huratoxin (Figure 10a) (Sakata et al. 1971), but there is the additional possibility of concomitant action by one or more lectins. It has been shown that the seeds and sap of this plant contain similar lectins having mitogenic and hemagglutinating properties (Falasca et al. 1980; Barbieri et al. 1983). Esters of various phorbol derivatives are known to occur in certain Asian species of *Sapium* (Evans and Taylor 1983), but South American members of the genus have not yet been investigated for this type of compound.

The co-carcinogenic properties of the tigliane derivatives have stimulated research into the mechanism of chemical carcinogenesis and paradoxically it has been found that mezerein (Figure 10b), from the fruits of *Daphne mezereum* L. (Thymelaeaceae), has potentially useful antileukemic properties (Kupchan and Baxter 1975; see Kupchan et al. 1976). Another *Daphne* species, *D. kamtschatica* Maxim. var. *yezoensis* (Maxim.) Ohwi, was at one time a component of an arrow poison used by the Ainu on the island of Hokkaido (Ishikawa 1962).

Acokanthera G. Don (Apocynaceae)

An extensive review of African arrow poisons dealing with their plant sources and the chemistry and pharmacology of their active principles has been published by Neuwinger (1974). Many of the poisons are prepared from plants containing cardiotonic glycosides, and in East Africa, among the Apocynaceae used, are members of the genus *Acokanthera*. Two of the species utilized are the morphologically rather variable *A. schimperi* (A. DC.) Schweinf. and *A. oppositifolia* (Lam.) Codd.

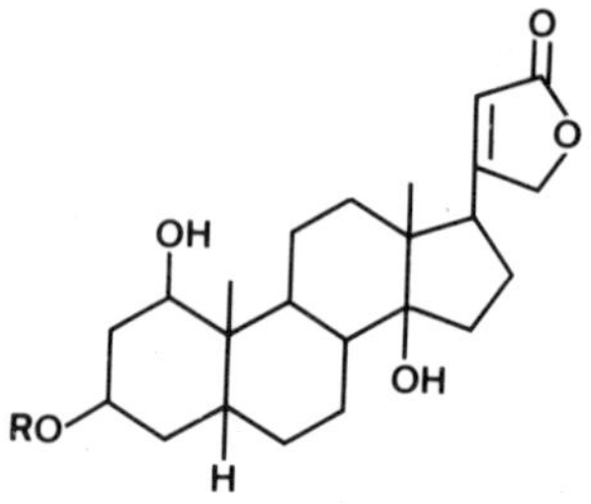

Acovenoside A R = L-Acovenose

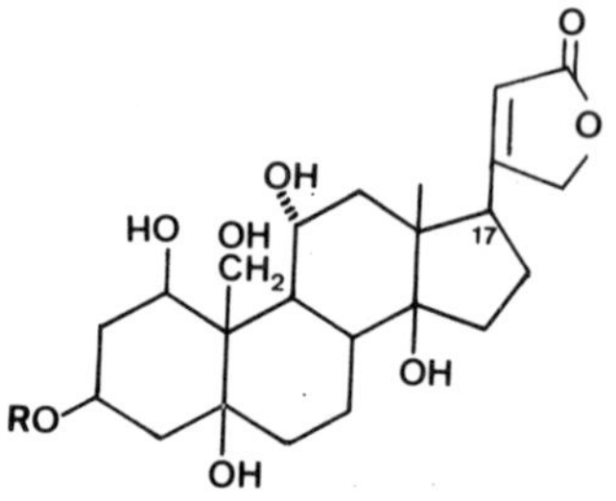

Ouabain R = L-Rhamnose

Acolongifloroside K R = 6-Deoxy-L-talose

Figures 11–12. Acovenoside, ouabain, and acolongifloroside K are powerful cardiotonic glycosides that have been found in plants used as arrow poisons in Africa. Ouabain, in particular, has been used medically to treat certain heart conditions.

In the past, *Acokanthera schimperi* has been considered to comprise a number of not very easily defined forms, among them *A. ouabaio* (Franch. & J. Poiss.) Cathel., *A. deflersii* Schweinf. ex L. Lewin, and *A. friesiorum* Markgr. The latest reviser of the genus, Kupicha (1982), did not maintain these forms but treats the taxon as a single entity. The principal glycosides present appear to be acovenoside A (Figure 11) and ouabain (Figure 12a), but the form described as *A. ouabaio* contains only the latter and is the plant from which the glycoside was originally obtained. [The principal source of ouabain, or strophanthin G, is the seed of *Strophanthus gratus* (Wall. & Hook. f.) Baill. The glycoside is readily obtained in crystalline form and in a pure state; and for this reason, in addition to being used medicinally, it has long been the international standard for the measurement of cardiotonic activity.] Ouabain, besides being more active than acovenoside A, is also more readily soluble in water and is therefore able to act very rapidly.

It is the branches and/or twigs of the *Acokanthera ouabaio* form of *A. schimperi* that the Wa Giriama of Kenya selected for making their arrow poison and as it is the form richest in ouabain it is also the most effective (Bally et al. 1958; Thudium et al. 1958). Of interest is its occurrence near the Somali coast where it may have been planted by Wa Giriama for use in their arrow poison (Kupicha 1982). It is thus evident the Africans were well able to distinguish the form of the plant that best suited their purposes—a form that is only weakly differentiated by its morphological characters, the chemical distinction of which made it of particular importance.

Kupicha (1982) has merged *Acokanthera longiflora* Stapf (*A. venenata* auct. non G. Don) with *A. oppositifolia*, both of which contain principally aconvenoside A, a considerable amount of acolongifloroside K, and very little or no ouabain. Most of the work on *A. schimperi* and *A. oppositifolia* (Reichstein 1965) was done on the seeds, the cardenolide composition of which is not necessarily an accurate guide to that of other parts of the plant. As already mentioned, however, Thudium et al. (1958) showed that ouabain (0.2 percent) was the main glycoside of *A. ouabaio*, the form of *A. schimperi* utilized by the Wa Giriama in making their poison.

About 80 percent of the poisoned arrows examined in the laboratory of the government chemist in Tanzania is reported to contain ouabain. This finding appears to be at variance with the observations of a number of explorers that *Acokanthera longiflora* is the chief species used in making the poison, for Reichstein's findings—unfortunately, again based on an analysis of the seeds—indicated the presence of little or no ouabain in this species. Neuwinger (1974) attempted to explain this seeming discrepancy by postulating that the "ouabain" found in the analyses represented "ouabain equivalents" rather than simply ouabain. Work by Cassels (1985) appears to shed some light on the situation. From a Masai arrow poison of Kenyan origin, known to have been prepared from *Acokanthera schimperi*, he isolated acolongifloroside K (Figure 12b) which accounted for 50 percent of the total cardenolides; ouabain comprised a further 20 percent and acovenoside A only 1.2 percent. It is therefore likely that the cardenolide composition of this and other species of *Acokanthera* is more variable than previous work has suggested. The R_f values of acolongifloroside K and ouabain in the paper and thin-layer chromatographic systems that have been used in analyzing arrow poisons are very close, and it seems quite possible that one or both of these glycosides may have been present in the poisons examined. The LD_{50} values of the three glycosides on intravenous injection into the cat are acovenoside A 0.24 milligram per kilogram, acolongifloroside K 0.11 milligram per kilogram, and ouabain 0.12 milligram per kilogram. As the latter two compounds are both readily soluble in water, they would be equally effective in an arrow poison.

Aconitum L. (Ranunculaceae)

The genus *Aconitum* has probably been the most widely used source of arrow-poison plants. It comprises over 350 species of herbs and climbers distributed throughout the temperate Northern Hemisphere. Morphologically, they are rather variable and many subspecies and varieties have been described. Members of the genus appear to have been used over a large part of the genus's distribution range—from Europe through Asia to Alaska in North America—for hunting anything from human beings to wolves and whales (Bisset 1989). The active principles are highly toxic (LD_{50} ca. 0.1 milligram per kilogram) diterpene ester alkaloids of the aconitine (Figure 13a) type, although other types of alkaloid (e.g., isoquinolines) are also present.

In spite of the presence of these dangerous substances, *Aconitum* species have been, and are, extensively used as medicinal agents throughout the aforementioned vast region. In European medicine, the tubers, or occasionally the herb, have been employed chiefly as a liniment or tincture in the treatment of neuralgia, sciatica, and rheumatism

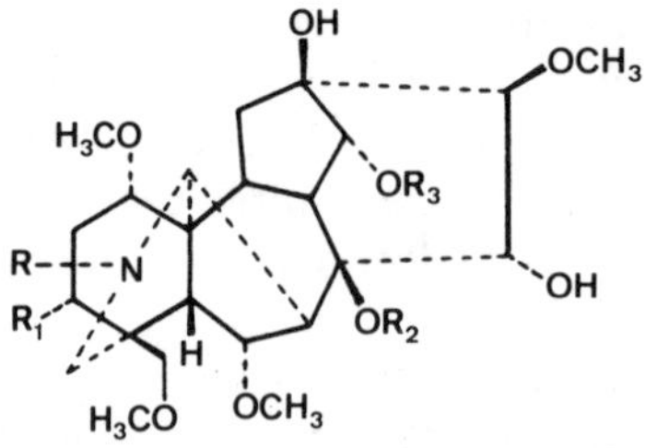

Aconitine R = C$_2$H$_5$, R$_1$ = OH, R$_2$ = CH$_3$·CO, R$_3$ = C$_6$H$_5$·CO

Mesaconitine R = CH$_3$, R$_1$ = OH, R$_2$ = CH$_3$·CO, R$_3$ = C$_6$H$_5$·CO

Figure 13. Aconitine and mesaconitine are highly toxic alkaloids present in various species of *Aconitum* (monkshood), but they also have medicinal properties that are utilized on a large scale in Chinese and other traditional medicines of the Orient.

and internally to alleviate fevers. The variability of the drug (see Katz and Staehelin 1979) and difficulties in its standardization have led to its abandonment except for internal use in homeopathic doses (Trease 1949, p. 269; Nelson 1951, pp. 381–382; Trease and Evans 1983, p. 607).

In marked contrast with the Western experience of the drug is its continuing widespread application in the traditional medicine of India and China and neighboring countries such as Tibet, Mongolia, Korea, and Japan. Besides its use in the raw state, the drug may be "processed" or "mitigated" in various ways to reduce the toxic properties and, at the same time, to allow dosage levels to be raised and to enable other therapeutic properties which the drug is believed to possess to be utilized. This specially pretreated form of the drug was never developed in the West. The unprocessed tubers are applied externally to bring about surface anaesthesia and internally to treat lumbar and leg pains, neuralgia, and rheumatoid arthritis. On the other hand, after "processing," the tubers provide medicines with cardiotonic, diuretic, and tonic properties.

In the 1970s and 1980s, very extensive chemical work on *Aconitum* species has been carried out in China and Japan. The late professor Hikino and his coworkers at Sendai, in particular, have been able to show that the raw tubers owe a number of their effects—analgesic (Hikino et al. 1979), antirheumatic (anti-inflammatory) (Hikino et al. 1980), CNS (including antipyretic) (Hikino et al. 1979) properties—to the aconitine-type alkaloids they contain. Aconite extracts, particularly mesaconitine (Figure 13b), also bring about a dose-dependent anabolic response in mice in which liver protein synthesis is increased. This is interpreted as providing an understanding for the use in Chinese traditional medicine of aconite to stimulate anabolic activity in debilitated patients (Hikino et al. 1983; Murayama and Hikino 1984). In spite of all the work that has been carried out, however, it is not yet clear how the various findings relate to the drug as prescribed and taken in the many traditional preparations of which it is a component. Moreover, it has been claimed that essentially alkaloid-free aqueous extracts of aconite exhibit many of the activities reported for the aconitine-type bases (Arichi and Uchida 1981).

Investigation of the source of the cardiovascular activity in *Aconitum* has brought to light the presence of several substances with such properties: coryneine chloride (dopamine methochloride) (Figure 14) (Konno et al. 1979), higenamine ((+)-demethylcoclaurine) (Figure 15) (Kosuge et al. 1978), and salsolinol (Figure 16) (Chen and Liang 1982). The cardiotonic properties of higenamine have been studied in considerable detail (Chou et al. 1978), but the substance's instability (being a triphenol) and short duration of action are disadvantages. In spite of this, the fact that it is active at about the same dosage level

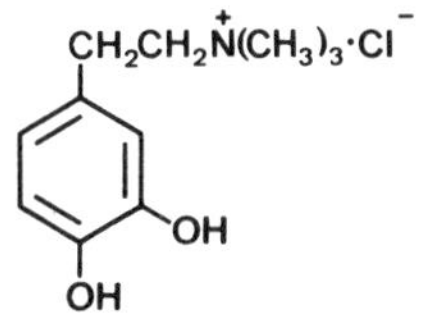

Coryneine chloride (Dopamine methochloride)

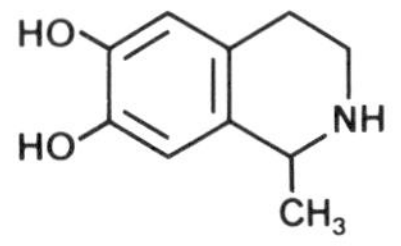

Salsolinol

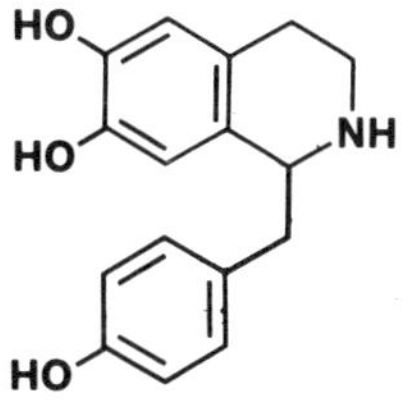

Higenamine ((±)-Demethylcoclaurine)

Figures 14–16. Coryneine chloride, higenamine, and salsolinol have been obtained from Asian species of monkshood. They have cardiotonic properties, and higenamine, especially, has been applied clinically and offers another possible approach to the treatment of certain heart conditions.

as the cardiotonic glycosides of the digitalis type, coupled with its very low chronic toxicity, points to another possible direction in the search for safer cardiotonic drugs.

Antitumor Activity of Arrow-Poison Plants

As part of a discussion dealing with plant folklore as a tool for predicting antitumor activity, from a "comprehensive but not exhaustive review of the literature" Spjut and Perdue (1976) compiled a list (not detailed) of plants used in arrow poisons. Of the seventy-six species representing sixty-three genera in twenty-nine families, forty-six species had been screened for anticancer activity. Fifty-two percent of the species and 75 percent of the genera had been found to be active. In quoting these figures, Farnsworth and Kaas (1981) pointed out that the high apparent correlation of arrow poisons with antitumor/cytotoxic activity is due probably to the fact that many arrow poisons are made from plants containing cardenolide glycosides and that almost all these substances are predictably cytotoxic and in some cases have *in vivo* antitumor activity as well. Unfortunately, their effect on the heart precludes possible use in treating human cancers.

The numbers of plants and families discussed by Spjut and Perdue are certainly a considerable underestimate. The indexes of the two major accounts of arrow poisons, the books by Perrot and Vogt (1913, pp. 335–350) and Lewin (1923), neither of which was included in the literature reviewed, between them include the names of approximately

160 genera, and much information has accumulated since these books were written. According to Schultes (pers. com. 1984), for example, the Kofán Indians of Colombia and Ecuador alone utilize more than seventy-five different species in the preparation of their various arrow and dart poisons. While most of the genera indexed in the two books mentioned are those of adjuncts rather than primary sources of active principles, about fifteen genera are known to contain cardiac glycosides and at least thirty-five contain biodynamically active alkaloids of various types. A complete reappraisal of the literature is required to determine whether the conclusions drawn and the inferences made regarding the antitumor properties of arrow- and dart-poison plants are indeed valid.

Summary

The role of certain arrow- and dart-poison plants in the development of medicinal agents is briefly reviewed: *Strychnos* (Loganiaceae), *Chondrodendron* (Menispermaceae), *Hippomane* and *Hura* (Euphorbiaceae), *Daphne* (Thymelaeaceae), *Acokanthera* (Apocynaceae), and *Aconitum* (Ranunculaceae). The suggestion that plants with antitumor activity are frequent particularly among those used in arrow and dart poisons is based on inadequate data and needs reevaluation.

LITERATURE CITED

Adolf, W., and E. Hecker. 1984. On the active principles of the spurge family. X. Skin irritants, cocarcinogens, and cryptic cocarcinogens from the latex of the manchineel tree. *Journal of Natural Products* 47: 482–496.

Angenot, L. 1971. De l'existence en Afrique Centrale d'un poison de flèche curarisant, issu du *Strychnos usambarensis* Gilg. *Ann. Pharm. Franç.* 29: 353–364.

Arichi, S., and Y. Uchida. 1981. Aconite root extract. U. K. Patent Appl. GB 2 053 680 A.

Bally, P. R. O., F. Thudium, K. Mohr, O. Schindler, and T. Reichstein. 1958. Giriama-Pfeilgifte. *Helvetica Chimica Acta* 41: 446–459.

Barbieri, L., A. Falasca, C. Franceschi, F. Licastro, C. A. Rossi, and F. Stirpe. 1983. Purification and properties of two lectins from the latex of the euphorbaceous plants *Hura crepitans* L. (sand-box tree) and *Euphorbia characias* L. (Mediterranean spurge). *Biochemistry Journal* 215: 433–439.

Bassleer, R., M.-Cl. Depauw-Gillet, B. Massart, J.-M. Marnette, P. Williquet, M. Caprasse, and L. Angenot. 1982. Effets de trois alkaloïdes extraits du *Strychnos usambarensis* sur des cellules cancéreuses en culture. *Planta Medica* 45: 123–126.

Bisset, N. G. 1980. Alkaloids of the Loganiaceae. In *Indole and Biogenetically Related Alkaloids*. Eds. J. D. Phillipson and M. H. Zenk. London/New York: Academic Press. 27–61.

————. 1989. Arrow and dart poisons. *Journal of Ethnopharmacology* 25: 1–41.

————. 1990. The ethnographic approach to ethnopharmacology. A critique. *Actes 1er Colloque Europ. Ethnopharmacol.* (March): 87–94.

————. 1991. One man's poison, another man's medicine? *Journal of Ethnopharmacology* 32: 71–81.

————. 1992. [need article title]. In *Alkaloids: Chemical and Biological Perspectives*, vol. 8. Ed. S. W. Pelletier. New York: Springer-Verlag. 1–150.

Bisset, N. G., and M. C. Woods. 1966. The arrow and dart poisons of South-East Asia, with particular reference to the *Strychnos* species used in them. Part II. Burma, Thailand, and Malaya. *Lloydia* 29: 172–195.

Bovet, D. 1959. Rapports entre constitution chimique et activité pharmacodynamique dans quelques séries de curares de synthèse. In *Curare and Curare-like Agents*. Eds. D. Bovet, F. Bovet-Nitti, and G. Marini-Bettòlo. New York: Elsevier. 252–287.

Bovet-Nitti, F. 1959. Les curares à brève durée d'action. In *Curare and Curare-like Agents*. Eds. D. Bovet, F. Bovet-Nitti, and G. Marini-Bettòlo. New York: Elsevier. 230–243.

Bowman, W. C., and M. J. Rand. 1980. *Textbook of Pharmacology*. 2nd ed. Oxford: Blackwell. 13.15, 17.33–49, 21.21.

Cassels, B. K. 1985. Analysis of a Masai arrow poison. *Journal of Ethnopharmacology* 14: 273–281.

Chen, D. H., and X. T. Liang. 1982. Studies on the constituents of lateral root of *Aconitum carmichaelii* Debx. (Fu Zi), a traditional Chinese medicine. I. Isolation and structural determination of salsolinol (in Chinese). *Yaoxue xuebao* 17: 792–794.

Chou, Y. P., L. L. Fan, L. Y. Chang, K. Y. Tseng. 1978. Pharmacological studies on aconite. I. Effect of higenamine on the cardiovascular system (in Chinese). *Chung Hua Yi Hsüeh Tsa Chih* 58: 664–669. Eng. summary in *Chin. Med. J.* (1979) 92: 292.

Dodonaeus, R. 1644. *Herbarivs oft Crvydt-boeck*. Antwerp: Balthasar Moretus.

Edwards, M. C., F. J. Evans, M. L. Barrett, and D. Gordon. 1985. Structural correlations of phorbol-ester-induced stimulation of PGE_2 production by human rheumatoid synovial cells. *Inflammation* 9: 33–38.

Evans, F. J. 1989. Protein kinase C and Rx-kinase: Two receptors for the tumor-promoting and pro-inflammatory phorbol/daphnane esters. *Proceedings of the International Symposium on New Drug Development from Natural Products*, Seoul, May 1989. 161–171.

Evans, F. J., and S. E. Taylor. 1983. Pro-inflammatory, tumor-promoting and anti-tumor diterpenes of the plant families Euphorbiaceae and Thymelaeaceae. *Progr. Chem. Org. Nat. Prod.* 44: 1–99.

Evans, W. C. 1989. *Trease and Evans' Pharmacognosy*. 13th ed. London: Baillière Tindall.

Falasca, A., C. Franceschi, C. A. Rossi, and F. Stirpe. 1980. Mitogenic and hemagglutinating properties of a lectin purified from *Hura crepitans* seeds. *Biochem. Biophys. Acta* 632: 95–105.

Farnsworth, N. R., and C. J. Kaas. 1981. An approach utilizing information from traditional medicine to identify tumor-inhibiting plants. *Journal of Ethnopharmacology* 3: 85–99.

Ford, R. I. 1978. Ethnobotany: Historical diversity and synthesis. In *The Nature and Status of Ethnobotany*. Ed. R. I. Ford. Anthropological Papers, no. 67. Ann Arbor, MI: University of Michigan Museum of Anthropology. 33–49.

Hikino, H., C. Konno, H. Takata, Y. Yamada, C. Yamada, Y. Ohizumi, K. Sugio, and H. Fujimura. 1980. Antiinflammatory principles of *Aconitum* roots. *J. Pharmaco-Biodyam.* 3: 514–525.

Hikino, H., H. Sato, C. Tamada, C. Konno, Y. Ohizumi, and K. Endo. 1979. Pharmacological actions of *Aconitum* roots (in Japanese). *Yakugaku Zasshi* 99: 252–263.

Hikino, H., H. Takata, and C. Konno. 1983. Anabolic principles of *Aconitum* roots. *Journal of Ethnopharmacology* 7: 277–286.

Ishikawa, M. 1962. Poisons and venoms used for hunting by the Ainu in Yeso (Hokkaido): A study of ethnotoxicology (in Japanese). *Jinruigaku Zasshi* 69: 141–153.

James, R. 1764. *Pharmacopoeia Universalis* or *A New Universal English Dispensatory*. 3rd ed. London.

Katz, A., and E. Staehelin. 1979. DC-Untersuchung der Europäischen Aconitum-napellus-Gruppe. *Pharm. Acta Helv.* 54: 253–265.

King, H. 1935. Curare alkaloids. Part I. Tubocurarine. *J. Chem. Soc.*: 1381–1389.

Konno, C., M. Shirasaka, H. Hikino. 1979. Cardioactive principle of *Aconitum carmichaelii* roots. *Planta Medica* 35: 150–155.

Kosuge, T., M. Yokota, and M. Nagasawa. 1978. Studies on cardiac principle in aconite roots. I. Isolation and structural determination of higenamine (in Japanese). *Yakugaku Zasshi* 98: 1370–1375.

Kupchan, S. M., and R. L. Baxter. 1975. Mezerein: Antileukemic principle isolated from *Daphne mezereum* L. *Science* 187: 652–653.

Kupchan, S. M., I. Uchida, A. R. Branfman, R. G. Dailey, Jr., and B. Yu Fei. 1976. Antileukemic principles isolated from Euphorbiaceae plants. *Science* 191: 571–572.

Kupicha, F. K. 1982. Studies on African Apocynaceae: The genus *Acokanthera*. *Kew Bulletin* 37: 41–67.

Leenhouts, P. W. 1962. *Strychnos* L. In *Flora Malesiana* [i] 6. Ed. C. G. G. J. van Steenis. 343–361.

Lewin, L. 1923. *Die Pfeilgifte*. Nach eigenen toxikologischen und ethnologischen Untersuchungen. 2nd ed. Leipzig: J. A. Barth. 498–517.

Massiot, G., and C. Delaude. 1988. Alkaloids of African *Strychnos* species. In *The Alkaloids. Chemistry and Pharmacology*, vol. 34. Ed. A. Brossi. London/New York: Academic Press. 211–329.

Murayama, M., and H. Hikino. 1984. Stimulating actions on ribonucleic acid biosynthesis of aconitines, diterpenic alkaloids of *Aconitum* roots. *Journal of Ethnopharmacology* 12: 25–33.

Nelson, A. 1951. *Medical Botany*. Edinburgh: E. & S. Livingstone.

Neuwinger, H. D. 1974. Afrikanische Pfeilgifte. *Naturwiss. Rundschau* 27: 340–359, 385–402.

Ohiri, F. C., R. Verpoorte, and A. Baerheim Svendsen. 1983. The African *Strychnos* species and their alkaloids: A review. *Journal of Ethnopharmacology* 9: 167–223.

Pelletier, J., and J. B. Caventou. 1819. Sur un nouvel alcali végétal (la *strychnine*) trouvé dans la fève de Saint-Ignace, la noix vomique, etc. *Ann. Chim. Phys.* 10: 142–176.

Pelletier, J., and H. Petroz. 1829. Examen chimique du curare. *Ann. Chim. Phys.* 40: 213–219.

Pereira, J. 1850. *The Elements of Materia Medica and Therapeutics,* vol. 2/1. 3rd ed. London: Longman, Brown, Green, & Longmans.

Perrot, É., and É. Vogt. 1913. *Poisons de flèches et poisons d'épreuve.* Paris: Vigot.

Quetin-Leclercq, J. 1990. Des poisons de flèches aux substances naturelles antimitotiques. *Actes 1er Colloque Europ. Ethnopharmacol.* (March): 279–290.

Quetin-Leclercq, J., L. Angenot, and N. G. Bisset. 1990. South American *Strychnos* species. Ethnobotany (except curare) and alkaloid screening. *Journal of Ethnopharmacology* 28: 1–52.

Reichstein, T. 1965. Chemische Rassen in *Acokanthera. Planta Medica* 13: 382–399.

Reynolds, J. E. F., ed. 1989. *Martindale: The Extra Pharmacopoeia.* 29th ed. London: Pharmaceutical Press.

Roulin, F., and (J. B. J. D.) Boussingault. 1828. Examen chimique du curare. *Ann. Chim. Phys.* 39: 24–30.

Sakata, K., K. Kawazu, T. Mitsui, and N. Masaki. 1971. Structure and stero-chemistry of huratoxin, a piscicidal constituent of *Hura crepitans. Tetrahedron Letters* 1141–1144.

Schlittler, E. 1971. Pharmacologically interesting and clinically useful alkaloids. In *The Alkaloids.* Specialist periodical reports. London: The Chemical Society. 1: 478–480.

Schultes, R. E. 1967. The place of ethnobotany in the ethnopharmacologic search for psychotomimetic drugs. In *Ethnopharmacologic Search for Psychoactive Drugs.* Eds. D. H. Efron, B. Holmstedt, and N. S. Kline. U.S. Department of Health, Education and Welfare Publication no. 1645. Washington, DC: Government Printing Office. 33–57.

Spjut, R. W., and R. E. Perdue, Jr. 1976. Plant folklore: A tool in predicting sources of antitumor activity? *Cancer Treatment Report* 60: 979–985.

Thudium, F., K. Mohr, O Schindler, and T. Reichstein. 1958. Die Glykoside von *Acokanthera schimperi* (A. DC.) Benth. et Hook. 2. Mitteilung. Untersuchung des Wurzel-und Zweig-Holzes der Form, die von den *Wa-Giriama* zur Pfeilgiftbereitung verwendet wird. *Helvetica Chimica Acta* 41: 604–614.

Trease, G. E. 1949. *A Textbook of Pharmacognosy.* 5th ed. London: Baillière Tindall & Cox.

Trease, G. E., and W. C. Evans. 1983. *Pharmacognosy.* 12th ed. London: Baillière Tindall & Cox.

Verpoorte, R., E. W. Kode, H. van Doorne, and A. Baerheim Svendsen. 1978. Antimicrobial effect of the alkaloids from *Strychnos afzelii* Gilg. *Planta Medica* 33: 237–242.

Woodville, W. 1794. *Medical Botany.* Part II. London: J. Phillips.

———. 1832. *Medical Botany.* 3rd ed. London: J. Bohn.

New Concepts in Medical and Dental Ethnobotany

MEMORY ELVIN-LEWIS AND WALTER H. LEWIS

In the last half of the twentieth century, books on plants in relation to human health began appearing in ever-increasing numbers. Although some contained well-documented herbal recipes from traditional medicine, many others were merely lists of plants accompanied by supposed efficacies. Little attempt was made to relate their contents to scientific data and only rarely were they correlated with current use in modern medicine (Lewis and Elvin-Lewis 1983). Few were based on new, objectively obtained, quantitative data, which should be fundamental to ethnobotanical surveys as they apply to current biomedical experimentation and to medical-dental clinical practices. The following discussion outlines several new approaches to medical and dental ethnobotany, including organization of data, field surveys, biomedical experimentation and an understanding of plant diversity.

Computerized Data Processing

Current concepts in data processing are revolutionizing the synthesis of data. In the past, researchers were dependent upon time-consuming techniques for arriving at a consensus, and, therefore, the in-put of large amounts of data was usually not feasible. There was a tendency to rely on a small number of informants, who were authoritative in one respect but who could be highly subjective in another. Impressions regarding the preference of a group of individuals for one type of plant over another were often just that. It was rarely possible to understand, after analyzing such data, whether age, sex, religion, or other factors influenced the choices available; the making of correlations was frequently too complex to be considered worthwhile.

To overcome the deficiencies inherent in the gathering of qualitative data, sampling sizes sufficient to allow significant statistical analysis of subsets of populations must be generated. If greater objectivity is a goal of further ethnobotanical surveys, a quantitative approach is essential, and, through the use of computers, the synthesis of multifaceted information involving many informants should be no longer a seemingly hopeless task.

Medical-Dental Ethnobotanical Surveys

The protocol of a dental ethnobotanical study followed by Adu-Tutu et al. (1979) and Elvin-Lewis et al. (1980), and further refined by Elvin-Lewis (1982), may serve as a model. The study, involving numerous parameters as well as many individuals, was designed to examine the reasons for chewing-stick preferences among a number of Ghanian tribes and how these choices might correlate with a crude analysis of dental health and other cultural variations. The whole was subjected to computer analysis. Data from 887 subject interviews were analyzed for information related to the population's profile (sex, age, education, job, tribe), territory (region, town size, forest type), chewing-stick (species, how acquired, reasons for adult and childhood choices), effect on dental health (tooth condition, absence of teeth), and type of sugar diet preferred (refined or cane, amount).

Results showed that only a few species were preferred and that the rationale behind their use was based on efficacy in dental hygiene and in preventing or curing dental disease. Furthermore, examination of these species in the laboratory indicated that most had a wide range of antibacterial activity against a number of odontopathic bacterial species and that many also contained a number of healing and/or analgesic compounds (Elvin-Lewis 1983, 1985). In spite of the limited type of clinical data acquired, the study provided information sufficient to enable Ghanian dentists to understand better how chewing-stick and sugar cane use were affecting the dental health of their patient populations and also to determine which species could be recommended for dental hygienic purposes, should the economy or availability dictate this necessity. The data also suggested where additional clinical studies would be useful. For example, a periodontal evaluation of persons who preferred certain species when they had bleeding or sore gums would be helpful in determining the therapeutic value of plants used under these circumstances.

Meriting serious evaluation is the ambiguity surrounding medical and dental descriptors that continue to be used and incorporated into modern writings. Often they are obscure at best, and many have no meaning in a modern biomedical context. For instance, *alternative* is an obsolete term used for a drug thought to reestablish healthy functions of the system; whereas *panacea* stands for a cure-all that supposedly maintains normal healthy functioning. Without a scientific basis for either of these terms, is there any value in continuing to use them except in the absence of more precise ones? Modern published herbals still utilize a multitude of terms inherited from the Middle Ages and earlier that lack the precision needed in a modern therapeutic context. Clearly, to ensure that biomedical scientists are able to interpret information obtained in ethnobotanical surveys, collaboration must be sought to develop a language for describing medical and dental aspects.

Biomedical Experimentation

In the next phase of such ethnobotanical studies, the collaboration of laboratory scientists such as natural chemists, pharmacologists and microbiologists is essential. By means of *in vitro* and *in vivo* screening, they can establish the therapeutic basis for plant products identified as promising through ethnobotanical surveys. The position of a particular species in a family or higher taxon often provides information regarding types of compounds and potential effacacies associated with that species. For example, similar antibiotics and analgesic substances are often found among members of plants belonging to closely related groups (Elvin-Lewis 1983, 1985), just as toxic factors may be a feature of other taxa (Kingsbury 1983; Lewis and Elvin-Lewis 1983).

Following the identification of promising compounds in the laboratory, a protocol that includes extensive toxicological trials in animals must be followed before human trials are initiated. Clearly, the temptation exists to circumvent these procedures when traditional use suggests low toxicity. Purified substances, however, although more potent in their activity, may also present potential problems that are not as apparent in weaker, herbal mixtures. As toxicological tests become more refined, results derived through older methods may not have the same value.

Sometimes, too, the selection of a plant for use in a proprietary remedy may have been made without careful review of known side effects associated with its traditional use among indigenous populations. Thus, the habitual use of a modern plant product could result in eliciting serious reactions that are not readily apparent in its casual use for other purposes. Moreover, according to some natural product scientists, toxicological tests appropriate to accounting for such contingencies may have yet to be undertaken (Persinos 1990; Sears 1991; Elvin-Lewis et al., in prep.).

Such is the current status regarding proof of safety of the daily use of low concentrations of root extracts of *Sanguinaria canadensis* L. containing sanguinarine and other isoquinoline alkaloids in dental products for the purpose of plaque control (Southard et al. 1984; Griangrego and Mitchell 1985; Balanyk 1990). While industry has generated a wide range of toxicological studies to defend its case (Schwartz 1986; Becci et al. 1987; Lord et al. 1989; Keller and Meyer 1989), and employed a panel of toxicologists to review its findings (Richardson 1990; Frankos et al. 1990), the basis of continued concern is that the bulk of these data has yet to appear in toxicological journals where it could be objectively assessed. Therefore, general acceptance by toxicologists and pharmacologists of this proposed safety profile is unlikely until independent verification and careful peer review and publication of these and other studies have been conducted.

This check is necessary because the industry's conclusions are not altogether consistent with observations associated with folk use among Native Americans (Erichsen-Brown 1979), adverse reactions associated with old proprietary expectorating and emetic preparations (Osol and Farrar 1955, pp. 120–121), and decades of pharmacological research that suggests that sanguinaria alkaloids, and particularly sanguinarine, are potentially irritating, bioreactive, and poisonous (Casarett and Doull 1975, p. 284). At high concentrations these alkaloids affect functions of the cardiovascular system (Seifen et al. 1979; Whittle et al. 1980), liver (Dalvi 1985), and eyes (Hakim 1954, 1970; Hakim et al. 1961), and at the molecular level at relatively low concentrations they inhibit a wide variety of enzyme systems (Vallejos 1973; Straub et al. 1975; Cohen et al. 1978; Moore and Rabovsky 1979; Vaidya et al. 1980; Walterovia et al. 1981; Cala et al. 1982), interact in *in vivo* drug metabolism (Dalvi and Peeples 1981; Peeples and Dalvi 1982), intercalate with DNA (Faddejeva et al. 1980; Maiti et al. 1982; Nandi and Maiti 1985), and may also be carcinogenic (Farnsworth et al. 1976).

Since habitual use of these dental preparations is recommended, future investigations must address the potential toxic conditions that could arise should these alkaloids, while circulating in the body, accumulate or interact with medications, especially those utilized for glaucoma or the heart. In this regard the issue of sanguinarine-induced glaucoma in predisposed humans (Sachdev et al. 1988) or the primary test animal, the cat, remains poorly understood and warrants careful reevaluation. With all these parameters to consider, the choice of such powerful compounds for use in herbal preparations or proprietary medications will always be problematical until all risks are fully understood.

It is recognized that even herbal practitioners have difficulty detecting toxicity due to long-term or chronic exposure to such plants (Croom 1983) without the diagnostic tools and techniques of modern medicine. Clearly, even most dental practitioners are in a poor position to recognize many effects that would be more apparent to either medical specialists or toxicologists. It is for this reason that before general use is allowed, very precise animal toxicology studies must be done on herbal preparations. Information gleaned from many sources, including impressions concerning dosage, efficacy, and/or toxicity known from original indigenous uses (Erichsen-Brown 1979), should serve as a guide for evaluations needed. Moreover, it would be useful if regulating agencies such as the Food and Drug Administration would develop guidelines that could be followed when the use of a specific phytochemical is being proposed.

Many claims can be made regarding the efficacy of a plant remedy, and yet when eagerness to substantiate certain opinions is not metered with careful, objective scientific evaluation, faulty concepts that are difficult to eradicate can be perpetuated. The saga of amygdalin is one of the best modern examples of this phenomenon. Following its registration as laetrile in 1952, it completely eclipsed the use of other unorthodox therapies for treating cancers and all other diseases in our time. It was legalized for use in twenty-seven states of the United States and remained a major and unresolved public health problem for over a quarter of a century. Thousands of cancer patients were treated with laetrile, and, as animal model trials were not convincing, its widespread use posed a serious problem for medical practitioners.

To answer finally these humanitarian and scientific issues, a carefully designed clinical trial was conducted by Moertel et al. (1982) at the Mayo Clinic, UCLA Cancer Center, Memorial Sloan-Kettering Cancer Center, University of Arizona Cancer Center, and National Cancer Institute, among 178 cancer patients following recommended protocols and also employing a "metabolic therapy" program of diet, enzymes, and vitamins. No substantive benefit was observed in terms of cure, improvement, stabilization of cancer, cancer-related symptoms, or extension of life span. Moreover, the hazards of cyanide toxicity were evident in a number of patients. It was concluded that amygdalin (laetrile) was a toxic drug that had no effect as a cancer treatment.

Sometimes the use of controlled clinical trials is not possible because medical ethics of a culture dictate otherwise. For instance, placebo treatments cannot be introduced as a control if efficacy for the treatment is presumed. Often the only data available are those obtainable under such circumstances, and, again, assumptions of efficacy often are derived from this kind of source. So it is with many of the claims regarding *Panax* (ginseng), most of whose animal and clinical research has been conducted in Oriental countries where such a concept prevails. The value of experiments is further compromised when patients must answer difficult, even meaningless questions regarding their progress, and when ginseng of unknown composition is used as the basis for treatment. Moreover, stressful conditions in the laboratory *per se* usually are not considered, and neutral or negative data are rarely reported. Careful statistical evaluation should and must be a part of the protocol, since studies claiming the value of ginseng in cancer therapy have subsequently proven to be erroneous (Lewis 1986).

Other studies using animal models are frequently undertaken without following a double-blind protocol; with so much inherent bias involved, the double blind is essential to objectivity. Studies of *Panax* and its close relatives are very complicated, because a number of active compounds have been identified that may act in divergent ways and may be effective only after a prolonged course of treatment.

The activity of certain ginsenosides *in vitro* is not questioned; but, the significance of their effects *in vivo* is largely conjectural, since ginsenoside absorption, transport, degradation, and excretion in relation to specific tissues and organs is poorly known. Unfortunately, ginseng remains a medical enigma, since current evidence does not support original claims that this plant, or *Eleutherococcus*, has antineoplastic, immunologic, or reproductive effects, or even that it improves performance under stressful conditions (Lewis 1986). Clearly, more objective and precise experimentation in animals and in humans is needed.

Plant Diversity

It is recognized that genetic variation, including polyploidy, of plant species can influence the total content of their efficacious or detrimental principles. Although the formulation of certain remedies may appear similar from one part of the world to the next, care must be taken to ensure that cultivars employed in local remedies contain the same amount or type of substances desired. Chemical diversity, therefore, is a significant variable of plants that can affect their use in herbal remedies.

In obtaining formulas from indigenous medical practitioners, ethnobotanists should appreciate the importance of their plant-collecting methodology. Ayurvedic practitioners follow a set of regulations that govern the collection and preparation of plants to insure the greatest efficacy in treating various illnesses (Lewis and Elvin-Lewis 1979). Similarly, the ancient Greeks collected certain herbs by night and certain others by day, and some before the sun struck them.

Only recently has the significance of these collecting techniques been appreciated; now it is known that alkaloids and many other secondary metabolites are dynamic products that fluctuate in both total concentration and in rate of turnover (Lewis 1984). For example, alkaloid changes during the ontogeny of *Catharanthus roseus* (L.) G. Don are striking. Virtually no alkaloids are present in the seeds; they first appear during germination, and after three weeks they are present throughout the plant. Alkaloids then gradually disappear almost completely and finally reappear in about eight weeks (Mothes et al. 1965).

The seeds of *Papaver somniferum* L. are essentially free of alkaloids, but, at germination, seedlings produce narcotine in three days and then codeine, morphine, and papaverine, when the seedlings are about 7 centimeters (3 inches) high. Total alkaloid content slowly increases until flowering when there is a sharp increase lasting until the floral leaves dehisce. There are many other well-documented examples of ontogenic changes in alkaloid content that generally follow the principle of rapid increase at the time of cell enlargement and vacuolization followed by a slow decline of concentration during senescence (Lewis 1984).

Active metabolism of these alkaloids is further demonstrated by fluctuations in their concentration during a single day. Fairbairn and Suwal (1961) showed that amounts of coniine and γ-coniceine of *Conium maculatum* L. varied considerably during the day and that an increase in one alkaloid corresponds to a decrease in the other (Table 1). Thus, at 4 A.M. the fruit contains 226 micrograms of coniine and no γ-coniceine, but at 4 P.M. the γ-coniceine concentration is 21 micrograms and that of coniine only 8 micrograms per fruit, indicating not only a reversal in the concentration of these compounds by the afternoon but also a total decrease in the amount of poisonous principles during this time. In fact, according to these data, poison hemlock is hardly poisonous during the late afternoon!

Commercial harvest of some herbs may depend on other changes in their environment. For example, the tropane alkaloid content of *Datura* species is considerably less after a rainy period than after clear weather; the difference is so marked that collecting is done only after a period of clear days. Leaves of many plants dried in the shade contain more alkaloid than those dried in the sun, and leaves allowed to dry on the plant often contain more alkaloids than those dried after removal. Furthermore, the alkaloid yield of leaves is also increased when floral buds are removed. Apparently, methods of harvest, light exposure, and photoperiod are all factors important in secondary compound yields.

Since the production of some of these compounds may be applicable to the expanding field of tissue and cell suspension cultures, care must be taken to ensure that the variations that can occur do not affect the total yield of a compound. Cultural conditions,

Table 1. Amounts of coniine and γ-coniceine from *Conium maculatum* during four-hour samples in one day of week five following flowering (from Fairbairn and Suwal 1961 and courtesy of the publishers of *Phytochemistry*).

Time	Coniine (μg/fruit)	γ-Coniceine (μg/fruit)
4 A.M.	226	0
8 A.M.	130	2
Noon	174	9
4 P.M.	8	21
8 P.M.	200	0
Midnight	213	0

including the use of growth regulators, should be kept optimal if maximum yield of specific compounds is desired. Moreover, careful surveillance of stock cultures should be maintained throughout propagation, since frequencies of high-yielding clones are known to vary (Dougall 1979).

Summary

Ethnobotanical research involving medicine has been valuable in the past, but its reliability and importance will be much greater with improved field-collecting techniques and with wider use of computers for correlative analysis of data. This basic format should be followed in any collaborative biomedical research that includes experimentation and makes use of diverse plant populations.

ACKNOWLEDGMENTS

The research presented in this paper was supported in part by a grant from the National Science Foundation (BSR-850875).

LITERATURE CITED

Adu-Tutu, M., Y. Afful, K. Asanti-Appiah, D. Lieberman, J. B. Hall, and M. Elvin-Lewis. 1979. Chewing stick usage in southern Ghana. *Economic Botany* 33: 320–328.

Balanyk, T. E. 1990. Sanguinarine: Comparison of anti-plaque/anti-gingivitis reports. *Clin. Prev. Dent.* 12(3): 18–25.

Becci, P. J., J. Schwartz, H. H. Barnes, and G. L. Southard. 1987. Short-term toxicity studies of sanguinarine and two alkaloid extracts of *Sanginaria canadensis* L. *J. Toxicol. Environ. Health* 20: 199–208.

Cala, P. M., J. G. Norby, and D. C. Tosteson. 1982. Effects of the plant alkaloid sanguinarine on cation transport by human red blood cells and lipid bilayer membranes. *J. Membr. Biol.* 64: 23–31.

Casarett, L. F., and J. Doull, eds. 1975. *Toxicology, The Basic Science of Poisons.* New York: MacMillan Publishing Company.

Cohen, G., E. E. Seifen, K. D. Straub, C. Tiefenback, and F. R. Stermitz. 1978. Structural specificity of the NaK-ATPase inhibition of sanguinarine, an isoquinoline benzophenanthridine alkaloid. *Biochem. Pharm.* 27: 2555–2558.

Croom, E. M., Jr. 1983. Documenting and evaluating herbal remedies. *Economic Botany* 37: 13–27.

Dalvi, R. R. 1985. Sanguinarine: Its potential as a liver toxic alkaloid present in the seeds of *Argemone mexicana. Experientia* 41: 77–78.

Dalvi, R. R., and A. Pepples. 1981. In vivo effect of toxic alkaloid on drug metabolism. *J. Membr. Biol.* 64: 23–31.

Dougall, D. K. 1979. Factors affecting the yields of secondary products in plant tissue cultures. In *Plant Cell and Tissue Culture Principles and Applications*. Eds. W. R. Sharp, P. O. Larsen, E. F. Paddock, and V. Raghavan. Columbus: Ohio State University Press. 727–743.

Elvin-Lewis, M. 1982. The therapeutic potential of plants used in dental folk medicine. *Odontostomatol. Trop*. 55: 107–117.

———. 1983. The antibiotic and anticarcinogenic potential of chewing-sticks. In *The Anthropology of Medicine*. Eds. L. Romanucci-Ross, D. E. Moerman, and L. R. Tancredi. New York: Praeger. 201–220.

———. 1985. Therapeutic rationale of plants used to treat dental infections. In *Plants Used in Indigenous Medicine and Diet*. Ed. N. L. Etkin. South Salem, NY: Redgrave. 48–69.

Elvin-Lewis, M., R. Adams, C. Beecher, A. der Marderosian, N. R. Farnsworth, J. D. McChesney, M. Malone, J. Pezzuto, A. Spiegelman, and D. P. Waller. Is sanguinarine safe for dental purposes? (In preparation)

Elvin-Lewis, M., J. B. Hall, M. Adu-Tutu, Y. Afful, K. Asanti-Appiah, and D. Lieberman. 1980. The dental health of chewing stick users of southern Ghana: Preliminary findings. *J. Prev. Dent*. 6: 151–159.

Erichsen-Brown, C. 1979. *Uses of Plants for the Past 500 Years*. Aurora, Ontario: Breezy Creeks Press.

Fairbairn, J. W., and P. N. Suwal. 1961. The alkaloids of hemlock (*Conium maculatum* L.)—II. Evidence for a rapid turnover of the major alkaloids. *Phytochemistry* 1: 38–46.

Faddejeva, M. D., I. N. Baclyaeva, J. P. Novikov, and H. G. Shalabi. 1980. Possible binding of alkaloids sanguinarine and berberine. *I.R.C.S. Med. Sci. Biochemistry* 8: 612.

Farnsworth, N. R., A. S. Bingel, H. H. S. Fong, A. A. Saleh, G. M. Christenson, and S. M. Saufferer. 1976. Oncogenic and tumor-promoting spermatophytes and pteridophytes and their active principles. *Cancer Treatment Report* 60: 1171–1214.

Frankos, V. H., D. J. Brusick, E. M. Johnson, H. I. Maibach, I. Munro, and R. A. Squire. 1990. Safety of *Sanguinaria* extract as used in commercial toothpaste and oral rinse products. *J.A.D.A.* 56 (supplement 7): 41–47.

Griangrego, E., and E. W. Mitchell. 1985. Chemical agents for the control of plaque. *J.A.D.A.* 112: 18–28.

Hakim, S. A. E. 1954. *British Journal of Ophthal*. 38: 193.

———. 1970. Death, cardiopathy, symptomless glaucoma, and cancer from edible oils containing *Argemone*. *Maharashtra Med. J*. 17: 109–130.

Hakim, S. A. E., V. Mijovic, and J. Walker. 1961. Distribution of certain poppy-fumaria alkaloids and a possible link with the incidence of glaucoma. *Nature* 189: 198–201.

Keller, K. A., and D. L. Meyer. 1989. Reproductive and developmental toxicological evaluation of *Sanguinaria* extract. *J. Clin. Dent*. 1: 59–66.

Kingsbury, J. M. 1983. The evolutionary and ecological significance of plant toxins. In *Handbook of Natural Toxins*, vol. 1 of *Plant and Fungal Toxins*. Eds. R. F. Keeler and A. T. Tu. New York: Dekker. 675–706.

Lewis, W. H. 1984. Biosystematics and medicine. In *Plant Biosystematics*. Ed. W. F. Grant. Dan Mills, Ontario: Academic Press Canada. 561–578.

———. 1986. Ginseng: A medical enigma. In *Plants Used in Indigenous Medicine and Diet*. Ed. N. L. Etkin. South Salem, NY: Redgrave. 290–305.

Lewis, W. H., and M. P. F. Elvin-Lewis. 1979. Systematic botany and medicine. In *Systematic Botany, Plant Utilization and Biosphere Conservation*. Ed. I. Hedberg. Stockholm: Almquist and Wiksell. 24–31.

———. 1983. Contributions of herbology to modern medicine and dentistry. In *Handbook of Natural toxins*. Vol. 1, *Plant and Fungal Toxins*. Eds. R. F. Keeler and A. T. Tu. New York: Dekker. 785–815.

Lord, G., E. I. Goldenthal, and D. L. Meyer. 1989. Sanguinarine and the controversy concerning its relationship to glaucoma in epidemic dropsy. *J. Clin. Dent*. 1: 110–115.

Maiti, M., et al. 1982. Sanguinarine: a monofunctional intercalating alkaloid. *FEBS Letters* 142(2): 280–284.

Moertel, C. G., T. R. Fleming, F. Rubin, L. K. Kvols, G. Sarna, R. Koch, V. E. Currie, C. W. Young, S. E. Jones, and J. P. Davignon. 1982. A clinical trial of amygdalin (laetrile) in the treatment of human cancer. *New England Journal of Medicine* 306: 201–206.

Moore, R. D., and J. L. Rabovsky. 1979. Activation by sanguinarine of active sodium efflux from frog skeletal muscle in the presence of ouabain. *J. Physiol*. (London) 295: 1–20.

Mothes, K. I., I. Richter, K. Stolle, and D. Groger. 1965. Physiologische bedingungen der alkaloid-synthese bei *Catharanthus roseus* G. Don. *Naturwissenschaften* 52: 431.

Nandi, R., and M. Maiti. 1985. Binding of sanguinarine to deoxyribonucleic acids of differing base composition. *Biochem. Pharmacol.* 34: 321–324.

Osol, A., and G. L. Farrar, eds. 1955. *The Dispensatory of the United States of America*. 25th ed. Philadelphia: J. B. Lippincott.

Peeples, A., and R. R. Dalvi. 1982. Toxic alkaloids and their interaction with microsomal cytochrome P-450 in vitro. *Journal of Applied Toxicology* 2: 300–302.

Persinos, G. J. 1990. Scientists' persistence pays off: Company performs additional toxicity tests on *Sanguinaria* extract. *Washington Insight* 3(2): 6. North Bethesda, MD.

Richardson, D. I. 1990. Letter to the editor, *Sanguinaria* extract. *Washington Insight* 3(3): 2. North Bethesda, MD.

Sachdev, M. S., N. Sood, L. K. Berma, S. K. Gupta, and N. F. Jaffery. 1988. Pathogenesis of epidemic dropsy glaucoma. *Arch. Opthalmol.* 106: 1221–1223.

Schwartz, H. G. 1986. Safety profile of sanguinarine and sanguinaria extract. *Compen. of Cont. Educ. in Dent.* Suppl. 7: 212–217

Sears, C. 1991. Flak over plaque. *American Health* 10(1): 22.

Seifen, E., R. J. Adams, and R. K. Reimer. 1979. Sanguinarine: Its potential as a liver toxic alkaloid present in the seeds of *Argemone mexicana*. *Experientia* 41: 77–78.

Southard, G. L., R. T. Boulware, D. R. Walborn, W. J. Groznik, E. E. Thorne, and S. L. Yankell. 1984. Sanguinarine, a new anti-plaque agent: Retention and plaque specificity. *J.A.D.A.* 108: 338–341.

Straub, K. D., and P. Carver. 1975. Sanguinarine, inhibitor of Na/K-dependent ATPase. *Biochem. Biophys. Res. Comm.* 62: 913–922.

Vaidya, A. B., T. G. Rajagopalan, A. G. Kale, and R. J. Levine. 1980. Inhibition of human pregnancy plasma diamine oxidase with alkaloids of *Argemone mexicana* berberine and sanguinarine. *J. Postgrad. Med.* 26: 28–33.

Vallejos, R. H. 1973. Uncoupling of photosynthetic phosphorylation benzophenanthridine alkaloids. *Biochem. Biophys. Acta* 292: 193–196.

Walterovia, D., J. Ulricova, V. Preininger, V. Simanek, J. Lenfeld, and J. Lasovsky. 1981. Inhibition of liver alanine aminotransferase activity by some benzophenanthridine alkaloids. *J. Med. Chem.* 24: 1100–1103.

Whittle, J. A., J. K. Bissett, K. D. Straub, J. E. Doherty, and J. R. McConnell. 1980. Effect of sanguinarine on ventricular refractoriness. *Res. Commun. Pathol. Pharmacol.* 29: 377–380.

Medicinal Chemistry's Debt to Ethnobotany

ALBERT HOFMANN

Medicinal chemists working on plants in search of pharmacologically active principles will select for their investigations plants with known biological activity. To whom are they indebted for the information about pharmacological effects in particular plants? In nearly all cases they are unknown, nameless discoverers from long past ages. It may well be that this knowledge was acquired empirically in the course of searching for food in the Vegetable Kingdom, or, as it is often supposed, that the first human beings were endowed with instincts that enabled them to recognize the curative properties of plants and to make use of them.

This knowledge became the lore of folk medicine. In culturally more highly developed countries, these medicinal plants were recorded in comprehensive herbals and pharmacopeias; whereas in so-called primitive societies, the knowledge and use of vegetable drugs remained hidden in the hands of healers and shamans. In the last case it was, and still is, the ethnobotanist who is the mediator between healers or shamans and medicinal chemists and who provides medicinal chemists with information about the existence of those particular plants, their use, and biological effects. On this kind of information valuable projects in medicinal chemistry were based and continue to be based to this day. In this chapter I shall report examples of my own research showing the fruitful collaboration between ethnobotanist and medicinal chemist.

The Sacred Mushrooms of Mexico

The first example refers to investigations of the so-called sacred mushrooms of Mexico. The beginning of scientific research in this field is documented by two publications that appeared in 1939, one by ethnologist Jean Bassett Johnson and the other by ethnobotanist Richard Evans Schultes, who published a second paper on this subject in 1940. Johnson and his coworkers had discovered the existence of a mushroom cult in the Sierra Mazateca in the state of Oaxaca in southern Mexico. Schultes gave a botanical description and identified a mushroom used in this cult.

Based on this pioneering ethnobotanical research, the ethnomycologists Robert Gordon Wasson and his wife, Valentina Pavlovna, carried out systematic studies of the ceremonial use of the mushroom, its history, and its present form on several expeditions to the

Mazatec country between 1953 and 1956. The Wassons published the results of their investigations in a comprehensive monograph (Wasson and Wasson 1957). It turned out that the use and worship of hallucinogenic mushrooms, called by the Aztecs *teonanacatl* (meaning "divine flesh"), was very ancient. The Wassons likewise discovered the use of numerous other mushrooms in this context.

The botanical identification and classification of most of the sacred mushrooms was carried out by mycologist Roger Heim, director of the Laboratoire de Cryptogamie du Musée Nationale d'Histoire Naturelle in Paris, who had accompanied the Wassons on their 1956 expedition. These were foliate mushrooms (Agaricales) of the Strophariaceae, mostly new species, the greater part belonging to the genus *Psilocybe* (Heim and Wasson 1958) (Figure 1).

Figure 1. *Psilocybe mexicana* Heim. Photo A. Brack.

After unsuccessful attempts to isolate the active principles of the mushrooms in Paris and in two laboratories in the United States, Heim sent samples of *Psilocybe mexicana* to my laboratory in the Pharmaceutical Department of Sandoz, Ltd., in Basel, Switzerland, for chemical analysis. Attempts to follow the active components in the extraction procedures by animal tests gave inconclusive results. Thanks to the testing of the extract fractions on myself and several of my colleagues who volunteered to serve as guinea pigs, we succeeded in isolating the active principles and in crystallizing them in pure form. The main active constituent was named psilocybin, the accompanying alkaloid, psilocin (Hofmann et al. 1958) (Figure 2). The dried mushrooms contain 0.2 to 0.4 percent of psilocybin. Psilocin is present in trace amounts only.

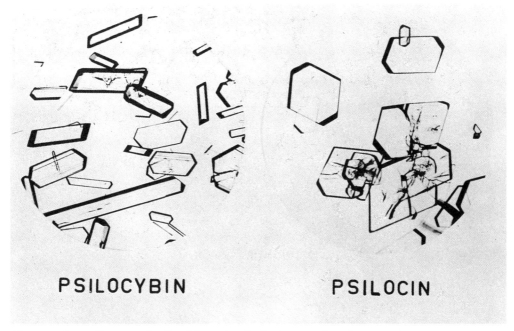

Figure 2. Crystallization of psilocybin and psilocin from methanol.

Elucidation of their structures led to the interesting result that these were novel indole derivatives. Degradation studies showed psilocybin to be 4-phosphoryloxy-N, N-dimethyltryptamine. Hydrolysis of psilocybin gave equimolecular amounts of phosphoric acid and psilocin, which proved to be 4-hydroxy-N, N-dimethyltryptamine. These structures were confirmed by synthesis of psilocybin and psilocin (Hofmann et al. 1959) (Figure 3).

The synthetic production of these compounds is much more rational than the extraction from mushrooms. Psilocybin is a stable compound, readily soluble in water, while psilocin is very sensitive to oxygen and nearly insoluble in water. Psilocybin is the first and hitherto only known natural indole compound containing a phosphoric acid radical. The two compounds are also novel in that they are substituted by a hydroxyl group in the 4-position of the indole nucleus. The availability of the active principles of the hallucinogenic mushrooms in the form of pure chemical compounds made it possible to study their pharmacological properties and mental effects.

PHARMACOLOGICAL PROPERTIES OF PSILOCYBIN

The first study of the effects of psilocybin on the whole animal and on isolated organs was carried out in the Pharmacological Department of Sandoz, Ltd., in Basel under the direction of Aurelio Cerletti. Psilocybin does not exhibit typical effects on isolated organs (intestine, uterus, heart), with the exception of a pronounced inhibiting action towards serotonin. On the entire animal, however, it shows characteristic autonomic effects,

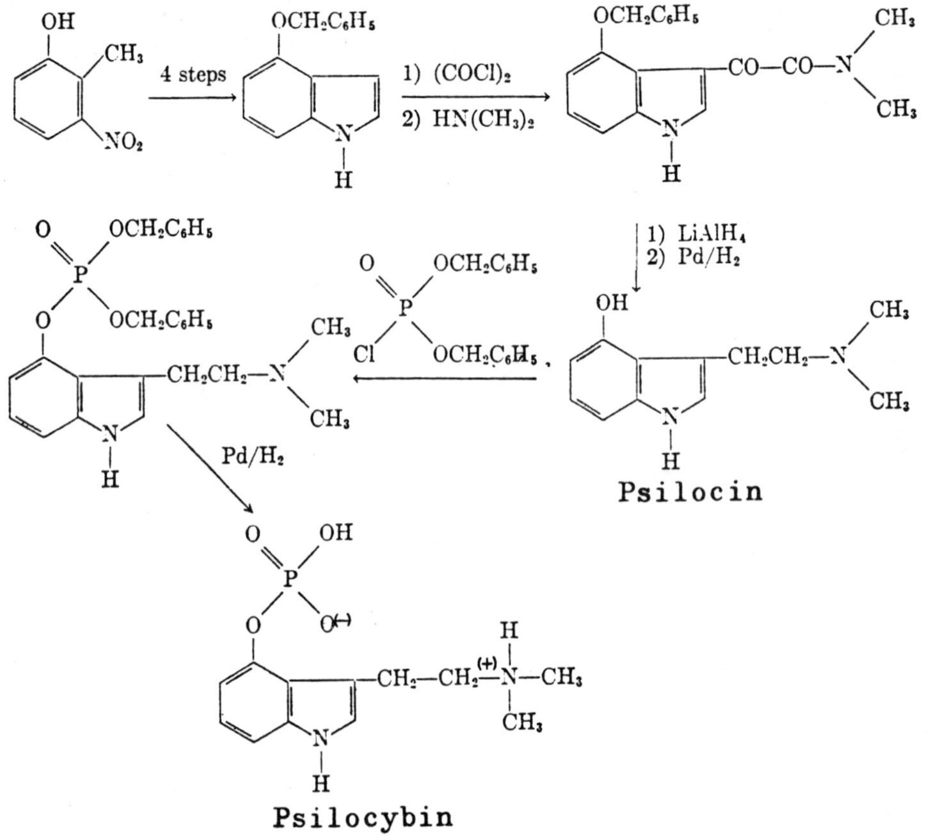

Figure 3. Synthesis of psilocybin and psilocin.

namely, dilation of the pupils, contraction of the nictitating membrane, piloerection, and temperature increase. This is an ergotropic excitation syndrome, which results mainly from a central stimulation of sympathetic structures (Weidmann et al. 1958; Cerletti et al. 1959). A very characteristic effect of psilocybin is the regular enhancement of mono-synaptic spinal reflexes (e.g., the patellar reflex in cats) (Weidmann and Cerletti 1959).

The pharmacological properties of psilocin amply correspond to those of psilocybin. The phosphoric acid radical, therefore, does not contribute to the pharmacological activity but, since psilocybin is much more stable than psilocin, it could act biologically as a protective group.

MENTAL EFFECTS OF PSILOCYBIN

Psilocybin and psilocin produce psychic effects in humans similar to those of mescaline or LSD. The medium oral dose is 6 to 10 milligrams, which elicits the same symptoms as the consumption of about 2 grams of dried *Psilocybe mexicana*.

The first systematic analysis of the effects of psilocybin in humans was carried out at the psychiatric clinic of the University of Basel and was based on personal studies made by several members of the staff of the Sandoz Research Laboratories (Gnirss 1959; Rümmele 1959). Oral doses of a few milligrams led, after 20 to 40 minutes, to changes in the psychic sphere. The mental symptoms produced by small doses of up to 4 milligrams comprise effects on mood and environmental contact; frequently there is a subjectively pleasant sensation of intellectual and physical relaxation and detachment from the environment. Not infrequently, these effects are associated with a pleasant feeling of physical tiredness and heaviness, but sometimes they are accompanied by a feeling of extraordinary lightness, a bodily hovering. With higher doses, 6 to 20 milligrams, more profound psychic changes are prominent and are associated with alterations in spatial and temporal perception and with changes in the awareness of self and body image. Visual and acoustic hypersensitivity are present and may lead to illusions and hallucinations. In this dreamlike state, long-forgotten memories, even from early childhood, are often recalled.

THE USE OF PSILOCYBIN IN BIOLOGY AND IN PSYCHIATRY

The close relationship in chemical structure between the mushroom principles and the brain factor serotonin (5-hydroxy-tryptamine), differing only by the position of the hydroxyl group and the methylation of the amino group, makes psilocybin and psilocin valuable tools in experimental neurology and brain research. The production of so-called model psychoses by psilocybin and other hallucinogens enables the study of biochemical and electrophysiological processes involved in mental disorders (Leuner 1962).

The application of these agents in drug-supported psychoanalysis and psychotherapy which, in the beginning, gave promising results, was interrupted by draconic legal restrictions regarding possession and use of hallucinogens, even by psychiatrists. This intervention of the health authorities became necessary because of the unfortunate misuse and abuse of hallucinogens in the drug scene.

From the Mexican Sacred Mushrooms to Visken

The natural active principles of the sacred mushrooms (i.e., psilocybin and psilocin) did not find application in the doctor's office, but they led to the origin of an important medicament. My coworker in the synthesis of psilocybin, Franz Troxler, became engaged in a project designed to develop new substances with an inhibiting effect on adrenergic β-receptors. This type of agent is employed therapeutically to regulate cardiac function. It had been found that β-receptor inhibiting substances are obtained if the isopropylamino-2-hydroxypropyl radical is connected with an aromatic system containing a phenol hydroxyl group by an etherlike linkage. Among the many phenols used in this research program was also 4-hydroxyindole, the phenolic nucleus of psilocybin and psilocin. This very rare substance was available in our laboratories only because it had been synthesized in a larger quantity for the synthetic preparation of the mushroom alkaloids. In no other laboratory of the world would it have been found.

The combination of 4-hydroxyindole with the isopropylamino-2-hydroxypropyl radical turned out to be a direct hit. Under the proprietary name Visken, this new drug has acquired a leading position among the β-receptor inhibitors, especially for the treatment of hypertension (Hofmann 1979). Without our investigations on the hallucinogenic Mexican mushrooms, which were based on discoveries and information from ethnobotanists, 4-hydroxyindole would not have been available and, consequently, Visken would not have been developed.

Ololiuqui, Sacred Drug of the Aztecs and the Eleusinian Mysteries

Ololiuqui is the Aztec name for the seeds of certain convolvulaceous plants that have been used since pre-Hispanic times by the Aztecs and related tribes, just as the sacred mushrooms and the cactus peyote have been used in their religious ceremonies and for magic healing purposes. Ololiuqui is still used by certain tribes, such as the Zapotecs, Chinantecs, Mazatecs, and Mixtecs, who live in the remote mountains of southern Mexico in comparative isolation, little or hardly influenced by Christianity.

The chemical investigation of this ancient sacred drug, which led to interesting results, was stimulated by ethnobotanists Schultes and Wasson, who already had provided me with fundamental information on the hallucinogenic Mexican mushrooms. An excellent review of the historical, botanical, and ethnological aspects of ololiuqui was published in 1941 by Schultes in his monograph "A Contribution to Our Knowledge of *Rivea corymbosa*: The Narcotic Ololiuqui of the Aztecs." Today *R. corymbosa* (L.) Hall. f. is known as *Turbina corymbosa* (L.) Raf. (Schultes and Hofmann 1980). The only report on chemical investigations with seeds of *T. corymbosa* mentioned in Schultes' review is that of Stockholm pharmacologist Santesson in 1937, who was, however, unsuccessful in isolating definite crystalline compounds.

In 1955, Canadian psychiatrist Humphrey Osmond conducted a series of experiments on himself. After taking sixty to one hundred ololiuqui seeds, he passed into a state of apathy and listlessness accompanied by increased visual sensitivity. After about four hours, he had a relaxed feeling of well-being that persisted for an extended period (Osmond 1955). In contrast to these results, Kinross-Wright in 1959 published experiments performed on eight male volunteers who had taken doses of up to 125 seeds without any noticeable effects.

Our investigations with authentic ololiuqui, which I received from Wasson, proved that Osmond was right and that Kinross-Wright probably had not experimented with true ololiuqui. Wasson sent me two different samples of ololiuqui seeds, collected by a Zapotec Indian near Oaxaca in southern Mexico. One sample consisted of brown seeds, which proved on botanical identification to be *Turbina corymbosa*. The second sample, black seeds, was identical with *Ipomoea violacea* L. (syn. *I. tricolor* Cav.). These black seeds, called *badoh negro*, are used particularly by the Zapotecs in conjunction with or instead of *badoh*, the brown seeds of *T. corymbosa* (MacDougall 1960).

Table 1. Percentage of alkaloids in the seeds of *Turbina corymbosa* and *Ipomoea violacea*.

Alkaloids	*Turbina corymbosa* (ololiuqui, badoh)	*Ipomoea violacea* (badoh negro)
d-Lysergic acid amide (ergine)	0.0065	0.035
d-Isolysergic acid amide (isoergine)	0.0020	0.005
Chanoclavine	0.0005	0.005
Elymoclavine	0.0005	0.005
Lysergol	0.0005	—
Ergometrine	—	0.005
Total alkaloid content	0.012	0.06

Chemical investigation revealed that the hallucinogenic principles of ololiuqui are ergot alkaloids (Hofmann and Tscherter 1960; Hofmann 1961, 1963). *Badoh* and *badoh negro* contain an alkaloid mixture of nearly the same composition (Table 1). The main component in both seeds is d-lysergic acid amide, also called ergine. In later investigations, it was found that ergine and isoergine were present in the seeds to some extent in the form of their condensation product with acetaldehyde (i.e., d-lysergic acid hydroxy-ethylamide and d-isolysergic acid hydroxyethylamide, respectively). The latter compounds are easily hydrolyzed in the course of the extraction process to provide ergine (respectively, isoergine and acetaldehyde) (Figure 4).

Ergine $R = NH_2$ Isoergine

d-Lysergic acid $R = N \begin{matrix} H \\ CHOHCH_3 \end{matrix}$ d-Isolysergic acid
hydroxyethylamide   hydroxyethylamide

Figure 4. Main alkaloids of *Turbina corymbosa* and *Ipomoea violacea* (ololiuqui) and of ergot from *Paspalum distichum* (kykeon of the Eleusinian Mysteries).

The hallucinogenic activity, on which the ceremonial use of ololiuqui by the Indians is based, is produced mainly by ergine, isoergine, and their hydroxyethyl derivatives. The hallucinogenic effects of these ololiuqui alkaloids are similar to those of the well-known LSD (d-lysergic acid diethylamide). The very close relationship in chemical structure between LSD and the ololiuqui alkaloids accounts for it. LSD, a synthetic compound, can be regarded as a chemical modification of an ancient, magic Mexican drug. The medium oral dose in humans of the ololiuqui constituents is 2 milligrams, whereas LSD is active in 0.05 to 0.1 milligram (Hofmann 1963).

It is a finding of extraordinary interest that the ololiuqui alkaloids ergine, isoergine, and their hydroxyethyl derivatives are also the main active constituents of an ergot species growing on the wild grass *Paspalum distichum* L., produced by the fungus *Claviceps paspali* Stev. et Hall. Paspalum grass is widely distributed in the Mediterranean basin (Arcamone et al. 1960).

From these facts and from studies of ancient Greek texts, the hypothesis was put forth that the hallucinogenic ingredient in Kykeon, the holy potion presented to the adepts in the Mysteries of Eleusis, was ergot of *Paspalum* (Wasson et al. 1978). The Eleusinian Mysteries were founded by Demeter, the grain goddess. Thus, ethnobotanists suggested to medicinal chemists an investigation of the active principles of ololiuqui that led to the discovery of the same alkaloids found also in ergot growing around Eleusis. This discovery allowed scientists to build a bridge between an ancient, still existing cult of Mexican Indians and the most important mysteries of Greek antiquity.

Summary

These examples show the incredible ramifications of ethnobotanical data into apparently unrelated fields of science, data gathered from the knowledge of plant properties possessed in primitive societies—knowledge that in many regions is fast disappearing.

LITERATURE CITED

Arcamone, F., C. Bonino, E. B. Chain, A. Ferretti, P. Pennella, A. Tonolo, and L. Vero. 1960. Production of lysergic acid derivatives by a strain of *Claviceps paspali* Stevens and Hall in submerged culture. *Nature* 187: 238–239.

Cerletti, A. 1959. In *Neuro-psychopharmacology*. Eds. P. B. Bradley, P. Deniker, and C. Radouco-Thomas. Amsterdam: Elsevier. 291.

Gnirrs, F. 1959. Schweiz. *Arch. Neurol. Neurochi. Psychiat.* 84: 346.

Heim, R., and R. G. Wasson. 1958. *Les Champignons Hallucinogènes du Mexique*. Paris: Editions du Musée National d'Histoire Naturelle.

Hofmann, A. 1961. Der Wirkstoffe der mexikanischen Zauberdroge "ololiuqui." *Planta Medica* 9: 354–367.

———. 1963. The active principles of the seeds of *Rivea corymbosa* and *Ipomoea violacea*. *Botanical Museum Leaflets* (Harvard University) 20: 194–212.

———. 1979. Planned research and chance discovery. *International SANDOZ Gazette* N3 (July) 23: 3–10.

Hofmann, A., R. Heim, A. Brack, and H. Kobel. 1958. Psilocybin, ein psychotroper Wirkstoff aus dem mexikanischen Rauschpilz *Psilocybe mexicana* Heim. *Experientia* (Basel) 14: 107–113.

Hofmann, A., R. Heim, A. Brack, H. Kobel, A. Frey, H. Ott, T. Petrzilka, and F. Troxler. 1959. Psilocybin und Psilocin, zwei psychotrope Wirkstoffe aus mexikanischen Rasuchpilzen. *Helvetica Chimica Acta* 42: 1557–1572.

Hofmann, A., and H. Tscherter. 1960. Isolierung von Lysergsäure-Alkaloiden aus der mexikanischen Zauberdroge Ololiuqui (*Rivea corymbosa* [L.] Hall.f.). *Experientia* (Basel) 16: 414–416.

Johnson, J. B. 1939. *Elements of Mazatec Witchcraft*. Ethnol. studies (Gothenburg), no. 2. 128–150.

Kinross-Wright, V. J. 1959. Research on ololiuqui: The Aztec drug. In *Neuro-Psychopharmacology*. Eds. P. B. Bradley, P. Deniker, and C. Radouco-Thomas. Amsterdam: Elsevier. 452–456.

Leuner, H. 1962. *Die experimentelle Psychose*. Berlin: Springer-Verlag.

MacDougall, T. 1960. *Ipomoea tricolor*, a hallucinogenic plant of the Zapotecs. *Boletín Centro Inv. Antrop. Mex.*, no. 6: 6–8.

Osmond, H. 1955. Ololiuqui: The ancient Aztex narcotic. Remarks on the effects of *Rivea corymbosa* (ololiuqui). *J. Ment. Sci.* 101: 526–537.

Rümmele, W. 1959. Schweiz. *Arch. Neurol. Neurochi. Psychiat.* 84: 348.

Schultes, R. E. 1939. The identification of teonanacatl, a narcotic Basidiomycete of the Aztecs. *Botanical Museum Leaflets* (Harvard University) 1: 37–54.

———. 1940. Teonanacatl, the narcotic mushroom of the Aztecs. *American Anthropologist* 42: 429–443.

———. 1941. A contribution to our knowledge of *Rivea corymbosa*, the narcotic ololiuqui of the Aztecs. *Botanical Museum Leaflets* (Harvard University).

Schultes, R. E., and A. Hofmann. 1980. *The Botany and Chemistry of Hallucinogens*. 2nd ed. Springfield, IL: Charles C. Thomas.

Wasson, R. G., A. Hofmann, and C. A. P. Ruck. 1978. *The Road to Eleusis: Unveiling the Secret of the Mysteries*. With a new translation of the *Homeric Hymn to Demeter* by D. Staples. New York: Harcourt Brace Jovanovich.

Wasson, V. P., and R. G. Wasson. 1957. *Mushrooms, Russia and History*. New York: Pantheon Books. 2: 215–322.

Weidmann, H., and A. Cerletti. 1959. Zur pharmakodynamischen Differenzierung der 4-Oxyindolderivate Psilocybin und Psilocin im Vergleich mit 5-Oxyindolköroerb (Serotonin, Bufotenin). *Helvetica Physiol. Acta* 17, C 46–48.

Weidmann, H., M. Taeschler, and H. Konzett. 1958. Zur Pharmakologie von Psilocybin, einem Wirkstoff aus *Psilocybe mexicana* Heim. *Experientia* (Basel) 14: 378–379.

Historical Perspective and Future of Ethnopharmacology

BO R. HOLMSTEDT

Discovery by Luck

On David Livingstone's expedition to the Zambezi between 1858 and 1864 he was accompanied by Dr. John Kirk and, among other things, they collected from the natives the arrow poison from *Strophanthus kombe*. Livingstone writes as follows:

> The poison used here, and called *kombi* [sic], is obtained from a species of *Strophanthus*, and is very virulent. Dr. Kirk found by an accidental experiment on himself that it acts by lowering the pulse. In using his tooth-brush which had been in a pocket containing a little of the poison, he noticed a bitter taste, but attributed it to his having sometimes used the handle in taking quinine. Though the quantity was small, it immediately showed its power by lowering his pulse which at the time had been raised by a cold, and next day he was perfectly restored. Not much can be inferred from a single case of this kind, but it is possible that the *kombi* may turn out a valuable remedy; and as Professor Sharpey has conducted a series of experiments with this substance, we look with interest for the results (D. Livingstone and C. Livingstone, 1865, pp. 466–467).

This is a perfect case of serendipity, but mind that the two explorers brought the drug back for serious pharmacological testing. They established a connection between the use of an arrow poison by an ethnic group in Africa and a person in a laboratory who studied its pharmacodynamics.

Early Studies

Ethnopharmacology can be defined as: "... the interdisciplinary scientific exploration of biologically active agents traditionally employed or observed by man. ..."

During the time of exploration and colonialism, pharmacologists in Europe had access to a multitude of crude drugs. In fact, one of the best surveys of these drugs was written during this time by Hartwich (1911).

Interest in traditional drugs is thus not new but has been spurred in recent years by methodological advances in phytochemistry, a growing number of ethnobotanical studies, and an upsurge of interest in renewable resources and traditional medicine (Schultes and Raffauf 1990). As a subject it is comparatively new although, of course, collaborative studies have taken place in the past. It constitutes mostly studies made on plants and plant products and, to a minor degree, products from animals.

The really basic standardization of modern botany can be said to have started with the work of the Swedish naturalist-physician, Carl von Linné or, as he is better known by his Latinized name, Carolus Linnaeus. The starting point of modern botanical nomenclature is taken as 1753, when he published his book *Species Plantarum*, in which he gave Latin binomials to some 10,000 species of plants. A professor at the University of Uppsala, Linné had a large number of students, many of whom presented an academic dissertation: he supervised no fewer than 509 students to whom are attributed 186 dissertations (Strandell 1982). These many dissertations have been preserved and copies are available. It is believed that Linné himself wrote wholly or in part some of his pupils' theses (Schultes and Holmstedt 1989).

Although Linnaeus, except for his native country, only traveled in Holland, England, France and Germany, he sent his pupils to study and collect all over the world. Two of his pupils, Daniel Solander and Anders Sparrman, sailed on different explorations with Captain Cook. C. P. Thunberg wrote *Flora Japonica* and Pehr Löffling was the first modern botanist in South America. In not a few cases, the scientists did not return from their travels but perished abroad. However, usually their collections reached Linnaeus. One example of this is the case of Peter Forsskål (1732–1763), who died in what is now called Yemen.

The First Successful Multidisciplinary Attack on an Ethnopharmacological Problem

A French scientific expedition starting in 1800 under the leadership of Captain Baudin went to the Southseas and what was then called New Holland (Australia). The chief naturalist on this expedition was Leschenault de la Tour.

Louis Théodore Leschenault de la Tour (1773–1826) was born in Chàlon-sur-Saône, France. He was a botanist and naturalist and took part in Capt. Baudin's last exploring expedition. In June 1803, he fell ill on Timor and stayed behind when the expedition left that island; after recovery he went to Java, making botanical investigations on that and other islands, with the support of Nicolaus Engelhard, then Governor of the northeast coast of Java. In 1806, he sailed for America, returning to France in July 1807. In 1816, he came to Pondicherry as *"Inquisitor rerum naturae"* and in the next years made extensive travels in British India, Bengal, Ceylon, etc., returning to Nantes towards the end of May 1822. Finally, he made a voyage to Brazil, Guiana and Surinam in 1823–1824. He is commemorated in several Malaysian plant species and in the genus *Leschenaultia* R.Br. (Jeandet 1883).

Leschenault was a typical field worker and an ethnobotanist. He traveled widely on the islands and became curious about one of the poisons (*Upas tieuté*) with which the natives tipped their arrows. During his many years in Java, where he had learned the local language, he made friends with one of the natives and persuaded him to show him how to prepare this poison. The essential ingredient proved to be the root of a plant. Scrapings of the bark of the root were put into water and boiled with some admixtures, the liquor was decanted off, and the final mixture was boiled down until it was of the consistency of thick molasses. Sharp slivers of bamboo were dipped in the mixture and shot into a hen. The hen died in two minutes (Bisset 1966, 1985, 1989).

Leschenault brought back to Paris samples of this gummy substance. The botanist A. L. de Jussieu (1748–1836) identified it as belonging to the genus *Strychnos* and pointed out its relation to the St. Ignatius bean and particularly to *Nux vomica*, the extract of which had been known since 1683 to produce vomiting and convulsions (Leschenault de la Tour 1810).

Leschenault then gave the poison to two young and aspiring medical experimenters, François Magendie (1783–1855) and Raffeneau-Delile (Olmsted 1944).

CHRONOLOGY STRYCHNINE

1805   Leschenault describes the preparation of the Javanese dart poison *Upas Tieuté*. Published only 1811.

1809   Magendie and Delile publish experiments on mechanism of action of the poison.

1819   Pelletier and Caventou isolate strychnine from other sources. Magendie uses strychnine in clinical medicine.

1824   Pelletier and Caventou isolate strychnine from *Upas Tieuté*.

1963   Total synthesis of strychnine by Woodward et al.

Alire Raffeneau-Delile (1778–1850) lived a varied life. Son of a court official, who countered difficulties during the revolution, he embarked upon medical studies which, however, were interrupted by his participation in Napoleon's Egyptian campaign. At the age of 20, he became one of the "learned soldiers" whom Napoleon took with him. During his time in Egypt, Raffeneau-Delile was instrumental in founding the Botanical Garden in Cairo—his interest in botany becoming apparent very early. Having returned to France, Raffeneau-Delile was sent as a consul by Napoleon to Wilmington in North Carolina, where he remained for some years fulfilling his task, it is said, with both harmony and good nature. President Jefferson was impressed by Raffeneau-Delile. The latter could not resist studying the natural flora of the region and sent back specimens to France. Some of these specimens were grown in the garden of Malmaison. However, he interrupted his stay in North Carolina in 1806 and went to New York where he became the disciple of one Dr. Hosak and where he continued his medical activities and in 1807 defended a thesis on pulmonary consumption, which he passed with honours. He had plans to settle in New Orleans but was called back by Napoleon. In Paris, he continued his medical activities and came to know Magendie, working with him for some years before taking up the position as head of the Botanical Gardens in Montpellier (Joly 1839; Pascallet 1847; Perraud de Thoury 1853).

Magendie and Raffeneau-Delile proceeded to administer the poison they had obtained (by wounding animals with homemade arrows) to hens, rabbits, dogs, and a horse. The symptoms were characteristic. For a few moments, there appeared to be no effect; then, suddenly, the animal was thrown into a violent convulsion, all its muscles contracting vigorously. A period of calm followed this seizure, then there was another period of convulsion, and the chest muscles were now held so rigidly tight that breathing was suspended, and the animal passed into a state of asphyxia and ultimately died. Death occurred in five minutes and at autopsy the blood was black as if the animal had died from asphyxiation. The experimenters concluded that there was little effect on the brain, the chief action being on the spinal cord. When the cord was severed from the brain there were still convulsions of the limbs on administration of the drug, but if the spinal cord had been destroyed there were no convulsions. This is the first time that the action of a poison has been shown to act upon a specific organ, a landmark in pharmacodynamics. It also led to Magendie's work on the absorption and distribution of poisons and other substances and, most importantly, to the discovery that the dorsal roots are sensory and that the ventral roots motor, "the law of Magendie" (Grmek 1973).

Today, we recognize that the symptoms described by Magendie and Raffeneau-Delile are typical of strychnine poisoning. It was not, however, until about a decade later that the chemical substance responsible for these symptoms, the alkaloid strychnine, was isolated by Pelletier (1788–1842) and Caventou (1795–1877). In 1819, they isolated strychnine from *Nux vomica* (Pelletier and Caventou 1819); and, in 1824, from *Upas tieuté* (Pelletier 1824; Pelletier and Caventou 1824). Thus, we have here for the first time an inter-disciplinary study starting with fieldwork among natives on botany, pharmacology, and up to the isolation of the active principle. Later on, this was also followed by its introduction into clinical medicine. In 1809, Raffeneau-Delile presented his second thesis which set down the action of *Upas tieuté* and several species of *Strychnos* (Raffeneau-Delile 1809).

In their 1819 paper, Pelletier and Caventou had repeated the animal experiment, and Magendie added a comment to this work (Magendie and Delille 1809). He suggested that one should call the compound isolated tetanine instead of strychnine, by analogy with morphine and emetine. In 1819, he also made an experiment which should certainly not have been carried out. I quote:

> I have administered a quarter of a grain of strychnine to a sick person, 67 years of age. He was suffering from muscular weakness following a disease of the central nervous system.
>
> I perceived a state of unquestionable tetanic convulsions. After eight days of this treatment I obtained a remarkable improvement in his muscular strength.

Thus strychnine was introduced in clinical medicine, where it has been used for almost everything (e.g., cholera, epilepsy, tuberculosis and "to strengthen the pelvic organs of young girls entering the stage of puberty").

Strychnine has an undeserved reputation as a useful therapeutic agent. To the drug have been ascribed properties that it does not possess or that it exhibits only when administered in toxic doses. There is no rational basis for the use of strychnine in therapy and, consequently, no justification for the presence of strychnine in any proprietary medication.

Strychnine has been the subject of many experimental investigations and is indeed an important experimental tool in pharmacodynamics. The convulsant action of the drug is due to interference with postsynaptic inhibition that is mediated by glycine. Glycine is an important inhibitory transmitter to motoneurons and interneurons in the spinal cord, and strychnine acts as a selective, competitive antagonist to block the inhibitory effects of glycine at all glycine receptors (Kuno and Weakly 1972).

From Cat to Cathinone

Peter Forsskål (1732–1763) was one of Linnaeus' most intelligent pupils. He was born in Finland (Finland then being part of Sweden) and was politically active during his university years in Uppsala. He had also studied in Göttingen. His political activities were contributing factors to his taking part in a Danish expedition to Arabia Felix, now Yemen. Only one man—Carsten Nieburg—returned from this expedition, but the collections arrived later, together with the manuscripts. Carsten Nieburg eventually published in 1775 Forsskål's *Flora Egyptica-Arabica* in which Forsskål characterized the *cat* plant which he called *Catha edulis*. The voucher specimens are still available in the Copenhagen herbarium.

CHRONOLOGY *CAT*

1763   P. Forskål describes *Catha Edulis*. Publication only in 1775.

1856   Burton gives a detailed scholarly account of cat and its use.

1880   Botta gives description of effect of cat.

1900   Beitter isolates alkaloids from *Catha*.

1930   Wolfe et al. isolate *d*-norpseudoephedrine from *Catha*.

1941   Von Brücke deduces from pharmacological experiments that the chemistry of *Catha* is complex and not just *d*-norpseudoephedrine.

1961   Alles et al. imply other components.

1964   WHO warns for craving and psychic dependence.

1980   Szendrei et al. identify cathinone as the main pharmacologically active component.

Inhabitants of parts of East Africa, Aden, and Yemen have, from very early times, employed the fresh leaves and stem tips of a bush or small tree *Catha edulis* Forsk. (family Celastraceae) as a stimulating drug. These succulent plant parts are either chewed and then swallowed or, less commonly, steeped in water to prepare a tea, or mixed with honey to form an edible paste. Variously known as tschat, cat, gat, or Abyssinian or Arabian tea, the plant has been used in the Ethiopian highlands around Harrar for hundreds of years, and its cultivation there antedates that of the coffee plant (Tyler 1966). Detailed accounts of the history and pharmacology of cat have recently been published (Getahun and Krikorian 1973, I and II; Kalix 1990).

Mastication of the leaves results in varying degrees of exhilaration and stimulation. Cat chewers may talk to each other all night without ceasing, eating little and drinking excessively. Used in moderation no ill effects are noted, but excessive use results in subsequent depression. Habitual and immoderate use is said to cause the cat eater to become withdrawn from reality and to undergo a deterioration of character.

Authorities are not in complete agreement as to the deleterious nature of cat chewing. Some governments consider it a narcotic. Regardless, cat chewing is a theologically accepted and lawful custom in Arabian and African countries at the present time, being employed by many to alleviate the sensations of hunger and fatigue.

The numerous studies carried out with this plant have been described (Krikorian 1984). It is appropriate here to deal only with the long-lasting efforts to characterize the active components in this plant responsible for its action on the central nervous system.

Beitter in 1900 performed the first comprehensive chemical investigations on cat. In addition to attempting a quantitative investigation of the basic fractions, he noted the presence of an abundance of tannins. Beitter also found that the alkaloid content was chiefly in the leaves and bark of young cat twigs and branches.

Early chemical studies of *Catha edulis* leaves yielded several bases or mixtures of bases which remained uncharacterized. Finally, Wolfe et al., in 1930, succeeded in establishing that what was called cathine was actually *d*-norpseudoephedrine. Whether this compound alone might account for the use of cat remained in doubt. In 1941 von Brücke, based upon pharmacological experiments, concluded that *d*-norpseudoephedrine alone could not be the only active component in cat extracts.

This belief lasted up to very recent years, until the true composition of cat was discovered; and this most certainly had to do with the modern techniques of separation and identification.

In an analytical HPLC-method for separation and quantitative analysis of the CNS-active components, cathinone (CA), (+)-norpseudoephedrine (NPE) and (–)-norphedrine (NE) (Fig. 1) was isolated (Schorno et al. 1982).

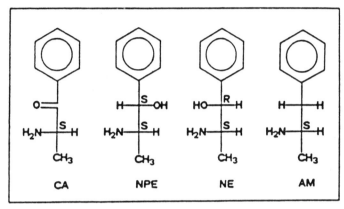

CA = (-)-Cathinone
NE = (-)-Norephedrin
NPE = (+)-Norpseudoephedrine
AM = (+)-Amphetamine

Figure 1. Structural formulas and configuration of compounds contained in *Catha*.

By now, we also know something about the metabolism of these compounds. S-(–)-Cathinone (S-(–)α-aminopropiophenone) is the major active principle of cat leaves. After oral administration of synthesized cathinone isomers (racemate), 22–52 percent was recovered in 24-hr urine samples, mainly as aminoalcohol metabolites. With a gas chromatograph/mass spectrometer (GC/MS), HPLC, and CD, the main metabolite of S-(–)-cathinone was identified as R/S-(–)-norephedrine and the main metabolite of R-(+)-cathinone as R/R-(–)-norpseudoephedrine. Both aminoalcohols are formed by a stereospecific keto reduction (Brenneisen et al. 1986).

The Rediscoveries of Ephedrine

Ephedra is the plant from which the alkaloid ephedrine was first isolated and characterized by the Japanese N. M. Nagai, who also arranged a preliminary study of its pharmacological action.

Nagayoshi Nagai (1844–1929) studied medicine at the Dutch Medical Academy in Nagasaki and completed these studies in Tokyo. Shortly after the abolition (in 1869) of the refusal to allow foreigners to visit Japan, young Japanese scientists were also sent to Europe in order to get acquainted with cultural advances made in western countries; Nagai was one of a group that in 1871 came to Germany. He intended to start further medical work in Berlin, but after hearing the lectures of A. W. Hofmann, professor of chemistry and well known as a pioneer with regard to organic dye substances, he decided to devote himself to chemistry. He stayed in Berlin for thirteen years, five of them as the personal assistant of Hofmann. He worked with great energy, publishing papers on eugenol and on ferulic acid (from asafoetida) and took an active part in several important investigations by Hofmann (such as the discovery of aminophenylmercaptan, and papers on primary aliphatic amines, and on piperidine). After his return to Japan (1883), he was attached to the University of Tokyo as professor of chemistry, supervising education in that discipline and in pharmacy. He retired at the age of seventy-three. He was also the leader of a school for studying German, which he considered necessary in those days for students of chemistry, medicine or pharmacy. Nagai took a great part in the development of chemistry and pharmacy in his fatherland. He was especially interested in the drugs of China and Japan and their active constituents. To this work belong his contributions to our early knowledge of ephedrine, which he discovered in 1885 and about which he reported briefly as follows:

> As far as these investigations are concerned I would like to mention only one of them. I analyzed a drug which for thousands of years has been appreciated as an antipyretic, and found that it contained ephedrine. I elucidated the structure of this substance and was able to confirm it by synthesis.

$$C_{10}H_{15}NO = \begin{array}{c} C_6H_5 \bullet CH(OH) \bullet CH \bullet CH_3 \\ | \\ NH \bullet CH_3 \end{array}$$

The synthesis yields an optically inactive mixture which consists of two racemic substances with melting points of 40° and 170°, respectively. I was able to resolve the mixture into two dextro- and two laevo-ephedrines by using tartaric acid and to obtain, thus, six different isomers. I chose the name isoephedrine for the pair with the higher melting point. A friend of mine determined its physiological activity and found that it was comparable to that of atropine; the action lasted only for two hours and the substance is, therefore, particularly useful in investigations on the eye. Recent findings show that ephedrine has actions similar to those of adrenaline and can be recommended for the treatment of asthma.

CHRONOLOGY EPHEDRINE

500	First edition of Chinese dispensatory *Pen Ts'ao*.
1885–1887	Nagai isolates and synthesizes ephedrine and resolves optical antipodes. It was found to dilate the pupils.
1920	Späth synthesizes ephedrine.
1923–1924	Chen and Schmidt isolate ephedrine from *ma huang* and study its pharmacology.
1927	Alles synthesizes amphetamine.
1975	Majno draws attention to the description of ephedra in Plinius.

The identity and formula of the alkaloid were confirmed by the German firm E. Merck. The early experiments on its pharmacology were performed by K. Miura, who studied some of its toxic effects and also observed that, in a rather high concentration applied locally, it caused in man moderate dilatation of the pupil without affecting accommodation. Then ephedrine was forgotten for many decades.

Ephedrine as *ma huang* had been used in China for over 5000 years before being introduced into Western medicine in 1924. In fact, the introduction of ephedrine from China is a classic episode in the recent history of medicine. Between 1922 and 1924, two young American physicians, Carl Frederic Schmidt and K. K. Chen, worked at the Peking Union Medical College, testing some of the most popular Chinese traditional herbs in the hope of discovering some new active principle. Of the five drugs selected out of nearly two thousand, one gave significant results: an extract of *Ephedra*, when injected intravenously in dogs, caused a spectacular increase in blood pressure. Thus began the modern career of ephedrine. (Chen 1974, 1981). It was again synthesized in 1927.

Schmidt and Chen found out that ephedrine has effects closely resembling those of adrenaline, but with two advantages: it can be given by mouth and its effects, though weaker, last longer. Its classical uses are for asthma, allergic cough, and hemorrhage. It also has a stimulating action on the CNS and is known to have caused dependence. As Schmidt and his Chinese collaborator were wondering what to call their newly discovered substance, they found out that a crystalline substance extracted from *ma huang* had already been prepared in 1885; it had been called ephedrine in 1887, and even synthesized by Nagai.

The story of ephedrine is loaded with irony. In 1975 a pathologist, Guido Majno, wrote a book covering the world history of wound healing. When sifting through the works of Pliny, he found that the author talks about a plant that some call *ephedron*:

> The Greeks hold various views about this plant . . . assuring us that so wonderful is its nature, its mere touch staunches a patient's bleeding ... its juice kept in the nostrils checks hemorrhage . . . and taken in sweet wine it cures cough.
>
> One cannot help but be startled by the association of a plant named "ephedron" with stopping hemorrhage and curing cough: for these are the two main effects of a powerful drug called *ephedrine*. A surgical incision in skin injected with ephedrine is almost bloodless; and a spell of asthmatic cough can be relieved almost as if by miracle. Coincidence? The data are repeated elsewhere in Pliny with a description of the plant: "*Ephedra* ... has no leaves, but numerous rush-like, jointed tufts.... For coughs, asthma and colic it is given pounded in dark-red, dry wine. (Majno 1975).

This is an accurate description of both the plant *Ephedra* and its product, ephedrine, the precious drug supposedly discovered by the Chinese alone.

Majno speculates about the possibility that Pliny somehow inherited the notion of *Ephedra* from China. It is extremely unlikely. His immediate source could have been Dioscorides, who has essentially the same information. There is no record of anyone from China going to Rome in antiquity. The Romans knew little about China, except that there were "people called Seres famous for the woolen substance obtained from their forests." If the use of ephedrine has come from China, its use should have increased in Europe as contacts developed; but in fact, it was completely forgotten. In ancient India, Sushruta and Charaka do not mention it. There are many *Ephedrae* in Europe, including an *Ephedra helvetica*; among the Italian varieties, some from Sardinia were studied even in 1940 for possible industrial exploitation. So we are probably dealing with an independent Mediterranean discovery, not related to *ma huang*.

Further development in this field is due largely to Gordon Alles.

Gordon Alles (1901–1963) was an eminent chemist and pharmacologist holding faculty appointments at the University of California and at California Tech. in Pasadena. He spent most of his life in California but had previously taken part in the isolation and properties of insulin with John Abel at the Johns Hopkins University, and in the work on the pernicious anaemia principle with George Minot at Harvard.

Alles had an interest in centrally acting compounds such as anticonvulsants, trypta-mines, and compounds related to cannabinol. It is less well known that he also had an interest in ethnopharmacology and did field work. This interest in central stimulants led him to many parts of the world—to Ethiopia to study the use of cat, to Mexico to learn about peyote and other cacti, and to Tahiti and Fiji to see why the kava, taboo in the former, is used so freely in the latter. He was a consultant to the World Health Organization and to many national, state, and local groups on problems of habituation or addiction to many compounds.

Alles had published on cat and peyote and was well aware of the work of K. K. Chen, who had introduced ephedrine as an orally effective compound for the relief of asthma. The by then diminishing supplies of natural plant sources from China and the absence of ephedrine in these plants when they were grown in the deserts of the United States led to the work which resulted in Alles' report on phenylethanolamine, used by his clinical colleagues, George Piness and Hyman Miller, as a possible compound for topical application to shrink nasal mucous membranes. This led him to work on other phenylethanol amines. In 1962, he sent the present author a letter about the development of psychoactive properties of benzedrine, from which I quote as follows:

> There is no simple first report on the psychoactive properties of benzedrine, since the development evolved in unorthodox ways.
>
> In the fall of 1928, I returned from Harvard Medical School to this problem and set out to find why *phenylamineoethanol had a shorter duration of action than ephedrine— and was not active when administered orally*. The structural comparisons were worked out with compounds not having an hydroxyl group on the side chain in order to minimize the complications of stereoisomeric compounds. I soon found phenylisopropylamine had the duration of action I was looking for but was not notably active as a bronchodilator on perfused guinea pig lungs, so I set the compound aside and worked on syntheses of ring substituted hydroxy derivatives.
>
> *In June 1929 I tried the oral activity of phenylisopropylamine hydrochloride* out by swallowing 50 mg, since I was at that time well calibrated to the effects of 50 mg ephedrine hydrochloride and had noted the lack of activity of 50 mg each of phenylamino-ethanol sulfate and phenethylamine and phenethyl methylamine hydrochlorides.
>
> *Almost needless to tell you I discovered the marked central effect of β-phenylisopropyl-amine when I took 50 mg of its salt!*

Arthur Heffter and Self-experiments with Peyote

John Raleigh Briggs (1851–1907), a Texan physician, was the first to experiment with peyote (then called mescal buttons) in 1886. Briggs wrote a short article which was the opening of an intensive era of peyote research.

In 1887 Louis Lewin (1850–1929) visited the United States; he arrived in Detroit on 16th September and wrote as follows: "My first errand, of course, was a visit to Parke-Davis in Detroit." Lewin obtained from this drug company a "peyote" sample, which he promised to investigate. After his return to Germany, he extracted the drug in various ways and obtained a basic syrupy substance which he called anhalonine. He studied this "substance," which was in fact a crude mixture of alkaloids, in animal experiments and, to his surprise, found that it was intensely poisonous. Shortly afterwards, Lewin published the first account of his chemical and pharmacological studies. He said in his summary: "It has been proven for the first time that a cactus can possess an extraordinary high toxicity."

Arthur Heffter (1860–1925) first studied chemistry and received his Ph.D. in Greifswald in 1883. After a few years as a chemist, he switched over to the study of medicine in Leipzig and received his M.D. there in 1890. In 1908, he became professor of pharmacology at the University of Berlin. He is known for his studies of arsenic and strophanthine, but most of all "peyote" (Holmstedt and Liljestrand 1981).

Heffter crystallized two alkaloids from his plant material. He characterized them in animal experiments and also tested on himself whether the new compounds could explain the effects of the crude drug. In this way he found that one of the isolated alkaloids, pellotine, was an active sedative and hypnotic.

In the meantime, Lewin had studied other problems, but when Heffter's first paper was published in 1894, Lewin rushed to vindicate his position in this field. In two papers, he discussed his own and Heffter's results, as well as botanical aspects of "peyote." This brought him into conflict with the renowned cactus botanist Karl Schumann, and Heffter's work was thereby overlooked.

In a letter to the editor, published on 6 April 1895, in the *Pharmaceuticishe Zeitung*, Heffter recognized Lewin's priority, but also showed that he, Heffter, had been first to isolate a pure alkaloid from "peyote." Lewin sent an angry reply to the editor, and this animated exchange of letters developed into a life-long grudge between the two men. Lewin never worked with "peyote" again.

The drug had now been studied in human experiments in the United States and had been found to produce hallucinations. Attempts were also made to isolate the alkaloids, but Heffter's lead was too great. In 1896, he described four new pure alkaloids—anhalonine, anhalonidine, lophophorine and mescaline. He was also very close to elucidating the structural formula of mescaline (Fig. 2).

The following year, he performed a series of self-experiments to find the active constituent of "peyote." On 5 June, he tried the crude drug to learn about its effects and confirmed the reported symptoms: color visions, pupillary dilation, loss of appreciation of

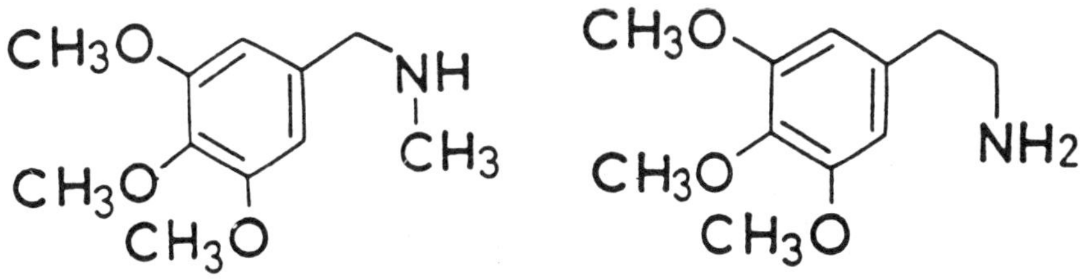

Figure 2. Structural formulas of mescaline: (*left*) as suggested by Heffter and (*right*) as confirmed by Späth.

time, nausea, and headache. He then extracted the alkaloids from the drug and ingested the extraction residue, a brown resin. This experiment was performed on 21 July and clearly established that the resin was ineffective.

Two days later, it was time to investigate the activity of the combined alkaloids. These produced the same effects as the whole drug. Heffter concluded: "The peculiar actions of peyote on the visual apparatus must, therefore, be produced by one of its alkaloids." All of the four new alkaloids were now tested separately by the thorough Heffter. Anhalonine and anhalonidine made him sleepy, lophophorine caused his face to flush and gave him an occipital headache.

On 23 November 1897, he took 150 mg of mescaline hydrochloride. In his laboratory notes, we read:

> 2:00 pm. Violet and green spots appear on the paper during readings. When the eyes are kept shut, the following visual images occur . . . carpet patterns, ribbed vaulting, etc. . . . Later on landscapes, halls, architectural scenes (e.g., pillars decorated with flowers) also appear. The images can be observed until about 5:30 pm. Nausea and dizziness are at times very distressing. . . . In the evening, well-being and appetite are undisturbed and there is no sign of sleeplessness.

Heffter concluded: "The results described above show that mescaline is exclusively responsible for the major symptoms of peyote poisoning."

Ten years had then passed since the publication of Briggs' report. A detailed description of early peyote research can be found in Bruhn and Holmstedt (1974).

Stability of Ethnobotanical and Archeological Material

The work carried out during the 19th century on ethnobotanical material was mostly done in Europe. Transportation to destinations was long and the time taken for samples to be analyzed was also long. One can only speculate on how much of the active components were thus lost. However, recent examples have shown the following: If a plant is kept dry and the cells are not broken, alkaloids can remain for thousands of years. Holmstedt and Lindgren in 1972 found caffeine in well-preserved leaves of *Ilex guayusa*. In the same archeological material, nicotine could be identified in fragments of leaves from *Nicotiana glauca* (Bruhn et al. 1976). Both materials were more than 1000 years old. Bruhn et al. (1976) again found mescaline and related alkaloids in a prehistoric specimen dated about a thousand years.

It has also been pointed out that loss of psychoactivity from stored cannabis samples does occur, and Harvey (1990) indicated that THC decomposes relatively rapidly in an ethanolic solution left in the light. Dryness and darkness favour stability of the active components.

Some components, however, may have disappeared more easily than others—as pointed out by Schultes et al. (1977) in the analyses of samples of *Anadenanthera peregrina*. In an ethnobotanical collection made by Spruce more than 100 years ago, it has been possible to identify alkaloidal material by the use of modern analytical techniques never dreamed of by this intrepid plant explorer. Of several alkaloids found in freshly collected reference material, only one remained in the Spruce collection: bufotenine (5-OH-DMT). Storage of freshly collected material over a period of 2 years resulted in the disappearance of all alkaloids except 5-OH-DMT. This may raise speculation as to whether or not they were originally contained in the Spruce material. The observation stresses the importance of storage time in addition to the knowledge of plant part, soil, season, and climatic conditions, when alkaloid analysis is carried out on seeds and on the snuffs prepared from them.

The first specimens of *Banisteriopsis* were sent home from the Amazon by Richard Spruce 125 years ago. Like so many collectors, even in modern times, Spruce was frustrated in his attempt to get things back to destination.

> I obtained a good many pieces of stem (from the type plant of *Banisteriopsis caapi*), dried them carefully, and packed them in a large box, which contained the botanical (herbarium) specimens, and dispatched them down the river for England in March 1853. The man who took that box and four others on freight, in a large new boat he had built on the Uaupés, was seized for debt when about half-way down the Rio Negro, and his boat and all its contents confiscated. My boxes were thrown aside in a hut, with only the damp earth for floor, and remained there many months, when my friend Senhor Henrique Antonij of Manáos . . . succeeded in redeeming them and getting them sent to the port of Pará. When Mr. Bentham came to open them in England, he found the contents somewhat injured by damp and mould, and the sheets of specimens near the bottom of the boxes quite ruined. The bundle of the *caapi* would presumably have quite lost its virtue from the same cause, and I do not know that it was ever analyzed chemically, but some portion of it should be in the Kew Museum at this day.

However, some of this material collected by Spruce reached the Department of Toxicology, Karolinska Institutet, Stockholm, in April 1968. It consisted of five pieces weighing in total 26.7 g. It was worked up for analysis by gas chromatography-mass spectrometry, and the yield of alkaloids was found to be 0.4 percent. A newly collected botanically verified specimen of *Banisteriopsis caapi* analyzed at the same time was found to contain 0.5 percent alkaloids. The latter material contained, as described by many authors, the main alkaloids harmine, harmaline, and tetrahydroharmine. In addition, it contained two minor components. By contrast, the alkaloid content of the Spruce material consisted exclusively of harmine. It is open to question whether the stems sent home by Spruce in 1853 contained from the beginning only harmine or, perhaps more likely, that harmaline and tetrahydroharmine have with time been transformed into the chemically more stable aromatic beta-carboline, harmine (Schultes et al. 1969).

Under any circumstances, it is remarkable that Spruce's query about the chemical analysis of the material that fared so badly on its way from the Amazonian rain forest to the Royal Botanic Gardens at Kew, has been answered by modern analytical microtechniques 115 years later.

On the whole, it is encouraging that modern analytical, sensitive techniques can elucidate the components in archeological material and material kept in the museums and herbaria; but, of course, it would be preferable to be able to analyze on the spot. However, although this is rarely possible, the two *Alpha Helix* expeditions in the Amazon are examples of the possibility of on-site analysis of fresh material.

R. V. Alpha Helix

Professor Per P. Scholander (1905–1980) of the University of California envisaged that many fundamental biological problems could best be investigated where they occur by teams of competent scientists supported by the advanced technology and equipment of a floating laboratory. The R. V. *Alpha Helix* (named to honor the helical configuration of proteins and genetic material) became the reality of this vision. This ship constitutes laboratory facilities singularly capable of supporting investigations of experimental biology and medicine in various geographical regions of the earth.

This vessel was designed and built under a grant from the National Science Foundation. It was delivered to the Scripps Institution of Oceanography in February 1966 and sailed the following month for a one-half year program at the Great Barrier Reef of Australia. Since then, the ship has been engaged in extended studies in the Amazon River basin during 1967 and again in 1977 (Figs. 3 and 4).

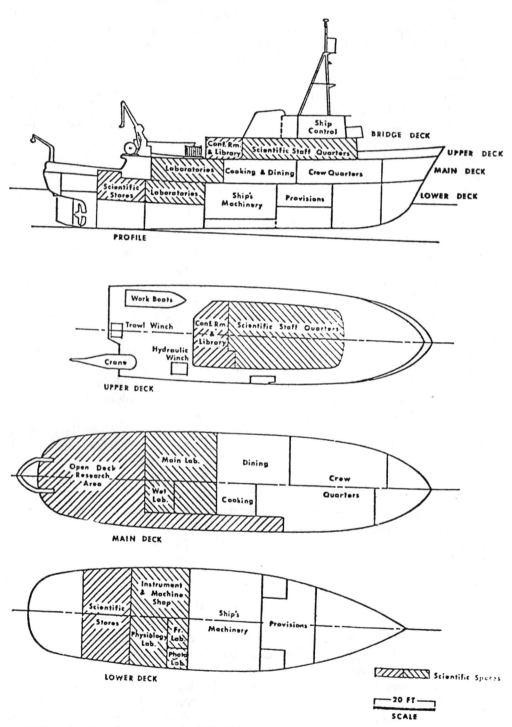

Figure 3. Functional arrangement of the R.V. *Alpha Helix*.

Figure 4. The R.V. *Alpha Helix.*

The R. V. *Alpha Helix* is a 133-ft ship built to accommodate twelve scientists. The ship has extraordinary laboratory space and is specially adapted to studies of biology, biochemistry and physiology.

Valuable data were gathered in the field during the two Amazonian *Alpha Helix* expeditions. The 1967 expedition was centered towards entomological studies in the Brazilian Amazon, but it also included one ethnobotanist and one ethnopharmacologist. By contrast, phase VII of the 1977 expedition had a scientific crew composed to study primarily the local Indian drug lore in the Peruvian Amazon. The chief scientists of this expedition were R. E. Schultes and the author.

The scientific crew of Phase VII, interdisciplinary and international in composition, was carefully chosen with a view to areas of mutual interest in their specialties. While each scientist had his own particular focus of research, the cooperative nature and interaction of research that characterized most of the problems under investigation were wide and general. While many of the research problems were resolved aboard during the period of Phase VII, some—or certain aspects of others—required further attention in the laboratory or in the literature following termination of the phase. Furthermore, in many instances, materials for additional work or wholly new research at the home institutions were collected.

During Phase VII, a productive research was conducted. Many factors contributed to the success of this expedition. Not the least of these factors was the willingness of all the scientific members to engage in cooperative and collaborative work, sometimes at the risk of sacrificing time for projects of closer personal interest. Cooperative help from the local Indians enabled the five botanists and zoologists, whose activity was mainly in the field, and some of the chemists and pharmacologists, tied necessarily to the laboratory most of the time, to work sympathetically and with understanding with the aboriginal populations—most certainly an important factor. Most important of all, however, was the fact that this phase of the *Alpha Helix* expedition was provided with a fully equipped laboratory. The presence on board of an LKB gas chromatograph-mass spectrometer (GC-MS) 2091, and an HPLC, greatly increased the amount and sophistication of the research.

The botanists collected a total of 960 plants, representing approximately 3500 specimens, mostly from the vicinity of Pebas, the Río Ampiyacu and Río Yaguasyacu in the Peruvian Amazon. Although general collecting was done, most of the time was spent in procuring material for chemical and pharmacological research and in gathering ethnobotanical data from Witoto and Bora Indians. One hundred and fifty species were submitted for chemical study. Of the results obtained only two will be mentioned here.

Experimental evidence on coca chewing gathered scientifically in the field has not previously been substantiated by measurements of blood levels of cocaine. During Phase VII, two methods of administration of coca leaves were studied, and the amount of cocaine in blood versus time was determined by an unequivocal method of analysis worked out aboard. Whole coca leaves (*Erythroxylum coca*) were obtained from Pisac, Department of Cuzco, Peru. Coca powder locally called ipadú (pulverized leaves of *E. coca* mixed with *Cecropia* leaf ash) was prepared by Witoto Indians of the Río Ampiyacu, Department of Loreto, Peru, according to the customs of the region. Coca leaves and the powdered coca-*Cecropia* preparation were taken orally by scientists on board in the same way as the Peruvian natives used the plant material. The cocaine, as measured by mass fragmentography, persisted in the plasma for more than 7 hr and reached concentrations from 10 to 150 mg/ml at 0.38–1.95 hr. Half-lives of the elimination of cocaine have been calculated ranging from 1.0 to 1.9 hr. The absorption half-lives ranged from 0.2 to 0.6 hr.

The stimulating effect obtained seems to be well correlated with the rising concentrations of cocaine in the blood. The shape of the curves fits with the subjective effects reported (Fig. 5). The differences in stimulation between using whole coca leaves or coca powder and taking cocaine by local application in the nose or by intravenous injections seems to be essentially a difference in means of administration and dosage. Consequently, there is no reason to believe that the stimulating effect achieved by the use of either coca leaves or powder is not due to cocaine (Holmstedt et al. 1979).

Another experiment, which could only have been performed in the field, concerned the coca-feeding moth (*Floria*). *Floria* is the principal lepidopterous pest that apparently lives in all its stages on leaves of the coca plant. The fate of the chemical compounds present in coca and the metabolism in the insect have never been studied. Cocaine appears to be sequestered by all stages in *Floria*. The alkaloid cocaine was detected and quantified by GC-MS, using deuterated standards in larval blood, prepupae, pupae, and adults of both sexes. Large amounts of cocaine are excreted by the larvae.

Analysis of the defensive behaviour of *Floria* larvae demonstrated that they were highly adaptive against predators. It was also proven that cocaine is an outstanding feeding deterrent for ants (Ethnopharmacological studies of the flora and fauna of the Pebas region of the Peruvian Amazon. Final report Phase VII *Alpha Helix* Amazon Expedition, 1976–77, unpublished).

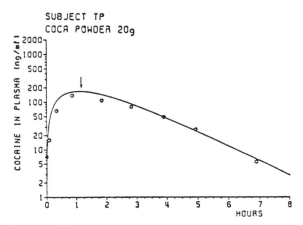

Figure 5. Blood concentrations of cocaine, as
determined by mass fragmentography, in the plasma
of a volunteer chewing 20 g of coca powder (voucher
specimen 6663A) corresponding to 48 mg of cocaine
(from Holmstedt et al. 1979). The arrow indicates time
of disappearance of the quid in the mouth. The curve
was generated by use of the parameters derived for the
pharmacokinetic model.

The two examples given above demonstrate the advantage of having a fully equipped
laboratory at hand in the field. Such undertakings are expensive but rewarding, and will
hopefully be carried out increasingly in the future.

ACKNOWLEDGMENTS

The inaugural lecture of the First International Congress on Ethnopharmacology, Strasbourg, 5–9 June 1990. Republished with adaptations from *The Journal of Ethnopharmacology* 32
(1991): 7–24, by permission of Elsevier Science Publishers BV, Academic Publishing Division.

REFERENCES

Bisset, N. G. 1966. The arrow and dart poisons of South-East Asia, with particular reference to
    the *Strychnos* species used in them. Part I. Indonesia, Borneo, Philippines, Hainan, and
    Indo-China. *Lloydia* 29: 1–18.
———. 1985. Plants as a source of isoquinoline alkaloids. In *The Chemistry and Biology of Iso-
    quinoline Alkaloids*. Eds. J. D. Phillipson, M. F. Roberts, and M. H. Zenk. Berlin/Heidelberg:
    Springer-Verlag.
———. 1989. Arrow and dart poisons. *Journal of Ethnopharmacology* 25: 1–41.
Brenneisen, R., S. Geisshusler, and X. Schorno. 1986. Metabolism of cathinone to (-)-
    norephedrine and (-)-norpseudoephedrine. *Journal of Pharmacy and Pharmacology* 38: 298–
    300.
Brücke, Franz Th. von. 1941. Über die zentral Wirkung des Alkaloides Cathin. *Naunyn-
    Schmiedeberg's Archiv fHr Experimentelle Pathologie und Pharmakologie* 198: 100–106.
Bruhn, J. G., and B. Holmstedt. 1974. Early peyote research—an interdisciplinary study. *Eco-
    nomic Botany* 28: 353–390.
Bruhn, J. G., B. Holmstedt, J.-E. Lindgren, and S. H. Wassén. 1976. The tobacco from Nino
    Korin: Identification of nicotine in a Bolivian archaeological collection. *Göteborgs Etno-
    grafiska Museum Årstryck* 1976: 45–48.
Bruhn, J. G., J.-E. Lindgren, B. Holmstedt, and J. M. Adovasio. 1978. Peyote alkaloids: Identi-
    fication in a prehistoric specimen of *Lophophora* from Coahuila, Mexico. *Science* 199:
    1437–1438.

Chen, K. K. 1974. Half a century of ephedrine. *American Journal of Chinese Medicine* 2: 359–365.
———. 1981. Two pharmacological traditions: Notes from experience. *Annual Review of Pharmacology and Toxicology* 21: 1–6.
Featherstone, R. M. 1963a. Obituary: Gordon Albert Alles 1901–1963. *Western Pharmaceutical Society, Proceedings* 6: i–v.
———. 1963b. Obituary: Gordon Albert Alles. *Pharmacologist* 5: 120–122.
Getahun, A., and A. D. Krikorian. 1973a. *Chat*: Coffee's rival from Harar, Ethiopia. I. Botany, cultivation and use. *Economic Botany* 27: 353–377.
———. 1973b. *Chat*: Coffee's rival from Harar, Ethiopia. II. Chemical composition. *Economic Botany* 27: 378–389.
Grmek, M. D. 1973. *Raisonnement expérimental et Recherches toxicologiques chez Claude Bernard.* Geneva: Librairie Droz.
Hartwich, C. 1911. *Die menschlichen Genussmittel, ihre Herkunft Verbreitung, Geschichte Anwendung, Bestandteile und Wirkung.* Leipzig: Tauchnitz.
Harvey, D. J. 1990. Stability of cannabinoids in dried samples of cannabis dating from around 1896–1905. *Journal of Ethnopharmacology* 28: 117–128.
Holmstedt, B., and H. Liljestrand, eds. 1981. *Readings in Pharmacology.* New York: Raven Press.
Holmstedt, B., and J. E. Lindgren. 1972. Alkaloid analyses of botanical material more than a thousand years old. *Etnologiska Studier* 32: 139–144.
Holmstedt, B., J. E. Lindgren, L. Rivier, and T. Plowman. 1979. Cocaine in blood of coca chewers. *Journal of Ethnopharmacology* 1: 69–78.
Holmstedt, B., and R. E. Schultes. 1989. INEBRIANTIA: An early interdisciplinary consideration of intoxicants and their effects on man. *Botanical Journal of the Linnean Society* 101: 181–198.
Jeandet, A. 1883. Notice sur la vie et les travaux de Leschenault de la Tour. *Extraits des Bulletins de la Société des Sciences Naturelles de Saône-et-Loire.* Chàlon-sur-Saône, impr. Marceau in 4°. 1–31.
Joly, N. 1839. Éloge historique d'Alyre Raffeneau Delile. *Extraits des Mémoires de l'Académie impériale des Sciences de Toulouse,* 5e Série, Tome III: 63–98.
Kalix, P. 1990. Pharmacological properties of the stimulant Khat. *Pharmac. Ther.* 48: 397–416.
Krikorian, A. D. 1984. Kat and its use: An historical perspective. *Journal of Ethnopharmacology* 12: 115–178.
Kuno, M., and J. N. Weakly. 1972. Quantal components of the inhibitory synaptic potential in spinal motoneurones of the cat. *Journal of Physiology* (London) 224: 287–303.
Leschenault de la Tour. 1810. Mémoire sur le *Strychnos tieuté* et l'*Antiaris toxicaria. Annales du Muséum d'Histoire Naturelle* (Paris). 459–467.
Livingstone, D., and C. Livingstone. 1865. *Narrative of an Expedition to the Zambesi and its Tributaries: and of the Discovery of the Lakes Shirwa and Nyass 1858–1864.* London: John Murray.
Magendie and Delile. 1809. *Nouveau Bulletin des Sciences par la Société philomatique,* 568. A. Raffeneau-Delile, *Dissertation sur les Effets d'un Poison de Java, appelé* l'Upas tieuté, *et sur la* Noix vomique, *la* Fève de St.-Ignace, *le* Strychnos potatorum, *et la* Pomme de Vontac, *qui sont du même genre des plantes que* l'Upas tieuté. Paris. (See also: *Procès-Verbaux des Séances de l'Académie des Sciences.* 1808–1811. Hendaye 1913, IV: 196, 208.)
Majno, G. 1975. *The Healing Hand—Man and Wound in the Ancient World.* Cambridge, MA: Harvard University Press.
Olmsted, J. M. D. 1944. *François Magendie.* New York: Schuman's.
Pascallet, M. E. 1847. Notice biographique sur M. Raffeneau-Delile (Alire). *Extrait de la Revue générale biographique et nécrologique.* Paris. 1–7.
Pelletier. 1824. Exam chim des *Upas. Archives générales de Médecin* (Paris) VI: 173–178.
Pelletier and Caventou. 1819. Mémoire—Sur un nouvel alcali (la strychnine) trouvé dans la *Fève de St.-Ignace,* la *Noix vomique,* etc. *Annales de Chimie et de Physique* 10: 142–177.
———. 1824. *Upas tieuté. Annales de Chimie et de Physique* (Paris) XXVI: 44.
Perraud de Thoury, M. E. 1853. *Notice Biographique sur M. Raffeneau (Delile).* Extrait du Panthéon biographique universel.
Raffeneau-Delile, A. 1809. *Dissertation sur les Effets d'un Poison de Java, appelé* l'Upas tieuté, *et sur la* Noix vomique, *la* Fève de St.-Ignace, *le* Strychnos potatorum, *et la* Pomme de Vontac, *qui sont du même genre des plantes que* l'Upas tieuté. Paris.
Schorno, X., R. Brenneisen, and E. Steinegger. 1982. Qualitative und quantitative unter-

suchungen über das vorkommen ZNS-aktiver phenylpropylamine in handelsdrogen und über deren verteilung in verschiedenen organen von *Catha edulis* Forsk. (Celastraceae). *Pharmaceutica Acta Helvetica* 57: 5–6, 168–175.

Schultes, R. E., and B. Holmstedt. 1989. Trans. of *Inebriantia*, an early consideration of intoxicants and their effects on man by a student of Linnaeus, by O. R. Arlander, 1761. *Bot. Journ. Linn. Soc.* 101: 181–198.

Schultes, R. E., B. Holmstedt, and J. E. Lindgren. 1969. Phytochemical examination of Spruce's original collection of *Banisteriopsis caapi*. De plantis toxicariis e mundo novo tropicale commentationes. III. *Botanical Museum Leaflets* (Harvard University) 22: 4, 121–132.

Schultes, R. E., B. Holmstedt, J. E. Lindgren, and L. Rivier. 1977. Phytochemical examination of Spruce's original collection of *Anadenanthera peregrina*. De plantis toxicariis e Mundo Novo tropicale commentationes. XVIII. *Botanical Museum Leaflets* (Harvard University) 25: 10, 273–287.

Schultes, R. E., and R. F. Raffauf. 1990. *The Healing Forest: Medicinal and Toxic Plants of the Northwest Amazonia*. Portland, OR: Dioscorides Press.

Strandell, B. 1982. Linnés lärjungar. Svenska Linnésällskapets Årsskrift (1982): 105–143.

Tyler, V. E., Jr. 1966. The physiological properties and chemical constituents of some habit-forming plants. *Lloydia* 29: 4, 275–292.

Woodward, R. B., M. P. Cava, W. D. Ollis, A. Hunger, H. U. Daeniker, and K. Schenker. 1963. The total synthesis of strychnine. *Tetrahedron* 19: 247–288.

Ethnopharmacology—A Challenge

BO R. HOLMSTEDT AND JAN G. BRUHN

Scientists like Rudolf Buchheim, Thomas Fraser and Rudolf Kobert devoted a great part of their active life to the study of the pharmacological effects of natural products. During the time of exploration and colonialism, they had access to a multitude of crude drugs. In fact, one of the best surveys of these drugs was written during this time by Hartwich (1911). It is to men like these that we owe many of the drugs we use today.

It is questionable if any of these prominent research workers can be regarded as ethnopharmacologists in the modern sense. Truly, they spent decades investigating ethnobotanical materials and their active principles, but they did not carry out field studies among different ethnic groups, nor did they in all cases realize the utmost importance of authenticating plant and drug materials. Whereas, at that time, it took months for dried leaves to arrive in Europe for analysis, it is now possible to take a fully equipped floating laboratory up the Amazon, not only for studies of accurately identified fresh botanical material, but also for the analysis of active components in blood.

Interest in traditional drugs is thus not new but has been spurred in recent years by methodological advances in phytochemistry, a growing number of ethnobotanical studies, and an upsurge of interest in renewable resources and traditional medicine.

Defining Ethnopharmacology

The observation, identification, description and experimental investigation of the ingredients and the effects of indigenous drugs is a truly interdisciplinary field of research. The term *ethnopharmacology* has been used loosely to describe this field (Efron et al., 1967; Schultes and Swain, 1976), but so far little attempt has been made to define the aims and scope of this discipline. Ethnopharmacologic research is based on botany, pharmacology and chemistry, but other disciplines have made vital contributions. Based on these considerations, we have recently defined ethnopharmacology as "the interdisciplinary scientific exploration of biologically active agents traditionally employed or observed by man" (Bruhn and Holmstedt 1981). This study of traditional drugs is not meant to advocate a return to the use of these remedies in their aboriginal form, nor to exploit traditional medicine.

The objectives of ethnopharmacology are to rescue and document an important cultural heritage before it is lost, and to investigate and evaluate the agents employed.

Ethnopharmacological Research

Field observations and descriptions of the use and effects of traditional remedies, botanical identification and phytochemical and pharmacological studies, are all within the scope of ethnopharmacology. It is essential, however, that anthropologists interested in ethnopharmacology seek contact and collaboration with experts in botany, chemistry and pharmacology. That such a multidisciplinary approach presents added advantages is not always realized. Even in recent times an anthropologist can give an utterly detailed description of an African poison ordeal without bothering about the chemical composition of the poisonous drink used or even its plant origin. Indeed remarkable in a time that favours teamwork and the involvement of many disciplines!

The first successful multidisciplinary attack on an ethnopharmacological problem was intiated by the French naturalist Leschenault de la Tour in 1803. He collected samples of an arrow poison in Java, as well as detailed first-hand information from the natives about the ingredients and preparation. In France a specimen of the major plant ingredient was studied by the botanist de Jussieu, who identified it as a *Strychnos* species. Leschenault then gave the poison to Magendie and Raffeneau-Delile, who studied the effects in hens, rabbits, dogs and a horse. They observed violent convulsions, asphyxia and death in 5 min and discovered that the chief action was on the spinal cord. This finding is a landmark in pharmacodynamics, since it represents the first time that the action of a poison was shown to act on a specific organ. A decade later the alkaloid responsible for these symptoms, strychnine, was isolated by Pelletier and Caventou. Thus, we have here for the first time an interdisciplinary study starting with fieldwork among natives and continuing with botany, pharmacology and the isolation of the active principle. Later on this was followed by the introduction of strychnine into clinical medicine.

The identification of medicinal plants and other traditional drugs is of course a crucial point, and good ethnopharmacological research can only be based on properly prepared voucher specimens, carefully authenticated by experts. There are many examples in the past, however, of pharmacologists neglecting this point, only to find that their results are difficult to repeat. Wherever possible, phytochemical studies on medicinal plants should be followed by a careful search for the biological activities of the compounds isolated. When biologically active principles have been found, the findings must be interpreted in the light of the traditional use.

It is impossible, however, to establish a dose-effect relationship unless the original drug preparations (water infusions etc.) are analysed and evaluated chemically and pharmacologically (Malone, 1977; Landgren et al. 1979). As a result we must, in ethnopharmacological research, have proper sampling and analysis methods, and this necessity requires close cooperation by pharmacologists, with anthropologists and ethnobotanists on the one hand and specialists in chemical analysis on the other. A recent example from our department of this approach is *Duboisia myoporoides*, used by the natives of New Caledonia as an antidote in ciguatera fish poisoning. Since a water infusion of *Duboisia myoporoides* is used for the treatment of ciguatera poisoning, such an infusion was prepared and the alkaloid content quantitated. The water infusion contains a powerful mixture of nicotine and scopolamine. Prepared in the traditional way, two mouthfuls would equal roughly 50 mg of nicotine and 20 mg of scopolamine. The potential of this preparation in the treatment of ciguatera poisoning became evident when modern knowledge of the poisoning was examined. Based on in vitro experiments with the toxin, two well-known anticholinesterase antidotes have been suggested: atropine and pyridinium aldoxime methiodide. Although they have been used with only moderate success, the recommendation of a mixture of pyridinium and tropane alkaloids is a striking parallel to the native use of chemically similar components.

Aqueous decoctions of *Oldenlandia affinis* are traditionally used to accelerate childbirth by the Lulua population in Zaire. In his capacity as medical missionary Gran (1973) observed increased uterine activity in women receiving these preparations. He also isolated serotonin from the plant. However, the usual decoction from about 100 g of dried plant would not give a higher total intake than about 2 mg of serotonin, and since serotonin is rapidly destroyed when taken orally it could have no action. In in vitro experiments with isolated organs, Gran showed that the increased uterine contractions produced by the water extract also persisted after treatment with methysergide. Therefore another, or several other compound(s) must be the active principle(s). In his research, which is a fine example of modern ethnopharmacology, Gran went on to isolate the active principles and found oxytocic peptides.

Most traditional drugs are administered as mixtures of many components, and with today's knowledge of the many possible interactions between drugs, and between food and drugs, ethnopharmacological research must deal with this aspect too. Additive, synergistic, or antagonistic effects are all possible. Various admixtures have also been shown to affect the bio-availability of pharmacologically active principles.

We believe that pharmacological studies of traditional medicinal agents should be initiated prior to, or in parallel with chemical research and should guide the isolation of active principles. Field observations of traditional therapies and the pharmacological effects in humans should be carried out by trained pharmacologists, and when interesting activity is found, controlled experiments should be initiated (Landgren et al., 1979).

Ethnopharmacology and Modern Medicine

Recent surveys have shown that the percentage of natural products in the modern drug armamentarium is considerable, estimates varying from 35% to 50%. Almost every class of drug includes a model structure derived from nature, exhibiting the classical effects of that specific pharmacological category. A great number of these natural products have come to us from the scientific study of remedies traditionally employed by various cultures. Most of them are plant-derived, and pilocarpine, vincristine, emetine, physostigmine, digitoxin, quinine, atropine and reserpine are a few well-known examples (Farnsworth and Bingel, 1977).

Evidently, the ethnopharmacological impulse to modern medicine can lead to many novel useful drugs, but modern and traditional uses may be entirely different. For example, the plant material studied at the National Cancer Institute (Bethesda, Maryland) has been collected at random, but the analysis performed by Spjut and Perdue (1976) (summarized in Table 1) shows that if antitumor screening had been guided by the knowledge of medicinal folklore and poisonous plants, the yield of active species would have been greatly increased. In this study, plants were classified as "active" regardless of tumor system or whether the results were obtained from in vivo or in vitro studies. The validity of the data in Table 1 should therefore be further analyzed. Another example is the antileukemic activity of vincristine from *Catharanthus roseus* which was found when the plant was investigated because of its folk use as a diabetes remedy.

It is generally accepted that ergot has constituted a gold-mine for finding therapeutically active components (naturally occurring or modified chemically) against many different diseases. It is by no means inconceivable that the cannabinoids may play a similar role in the future. Derivatives are already being tested for various purposes in clinical medicine.

The above data seem to support the conclusion that traditional medicine is a general, powerful source of biological activity.

Table 1. Plant folklore: a tool for predicting sources of antitumor activity?

Plant types	Percent active[1]
Plant collected at random	10.4
Plants used against cancer	19.9
Anthelmintics	29.3
Fish poisons	38.6
Plants poisonous to people	50.0
Arrow poisons	52.2

[1]Extracts show a significant inhibitory effect in experimental tumor systems (National Cancer Institute, Bethesda, MD) (Spjut and Perdue 1976).

Ethnopharmacology and Traditional Medicine

The field of medicinal plants is far from exhausted. The flora of the Amazon has been estimated at 73,000 species. During 30 years of ethnobotanical and ethnopharmacological research in this area, Schultes has collected information on the use of over 1300 species as poisons, narcotics or "medicines." This, and many other treasure-houses of human knowledge are waiting for pharmacologists to take up the challenge (Schultes and Swain, 1976; Perry, 1980).

The ultimate aim of ethnopharmacology is the validation (or invalidation) of these traditional preparations, either through the isolation of active substances or through pharmacological findings. The information gathered about indigenous drugs will permit a feed-back to traditional medicine. Harmful practices can be discouraged, such as the use of plants containing tumor-producing pyrrolizidine alkaloids. Knowledge of active constituents in indigenous drugs may lead to substantial improvements in traditional therapy. In recent years, WHO has emphasized the importance of scientific investigations into indigenous herbal medicines (WHO, 1978). Many Third World countries look upon native medicinal plants as possible additions to the WHO list of "essential drugs," once their value has been clinically proven.

Ethnopharmacology is not just a science of the past utilizing an outmoded approach. It still constitutes a scientific backbone in the development of active therapeutics based upon traditional medicine of various ethnic groups. Although not highly esteemed at the moment, it is a challenge to modern pharmacologists.

ACKNOWLEDGMENTS

Republished with changes by permission of *Trends in Pharmacological Sciences* 3, No. 5 (May 1982). Original title: "Is there a place for ethnopharmacology in our time?" Also published in *Journal of Ethnopharmacology* 8 (1983): 251–256 with the title "Ethnopharmacology—A challenge."

LITERATURE CITED

Bruhn, J. G., and B. Holmstedt. 1981. Ethnopharmacology: Objectives, principles, and perspectives. In *Natural Products as Medicinal Agents*. Eds. E. Reinhard and J. L. Beal. Stuttgart: Hippokrates. 405–430.
Efron, D. H., B. Holmstedt, and N. S. Kline, eds. 1967. *Ethnopharmacologic Search for Psychoactive Drugs*. Public Health Service Pub. no. 1645. Washington, DC: U.S. Government Printing Office. (Reprinted by Raven Press, New York, 1979.)

Farnsworth, N. R., and A. S. Bingel. 1977. Problems and prospects of discovering new drugs from higher plants by pharmacological screening. In *New Natural Products and Plant Drugs with Pharmacological, Biological or Therapeutic Activity*. Eds. H. Wagner and P. Wolff. Berlin: Springer. 1-22.

Gran, L. 1973. The uteroactive principle of "Kalata-Kalata" (*Oldenlandia affinis* DC.). Dissertation, University of Bergen.

Hartwich, C. 1911. *Die menschlichen Genussmittel, ihre Herkunft Verbreitung, Geschichte Anwendung, Bestandteile und Wirkung*. Leipzig: Tauchnitz.

Landgren, B. M., A. R. Aedo, K. Hagenfeldt, and E. Diczfalusy. 1979. Clinical effects of orally administered extracts of *Montanoa tomentosa* in early human pregnancy. *American Journal of Obstetrics and Gynecology* 135: 480–487.

Malone, M. H. 1977. Pharmacological approaches to natural product screening and evaluation. In *New Natural Products and Plant Drugs with Pharmacological, Biological or Therapeutic Activity*. Eds. H. Wagner and P. Wolff. Berlin: Springer. 23–53.

Perry, L. M. 1980. *Medicinal Plants of East and Southeast Asia* Cambridge, MA: MIT Press.

Schultes, R. E., and T. Swain. 1976. The plant kingdom: A virgin field for new bio-dynamic constituents. In *The Recent Chemistry of Natural Products, Including Tobacco: Proceedings of the Second Philip Morris Science Symposium*, New York. Ed. N. J. Fina. 133-171.

Spjut, R. W., and R. E. Perdue, Jr. 1976. Plant folklore: A tool for predicting sources of antitumor activity? *Cancer Treatment Reports* 60: 979–985.

WHO. 1978. *The Promotion and Development of Traditional Medicine*. Technical Report Series, 622. Geneva: WHO.

Amazonian Psychoactive Indoles: A Review

JAN-ERIK LINDGREN

In this chapter, only research concerning Amazonian psychoactive indoles will be treated. Collections were made during several expeditions. The work will be commented upon historically and with regard to analysis carried out. The description of botanical identification and chemical analysis is given in the reference list, where only work summarizing this research has been quoted.

LIST OF CHEMICAL ABBREVIATIONS USED

DMT	= N,N-Dimethyltryptamine
MMT	= N-Methyltryptamine
5-OH-DMT	= 5-Hydroxy-N,N-dimethyltryptamine or bufotenine
5-MeO-DMT	= 5-Methoxy-N,N-dimethyltryptamine
5-MeO-MMT	= 5-Methoxy-N-methyltryptamine
Harmine	= 1-Methyl-7-methoxy-β-carboline
Harmaline	= 1-Methyl-7-methoxy-3,4-dihydro-β-carboline
Tetrahydroharmine	= 1-Methyl-7-methoxy-1,2,3,4-tetrahydro-β-carboline
MTHC	= 2-Methyl-1,2,3,4-tetrahydro-β-carboline

Malpighiaceae

There is abundant evidence of the use of plants containing alkaloids by the South American Indians. This was first described in scientific detail by the explorer Richard Spruce (1817–1893), who became one of the outstanding plant explorers of South America of all time. Spruce was far ahead of his day in scientific thought and method. For a number of plants that have later attracted extensive phytochemical and pharmacological attention— and which are still claiming serious studies—it was Spruce who gave us detailed, accurate, pioneer information.

One of these plants was a jungle liana, source of an extraordinary hallucinogenic drink called *caapi* in Brazil, *ayahuasca* in Ecuador and Peru. Spruce first learned of *caapi* among the Tukanoan tribes of the upper Río Negro in 1852. The drink was employed to induce, for prophetic and divinatory purposes, an intoxication characterized, among other strange symptoms, by frighteningly realistic, colored visual hallucinations and a feeling of extreme and reckless bravery. Unlike many early reports of newly discovered narcotics, Spruce's contribution included a precise determination of the botanical source of the drug.

The botanical and chemical components of *caapi, ayahuasca,* and admixtures to the narcotic drink have been discussed in detail by Rivier and Lindgren (1972). More recently the monoamine oxidase (MAO) activity of *ayahuasca* and reference substances has been studied in detail by McKenna et al. (1984a). *Ayahuasca* was found to be an extremely effective inhibitor of MAO *in vitro*, and the degree of inhibition was directly correlated with the concentration of MAO-inhibiting β-carbolines. Inhibition experiments using mixtures of β-carbolines indicated that their effect in combination is additive.

Leguminosae

Another of the classical hallucinogens of the Americas is the snuff prepared from beans of the leguminous tree *Anadenanthera peregrina* (L.) Speg., better known in the literature by its former name *Piptadenia peregrina* (L.) Benth.

Long known from the Orinoco River basin of Colombia and Venezuela, this psychoactive drug has been mentioned by virtually all the early scientific explorers of the area. In 1916 it was identified by Safford as the source of the enigmatic *cohoba*, the narcotic snuff of the West Indies, the use and effects of which were seen among the Taino Indians of Hispaniola by early Spanish explorers in 1496.

Although the drug is no longer employed anywhere in the Caribbean islands, the extent of the use of *Anadenanthera peregrina* has still not been clearly defined. It may be that in isolated localities in the southern part of the Amazon Valley this tree was the source of a snuff. There is circumstantial evidence, too, that the very closely allied *Anadenanthera colubrina* was employed in preparing an intoxicating snuff known as *cébil* or *huilca*, used in former times in parts of northern Argentina, Paraguay and possibly in Bolivia and Peru. A narcotic snuff from this species is still employed by the Mataco Indian shamans of Argentina.

Our earliest botanical knowledge of *niopo* or *yopo*, as the snuff is called in the Orinoco basin, goes back to 1801 when von Humboldt and Bonpland encountered its use in Colombia and Venezuela. The next major botanical encounter with the drug was that of Spruce, who met with its use in June 1854 among the Guahibos of the upper Orinoco. Spruce wrote that his

> specimens of the leaves, flowers, and fruit agree so well with Kunth's description of *Acacia Niopo* that I cannot doubt their being the same species; especially as I have traced the tree all the way from the Amazon to the Orinoco, and found it everywhere identical.

An important point in Spruce's meticulous observation of the preparation of the snuff, however, is his statement that "there is no admixture of quicklime." Spruce found

> a wandering horde of Guahibo Indians . . . encamped on the savannas of Maypures (on the Orinoco) and . . . an old man grinding Niopo seeds. . . . The seeds, being first

roasted, are powdered on a wooden platter, nearly the shape of a watchglass. . . . For taking the snuff, they use an apparatus made of the leg-bones of herons . . . in the shape of the letter Y, or something like a tuning fork, and the two upper tubes are tipped with small black perforated knobs (the endocarps of a palm). The lower tube being inserted in the snuff box and the knobs in the nostrils, the snuff is forcibly inhaled, with the effect of thoroughly narcotising a novice or, indeed, a practiced hand, if taken in sufficient quantity.

There are also at Kew, in the Sir Joseph Banks Centre for Economic Botany, specimens of the pods of *Anadenanthera peregrina* which Spruce collected in 1854 on the Colombo-Venezuelan border at the savannahs of Maypures. These pods were purchased from an old Guahibo Indian who was grinding the seeds for preparation of the snuff.

I examined the material collected by Spruce, seeds of the collection *Spruce 119* of *Anadenanthera peregrina,* and compared my results with the analysis of similar freshly collected material (Schultes et al. 1977). The gas chromatographic (GC) trace of the chloroform soluble bases from the seeds collected in 1854 by Spruce gave a single peak. Its mass spectrum was identical to that of 5-OH-DMT (bufotenine). The extract of mature seeds freshly collected in Puerto Rico, seeds "March 1975," showed several GC peaks. Beside 5-OH-DMT, they have been identified by GC-MS (mass spectrometer) and are DMT, MTHC and 5-MeO-DMT.

The finding of 5-OH-DMT as the only alkaloid in the Spruce material is significant for several reasons. First, it again indicates that, with modern analytical tools, it is possible to detect and identify alkaloids in plant materials more than 100 years old. Second, identification of the botanical specimen is strengthened by the results of the chemical analyses, because the same compound has been found in both old and the freshly collected seeds.

Of several alkaloids found in freshly collected reference material, only one remained in the Spruce collection: bufotenine (5-OH-DMT). Storage of freshly collected material for 2 years resulted in the disappearance of all alkaloids except 5-OH-DMT. This may raise speculation as to whether or not they were originally contained in the Spruce material. My observation stresses the importance of storage time in addition to knowledge of plant part, soil, season, and climatic conditions, when alkaloid analysis is carried out on seeds and on the snuffs prepared from them.

Myristicaceae

Indians of the northwest Amazon and of adjacent parts of the Orinoco have employed various species of the myristicaceous genus *Virola* for many years as the base of hallucinogenic preparations. A resinlike liquid of the inner bark of the trees is elaborated into an intoxicating snuff and is prepared in pellets for oral consumption; on occasion it is ingested raw without any preparation (Holmstedt et al. 1980).

Although use of *Virola* is undoubtedly a custom of great age, discovery of its use as an important hallucinogen is recent. In the early part of the 20th century, German anthropologist Theodor Koch-Grünberg (1923) reported that the Yekwana Indians of the Orinoco of Venezuela utilized an intoxicating snuff prepared from the "bark of a tree." In 1938, Brazilian botanist Adolpho Ducke indicated that Indians of the Río Negro in Brazil made a snuff called *paricá* from the leaves of *Virola theiodora* and *V. cuspidata* (Schultes 1989).

The source of the narcotic snuff, however, was not definitely identified until 1954, when voucher specimens and field studies indicated that the leaves are not used but that a bark exudate is the real source of the snuff. At that time, the Indians of the Río Apaporis of Amazonian Colombia were found to be using in ritual ceremonies of the medicine men an intoxicating snuff made from the red "resin" of the inner bark or from scrapings of the inner bark itself of *Virola calophylla, V. calophylloidea,* and possibly *V. elongata.*

Later it was learned that the Witoto Indians of Colombia did not utilize the red "resin" of the bark in the form of snuff but valued it in pellets orally ingested as a magic and ceremonial hallucinogen. Subsequent field work among the Boras and Witotos of Peru has indicated the use of *Virola pavonis* and *V. elongata* as well as possibly *V. surinamensis* and *V. loretensis.* The Boras likewise point out *Iryanthera macrophylla,* of a related myristicaceous genus, as the source of a narcotic paste. This represents the first time that a genus other than *Virola* has been known to be involved in the myristicaceous hallucinogens of tropical South America.

Reports from the field work of anthropologist Peter L. Silverwood-Cope inform us that the primitive nomadic Makú Indians of the Río Piraparaná in Amazonian Colombia drink the "resin" directly, with no preparation and no admixture, for its hallucinatory effects.

There are suggestions that the bark of *Virola sebifera* may be smoked in Venezuela. Several herbarium collections note that the inner bark is smoked by witch doctors at dances when curing fevers and that they boil the bark "to drive away evil spirits."

It is now known that a number of *Virola* species are employed hallucinogenically in Brazil, Colombia, Peru, and Venezuela: *V. calophylla, V. calophylloidea, V. cuspidata, V. elongata, V. pavonis, V. surinamensis,* and *V. theiodora.* Native names for these species are many. Among those most frequently mentioned in the literature (mainly because of the use of the plants as hallucinogens) are *yakee* (Puinave) and *yato* (Kuripako) in Colombia; *epena* or *nyakwana* (Waika) in Brazil and Venezuela; and *paricá* (Tukano) in Brazil and Colombia.

It appears that the most important species of *Virola* for the preparation of the hallucinogenic snuff or paste is *Virola theiodora.* This species occurs mainly in the western Amazonia of Brazil and Colombia, possibly also in adjacent parts of Peru and Venezuela. Normally found in well-drained forests, the tree is especially abundant in the Río Negro drainage area.

Fifty-three voucher collections of *Virola* and related genera made by various botanists over several decades have been analyzed for alkaloids. Of these, 18 collections proved negative when analyzed for alkaloids in various parts of the plants. Occasionally, different collections representing the same species have proven to be alkaloid positive in some cases and negative in others. Analyses of 4 different collections of *Virola surinamensis,* however, all proved to be negative.

To our knowledge, only the bark and/or constituents of bark are used in the preparation of the intoxicating snuffs employed by the Indians. It is apparent that the species used are usually rich in alkaloids. *Virola rufula* is not known to be employed and, if it is not utilized, it would appear that the Indians have missed an alkaloid-rich species. In addition to the bark, the leaves and flowering shoots seem to be unusually rich in alkaloids. The simple alkaloids MMT, DMT, 5-MeO-MMT, and 5-MeO-DMT abound in the species used; they are also present in other species.

The *nyakwana* snuff analyzed by Agurell et al. in 1969 proved to be extraordinarily rich in base content (11 percent). This might explain why the resin of *Virola theiodora* is also employed as an arrow poison. The high content of bases in this snuff has surprised some research workers (McKenna et al. 1984a). An alternative snuff preparation from *Anadenanthera* has been proven to contain no less than 7.4 percent alkaloids (Chagnon et al. 1971). As has been pointed out by Gottlieb (1979), there exist appreciable differences in the base composition of different parts of a single plant; this is true also of different species and even in analyses of different specimens representing the same species. When examining the data with respect to use, it must be kept in mind that the preparation of snuffs, pellets, and arrow poison involves concentration of the resinous bark exudate to a thick syrup which is subsequently dried, powdered, or rolled into pellets which are then coated with the residue of leachings from ashes. Such treatment, not to speak of storage, would be expected to alter the original base composition.

The hallucinogenic myristicaceous snuffs of the South American Indians owe their biological activity to the simple methylated indoles mentioned above. They are not inhibitors of MAO (McKenna et al. 1984b). Bufotenine, a component of the snuff made from *Anadenanthera peregrina*, is not present in the species of *Virola* investigated; neither is it present in the snuff made from them.

When analyzing Indian snuffs in 1967, Holmstedt and Lindgren noted the presence of harmala alkaloids in several preparations of uncertain botanical origin. In one case, both the simple indoles and harmine were present in the same preparation. This observation led to the following conclusion:

> In South American botany, β-carbolines (harmine, harmaline and tetrahydroharmine) are usually associated with the species of *Banisteriopsis*, wherefore it is very likely that this is their origin in the snuffs. Very likely this is an admixture to the snuff, although definite botanical proof for it is lacking at the moment. To the knowledge of the authors, simple indoles and β-carbolines have not yet been isolated *from the same plant*.
> The occurrence of both tryptamines and β-carbolines in the South American snuffs is pharmacologically interesting. The β-carbolines such as harmine and harmaline are monoamine-oxidase inhibitors (Udenfriend et al. 1958) and could potentiate the action of the simple indoles. The combination of β-carbolines and tryptamines would thus be advantageous. However, pharmacological action of the β-carbolines unrelated to monoamine-oxidase inhibition has also been proven to exist (Schievelbein et al. 1966). Further botanical and chemical studies are obviously needed to see if the two groups of compounds in the snuff are derived from one plant or a mixture of plants (Holmstedt and Lindgren 1967).

This observation has often been quoted and even misquoted. Additional phytochemical and enzymatic evidence is now available. Trace amounts of β-carbolines have been found to be present in *Virola calophylla*, *V. theiodora*, and *V. elongata*. In one species, which is not known to be used hallucinogenically, *V. cuspidata*, harman bases have been found to be the main alkaloids (Cassady et al. 1971). Their existence has been unequivocally proven by isolation, spectroscopy, and mass spectrometry, as compared to synthetic reference compounds. Interestingly, these authors have also observed aromatization due to heat treatment such as that practiced by the Indians when preparing snuff from other species. They also point out the possibility of increased potency of enzyme inhibition due to the aromatization.

Although the MAO-inhibiting properties of harmine were observed indirectly before the enzyme was known to exist (Marinesco et al. 1930), it was only through Udenfriend et al. (1958) that these properties of the harmala alkaloids could be quantitated. Subsequently, the structure-activity relationship was worked out for a large number of β-carbolines (Buckholtz and Boggan 1977). From this work, it is clear that the β-carbolines contained in *Banisteriopsis caapi* and *Peganum harmala* are far superior MAO inhibitors than the compounds contained in usually trace amounts in *Virola*. Buckholtz and Boggan (1977) compared enzyme inhibitory power.

The occurrence of trace amounts of 6-methoxy-β-carbolines in some species of *Virola* is not surprising. It might be expected from the point of view of biosynthesis and workup procedure, and is pharmacologically of no importance. The occurrence of a mixture of simple methylated indoles and harmine in one Indian snuff of unknown origin, or of harmine and related compounds alone in other snuffs, justifies, however, the statement made in 1967 and quoted above, and should encourage further research on this interesting group of indigenous drugs and on the plants from which they are derived.

LITERATURE CITED

Agurell, S., B. Holmstedt, J.-E. Lindgren, and R. E. Schultes. 1969. Alkaloids in certain species of *Virola* and other South American plants of ethnopharmacologic interest. *Acta Chem. Scand.* 23: 903–916.

Buckholtz, N. S., and W. O. Boggan. 1977. Monoamine oxidase inhibition in brain and liver produced by β-carbolines: Structure activity relationships and substrate specificity. *Biochem. Pharmacol.* 26: 1991–1996.

Cassady, J. M., G. E. Blair, R. F. Raffauf, and V. E. Tyler. 1971. The isolation of 6-methoxyharmalan and 6-methoxyharman from *Virola cuspidata*. *Lloydia* 34: 161–162.

Chagnon, N. A., P. Le Quesne, and J. M. Cook. 1971. Yanomamó hallucinogens: Anthropological, botanical, and chemical findings. *Curr. Anthrop.* 12: 72–74.

Gottlieb, O. R. 1979. Chemical studies on medicinal *Myristicaceae* from Amazonia. *Journ. Ethnopharmacol.* 1: 309–323.

Holmstedt, B., and J.-E. Lindgren. 1967. Chemical constituents and pharmacology of South American snuffs. In *Ethnopharmacologic Search for Psychoactive Drugs*. Eds. D. Efron, B. Holmstedt, and N. S. Kline. U.S. Public Health Service Publ., Washington, DC, no. 1645: 339–373; 2nd ed., 1979, New York: Raven Press, 339.

Holmstedt, B., J.-E. Lindgren, T. Plowman, L. Rivier, R. E. Schultes, and O. Tovar. 1980. Indole alkaloids in Amazonian Myristicaceae. Field and Laboratory Research. *Botanical Museum Leaflets, Harvard University* 28(3): 215–234.

Marinesco, G., A. Kreindler, and A. Scheim. 1930. Klinische und experimentelle Beiträge zur Pharmakologie des Harmins. *Naunyn-Schmiedeberg's Arch.* 154: 301–312.

McKenna, D. J., G. H. N. Towers, and F. S. Abbott. 1984a. Monoamine-oxidase inhibitors in South American hallucinogenic plants: Tryptamine and β-carboline constituents of *ayahuasca*. *Journ. Ethnopharmacol.* 10: 195–223.

———. 1984b. Monoamine-oxidase inhibitors in South American hallucinogenic plants. Part 2: Constituents of orally active *Myristicaceous* hallucinogens. *Journ. Ethnopharmacol.* 12: 179–211.

Rivier, L. and J.-E. Lindgren. 1972. "Ayahuasca," the South American hallucinogenic drink: An ethnobotanical and chemical investigation. *Economic Botany* 26: 101–129.

Schievelbein, H., H. Peter, I. Trautschold, and E. Werle. 1966. Freisetzung von 5-Hydroxytryptamin aus thrombocyten durch harmalin. *Biochem. Pharmacol.* 15: 195–197.

Schultes, R. E. 1989. The "social-chemistry" of pharmacological discovery—the *Virola* story. *Social Pharmacology* 3(4): 297–314.

Schultes, R. E., B. Holmstedt, J.-E. Lindgren, and L. Rivier. 1977. De plantis toxicariis mundo novo tropicale commentationes. XVIII. Phytochemical examination of Spruce's ethnobotanical collection of *Anadenanthera peregrina*. *Botanical Museum Leaflets, Harvard University* 25(10): 273–287.

Udenfriend, S., B. Witkop, B. G. Redfield, and H. Weissbach. 1958. Studies with reversible inhibitors of monoamine oxidase: Harmaline and related compounds. *Biochem. Pharmacol.* 1: 160–165.

Biodynamic Constituents in Ayahuasca Admixture Plants: An Uninvestigated Folk Pharmacopeia

DENNIS J. MCKENNA, L. E. LUNA, AND G. N. TOWERS

From 1960 to 1990 scientific knowledge of the botany and chemistry of plant hallucinogens has expanded enormously. This has been accomplished through the cooperative efforts of ethnobotanists, working to collect and identify the source-plants utilized by aboriginal peoples, and phytochemists, who have isolated and characterized the biodynamic constituents responsible for these properties (Schultes 1970; Schultes and Hofmann 1980). The number of higher plant species is estimated at between 400,000 and 800,000; of these, only an insignificant number—somewhat fewer than 100—are known to be exploited as hallucinogens, and fewer than 20 of these species may be described as major (Schultes and Hofmann 1980).

Nowhere else in the world has the knowledge and use of endemic hallucinogenic plants developed to the extent found in the western part of the Amazon basin of South America. And of various hallucinogens utilized by indigenous populations in that region, none is as interesting or as complex botanically, chemically, or ethnographically as the hallucinogenic beverage known variously as *ayahuasca*, *caapi*, or *yagé*. Far from being "simply" a hallucinogenic plant or preparation, ayahuasca (Quechua for "vine of the soul") occupies an integral position in mestizo folk medicine.

Contemporary use of ayahuasca in Amazonian mestizo populations appears to be an amalgam of diverse tribal traditions. The large urban settlements have become melting pots; people of many different cultural backgrounds have migrated to these centers in search of employment in the lumber, petroleum, and other resource-based industries, bringing with them tribal traditions and belief systems (usually syncretically fused with Christianity due to prior contact with missionaries). The cultural background of these migrant laborers often extends to a knowledge of the medicinal plants valued in their own culture; over the years this drug-plant lore derived from diverse sources has gradually diffused through the larger mestizo society and become assimilated into mestizo folk medicine. This ethnomedical tradition is unique to the mestizo social class, although it incorporates elements of its diverse tribal origins.

This process of cultural assimilation has occurred over the same period of time in which the antecedents of mestizo folk medicine have disintegrated or disappeared from most tribal societies. As a result, mestizo folk medicine, as it is practiced in urban centers of the Amazon, is a living system of traditional medicine based on the ethnomedical lore of many cultures; in many cases these centers are the only places where such knowledge has been preserved. Hence it is important, even urgent, that mestizo folk medicine and the plants that form its basis be studied by investigators with backgrounds in medicine, pharmacology, phytochemistry, and botany while the opportunity still exists. This chapter presents phytochemical and ethnobotanical information on approximately fifty genera of medicinal plants utilized as ayahuasca admixtures in contemporary mestizo ethnomedicine.

Ethnomedical, Botanical, and Pharmacological Aspects of Ayahuasca

THE ROLE OF THE AYAHUASQUERO IN MESTIZO FOLK MEDICINE

In contemporary countercultural circles in Western society, hallucinogens are employed idiosyncratically; that is, they are usually self-prescribed and the individual consuming the drug does so outside the context of any magical, ritual, or metaphysical belief systems designed to accommodate the phenomenology of the drug experience. By contrast, the use of ayahuasca in mestizo folk medicine always takes place within a ritual and therapeutic context. Dispensation of the drug and the progress of the intoxication is under the control of the *ayahuasquero*, who uses various techniques, including singing, whistling, blowing of tobacco smoke, and making passes over the patient's body, to influence the content and course of his[1] patient's drug experience. By this means the set and setting of the ayahuasca experience is carefully controlled and manipulated by the *ayahuasquero*, and there is usually a specific purpose for consuming this drug—for divination, to discover the cause of an illness, or to communicate with the spirit world.

In traditional cultures, the boundaries between religion, magic, and medicine are not clearly delineated; the function of the *ayahuasquero* or traditional healer amalgamates the Western roles of priest, doctor, and psychotherapist; illness may be precipitated by physical, psychological, or supernatural causes, or a combination of these, and all are amenable to treatment with the methods available to the *ayahuasquero*. In this sense the trend in modern medicine toward "holistic" therapies is not that different from the therapeutic methods practiced by the traditional healer. Both proceed from the recognition that mind and body are an integrated unit, and that the most effective therapies are those directed at improving both physical and mental health. Thus it is not surprising that ayahuasca, which profoundly affects both the mind and the body, and affords access to and a certain degree of manipulation of [real or imagined] supernatural dimensions, should occupy such a prominent position in the pharmacopoeia of Indian and mestizo folk medicine.

The *ayahuasquero* employs ayahuasca as a diagnostic and therapeutic tool to uncover the causes of illness, rather than as a palliative for specific ailments. Through the interpretation of his own or his patient's visions, the *ayahuasquero* feels he is able to divine the source of the illness or misfortune which has precipitated the patient's visit; he is then able to recommend appropriate remedies. In some cases this may require the neutralization of malevolent supernatural forces directed to the patient by a sorcerer or *brujo*, and in other instances it may entail pharmacological intervention involving the use of various medicinal plants. In most instances, both magical and medicinal remedies will be employed. This use of ayahuasca in contemporary mestizo folk medicine has been previously described (Dobkin de Rios 1970, 1972).

Besides functioning as an important diagnostic tool in the medical practice of the *ayahuasquero*, use of the drug is also an intrinsic part of his shamanic training. As in most shamanic traditions, the apprentice *ayahuasquero* must undergo an initiatory period of training. During this time, which lasts for a minimum of six months but may extend for several years (depending on the degree of power he wishes to acquire), the *ayahuasquero* consumes ayahuasca frequently while adhering to a strict diet in which no salt, sugar, fat, alcoholic or cold beverages may be consumed; sexual abstinence is also a requirement. During this initiatory period the *ayahuasquero* acquires the magical songs, objects, and helping spirits which he will later use in curing ceremonies; he also learns the properties and uses of numerous medicinal plants, often by consuming them in the form of admixtures to ayahuasca. The assertion is nearly universal among *ayahuasqueros* that this shamanic knowledge is transmitted directly by ayahuasca and other "plant-teachers"; it is not acquired through instruction by an elder *ayahuasquero* or other human teacher. Luna (1984) has provided a detailed account of the initiatory training and practices of *ayahuasqueros* in Iquitos, Peru (Luna 1984, 1986, 1992).

The system of ethnomedicine practiced by the mestizo healer can in some sense be regarded as an alternative health-care system. The urban mestizo who is poor, barred by economic factors from all but the barest access to health-care based on Western medicine, looks to the *ayahuasquero* and his magical and botanical remedies for medical, psychiatric, and spiritual support. Although the health-care system of the *ayahuasquero* incorporates magical, religious, and psychotherapeutic elements, it also is largely based on pharmacology because of its reliance on numerous biodynamic plants. In that respect it is more akin to Western medicine than to other shamanic, quasi-medical systems of traditional healing.

BOTANY, CHEMISTRY, AND PHARMACOLOGY OF AYAHUASCA

Botanical sources of ayahuasca. The liana *Banisteriopsis caapi* (Malpighiaceae) forms the basis of ayahuasca. Although *B. caapi* is used normally, *B. inebriens*, *B. quitensis*, and *Tetrapterys methystica* have all been reported as sources of the drink (Schultes 1957). On rare occasions ayahuasca is prepared from the boiled bark or stems of one of these malpighiaceous species without the addition of any other botanical ingredients. More commonly, however, the leaves or bark of various admixture plants are added to the brew to strengthen or modify the effect (Pinkley 1969). The admixtures used most frequently are *Diplopterys cabrerana* (Cuatrecasas) (formerly known as *Banisteriopsis rusbyana*) and the rubiaceous species *Psychotria viridis* and *Psychotria carthaginensis*. Solanaceous admixtures are also common, including tobacco (*Nicotiana* species), *Brugmansia* species, and *Brunfelsia* species.

Chemistry and pharmacology of ayahuasca. The most detailed chemical study to date of ayahuasca and its botanical ingredients is that of Rivier and Lindgren (1972). Using GC/MS (gas chromatograph/mass spectrometer) analysis, these investigators found that the major active constituents of ayahuasca are the beta-carboline alkaloids harmine, harmaline, and tetrahydroharmine, and N,N-dimethyltryptamine (DMT). The beta-carbolines are constituents of *Banisteriopsis caapi* (Rivier and Lindgren 1972), while DMT has been isolated as a constituent of *Diplopterys carbrerana* (Agurell et al. 1968) and has also been detected in *Psychotria viridis* and *P. carthaginensis* (Rivier and Lindgren 1972).

The compound DMT is a potent hallucinogen and is probably responsible for the hallucinogenic activity of ayahuasca. A peculiarity of the pharmacology of DMT is that it is *not* orally active, possibly due to oxidative deamination in peripheral tissues by the enzyme monoamine oxidase (MAO). The beta-carbolines, although having some limited hallucinogenic activity themselves (Naranjo 1967), are extremely active, reversible inhibitors of MAO and thus may protect the DMT from degradation and render it orally active. This mechanism was postulated (Pinkley 1969; der Marderosian et al. 1968; Schultes 1972) to underlie the oral activity of ayahuasca long before it was experimentally investigated (McKenna et al. 1984).

Ayahuasca admixtures. The utilization of admixture plants in conjunction with ayahuasca has reached a rather high degree of botanical and pharmacological sophistication. Besides the rubiaceous admixtures almost always included in ayahuasca, a virtual pharmacopoeia of admixtures is used occasionally, depending on the magical, ritual, or medical purposes for which the drug is being made and consumed (Schultes 1957; Pinkley 1969; der Marderosian et al. 1968; Rivier and Lindgren 1972; Schultes 1972; Luna 1984; McKenna et al. 1984). Many of these admixtures have not been botanically identified, much less chemically characterized, but of those identified, many of them are known to contain biodynamic constituents.

Phytochemical data on ayahuasca admixtures. The extant chemical information on plants in approximately fifty genera utilized as admixtures to ayahuasca has been compiled in Table 1 with its accompanying references (see end of chapter). This information was assembled from a computer search of the *Biological Abstracts* database and the American Chemical Society database, covering the years 1970 to the present. The references are not intended to be exhaustive but rather to be indicators of the existence or nonexistence of information regarding biodynamic constituents in the genera listed. The chemistry of certain genera (e.g., *Alchornea, Erythrina, Ficus, Maytenus, Ocimum, Tabebuia, Tabernaemontana,* and *Uncaria*) has been extensively investigated, and the number of available references runs well into the thousands; in these instances only a limited number of key references are cited. In the many instances where phytochemical data are not available on a particular species used as an admixture to ayahuasca, the references cited refer to closely related species in the same genus.

The contribution of most of these admixtures to the pharmacological activity of ayahuasca is, at this time, a mystery and an area well deserving of further investigation by ethnopharmacologists. Information can be found in the literature on the chemical or biodynamic properties of about half the genera listed in the table; the corollary to this is that virtually *nothing* is known about the pharmacologically active constituents in the remaining genera listed. These uninvestigated genera form part of a neglected folk pharmacopoeia that potentially is of great interest to Western science. Because many of these genera have long been valued as medicinal agents by the traditional mestizo healers who employ them, and because a high proportion of these traditional medicines *has* yielded biodynamic compounds of medicinal value, there seems a strong likelihood that further biochemical investigations of these admixtures will more than repay the efforts involved.

Uses of Admixture Plants in Mestizo Folk Medicine

The idea that certain plants, animals, and inanimate objects, such as mountains, lakes and rivers, have a spirit, is implicit in the cosmology of many Amazonian people, including that of mestizo practitioners of the Peruvian Amazonas. These spirits, sometimes called the *madres* ("mothers") of the corresponding plants, animals, or objects (Deltgen 1978–1979; Chevalier 1982; Chaumeil 1983) may be contacted for the purpose of acquiring from them knowledge or certain powers. Intelligence is not considered to be a prerogative of the human species. Being in constant contact with nature, local people have learned to respect and fear certain species of plants and animals as well as natural phenomena.

Reality for these people has a twofold character, a secular and a sacred one. These two aspects are not divorced from each other, however. Some of the qualities attributed to the "spirit" of certain plants or animals are in fact based on accurate observation and experimentation. Much can be learned about the Amazonian people's knowledge of the natural world by studying their cosmological and religious ideas.

There is no doubt that nonliterate people possess an impressively comprehensive, scientifically accurate knowledge of their environment. Taxonomic recognition of species may be extremely sophisticated (Berlin and Berlin 1983). Knowledge of the effect on the human organism of certain species of plants and animals seems to be at least as important as the recognition of morphological differentiation. In the context of an animistic world view, it is not strange that plants possessing biodynamic compounds are considered to have particularly strong mother spirits, and those with psychotomimetic constituents are regarded as powerful plant-teachers.

Dietary prescriptions, which might also have symbolic connotations (Chevalier 1982) probably reflect accurate observations of the incompatibility of ingesting specific foods together with certain plants. It is well known, for instance, that when ingesting *chuchuhuasa*, a beverage made of the bark of *Maytenus ebenifolia* and alcohol, one should avoid eating peccary (Gunther Schaper, pers. com.). The combination produces an intermittent high fever, similar to malaria. Compatibility and incompatibility of plants is often explained in terms of friendship or enmity between the spirits of the plants.

Access to the sacred dimension of reality happens through consumption of psychotropic plants and the dietary prescriptions mentioned above. The initiation occurs usually through the mastering of the use of tobacco and ayahuasca. The personal disposition of the individual and his ability to stand the hard training and the dangers involved in the shamanic initiation will determine the degree of his development. He may continue to add new additives to the basic ayahuasca brew or consume other plant-teachers to increase his knowledge and abilities. Each plant taken means entering a new dimension where the initiate encounters beings who give him new powers to manipulate the environment, often through magic melodies or *icaros* and incantations. Each plant has its cluster of zoomorphic and anthropomorphic spirits with which it is associated. By establishing contact with these beings, the shaman acquires more knowledge and power.

The use of these plants is not without dangers. The *vegetalistas* are aware that sometimes they are dealing with very powerful, even highly toxic compounds. Dosage is then of crucial importance. Another important factor is the strict observation of the diet, which seems to have at least two functions: to "cleanse" the organism so that the initiate is able fully to experience the effect of the plants, and to protect the initiate against the adverse effects of certain foods when consuming some of these plants.

Practitioners often claim that some of these plants are very "jealous" (Dobkin de Rios 1973). The sexual continence and the diet should not be broken, as the person may be "punished" by the spirits of the plants with sickness or even death. This is the case of such plants as *Brunfelsia grandiflora* subsp. *schultesii* (*chiric sanango*), *Capirona decorticans* (*capirona negra*), *Chorisia speciosa* (*lupuna*), *Couroupita guianensis* (*ayahúma*), *Hura crepitans* (*catahua*), *Tabebuia* species (*tahuari*) and others, although symbolic ideas are also important. Chevalier (1982), for example, claims that the reason for the strict diet is the ritual transformation of the patient into a plant spirit. Most probably, however, there are reasons of a biological nature for most of these prescriptions. The possibility that through the diet the initiate is capable of maintaining the effect of the psychotropic plants for a longer period of time should not be excluded *a priori*.

By ingesting these plants and keeping the prescribed diet, the initiate is supposed to be in the appropriate state of consciousness for learning the body of knowledge necessary for his future shamanistic practices. These plants "open the mind" of the initiate, so that he can effectively explore the flora, fauna, and geographical setting which surrounds him and will be able to remember it all in the future. Much of this learning process takes place in dreams, which are said to be especially vivid during the period of initiation.

At the same time, these plants strengthen the body of the initiate by giving him some of the physical qualities of the plants: for instance, the ability to withstand heavy rains,

354   D. J. MCKENNA ET AL.

winds, and floods. Plant-teachers or *doctores*, as these plants are known, have a twofold aspect: they give both "strength and wisdom." *Vegetalistas*, when asked why they consume plant-teachers, say that they do it to "cure" (*curarse*) themselves. This implies that they consume plant-teachers not only to heal themselves of illness or to recover the energies of their youth, but also to "awaken" their minds.

Some of these plants, such as *Couroupita guianensis* (*ayahúma*), are also given to dogs, with the same aim: to make them stronger and to increase their hunting abilities. The idea that certain plants are teachers is even found in highly syncretic, modern rural-urban cults. In Brazil, in the state of Acre, there are groups that use the beverage prepared of *Banisteriopsis caapi* and *Psychotria viridis* under the name *Santo Daime* (Monteiro 1983), because it is believed that these plants heal both the body and the soul and teach the doctrine of Jesus Christ.

Adding admixtures to the ayahuasca beverage is a way of studying their properties, so that some *ayahuasqueros* continually are expanding their pharmacopoeia. Similar findings were made by Bristol (1966) among the Sibundoy Indians of southwestern Colombia and by Chaumeil (1983) among Yagua Indians. The latter use psychotropic plants with specific goals; some plants make you "see," others make you travel, teach you how to heal or to harm, give you strength, and so forth (Chaumeil 1983).

Our informants believe that the spirits of these admixtures present themselves either during the hallucinations elicited by the beverage or in the dreams following the intoxication, and that they disclose to the initiate their pharmacological properties. Our informants also recognize the synergistic effect that sometimes occurs when several plants are taken together. This concept is based on the idea that these plants "know each other" or "go well together," while other plants "do not like each other."

Each of the admixture plants is associated with a magic melody or *icaro*, which is individually revealed to the initiate when taking the ayahuasca beverage with that specific admixture. The number and quality of the magic chants increase when the diet is prolonged and new admixtures are added, one at a time, to the ayahuasca beverage. The *madre* spirit of these plants may be called by singing or whistling the appropriate *icaro*. New knowledge is first of all expressed through magic melodies (Luna 1992). A similar idea has been found among the Sharanahua (Siskind 1973).

NOTES

1. Although the masculine pronoun is used here, *ayahuasqueros* can be, and frequently are, women.

ACKNOWLEDGMENTS

This chapter was translated from the original "Ingredientes biodinámicos en las plantas que se mezclan al ayahuasca. Una farmacopea tradicional no investigada." *America Indígena* 46, no. 1 (1986). Published with the permission of Interamerican Indian Institute, Mexico.

LITERATURE CITED

Agurell, S., B. Holmstedt, and J.-E. Lindgren. 1968. Alkaloid content of *Banisteriopsis Rusbyana*. *American Journal of Pharmacy* 140: 148–151.
Berlin, B., and E. A. Berlin. 1983. Adaptation and ethnozoological classification: Theoretical implication of animal resources and diet of the Aguaruna and Huambiza. In *Adaptive Responses of Native Amazonians*. Eds. R. B. Hames and W. T. Vickers. New York: Academic Press. 65–111.
Bristol, M. L. 1966. The psychotropic *Banisteriopsis* among the Sibundoy of Colombia. *Botanical Museum Leaflets* (Harvard University) 21: 113–140.

Chaumeil, J. P. 1983. Voir, savoir, pouvoir. Le chamanisme chez les Yaguas du nordest peruvien. In *Adaptive Responses of Native Amazonians*. Eds. R. B. Hames and W. T. Vickers. New York: Academic Press.

Chevalier, J. M. 1982. *Civilization and the Stolen Gift: Capitol, Kin, and Cult in Eastern Peru*. University of Toronto Press.

Deltgen, F. 1978–1979. Culture, drug, and personality—a preliminary report about the results of a field research among the Yebasama Indians of Río Piraparaná in the Colombian Comisaría del Vaupés. *Ethnomedicine* V, no. 1/2.

Dobkin de Rios, M. 1970. *Banisteriopsis* used in witchcraft and folk healing in Iquitos, Peru. *Economic Botany* 24(35): 297–300.

———. 1971. *Ayahuasca:* the healing vine. *International Journal of Social Psychiatry* 17: 256–269.

———. 1972. *Visionary Vine: Healing in the Peruvian Amazon*. San Francisco: Chandler.

———. 1973. Curing with *ayahuasca* in an urban slum. In *Hallucinogens and Shamanism*. Ed. M. J. Harner. New York: Oxford University Press.

Deulofeu, V. 1967. Chemical compounds isolated from *Banisteriopsis* and related species. In *Ethnopharmacologic Search for Psychoactive Drugs*. Eds. D. H. Efron, B. Holmstedt, and N. S. Kline. U.S. Department of Health, Education and Welfare Publication no. 1645. Washington, DC: Government Printing Office.

Friedberg, C. 1965. Des *Banisteriopsis* utilisés comme drogue en Amérique du Sud. *Journal d'Agriculture tropicale et Botanique Appliquée*. Vol. 12: 403, 550, 729.

Luna, L. E. 1984. The concept of plants as teachers among four mestizo shamans of Iquitos, Peru. *Journal of Ethnopharmacology* 11: 135–156.

———. 1986. *Vegetalismo: Shamanism among the Mestizo Population of the Peruvian Amazon*. Stockholm: Almquist & Wiksell International.

———. 1992. Function of the magic melodies or *icaros* of some mestizo shamans of Iquitos in the Peruvian Amazonas. In *Portals of Power: Shamanism in South America*. Eds. J. M. Langdon and G. Baer. University of New Mexico Press. 231–253.

der Marderosian, A., H. V. Pinkley, and M. F. Dobbins. 1968. Native use and occurrence of N,N-dimethyltryptamine in the leaves of *Banisteriopsis rusbyana*. *American Journal of Pharmacy* 140: 137–147.

McKenna, D., G. H. N. Towers, and F. S. Abbott. 1984. Monoamine oxidase inhibitors in South American hallucinogenic plants: Tryptamine and β-carboline constituents of ayahuasca. *Journal of Ethnopharmacology* 10: 195–223.

Monteiro da Silva, C. 1983. *O Palacio de Juramidan. Santo Daime: Um ritual de transcendencia e despoluiçao*. Dissertaçao de mestrado. Recife, Pernambuco. Março 1983.

Naranjo, C. 1967. Psychotropic properties of the harmala alkaloids. In *Ethnopharmacologic Search for Psychoactive Drugs*. Eds. D. H. Efron, B. Holmstedt, and N. S. Kline. U.S. Department of Health, Education and Welfare Publication no. 1645. Washington, DC: Government Printing Office.

Pinkley, H. V. 1969. Plant admixtures to *ayahuasca*, the South American hallucinogenic drink. *Lloydia* 32: 305–314.

Rivier, L., and J. E. Lindgren. 1972. Ayahuasca, the South American hallucinogenic drink: An ethnobotanical and chemical investigation. *Economic Botany* 29: 101–129.

Schultes, R. E. 1957. The identity of the Malpighiaceous narcotics of South America. *Botanical Museum Leaflets* (Harvard University) 18: 1–56.

———. 1970. The botanical and chemical distribution of hallucinogens. *Annual Review of Plant Physiology* 21: 571–598.

———. 1972. Ethnotoxicological significance of additives to New World hallucinogens. *Plant Science Bulletin* 18: 34–41.

Schultes, R. E., and A. Hofmann. 1980. *The Botany and Chemistry of Hallucinogens*. 2nd ed. Springfield, IL: Charles C. Thomas.

Siskind, J. 1973. Visions and cures among the Sharanahua. In *Hallucinogens and Shamanism*. Ed. M. J. Harner. New York: Oxford University Press.

Soukup, J. 1970. *Vocabulario de los Nombres Vulgares de la Flora Peruana*. Lima: Colegio Salesiano.

Tessmann, G. 1930. *Die Indianer Nordost-Perus, Grundlegende Forschungen für eine Systematische Kulturkunde*. Hamburg: Friederichsen, De Gruyter & Company.

Williams, L. 1936. *Woods of Northeastern Peru*. Field Museum of Natural History, vol. 15, no. 377. Chicago.

Table 1. Biologically active constituents in ayahuasca admixtures.

Family: Genus & species	Vernacular name	Biodynamic constituents	References
Acanthaceae: *Teliostachya lanceolata* Nees var. *crispa* Nees in Mart.	Toe negro	None reported	43, 61
Amaranthaceae: *Iresine* sp. P. Br. *Alternanthera lehmanii* Ilieron	– Picurullana-quina	Cinnamic acide amides None reported	43, 45 54, 60, 73

Apocynaceae:
Malouetia tamaquarina (Aubl.)
    A. DC.
Tabernaemontana sp. L.
 	
Himatanthus succuba (Spruce)
    Woods
	Cuchura-caspi
	Uchu-sanango
	
	Bellaco-caspi
		Steroid alkaloids, conopharyngine
		Bis-indole alkaloids, terpenoids,
		    cornaridine
		Fulvoplumieron, flavonoids
			31, 46, 60
			11, 32, 33, 43, 60, 72, 76
			
			43, 51, 52

Araceae: *Montrichardia arborescens* Schott	Raya balsa	None reported	43

Bignoniaceae:
Mansoa alliacea (Lem.) Gentry[1]
Tabebuia heteropoda (DC.)
    Sandwith.[2]
 	
Tynnanthus panurensis (Bur.)
    Sandwith.
	Ajo sacha
	
	Tahuari
	
	Clavo huasca
		None reported
		
		Dibenzoxanthines, lapachol
		    naphthoquinones
		
		None reported
			43
			
			34, 43, 56, 82
			
			43

Bombacaceae: *Ceiba pentandra* L.[3] *Cavanillesia hylogeiton* Ulb.	Lupuna Puca lupuna	None reported ?	43 41, 43
Cactaceae: *Opuntia* sp. Mill. *Epiphyllum* sp. Haw	Tchai Pokere	N-Me-tyramine, mescaline None reported	50, 57, 78 57
Caryocaraceae: *Anthodiscus pilosus* Ducke	Tahuari	None reported	43

Celastraceae:
Maytenus ebenifolia Reiss[4]
	Chuchuhuasa	Sesquiterpene and nicotinoyl
		    alkaloids, triterpenes, maytensine,
		    tingonane, ansa macrolides, etc.
			19, 37, 39, 43, 49, 71, 81

Cyclanthaceae: *Carludovica divergens* Ducke	Tamshi	None reported	43
Cyperaceae: *Cyperus* sp.[5] *Cyperus digitatus* Roxb.	Piri-piri Chicorro	Quinones, essential oils, saponins, sesquiterpenes None reported	2, 24, 36, 43, 69 43

Euphorbiaceae:
Alchornea castaneaefolia (Willd.)
    Juss.
 	
Hura crepitans L.[6]
	Hiporuru
	
	Catahua
		Alchornine, imadazole and
		    corynanthe alkaloids, antifeedants
		Tigliane diterpenes, piscicidal
		    compounds, lectins
			22, 30, 43, 64, 72
			
			17, 43, 53, 59

Labiateae: *Ocimum micranthum* Willd.	Pichanga, abaca	Neolignans, sesquiterpenes, anthelmintics	13, 58, 60, 77

Family: Genus & species	Vernacular name	Biodynamic constituents	References
Lecythidaceae: *Couroupita guianensis* Aubl.[7]	Ayahuma	Indole alkaloids	6, 43, 65
Leguminoseae: *Calliandra angustifolia* Spruce	Bobinsana	Amino acids	12, 42, 43
Cedrelinga catenaeformis Ducke[8]	Huairacaspi	None reported	43
Erythrina glauca Willd.[9] *Erythrina poeppigiana* (Walp.) O. F. Cook	Amaciza, amasisa	*Erythrina* alkaloids	14, 26, 28, 43, 73
Pithecellobium laetum Poepp. & Endl.	Remo caspi	Phytomitogens, lupeopl spinasterol	21, 43, 83
Sclerolobium setiferum Ducke	Palosanto	None reported	43, 73
Vouacapoua americana Aubl.[10]	Huacapu	None reported	43
Lomariopsidaceae: *Lomariopsis japurensis* (Mart.) J. Sm.	Shoka	None reported	57
Loranthaceae: *Phrygilanthus eugenoides* var. *robustus* Glaz.	Miya	None reported	57
Phthirusa pyrifolia (H.B.K.) Eichler	Suelda con suelda	None reported	43
Marantaceae: *Calathea veitchiana* Veitch. èx Hooker	Pulma	Tryptophan	60, 74

Menispermaceae:
Abuta grandifolia (Mart.)
   Sandwith.[11]
	Abuta caimitillo,
	   sanango
		Tropolone isoquinolines,
		   palmatine, oxo-aporphines
			43, 44, 66, 67, 68, 70

Moraceae: *Coussapoa tessmannii* Mildbr.	Renaco	None reported	43
Ficus ruiziana Standl.	Renaco	Furocoumarins, triterpenes,	15, 16, 43, 79
Ficus insipida Willd.	Oje	biphenylhexahydroindolizines, phenanthroxindolizines	
Tovomita sp.[12]	Chullachaqui caspi	None reported	43

Myristicaceae:
Virola sp. Aubl.
	Cumala	Diaryl propanoids, tryptamines,
		   beta-carbolines, neolignans,
		   2-Me-ketones
			4, 5, 20, 43

Virola surinamensis Warb.	Caupuri	Neolignans	4, 43
Nymphaeaceae: *Cabomba aquatica* Aubl.	Murere, mureru	None reported	43, 73
Phytolaccaceae: *Petiveria alliacea* L.[13]	Mucura	Oligo sulfides, triterpenes, trithiolanes	1, 43, 62, 80
Polygonaceae: *Triplaris surinamensis* var. *chamissoana* Meisn.	Tangarana	None reported	43
Pontederiaceae: *Pontederia cordata* L.	Amaron borrachero	None reported	60, 61

(continued)

Table 1. Continued.

Family: Genus & species	Vernacular name	Biodynamic constituents	References
Rubiaceae: *Calycophyllum spruceanum* (Benth.) Hooker	Capirona negro	None reported	43
Guettarda ferox Standl.	Garabata	Cathenine, hetero-yohimbine alkaloids	25, 29, 43
Uncaria guianensis (Aubl.) Gmel.	Garabata	Spiro-oxindoles, bis-indoles, hetero-yohimbine alkaloids	3, 7, 11, 23, 38, 43, 72
Schizaceae: *Lygodium venustum* Sw.	Tachai del monte	Antifertility agents	18, 47, 57
Scrophulariaceae: *Scoparia dulcis* L.	Nuchu pichana	Triterpenes, 6-MeO-benzoxazolilinone	9, 40, 43, 48
Solanaceae: *Brugmansia suaveolens* (Willd.) Brechtold & Presl.[14]	Toe	Tropane alkaloids	10, 43, 54, 60, 61
Brunfelsia chiricsanango Plowman	Chiricsanango	Scopoletin, CNS depressants, anti-inflammatory compounds	8, 27, 43, 55, 60
Iochroma fuchsoides Meers in Hooker	Borrachero	None reported	54, 60, 61
Juanulloa ochracea Cuatr.	Ayahuasca	None reported	61
Verbenaceae: *Cornutia odorata* (Poepp. & Endl.) Poepp.	Shinguarana	None reported	44, 73
Vitex triflora Vahl.	Tahuari	Diterpene lactones, iridoid glycosides, flavonoid glycosides	35, 43, 63, 75

[1]Considered a very strong plant-teacher. Ingested, it is used for "learning medicine." Applied externally, it is used for rheumatic diseases and for good luck in work and love.

[2]A tree strong as iron (*"es un palo como fierro"*). A diet is required.

[3]A very strong *doctore*. It may kill you if the diet is not kept perfectly (Herbarium Amazonense, pers. com.).

[4]Taken by Lamisto shaman apprentices, a few weeks after ingesting *caapi*.

[5]Ayahuasqueros of Iquitos use several kinds of piri-piri as admixtures to ayahuasca. They are used to "teach medicine."

[6]A strong and dangerous *doctore*. It may kill you if the diet is not kept perfectly. Together with *patiquina* (*Dieffenbachia* species), it is used for destroying lakes inhabited by dangerous animals. It is also used as a strong purgative (informant Don Emilio Andrade, pers. com.).

[7]It cures strong sicknesses. Its spirit has no head. It is used to "fortify the body." The bark is cooked with *ayahuasca*. It requires a 30-day diet (informant Don Emilio Andrade, pers. com.).

[8]A big tree that can withstand tempests, lightning, and strong winds; thus it is a good "defense." A 30-day diet is required.

[9]*Vegetalistas* of Iquitos use the flower of this plant for stomach aches, and its bark for bathing persons with rheumatic diseases.

[10]A 30-day diet is required.

[11]In Puerto Maldonado, this plant is a remedy for lung sickness (tuberculosis?), stomach cancer, menstrual disturbances, and malaria (informant Don Manuel Pacherras, pers. com.).

[12]Four pieces of bark are cooked together with ayahuasca. The plant is used to strengthen the body. A 30-day diet is required.

[13]This plant is used as an admixture plant among Yagua shamans (Chaumeil 1983) and as a medicinal plant among the *vegetalistas* of Iquitos (Luna 1984b).

[14]Two or three leaves are required to see the witch who has harmed a patient.

REFERENCES FOR TABLE 1

1. Adesogan, E. K. 1974. Trithiolaniacin, a novel trithiolane from *Petiveria alliacea*. *Journal of the Chemical Society, Chemical Communications* 21: 906–907.

2. Allan, R. D., R. J. Wells, R. L. Correll, and J. K. MacLeod. 1978. The presence of quinones in the genus *Cyperus* as an aid to classification. *Phytochemistry* 17: 263–266.

3. Ban, Y., M. Seto, and T. Oishi. 1975. The synthesis of 3 spiro oxindole derivatives part 7: Total synthesis of alkaloids racemic rhynchophylline and racemic isorhynchophylline. *Chemical and Pharmaceutical Bulletin* 11: 2605–2613.

4. Barata, L. E. S., P. M. Baker, O. R. Gottlieb, and E. A. Ruveda. 1978. Neolignans of *Virola surinamensis*. *Phytochemistry* 17: 783–786.

5. Baruffoldi, R., E. Fedeli, and N. Cortesi. 1975. Study of the fat of *Virola surinamensis*. Part 1: Acidic and glyceride composition and chemical nature of some of the unsaponifiable components. *Revista de Farmacia e Bioquímica de Universidade de São Paulo* 13: 91–102.

6. Bergman, J., B. Egestad, and J.-O. Lindstrom. 1977. The structure of some indolic constituents in *Couroupita guianensis*. *Tetrahedron Letters* 30: 2625–2626.

7. Bindra, J. S. 1973. Oxindole alkaloids. In *The Alkaloids*, vol. 14. Ed. R. H. F. Manske. New York: Academic Press.

8. Chaubal, M., and R. P. Iyer. 1977. Carbon-13 NMR spectra of scopoletin. *Lloydia* 40: 618.

9. Chen, C.-M., and C.-T. Chen. 1976. 6-methoxybenzoxazolinone and triterpenoids from roots of *Scoparia dulcis*. *Phytochemistry* 15: 1997–1999.

10. Clarke, R. L. 1977. The tropane alkaloids. In *The Alkaloids*, vol. 16. Ed. R. H. F. Manske. New York: Academic Press.

11. Cordell, G. A., and J. E. Saxton. 1981. Bis-indole alkaloids. In *The Alkaloids*, vol. 20. Eds. R. H. F. Manske and R. G. A. Rodrigo. New York: Academic Press.

12. Dardenne, G. A. 1975. New free amino acids from Leguminoseae. *Phytochemistry* 14: 860.

13. Desai, D. G., N. S. Ambade, and R. R. Mane. 1982. Synthesis of ocimum, a new neolignan from *Ocimum americanum*. *Indian Journal of Chemistry*, section B, organic chemistry, 5: 491–492.

14. Dyke, S. F., and S. N. Quessy. 1981. *Erythrina* and related alkaloids. In *The Alkaloids*, vol. 20. Eds. R. H. F. Manske and R. G. A. Rodrigo. New York: Academic Press.

15. Eidler, Y. I., G. L. Genkina, and T. T. Shakirov. 1975. Quantitative determination of furocoumarins in *Ficus carica* leaves. *Khimiyo Prirodnykh Soedinenii* 3: 349–351.

16. Elgamal, M. H. A., B. A. H. El-tawail, and M. B. E. Payez. 1975. The triterpenoid constituents of the leaves of *Ficus nitida*. *Naturwissenschaften* 62: 486.

17. Evans, F. J., and C. J. Soper. 1978. The tigliane daphnane and ingenane diterpenes: Their chemistry, distribution, and biological activities; a review. *Lloydia* 4: 193–233.

18. Gaitonde, B. B., and R. T. Mahajan. 1980. Antifertility activity of *Lygodium flexuosum*. *Indian Journal of Medical Research* 72: 597–604.

19. Gonzales, J. G., G. Delle Monache, F. Delle Monache, G. B. Marini-Bettòlo. 1982. Chuchuhuasa, a drug used in folk medicine in the Amazonian and Andean areas of South America: A chemical study of *Maytenus laevis*. *Journal of Ethnopharmacology* 5: 73–78.

20. Gottlieb, O. R., and A. A. Loureiro. 1973. Distribution of diaryl propanoids in Amazonian *Virola* species. *Phytochemistry* 12: 1830.

21. Gunasekera, S. P., G. A. Cordell, and N. R. Farnsworth. 1982. Constituents of *Pithecellobium multiflorum*. *Journal of Natural Products* 46: 651.

22. Hankinson, B. L. H. 1982. Investigation of constituents and antifeedant activity of *Alchornea triplinervia*. *Dissertation Abstracts International*, sec. B, no. DA8217228.

23. Hemingway, S. R., and J. D. Phillipson. 1974. Alkaloids from South American species of *Uncaria* (Rubiaceae). *Journal of Pharmacy and Pharmacology* 26 (supplement): 1–113.

24. Hikino, H., and K. Aota. 1971. Structure and absolute configuration of α rotonol and β rotonol, sesquiterpenoids of *Cyperus rotundus*. *Tetrahedron Letters* 27: 4831–4836.

25. Husson, H. P., C. Kan-Fan, T. Sevener, and J. P. Vidal. 1977. Structure of cathenine, key intermediate in the biosynthesis of indole alkaloids. *Tetrahedron Letters* 22: 1889–1892.

26. Ito, K., M. Haruna, and H. Fukura. 1975. Studies on the *Erythrina* alkaloids. Part 10: Alkaloids of several *Erythrina* sp. plants from Singapore and Malaysia. *Yakuaaku Zasshi* 3: 358–362.

27. Iyer, R. P., J. K. Brown, M. G. Chaubal, and M. H. Malone. 1977. *Brunfelsia hopeana* I. Hippocratic screening and anti-inflammatory evaluation. *Journal of Natural Products* 40: 356–360.

28. Jackson, A. H., and A. S. Chawla. 1982. Studies of *Erythrina* alkaloids. Part IV: GC/MS investigations of alkaloids in the leaves of *E. poeppigiana*, *E. macrophylla*, *E. berteroana*, and *E. salviflora*. *Allertonia* 3: 39–45.

29. Kan-Fan, C., and H. P. Husson. 1978. Sterochemical control in the biomimetic conversion of heteroyohimbine alkaloid precursors: Isolation of a novel key intermediate. *Journal of the Chemical Society, Chemical Communications* 14: 618–619.

30. Khoung-Huu, F., and J. P. Le Forestier. 1972. Alchornine, isoalchornine, and alchorninone, products isolated from *Alchornea floribunda*. *Tetrahedron Letters* 28: 5207–5220.

31. Khoung-Huu, F., and M.-J. Magdelaine. 1973. Steroid alkaloids from *Malouetia brachylobe* and from *Malouetia heudelotii*. *Phytochemistry* 7: 1813–1816.

32. Kingston, D. G. 1978. Plant anti-cancer agents part 6: Isolation of voacangine, voacamine and epi-voacorine from *Tabernaemontana arborea* sap. *Journal of Pharmaceutical Sciences* 67: 271–272.

33. Kingston, D. G., B. B. Gerhart, F. Ionescu, M. M. Mangino, and S. K. Sami. 1978. Plant anti-cancer agents part 5: New bis-indole alkaloids from *Tabernaemontana johnstonii* stem bark. *Journal of Pharmaceutical Sciences* 67: 249–251.

34. Kingston, D. G., and M. M. Rao. 1980. Isolation, structure elucidation and synthesis of 2 new cytotoxic naphthoquinones from *Tabebuia cassinoides*. *Planta Medica* 40: 230–231.

35. Kodanda, R. U., R. E. Venkata, and R. D. Venkata. 1977. Phenolic constituents of the bark of *Vitex negundo*. *Indian Journal of Pharmacy* 39: 41.

36. Kokate, C. K., and K. C. Varma. 1977. Pharmacological study on essential oil of *Cyperus scariosus* Part I: Effect on central nervous system. In *Advances in Essential Oil Industry*. Eds. L. D. Kapoor and R. Krishnan. New Delhi: Today & Tomorrow's Printers & Publishers. 189–193.

37. Kupchan, S. M., and R. M. Smith. 1977. Maytoline, maytine, and matolidine, novel nicotinoyl sesquiterpenes alkaloids from *Maytenus serrata*. *Journal of Organic Chemistry* 42: 115–118.

38. Lavault, M., C. Moretti, and J. Bruneton. 1983. Alkaloids of *Uncaria guianensis*. *Planta Medica* 47: 244–245.

39. Leboeuf, M. 1978. Maytensine and maytensinoids. *Plantes Médicinales et Phytotherapie* 12: 53–70.

40. Li, J., Y. Li, R. Nie, and J. Zhou. 1981. Coixol and betulinic acid of *Scoparia dulcis* L. *Yun-nan Chih Wu Yen Chui* 3: 475–477.

41. Lopez Guillén, J. E., and I. K. De Cornelio. 1974. Medicinal plants of Peru, part 5. *Biota* 10: 76–104.

42. Lopez Guillén, J. E., and I. K. De Cornelio. 1974. Medicinal plants of Peru, part 4. *Biota* 10: 28–56.

43. Luna, L. E. 1984. The concept of plants as teachers among four mestizo shamans of Iquitos, Peru. *Journal of Ethnopharmacology* 11: 135–156.

44. McKenna, D., G. H. N. Towers, and F. S. Abbott. 1984. Monoamine oxidase inhibitors in South American hallucinogenic plants: Tryptamine and β-carboline constituents of ayahuasca. *Journal of Ethnopharmacology* 10: 195–223.

45. Martin Tanguy, J., F. Cabanne, E. Perdrizei, and C. Martin. 1978. The distribution of hydroxy-cinnamic acid amides in flowering plants. *Phytochemistry* 11: 1927–1928.

46. Medina, J. D., and R. Bracho. 1976. Constituents of the bark of *Malouetia glandulifera*. *Planta Medica* 29: 367–369.

47. Mehrotra, P. K., and V. P. Kamboj. 1978. Hormonal profile of coronaridine hydrochloride, an antifertility agent of plant origin. *Planta Medica* 33: 345–349.

48. Mohato, S. B., M. C. Das, and N. P. Sahu. 1981. The terpenoids of *Scoparia dulcis*. *Phytochemistry* 20: 171–173.

49. Nozaki, H., H. Suzuki, K.-H. Lee, and A. T. Mcphail. 1982. Structure and stereochemistry of maytenfolic acid and maytenfoliol, two new anti-leukemic triterpenes from *Maytenus diversifolia*: X-ray crystal structures. *Journal of the Chemical Society, Chemical Communications* 10: 1048–1051.

50. Pardanani, J., B. N. Meyer, and J. L. McLaughlin. 1978. Mescaline and related compounds from *Opuntia spinosior*. *Lloydia* 41: 286–288.

51. Paris, R. R., and S. Duret. 1974. The flavonoids of various Apocynaceae. *Plantes Médicinales et Phytotherapie* 8: 318–325.

52. Perdu, G. P., and R. N. Blomster. 1978. South American plants part 3: Isolation of fulvoplumieron from *Himatanthus succuba* (Apocynaceae). *Journal of Pharmaceutical Sciences* 67: 123–132.

53. Pere, M., D. Pere, and P. Rouge. 1981. Isolation and studies of the physicochemical and biological properties of lectins from *Hura crepitans*. *Planta Medica* 41: 344–350.

54. Pinkley, H. V. 1969. Plant admixtures to *ayahuasca*, the South American hallucinogenic drink. *Lloydia* 32: 305.

55. Plowman, T. 1977. *Brunfelsia* in ethnomedicine. *Botanical Museum Leaflets* (Harvard University) 25: 289–320.

56. Prakash, L., and R. Singh. 1981. Chemical examination of the leaves and stem heartwood of *Tabebuia pentaphylla* (Linn.) Hemsl. (Bignoniaceae). *Journal of the Indian Chemical Society* 58: 1122–1123.

57. Rivier, L., and J. E. Lindgren. 1972. Ayahuasca, the South American hallucinogenic drink: An ethnobotanical and chemical investigation. *Economic Botany* 29: 101–129.

58. Roy, R. G., N. M. Madesayaa, R. B. Ghosh, D. V. Gopalkrishnan, N. N. Murthy, T. J. Doraira, and N. L. Sitaraman. 1976. Study on inhalation therapy by an indigenous compound on *Plasmodium vivax* and *Plasmodium falciparum* infections: A preliminary communication. *Indian Journal of Medicinal Research* 10: 1451–1455.

59. Sakata, K., and K. Kawazu. 1971. Studies on a piscicidal constituent of *Hura crepitans* D. Part 1: Isolation and characterization of hura toxin and its piscicidal activity. *Agricultural and Biological Chemistry* 35: 1084–1091.

60. Schultes, R. E. 1972. Ethnotoxicological significance of additives to New World hallucinogens. *Plant Science Bulletin* 18: 34–41.

61. Schultes, R. E. 1979. New data on the malpighiaceous narcotics of South America. *Botanical Museum Leaflets* (Harvard University) 23: 137ff.

62. Segelman, F. P., and A. B. Segelman. 1975. Constituents of *Petiveria alliacea* (Phytolaccaceae). Part 1: Isolation of isoarborinol, isoarborinol acetate and isoarborinol cinnamate from the leaves. *Lloydia* 38: 537.

63. Sehgal, C. K., S. C. Taneja, K. L. Ohar, and C. K. Atal. 1982. 2'p-hydroxylbenzoyl musaenosidic acid, a new iridoid glycoside from *Vitex negundo*. *Phytochemistry* 2: 363–366.

64. Seigler, D. S. 1977. Plant systematics and alkaloids. In *The Alkaloids*, vol. 16. Ed. R. H. F. Manske. New York: Academic Press.

65. Sen, A. K., and S. B. Mahato. 1974. Couroupitine, a new alkaloid from *Couroupita guianensis*. *Tetrahedron Letters* 7: 609–610.

66. Setor de Fitoquimica, INPA, Manaus, Amazonas, Brazil. 1971. Chemical composition of Amazonian plants. *Acta Amazonica* 1: 83–86.

67. Shamma, M., and R. L. Castenson. 1973. The oxoaporphine alkaloids. In *The Alkaloids,* vol. 14. Ed. R. H. F. Manske. New York: Academic Press.

68. Silverton, J. V., C. Kabuto, K. T. Buck, and M. P. Cava. 1977. Structure of Imerubrine, a novel condensed tropolone isoquinoline alkaloid. *Journal of the American Chemical Society* 99: 6708–6712.

69. Singh, P. N., and S. B. Singh. 1980. A new saponin from mature tubers of *Cyperus rotundus*. *Phytochemistry* 19: 2056.

70. Skiles, J. W., J. M. Saa, and M. P. Cava. 1979. Splendidine, a new oxoaporphine alkaloid from *Abuta rufescens*. *Canadian Journal of Chemistry* 57: 1642–1646.

71. Smith, D. M. 1977. The Celastraceae alkaloids. In *The Alkaloids*, vol. 16. Ed. R. H. F. Manske. New York: Academic Press.

72. Snieckus, V. 1968. Distribution of indole alkaloids in plants. In *The Alkaloids*, vol. 11. Ed. R. H. F. Manske. New York: Academic Press.

73. Soukup, J. 1970. *Vocabulario de los Nombres Vulgares de la Flora Peruana*. Lima: Colegio Salesiano.

74. Splittstoesser, W. E., and F. W. Martin. 1975. The tryptophan content of tropical roots and tubers. *HortScience* 10: 23–24.

75. Taguchi, H. 1976. Studies on the constituents of *Vitex cannibolia*. *Chemical and Pharmaceutical Bulletin* 7: 1668–1670.

76. Talapatra, B., A. Patra, and S. K. Talapatra. 1975. Terpenoids and alkaloids of the leaves of *Tabernaemontana coronaria*. *Phytochemistry* 14: 1652–1653.

77. Terhune, S. J., J. W. Hoff, and R. W. Lawrence. 1974. Bicyclosesquiphellandrine and 1-epi-bicyclosesquiphellandrine: 2 new dienes based on the cadalene skeleton. *Phytochemistry* 13: 183–185.

78. Vander Veen, R. L., and L. G. West. 1974. N-methyl tyramine from *Opuntia clavata*. *Phytochemistry* 13: 866–867.

79. Venkatachalam, S. R., and N. B. Mulchandani. 1982. Isolation of phenanthroindolizine alkaloid from *Ficus hispida*. *Naturwissenschaften* 69: 287–288.

80. von Szczepanski, C., J. Heindl, G.-A. Hoyer, and E. Schroeder. 1977. Biologically active compounds from plants. Part 2: Synthesis and anti-microbial activity of some dissymmetric oligo sulfides. *European Journal of Medicinal Chemistry* 12: 279–284.

81. Wagner, H., and J. Burghart. 1977. Spermidine alkaloids and triterpenes from *Maytenus heterophylla* ssp. *heterophylla* and *Pleurostyla africana*: Chemical constituents of the Celastraceae. Part 4. *Planta Medica* 32A: 9-14.

82. Wong, R. Y., K. J. Palmer, G. D. Manners, and L. Jurd. 1976. The structure of guayacanin with acetone of crystallization, a naturally occurring dibenzoxanthine from *Tabebuia guayacan*. *Acta Crystallographica*, sec. B, 8: 2396–2400.

83. Yadav, M., and V. K. C. Ganaswaran. 1976. Phytomitogens of tropical legumes part I: Isolation from *Parkia* sp. and *Pithecellobium jiringa*. *Malaysian Journal of Science* 4: 25–26.

The Urgent Need for the Study of Medicinal Plants

PLUTARCO NARANJO

In the dawn of humankind, the art of curing was essentially magical (Pardal 1937; Lain-Entralgo et al. 1982; Naranjo 1984). Few plants were used and those employed were usually psychoactive ones, now known as magic or psychedelic plants. Later, empirical medicine arose, using many plants for the treatment of various afflictions. This tendency culminated, in the Old World, in the famous work *Materia Medica* by Dioscorides, published in the first century A.D., in which the characters and properties of numerous drugs are described, the majority of which come from the Plant Kingdom.

Scientific medicine arose much later. The seventeenth century, in particular, is that which opened the doors to true scientific understanding of medicine and, although the art of curing was refined, plants continued to occupy a preeminent position. Even though the first British pharmacopeia, that of the Royal College of Physicians, was published in 1618, the nineteenth century was the century of pharmacopeias. Several countries published their own pharmacopeias, incorporating drugs, the therapeutic effects of which were already "proven," at least according to the procedures and techniques of the period. Until 1930, around 90 percent of the official medicines were of plant origin (Swain 1972).

Chemistry had also realized important progress during the nineteenth century, and by the beginning of the twentieth century, the first artificial pharmaceuticals were obtained by synthesis, among these phenacetin, urea, barbital, and acetylsalicylic acid or aspirin. The fruitful period of chemotherapy began in the 1930s with the synthesis of the sulphonamides. The era of antibiotics began in the following decade, and when the Second World War ended, conditions became more favorable for the development of synthetic chemistry, to the point that in a few decades therapy was radically transformed. Since the 1960s, nearly 90 percent of all standard medicines are of synthetic origin or the product of fermentation, lowering medicines of plant origin to a secondary and reduced role.

Even though emphasis persists in research of synthetic compounds, a certain interest in medicinal plants has been reborn. Nonetheless, the attention that is given to this important field in science is totally insufficient. As previously analyzed by many authors (Farnsworth and Bingel 1977; von Reis Altschul 1973; OTA 1983; Balandran et al. 1985; Duke 1985), there are many powerful reasons for doing broad ethnobotanical, ethnopharmacological, and even clinical therapeutic research. Among these reasons, the following could be cited:

1. To rescue knowledge in imminent danger of being lost.

2. The utility of plants in current therapy.

3. To find new molecular models in plants.

4. The usefulness of plants in the development of physiopathology.

5. The wide use of plants in folk medicine.

6. To obtain intermediate chemicals.

To Rescue Knowledge in Imminent Danger of Being Lost

An inventory of medicinal plants compiled by the World Health Organization (WHO 1978) and encompassing only ninety member countries gave the large figure of 20,000 species, of which only 250 were of widespread use or had been analyzed to identify their main active chemical compound(s). That sample, even though a partial one, reveals the enormous empirical traditional knowledge about medicinal plants. Most of this knowledge is verbal and only incompletely incorporated in historical and folkloric works. The aboriginal knowledge is the fruit of centuries and, in some cases, millennia of plant use.

In this period of aerial transport, radio, and television, the accelerated process of acculturation that destroys ancient values, on the one hand, and epidemics caused by microorganisms (against which indigenous populations have no immune defenses), on the other hand, have brought about the extinction of certain tribes and with them their knowledge. If one reviews the texts of the first Latin-American histories, for example, it can be proven that since the discovery of the New World, hundreds of communities and tribes have disappeared. Currently, the progressive penetration of civilization into the immense Amazon region is causing the same effect. In other areas of the planet, acculturation has also relegated ancestral knowledge to partial or total oblivion. Finally, the uncontainable and at times irreversible transformation of ecosystems, the depredation of extensive natural areas, and other factors have led to the disappearance of many species. For example, of the thousands of Mexican medicinal plants cited by Hernández (1959) in the sixteenth century, many can no longer be found.

The Utility of Plants in Current Therapy

Despite the enormous availability of medicines and, above all, of pharmaceutical specialties, plants have a place in current therapy, as can be justified at least by the four following reasons.

First, there is renewed interest in using plants in therapy. Such is the case of *Artemisia* (Klayman 1985), a source of quinine. Behind the therapeutic success of chloroquine and its synthetic derivatives in the treatment of malaria, the use of quinine passed into a chapter in the history of medicine. Through a biological phenomenon that is now well studied, even at the molecular level, bacteria and parasites can develop resistance to chemotherapeutics; that is, they undergo selection that results in resistance to a particular chemical compound. This process has occurred, in part, with species of *Plasmodium*, the causative agent of malaria, to a point that synthetic antimalarial drugs have lost such a significant part of their efficiency in the last quarter of the twentieth century that it has often been necessary to return to the use of quinine. Currently, there is such a great demand for the plant alkaloid that extraction laboratories cannot satisfy the growing demand, maximized now that malaria has again become a great health risk in tropical areas. The demand has been accentuated further by the resistance of the insect vector, the *Anopheles* mosquito, to insecticides that were used in the twentieth century in sanitation campaigns.

Second, many plant-derived drugs have not yet been improved upon. Traditional medicine depends on a number of plants that are currently used in scientific medicine although they have not yet been improved upon. Such is the case of *Digitalis purpurea* L. and *D. lanata* Ehrh. Many other drugs exist to which therapeutic effects have been attributed. As is well known, synthetic chemistry has until now had little success in obtaining drugs effective in the treatment of various viral diseases; even though immunotherapy has achieved great success, we still do not have vaccines for all viral diseases. It is possible that plants may be useful to treat these diseases. An example from Ecuador is *La nigua* (*Margyricarpus setosus* Ruíz and Pavón), the roots of which, in infusion, are used in the symptomatic treatment of measles. Other examples are *nachag* (*Bidens humilis* Sessé and Moc.), the flowers or branches of which, in infusion, are used in the treatment of infectious hepatitis, or the latex of several species of Euphorbiaceae, especially of the genera *Croton* and *Euphorbia*, which in topical form are used in the treatment of common warts. Furthermore, numerous plants are known for certain antineoplastic effects (Cassady and Douros 1980), modifiers of fertility (Moreno and Schwartzman 1975), and other effects (Perdue and Hartwell 1969).

One must note, of course, that controlled studies are needed, for example, by the double blind system, to confirm the therapeutic effect of these plants. Nonetheless, in folk medicine these plants are employed with apparently favorable results, and above all without causing detectable unfavorable side effects.

Third, some plants are useful as coadjuvants of other treatments. These plants can help in the treatment of certain diseases, even though there already exists a basic recommended medicine. For example, *guayusa* (*Ilex guayusa* Loes.) is famous for being antidiabetic (Varea 1922) and is used as a coadjuvant of insulin derived from animal or synthetic sources. *Calaguala*, ferns of the genus *Polypodium*, constitutes another example. An infusion of the rhizomes is used as a coadjuvant in the treatment of some skin diseases, especially psoriasis, an affliction that still has no specific treatment. *Matico* (*Eupatorium glutinosum* Lamb.) is another example of a plant to which many therapeutic properties have been attributed, such as antiseptic and antiinflammatory; it has also been used in the treatment of gastric ulcers.

Fourth, plants are useful in the treatment of mild afflictions that are common and frequent, such as the simple common cold and mild diarrhea. These and other illnesses pass with little or no treatment; hygienic and dietary measures help, and infusions of some plants well known in folk medicine throughout the world can contribute to improving the state of the patient. Numerous species in this category could be named.

To Find New Molecular Models in Plants

Chemists have developed enormous technical capacity to obtain the basic molecular structure of tens or hundreds of derivatives with different side chains, all of which give the impression of an unlimited proliferation of modern drugs. For example, on the basis of a single sulfa drug with therapeutic properties, chemists have obtained hundreds of synthetic drugs, although not all have entered the therapeutic field. The fundamental problem is to develop new basic molecular structures with appropriate therapeutic properties and minimal side effects or toxicity.

Several chapters of pharmacology have developed from a basic molecular structure found in its natural form in drugs of plant origin. Such is the case of synthetic antimalarial drugs, based on the structure of quinine, or the synthetic antispasmodic (anticonvulsive) based on the structure of atropine, or synthetic analgesics based on the structure of cocaine, and so on.

The utility of plants in this period of fascinating technological revolution cannot be considered a closed book. On the contrary, while the capacity of chemists to modify a molecular structure is almost unlimited, the capacity to invent or create new structures with therapeutic properties has been limited. In the meantime the Plant Kingdom offers us thousands of new molecules (Evans et al. 1982; Gottlieb 1982), fruits of the most interesting and often bizarre nature. The study of those molecules identified as "active compounds" is indispensable. Phytochemical investigations carried out during the 1970s and 1980s have discovered a number of alkaloids and other pharmacologically active substances that are currently being studied and that can possibly serve as models for new synthetic compounds (Barz and Ellis 1980).

Among the plants that have become part of standard therapy, one can mention *Vinca rosea* L. (synonym *Catharanthus roseus* [L.] G. Don f.) from which numerous alkaloids with anticarcinogenic activity have been obtained, particularly antileukemic drugs. Its two principal alkaloids constitute a new molecular model, and synthetic chemistry is now trying to obtain similar compounds.

In pre-Columbian America, to cite another example, several species of the genus *Lippia*, among other plants, were used as sweeteners (Compadre et al. 1985, p. 417). The Spaniards, who learned sugar-making from the Arabs, brought sugar cane to the New World. Sucrose very quickly replaced the native plant and relegated it to total oblivion. Not until the twentieth century was *L. dulcis* Trevir chemically analyzed and a new substance discovered, for which the name *hernandulcin* has been suggested. Hernandulcin is a thousand times sweeter than sucrose, and its molecular structure will serve to synthesize new sweeteners for diabetics as well as many candies and other sweets for children, without the disadvantage of feeding bacteria that cause cavities.

Thus, plants are sources of dyes, flavors, aromas, perfumes, insecticides (Schmeltz 1971), antiparasitic drugs, and many other substances, so that the broadest and most exhaustive phytochemical study is justified.

The Usefulness of Plants in the Development of Physiopathology

The concept of intracellular or membrane "chemical receptors" has been very fertile in the fields of physiology and especially pharmacology. If a substance is to have a pharmacodynamic effect, it must bind to specific molecules located on the exterior or the interior of cellular membranes. This binding activates different biochemical systems that have, as a result, a physiological or pharmacological response. Compounds produced by the organism itself, such as hormones and other so-called chemical mediators, act through this initial mechanism.

In times past, we knew of the pharmodynamic effects of substances such as nicotine, obtained from tobacco, or muscarine, obtained from a mushroom. Now it is known that these substances are able to selectively block two different receptors of acetylcholine at the level of striated muscle-fiber and other cells. Likewise, it was discovered that strychnine, from *Strychnos*, blocks the effect of another chemical mediator, namely, gamma aminobutiric acid (GAMA) that acts at the level of certain neurons. Some alkaloids and other compounds of plant origin have served to clarify the normal and pathological mechanism of action of some chemical mediators.

In the last half of the twentieth century, the central nervous system has been the subject of thousands of studies in which it has been found that numerous chemical mediators of extremely variable molecular structure intervene (Naranjo 1970; Lynch and Baundry

1984). Among them are the endorphin polypeptides, the activity of which is blocked by morphine and other similar alkaloids, and serotonin, the receptors of which are blocked by many plant drugs derived from the tryptamine nucleus.

In conclusion, phytochemistry can contribute to a better understanding of physiology and pathology and perhaps secondarily to the synthesis of new drugs with therapeutic properties.

The Wide Use of Plants in Folk Medicine

All the wonderful progress of synthetic chemistry and of science in general, unfortunately, has not served to alleviate and cure all the sickness in the world. According to surveys and other research carried out in different countries, scientific (or standard) medicine in developing countries (Naranjo 1981) serves only a minority (estimated at 30 to 50 percent of the total population), while the rest of the population attends to its health needs through the process called traditional medicine, aboriginal medicine, or folk medicine, processes based essentially on the use of low-cost medicinal plants that are easily accessible to the entire population.

Since 1977, WHO has encouraged the study of traditional medicine with hopes of obtaining the benefits that it could possibly yield, while at the same time avoiding the irrational or harmful effects that this type of medicine can have. To reach its goal of "health for everyone in the year 2000," WHO has suggested that governments incorporate the favorable aspects of traditional medicine, especially the use of medicinal plants, in their primary health care procedures.

One positive aspect of the use of medicinal plants is their low cost compared to the high price of new synthetic drugs, which have become totally inaccessible to the vast majority of people. Another consideration in favor of the use of medicinal plants, when they are the only recourse available, is that they have no harmful side effects. Synthetic drugs, in general, have very potent pharmodynamic effects; but as they are active, many also have strong and possibly dangerous and harmful side effects. Between 3 and 5 percent of patient hospital admissions are attributed to side effects of pharmaceuticals. On the contrary, medicinal plants, with a few exceptions, do not have great therapeutic potency, but neither do they have intense or serious side effects. Therefore, their direct administration in folk medicine offers little risk.

Thus, there exists a wide field for research in the phytochemistry of those hundreds of plants that are used in folk medicine in each country, research confirming the presence of pharmacodynamic chemicals such as alkaloids, glucosides to a lesser degree, and essential oils and other substances, indispensable knowledge that justifies the practices of naturalist and folk medicine.

To Obtain Intermediate Chemicals

The discovery of cortisone and hydrocortisone as natural hormones that carry out multiple functions in organisms opened an enormous field in physiology, therapy, and synthetic and semisynthetic chemistry.

Synthesis of the steroid nucleus, from which different corticosteroids are derived, as well as estrogen and progesterone for therapeutic use, is not easy and, above all, is not economical. It was found that a basic nucleus existed in some plants in nature; thereafter the necessary studies were carried out, and on this basis a large industry for extraction and semisynthesis has been developed.

Going back to the traditions of Mexican medicinal plants it was discovered that yams (Kreig 1964), plants belonging to the genus *Dioscorea*, contained a steroid appropriate for the synthesis of hormones. Later this steroid was found in plants of the genus *Agave*. Of these, the most utilized species grows in Africa—*Agave sisalana* Perrine. Many other genera contain steroids and saponins.

In South America, there are numerous plants known for centuries for their capacity to produce suds and toxic effects. One of these, *Solanum marginatum* L.f. (Varea 1922), is known by the common name *huapag*. This plant has a high content of solasodine, another intermediate in the synthesis of corticosteroids. Merck has installed a factory in Ecuador for the extraction of such a substance and has developed the necessary culture techniques to obtain sufficient raw material. This is another important chapter opening to new research.

In conclusion, it would be worth reviewing work that has been carried out in species of *Agave*. Similarly, it would be interesting to study the saponins of quinoa (*Chenopodium quinoa* Willd.), *chocho* (*Lupinus mutabilis* Lindley), and other plants that could well constitute new sources of intermediate chemical compounds.

LITERATURE CITED

Balandran, M. F., J. A. Klocke, E. S. Wurtele, and H. Bollinger. 1985. Natural plant chemicals: Sources of industrial and medicinal materials. *Science* 228: 1154.

Barz, W., and E. Ellis. 1980. *Natural Products as Medicinal Agents*. Eds. J. L. Bel and E. Reinhard. Stuttgart: Hippokrates.

Cassady, J., and J. Douros, eds. 1980. *Anticancer Agents Based on Natural Product Models*. New York: Academic Press.

Compadre, C. M., J. M. Pezzuto, A. D. Kinghorn, and S. K. Klamath. 1985. Hernandulcin: An intensely sweet compound discovered by review of ancient literature. *Science* 227: 417–419.

Duke, J. 1985. Medicinal plants. *Science* 229: 1036.

Evans, D., J. Nelson, and T. Taber. 1982. Topics in stereochemistry. In *Interscience*. Eds. N. L. Alinger and E. L. Eliel. New York.

Farber, E. M., and A. J. Cox, eds. 1982. *Proceedings of the Third International Symposium on Ethnopharmacology*. Stanford University. New York: Grune & Stratton.

Farnsworth, N., and A. Bingel. 1977. *New Natural Products and Plant Drugs with Pharmaceutical, Biological, or Therapeutical Activity*. Eds. H. Wagner and P. Wolff. New York: Springer-Verlag.

Gottlieb, O. R. 1982. *Micromolecular Evolution, Systematics and Ecology*. Berlin: Springer-Verlag.

Hernández, F. 1959. *Historia Natural de Nueva España*. Universidad Nacional Autónoma de México.

Klayman, D. L. 1985. *Qinghaosu* (Artemisinin): An antimalarial drug from China. *Science* 228: 1049.

Kreig, M. B. 1964. *Green Medicine*. Skokie, IL: Bantam Books.

Lain-Entralgo, P., et al. 1982. *Historia Universal de la Medicina*. Vol. 7. Madrid: Salvat Edit.

Lynch, G., and M. Baundry. 1984. The biochemistry of memory: A new and specific hypothesis. *Science* 224: 1057.

Moreno, R., and B. Schwartzman. 1975. 268 plantas medicinales utilizadas para regular la fertilidad en algunos países de Sudamérica. *Reproducción* 2: 163.

Naranjo, P. 1970. Plantas psicotomiméticas y bioquímicas de la mente. *Terapia* 25: 87.

———. 1981. Farmacología y medicina tradicional. In *Fundamentos de Farmacología Médica*. Eds. E. Samaniego and R. Escaleras. Quito: Univ. Central.

———. 1984. La Medicina en el Ecuador preincaico. *Rev. Ecuat. de Medicina* 20: 93.

Office of Technology Assessment (OTA). 1983. *Plants: The Potentials for Extracting Protein, Medicines, and other Useful Chemicals*. Washington, DC: OTA.

Pardal, R. 1937. *Medicina Aborigen Americana*. Buenos Aires: J. Anesi.

Perdue, R. E., Jr., and J. L. Hartwell. 1969. The search for plant sources of anticancer drugs. *Morris Arboretum Bulletin* 20(3): 35–53.

von Reis Altschul, S. 1973. *Drugs and Foods from Little-Known Plants*. Cambridge, MA: Harvard University Press.

Schmeltz, I. 1971. *Naturally Occurring Insecticides*. Eds. M. Jacobson and D. G. Crosby. New York: Dekker.

Swain, T., ed. 1972. *Plants in the Development of Modern Medicine*. Cambridge, MA: Harvard University Press.

Varea, M. 1922. *Botánica Médica Nacional*. Latacunga: Imprenta León.

WHO (World Health Organization). 1978. *Promoción y Desarrollo de la Medicina Tradicional*. Ginebra.

Considerations in the Multidisciplinary Approach to the Study of Ritual Hallucinogenic Plants

PETER A. G. M. DE SMET

Ethnobotanical investigations have perhaps nowhere arrived at more scintillating results than in the domain of hallucinogenic plants. Only a few Western clinicians consider hallucinogens to have any therapeutic value.[1] According to a chapter in an authoritative pharmacological textbook, the use of LSD (lysergic acid diethylamide) has been abandoned, either because controlled studies have failed to demonstrate its therapeutic value, or because the elaborate precautions required to minimize adverse psychological reactions dampened enthusiasm and rendered its therapeutic use impractical (Jaffe 1980).

Nonetheless, scientific studies on ritual psychoactive drugs may have an impact on medical care in Western society. First, they may lead to an improved clinical management of careless Western youngsters who arrive in emergency wards after self-experiments with herbal "highs" derived from native practices (Hall et al. 1978; Stienstra et al. 1981; Young et al. 1982). Second, they may provide new pharmacological tools for neurochemical research. For instance, the *Banisteriopsis* alkaloid harmaline has turned out to be a valuable selective inhibitor of MAO-A enzymes (Fuller et al. 1981). Third, they may result in the discovery of synthetic substances with potentially therapeutic properties. An example is the development of specific agonists of the central GABA-ergic system from the fly agaric constituent muscimol (Falch et al. 1984). Fourth, might there never come another time when the present categorical condemnation of hallucinogenic drug therapy will be reevaluated (Grieco and Bloom 1981)?

Aside from any direct or indirect medical significance, ritual native intoxication is a fascinating area of research in its own right. It constitutes an important culture trait of primitive people, which, like other parts of our cultural heritage, deserves to be carefully documented and evaluated in a multidisciplinary approach (Bruhn and Holmstedt 1981; De Smet and Rivier 1989).

In various parts of the world, aboriginal peoples and tribes have taken intoxicating vegetal preparations as facilitating agents in religious trance induction, divination, witchcraft, and healing ceremonies (Efron et al. 1967; Furst 1976; Schultes and Hofmann 1980; Volger et al. 1981). The recovery of *Sophora secundiflora* from well-dated archaeological sites in northeastern Mexico and Trans-Pecos Texas suggests that such ritual plant uses may date back to millennia before our era (Adovasio and Fry 1976).

It is obvious that only the ethnological discipline can highlight the attitude of the aboriginals towards their sacred drugs. The essence of the Catholic mass for the churchgoer is certainly missed by saying that mass wine is prepared from *Vitis vinifera* L. (Vitaceae) and that it contains about 13 percent of the inebriating substance ethyl alcohol before it is diluted by the priest. There is an essential difference, however, between the Catholic priest and the indigenous shaman. The former has no intention whatsoever of becoming drunk, whereas the latter tries to reach an intoxicated state that will enable him to enter supernatural realms. Nobody has put this into words more eloquently than the peyotist Quanah Parker: "The white man goes into his church and talks *about* Jesus; the Indian goes into his tipi and talks *to* Jesus" (Grinspoon and Bakalar 1981).

Due to this fundamental difference, it is not merely allowed but even necessary to conduct experimental studies to validate or invalidate the alleged activity of drugs employed in indigenous drug rituals. A few of the factors that should be taken into consideration in this kind of approach are the subject of the this chapter. One of the most striking mistakes in some sources on ritual drugs is the inappropriate handling of pharmacological data. Actually, this does not come as a great surprise, because authors in this field often lack a pharmacological background. As a consequence, this chapter is meant particularly to provide a nonpharmacological audience with an easily readable outline of a few important pharmacological principles.

Some Botanical Considerations

In the multidisciplinary approach to indigenous ritual plants, botany forms the crucial hinge between field observations and laboratory results. By providing the scientific identity of the ritual plant, botany opens the way to chemical and pharmacological studies. It goes without saying that careful botanical recording requires the collection of a herbarium voucher specimen (Lipp 1989). Any field report that does not indicate voucher specimen numbers does not live up to modern scientific standards and therefore may be open to question. Especially in the early days, many explorers failed to do this, which usually makes it impossible to check the validity of their botanical information.

A voucher specimen in itself, however, is not the ultimate proof of reliability. Indigenous informants may deliberately supply wrong material because they do not want to disclose the botanical source of their sacred drug (Buckley et al. 1973), or because they like to "pull the investigator's leg." It is of course not the responsibility of the informants, but that of the botanist to collect accurate data. Undoubtedly, this problem and many other aspects concerning the botanical approach to hallucinogenic ritual plants will receive attention in other chapters of this book. Yet there are some specific points that should be outlined here, namely, geographical predilection, nonpharmacological terminology, and the usefulness of museum material.

GEOGRAPHICAL PREDILECTION

Botanical reviews on hallucinogens tend to offer relatively much information about the Western Hemisphere. This is hardly surprising in the case of proven hallucinogens, for nowhere in the world has the aboriginal use of truly hallucinogenic plants been more varied and extensive than in Middle and South America. The same tendency, however, is seen also with alleged hallucinogenic plants, which have not been properly evaluated by additional field studies, phytochemical analysis, and pharmacological experiments. Publications on Chinese herbals (Li 1977) and on indigenous peoples of New Guinea (De Smet 1983a) suggest that this is due not only to cultural and botanical differences between the Western and Eastern Hemispheres, but also to a scientific predilection for the New World.

For example, surveys on ritual hallucinogens usually fail to include the genus *Elaeagnus*. Yet there is evidence to suggest that the Gimi of the New Guinea Highlands pass into a trancelike state during divination rites by smoking a mixture of tobacco with leaves of an *Elaeagnus* species (Glick 1967). The genus *Elaeagnus* is rich in beta-carbolines such as tetrahydroharman (Hegnauer 1966; Allen and Holmstedt 1980), and such alkaloids are not likely to be pyrolyzed during smoking (Holmstedt, pers. com. 1983). Nothing appears to be known at present about the effects of tetrahydroharman in humans, but central neurochemical changes have been observed after intraperitoneal administration to mice (Buckholtz 1979). While these data are insufficient to draw conclusions, they are interesting enough to warrant closer examination.

The overlooking of the Eastern Hemisphere occurs not only when the hallucinogenic nature of a plant is still speculative, but it also happens to plant sources with unmistakable hallucinogenic potential. A good example of the latter is the amaryllidaceous plant *Boophone disticha*, which has been used in Africa as a medicine, an arrow poison, and as a means of suicide. The bulb contains alkaloids with atropinelike toxicity, and it is well documented that the use of the bulb as a decoction for recreational purposes can result in hallucinatory visions (Gordon 1947; Gelfand and Mitchell 1952; Goosen and Warren 1960; Haut and Stauffacher 1961; Laing 1979).

NONPHARMACOLOGICAL TERMINOLOGY

There is a tendency in the ethnological and botanical literature to disregard pharmacological definitions of certain terms. For instance, the pharmacological literature associates systemic oral therapy with gastrointestinal absorption. Consequently, it is somewhat of a misnomer from the pharmacological point of view to denote coca chewing as oral administration, as this practice involves significant buccal absorption (De Smet 1985a).[2] A more notable example is the custom to designate hallucinogenic plants as narcotic plants. In pharmacological parlance, the term *narcotic* does not refer to hallucinogens, but to stupefying agents in general and to morphinelike analgesics in particular (Jaffe and Martin 1980).

MUSEUM MATERIAL

Original data may be obtained not only by going out into the field but also by studying already collected material. The collections of many ethnographical museums comprise paraphernalia for ritual drug-taking, and sometimes the drug itself or its vegetal source is also present. Such materials have often been gathered by travellers, who merely recorded ethnological data because they lacked specific botanical interest or training. In such cases, botanical examination still may reveal the identity of the drug source, especially if it can be backed up by the results of chemical analysis.

The Museum for Ethnology in Vienna, for example, possesses well-preserved paricá seeds from the nineteenth century, which have been used as an enema and snuff ingredient by the Brazilian Maué Indians. These seeds seem to be the only ethnobotanical material from the Western Hemisphere that has ever been associated directly with ritual rectal intoxication. The seeds certainly look like they belong to the genus *Anadenanthera*, a view corroborated by the isolation of the *Anadenanthera* alkaloid bufotenine (De Smet and Rivier 1987). This example clearly illustrates the importance of museum material as a tool to extend our ethnobotanical knowledge on indigenous ritual practices.

Some Chemical Considerations

In the multidisciplinary studies on ritual plants, chemistry is an important link between the botanical and pharmacological disciplines. It is difficult to speak in scientific terms of the pharmacology of psychoactive plant material until the relevant constituents have been identified and made available for pharmacological studies.

Chemical reports on ritual plants must give careful attention to the investigated material. Voucher specimens are as essential for chemical laboratory data as they are for botanical field data. The material should come preferably from the area where the plant is ritually utilized. For example, suggestions that Native Americans may have taken *Acorus calamus* in a ritual context are frequently accompanied by statements that this plant has sedative properties due to its asarone fraction. Unfortunately, such statements are based on studies with samples from India. There is considerable evidence that a substantial asarone fraction cannot be expected in diploid plants of North America, but only in triploid and tetraploid specimens of the Old World (De Smet 1985a).

It is well known that the composition of a plant may vary with its parts. In other words, chemical research on ritual plants should include the plant parts used in the ritual. Less obvious factors that also may be relevant include the method of harvesting and the freshness of the sample studied. For instance, the scopolamine content of a peach-flowered form of *Brugmansia* varies with leaf age (Griffin 1976), and storage may alter the tryptamine spectrum of seeds of *Anadenanthera peregrina* (Schultes et al. 1977).

Chemists should not limit their research to the botanical sources of ritual dosage forms, but they must include also the ultimate dosage forms, as the original plant composition may alter during preparation. For example, one step in the preparation of *Virola* snuffs involves concentration of the exudate to a more viscose liquid. The effect of such a treatment has been assessed in the laboratory for a 6-methoxy-tetrahydroharman. This is the major alkaloid in *V. cuspidata*, a species that possibly could serve as a snuff source. Refluxing in water for eight hours results in partial aromatization to 6-methoxy-harmalan and 6-methoxy-harman (Cassady et al. 1971).

The usefulness of chemical results clearly depends on the specificity and sensitivity of the analytical method by which they have been obtained. Illustrative are the paper chromatographical studies on tree daturas (*Brugmansia* species) in the 1960s. Their failure to distinguish between l-hyoscyamine and atropine is pharmacologically relevant, since atropine is the racemic mixture of active l-hyoscyamine and the practically inactive d-hyoscyamine (De Smet 1983b).

Chemical data may be devalued not only by inaccuracy of the final assay, but also by the reactivity of the workup procedure. For instance, ecgonine methyl ester is not a genuine *Erythroxylum* alkaloid but an artifact arising from prolonged extraction with sulfuric acid or chloroform (Rivier 1981). Similarly, the cannabicyclol-type and cannabielsoin-type cannabinoids reported for *Cannabis sativa* are artificial (Turner et al. 1980).

Some Pharmacological Considerations

In the scientific approach to indigenous ritual drugs, there is a clear-cut distinction between ethnological field observations and pharmacological test results (Alger 1976). All ethnological data on the hallucinogenic activity of indigenous vegetal drugs require careful pharmacological evaluation, whereby due attention must be paid to differences between native and experimental drug administration.

A hallucinogen is often defined as a nonaddictive substance that consistently produces changes in perception, thought, and mood, occurring alone or in concert, without causing serious disabilities, like major disturbances of the autonomic nervous system; high doses may elicit disorientation, memory disturbances, hyperexcitation, stupor, or narcosis, but these reactions are not characteristic. This definition is widely accepted, but some authors apply it more rigorously than others (Hoffer and Osmond 1967; Brimble-combe and Pinder 1975; Diaz 1979; Schultes and Hofmann 1980; Grinspoon and Bakalar 1981), so that two types of hallucinogenic agents emerge from the literature:

1. Hallucinogens in a very strict sense, which are always classified as hallucinogens (e.g., indolalkylamines, such as psilocybin, and phenethylamines, such as mescaline).

2. Hallucinogens in a broader sense, which sometimes, but not always, are classified as hallucinogens (e.g., tropane derivatives, such as scopolamine, and dibenz-pyran derivatives, such as ^{9}-tetrahydrocannabinol).

At present, there appears to be no generally applicable method for detecting hallucinogenic activity, except by administering an agent to humans and observing its effects. The use of a universal animal model can already be rejected on the ground that there are several classes of hallucinogens that have different effects and different mechanisms of action. Only within a specific class of closely related agents may animal testing be useful for obtaining an impression of the hallucinogenic potency in humans.

A plausible example is the observed correlation between the human central potency of classical anticholinergic hallucinogens, such as the *Datura* alkaloid atropine, and the ability to block oxotremorine-induced tremors in laboratory animals. In the majority of cases, however, it is still too early to assess the predictive value of an animal model, or the method has already been found to have insufficient specificity (Brimblecombe and Pinder 1975). In other words, animal testing is not very useful when a compound does not belong to a group of well-established hallucinogens; and, even when this is the case, the results may be inconclusive. If a relationship with an established hallucinogenic class is lacking, animal studies can merely assess the somatic toxicity of the test compound prior to human experiments.

The type and degree of hallucinogenic symptoms depend on individual personality, mental condition, and experimental setting, and they may vary with the dose level (Szára 1961; Faillace and Szára 1968; Stark-Adamec et al. 1981). Perceptual changes are not a reliable criterion, as their origin may be peripheral rather than central, and they may be less prominent than alterations in thought and mood.[3] Nor can these latter symptoms be used indiscriminately as a diagnostic feature, since many nonhallucinogenic drugs are known to affect mood (Schultes and Hofmann 1980; Grinspoon and Bakalar 1981).

A simple and objective criterion for hallucinogenic activity would be cross-tolerance with LSD, but this phenomenon is not observed with hallucinogens in the broader sense (Fanchamps 1978; Jaffe 1980). All in all, the classification of a new compound as a hallucinogen will depend primarily on the integrity and experienced judgment of the clinical investigator and the test subjects. Obviously, observed effects must be carefully recorded to allow comparison with future results on the same compound and with clinical data on other substances that are already recognized as hallucinogens.

Ethnological reports on the trance-inducing effects of tobacco (Wilbert 1987) attest to the fact that a ritual plant may have effects that are based more on cultural pre-conditioning than on pharmacological activity. Methodologically, the only conclusive way to distinguish between pharmacological and psychological actions is the cross-over double-blind design. Hereby the subject receives the test drug and a dummy drug (placebo) on two different occasions, and neither the subject nor the investigator knows at the time of administration which drug is being taken.

In an experiment by Manno et al. (1974), subjects who had been given placebo cigarettes but thought they had received marihuana cigarettes indicated that they were high and some actually became stimulated. Only after marihuana cigarettes had been smoked in the next testing session did several subjects question whether they had received marihuana the first time.

Hochman and Brill (1971) tested cigarettes with marihuana extracts varying in strength from 0 to 7.5 milligrams of ^{9}-tetrahydrocannabinol (THC) in regular marihuana users. These volunteers reported the greatest intoxication from the most potent material, but they experienced a higher degree of subjective intoxication from the THC-free extract than from low-potency cigarettes. This might indicate that the THC-free extract contained some other psychoactive constituent, but it may also signify that only the most potent extract exceeded the threshold level needed for a pharmacological intoxication.

A similar pattern has been observed in a study on the activity of nasal cocaine in healthy recreational users of this alkaloid.[4] Central stimulation from 0.2 milligram per kilogram, 0.75 milligram per kilogram, and 1.5 milligrams per kilogram of cocaine Hcl was compared with that from 0.2 milligram per kilogram of lidocaine Hcl as a placebo. Although the latter local anaesthetic does not have euphorizing properties, it induced a more intense "high" than did the same dose of cocaine HCl, and it was only slightly less psychoactive than 0.75 milligram per kilogram of cocaine HCl (Van Dyke et al. 1982). Such data show that any vegetal drug or constituent that is only mildly psychoactive in controlled studies must be properly compared with a placebo, and that the placebo-controlled approach is needed for potent principles to determine the pharmacological threshold level of the drug.

The validity of clinical data is enlarged not only by the inclusion of a placebo in the experimental design, but also by testing a compound in more than one subject. It should be clear that individual experimentation is one thing and that a scientific dose-response study is something else. An experiment in a single individual seldom, if ever, will provide sufficient information. This may be particularly true when the subject is an experienced user of hallucinogenic drugs. It is well documented that nonhallucinogenic stimuli may precipitate a flashback phenomenon in individuals with a history of LSD-use (Abraham 1983).

Other relevant factors include the influence of concomitant drug use and the mental state of the test subjects. For example, the response to LSD is accentuated by pre-treatment with the *Rauwolfia* alkaloid reserpine and attenuated by pre-treatment with the monoamine oxidase inhibitor isocarboxazide (Resnick et al. 1967). It is also known that schizophrenics are less sensitive to the hallucinogenic effects of LSD than normal or neurotic individuals (Fanchamps 1978).

Another point of interest is the duration of the experimental administration. When pharmacological data do not relate to acute effects but to symptoms of prolonged use, they should not be applied to indigenous practices unless there is substantial ethnological evidence that the frequency and duration of the indigenous use are equal. For instance, alcohol can induce a psychotic syndrome in alcoholics (Jaffe 1980), but it would be absurd to extrapolate this chronic effect to nonalcoholics who indulge in drinking alcoholic beverages during occasional ritual festivities.

Pharmacological data mostly are not obtained by testing whole indigenous dosage forms via the indigenous route of administration. Most experiments are performed with isolated constituents, which may be taken in another dose and by another route of administration. Such differences between experimental and indigenous drug uses must be given careful consideration. It must be verified that the amount of active constituent present in the usual native dose exceeds the threshold level observed in the pharmacological experiment. Obviously, a native dose will be limited by the somatic toxicity of the active constituents, and the dose problem will be especially relevant for trace components that occur next to a potent major principle.

Among the numerous identified constituents of tobacco smoke besides nicotine, there are various compounds with suspected hallucinogenic properties, such as carbon dioxide, myristicin, nitrous oxide, and beta-carbolines, and there are also hydrocarbons and ketones with deliriant effects. All these compounds, however, appear present in such small amounts, at least in commercial tobaccos, that any suggestion of endogenous hallucinogens present in *Nicotiana* in behaviorally active amounts must be viewed with appropriate caution (Siegel et al. 1977).

Another example is the presence of beta-carbolines in *Virola* species. Since such compounds have monoamine oxidase inhibiting properties, it is sometimes suggested that they prevent the degradation of hallucinogenic tryptamines, which are the main alkaloids in endogenously utilized *Virola* species. The trace amounts in which the *Virola* beta-carbolines occur, however, are unlikely to be of pharmacological importance (Holmstedt et al. 1980).

It must be checked whether there are pharmacological data on the activity via the indigenous route of administration. The difference between oral ingestion in the aboriginal ritual and parenteral injection in the clinical setting could be of particular importance. Some drugs, such as quaternary ammonium compounds, show poor absorption from the gastrointestinal tract because of their low lipid solubility. Other drugs are inactivated by the acid gastric juices or undergo elimination by intestinal or hepatic enzymes before they reach the general circulation. This last phenomenon, which is commonly known as first-pass metabolism, is observed with many drugs, including the opium alkaloid morphine (Routledge and Shand 1979). A parenterally active amount will be effective only via the oral route, when the drug is absorbed and is not prematurely degraded.

In general, centrally active compounds can be expected to be absorbed well, since passage into the brain and absorption from the gastrointestinal tract are both processes that require sufficient lipid solubility. Consequently, poor absorption of hallucinogenic principles would not appear to be a general problem, although exceptions may occur. There is substantial evidence to suggest, however, that first-pass inactivation is not uncommon for naturally occurring hallucinogens. Extensive metabolism and route-dependent activity have been reported for the tryptamine derivative dimethyltryptamine, for the beta-carbolines harmine and harmaline, and for the tropane alkaloid scopolamine (De Smet 1983b).

Among the indigenous inhabitants of the Western Hemisphere, oral ingestion is certainly not the only way to elicit ritual intoxication. Major nonoral ways are rectal application (De Smet 1983b), snuffing (De Smet 1985a), and smoking (De Smet 1985b). An ethnopharmacological view of such practices among indigenous South American people is presented in Table 1 (see end of chapter). Rectal administration is not generally an adequate method to avoid first-pass metabolism, but nasal administration can undoubtedly provide a bypass, and this may also apply to smoking.

The picture for nasal administration is complicated by the possibility that the developing heat may destroy existing components (pyrolysis) and may form new ones (pyrosynthesis). For instance, the beta-carbolines harman and norharman in tobacco smoke are largely formed during smoking (Janiger and Dobkin De Rios 1976), whereas the [9]-tetrahydrocannabinol present in a marihuana cigarette only partially survives the burning process (Ohlsson et al. 1980; Turner 1980).

Another South American Indian practice that may lead to a direct passage into the systemic circulation is the holding of a chewing quid between the gum and the cheek or lip (Plowman 1981; Wilbert 1987). This is not a pure method to avoid first-pass metabolism, however, when the juice or the plant material is swallowed (De Smet 1985a).

The substantial influence of drug absorption, distribution, and elimination on drug action has long been recognized by pharmacologists, who have even devoted a special part of their discipline, namely, pharmacokinetics, to the investigation of such processes. By analogy, ethnopharmacology deserves a branch called *ethnopharmacokinetics*, which must be aimed at the fate of indigenous drug constituents in the body (De Smet and Rivier 1989). This branch must assess whether adequate concentrations of indigenous drug principles are reached and sustained at their appropriate sites of action.

In practice, it is quite laborious or even impossible to perform direct measurements at a specific site of action in the human body. Instead, most pharmacokinetic studies perform measurements in a readily accessible body fluid, such as blood, urine, or saliva. Although monitoring at these convenient sites has its limitations, it has proved to be very useful in many cases (Rowland and Tozer 1980). A good ethnopharmacokinetic example is the demonstration of substantial cocaine plasma levels in coca chewers (Holmstedt et al. 1979; Paly et al. 1980).[5] This puts an end to speculation that cocaine might be hydrolyzed in the mouth before it is absorbed.

The pharmacological activity of a naturally occurring mixture may be different from that of the most active isolated constituent or fraction. An interesting experiment on the difference between a crude hallucinogenic preparation and a pure active constituent was performed by List et al. (1969), who measured the permeation of l-hyoscyamine through the isolated small intestine of the rat. When an aqueous solution with pure l-hyoscyamine was compared with a fresh extract of *Atropa belladonna* leaves, the permeation rate was found to be from 70 to 80 percent higher in the case of the extract. Unfortunately, this experiment was not followed by *in vivo* investigations, so it remains difficult to assess the clinical significance of this *in vitro* result.

The most striking example of superiority of a whole indigenous dosage form over its isolated principles may well be the South American beverage ayahuasca, containing mostly *Banisteriopsis* beta-carbolines like harmine, together with dimethyltryptamine from *Psychotria*. Although conclusive clinical evidence has not been published, growing evidence suggests that the beta-carbolines may protect the dimethyltryptamine from first-pass inactivation by monoamine oxidase A enzymes, thus rendering the tryptamine derivative orally less inactive (De Smet 1983b).

The activity of a constituent may be modified also by other components of the indigenous dosage form, which do not have a systemic action by themselves. Many South American tribes prepare their intoxicating snuffs by mixing psychoactive vegetal ingredients with lime or plant ash. Such alkaline admixtures may facilitate the absorption of the psychoactive alkaloids through the nasal mucosa (De Smet 1985a).

It is clear from these data that the influence of accompanying substances should not be neglected, but it should not be exaggerated either. When Hofmann (1982) gave some pills with pure psilocybin to the famous Mexican curandera Maria Sabina, she attested that there was no difference in efficacy between the pills and her own psilocybin mushrooms. In contrast with this report, some ethnological and botanical publications tend to overemphasize the difference between isolated constituents and whole indigenous dosage forms. This may even reach the point where vegetal constituents are assumed out of hand to act in a synergistic manner. It should always be borne in mind, however, that antagonism may also occur and that in many cases the effects will be additive rather than synergistic. For example, Heimann (1965) showed that a mixture of the ololiuqui constituents d-lysergic acid amide, d-isolysergic acid amide, and lysergol in healthy volunteers merely elicits the combined symptoms of each separate alkaloid.

The pharmacological understanding of hallucinogenic drug rituals must not only consider the influence of accompanying ingredients in the aboriginal dosage form, but also the concomitant use of other separate drugs. This is well illustrated by the clinical finding that alcohol adds significantly to the subjective effects of marihuana smoking (Manno et al. 1974). As is the case with single drugs, folkloristic data on combinations should not be accepted without experimental confirmation. In India, the tamarind fruit is reputed to antagonize the effects of bhang (*Cannabis*), but Hollister (1976) found the fruit to be ineffective in humans in doses up to 180 grams, when it was given an hour before, simultaneously, or an hour after oral administration of ^{9}-tetrahydrocannabinol.

Summary

Many factors that have been discussed here are so obvious it may seem superfluous to review them. Nothing is less true, however, for various publications fail to afford them proper attention. Perhaps the most poignant example of why they never should be overlooked is the famous case of the reports by Castaneda (1970, 1972). This author described the use and hallucinogenic effects of a smoking mixture consisting of dried mushrooms with the addition of other dried plants as sweeteners. The mushrooms were vaguely suggested to be a *Psilocybe* species, possibly *P. mexicana*. There are substantial scientific reasons to believe that these reports are not authentic but fictional. For instance, no voucher specimens are cited, and Schultes (pers. com. 1991) was unable to ascertain from Castaneda whether voucher specimens had been collected. The suggestion that these mushrooms could have been *P. mexicana* must be refuted on the basis of their habitat (De Mille 1980). What is more, the smoking mixture is unlikely to have contained sufficient psilocybin to elicit the profound effects described by Castaneda. In addition, psilocybin is said to be largely degraded during smoking in a pipe, whereby its availability is further reduced by condensation and deposition on the inner surface of the bowl and stem of the pipe (Siegel 1981). In other words, if proper attention had been given immediately to dose and to way of administration, scientific doubts about the authenticity of Castaneda's accounts would have arisen sooner.

NOTES

[1] Although the term *hallucinogen* understates the importance of thought and mood changes by emphasizing the perceptual activity, it is still one of the most commonly used terms in the medical and pharmacological literature.

[2] Although coca is not hallucinogenic, data of this psychoactive plant and its main constituent, cocaine, have been included because of their illustrative value.

[3] When the symptoms seem to be largely autonomic in nature, it may be worthwhile to ascertain whether the tested compound readily enters the brain or not. If animal experiments fail to show a significant entry, the effects are likely to be peripheral rather than central.

[4] Although coca is not hallucinogenic, data on this psychoactive plant and its main constituent, cocaine, have been included because of their illustrative value.

[5] Although coca is not hallucinogenic, data on this psychoactive plant and its main constituent, cocaine, have been included because of their illustrative value.

ACKNOWLEDGMENTS

I am indebted to Professors Bo Holmstedt and Richard E. Schultes for their personal communications.

Table 1. Multidisciplinary overview of indigenous South American ritual snuffing, enema taking, and smoking.[1]

Scientific name	Snuffing	Enema taking	Smoking	Phytochemistry	Pharmacology
Anadenanthera sp. (Leguminosae)	The seeds are a common source of South American snuffs.	Various claims that South American tribes use the seeds as an enema source appear to be partially correct.	Smoking of preparations associated with Anadenanthera has been reported in British Guiana, Paraguay, and Argentina.	DMT; 5-OH-DMT; 5-MeO-DMT.[7]	Established hallucinogenic activity for DMT and 5-MeO-DMT, but questionable for 5-OH-DMT.
Banisteriopsis sp. (Malpighiaceae)	Evidence lacking, but alkaloids have been isolated from snuffs of the Surara and Piaroa Indians.[3]	The Aguaruna Indians of the Peruvian Montana reportedly use clysters prepared from ayahuasca mixed with tobacco syrup.[6]	—	Harmine; harmaline; tetrahydroharmine.	Established hallucinogenic activity.
Brugmansia sp. (Solanaceae)	—	Ritual use of enemas among the Jivaro Indians is well documented.	—	Scopolamine & related alkaloids.	Established deliriant activity.
Cannabis sativa (Cannabaceae)	—	—	The Tenetehara Indians of Brazil smoke the dried flower and leaf as a recreational stimulant.	Delta-9-tetrahydro-cannabinol & related cannabinoids.	Established mild hallucinogenic activity.
Dinorphandra parviflora (Leguminosae)	According to a 19th-century herbarium annotation, the seeds were used in Brazil to prepare a noted snuff.[4]	—	—	—	—
Erythroxylum sp. (Erythroxylaceae)	Coca is used occasionally as a snuff source in the Northwest Amazon.	—	It is vaguely claimed that the Omagua Indians of Peru smoke these species.	Cocaine.	Established psychostimulating properties.
Ilex guayusa (Aquifoliaceae)	Bundled leaves have been found together with snuff trays in a pre-Hispanic Bolivian grave.	Bundled leaves have been found together with enema syringes in a pre-Hispanic Bolivian grave.	—	Caffeine.	Established mild psychostimulating properties.
Justicia pectoralis (Acanthaceae)	Leaves are mostly taken as an admixture to Virola bark exudate, but occasionally are used alone.	—	—	Coumarin; umbelliferone; justicidin B[8]; swertisin & related flavones.[9]	—

Plant (family)				Constituents	Activity
Lobelia tupa (Campanulaceae)	—	—	The Mapuche Indians of Chile reportedly smoke the leaves as an intoxicant.	Lobeline; lobelanidine; norlobelanine.	Lobeline has peripheral effects similar to those of nicotine, but its central activity may be different.[5]
Maquira sclerophylla (Moraceae)	Circumstantial evidence indicates the bark[5] or fruit was used as a snuff source in the central part of the Brazilian Amazon.	—	—	The related *Maquira calophylla* has yielded the cardiac glycoside maquiroside A.[10]	Effects suggestive of amphetamine-like stimulation were found in rats, but further study is required to show that this was not due to toxic effects mimicking central stimulation.[5]
Nicotiana sp. (Solanaceae)	Tobacco leaves are used widely in South America as a snuff source.	The Aguaruna Indians of the Peruvian Montana use tobacco syrup, either alone or mixed with *Banisteriopsis*, as an enema.	Tobacco smoking is widespread in South America.	Nicotine.	Established psychostimulating properties.
Pagamea macrophylla (Rubiaceae)	Pulverized leaves are snuffed by the Barasana Indians of Colombia.	—	—	—	—
Piper interitum (Piperaceae)	The Kulina Indians of Peru reportedly prepare a snuff from the dried leaves and roots.	—	—	—	—
Tanaecium nocturnum (Bignoniaceae)	The Paumari Indians of Brazil snuff a mixture of roasted, ground leaves with tobacco powder.	—	—	—	—
Trichocereus pachanoi[2] (Cactaceae)	—	—	Archaeological excavations in Peru have yielded what appear to be cigars made from this cactus.	Mescaline.	Established hallucinogenic activity.
Trichocline sp. (Asteraceae)	—	—	The Chaco Indians of Argentina smoke the powdered root.	Isopimpinelline; phellopterin; trichoclin.	—
Virola sp. (Myristicaceae)	The bark exudate is a common source of snuffs in South America.	—	The bark reportedly is smoked by witch doctors in Venezuela and Brazil.	DMT; 5-MeO-DMT.[7]	Established hallucinogenic activity.

[1] Unless otherwise indicated, data for snuffing adapted from De Smet (1985a); for enema taking, De Smet (1983b), and for smoking, De Smet (1985b).
[2] This cactus also is known as *Echinopsis pachanoi*.
[3] De Smet and Rivier (1985).
[4] De Smet and Lipp (1987).
[5] De Carvalho and Lapa (1990).
[6] Wilbert (1987, pp. 47–48).
[7] DMT = N,N-dimethyltryptamine; 5-OH-DMT = 5-hydroxy-N,N-dimethyltryptamine; 5-MeO-DMT = 5-methoxy-N,N-dimethyltryptamine.
[8] Gleye et al. (1988).
[9] Joseph et al. (1988).
[10] Rovinski et al. (1987).

LITERATURE CITED

Abraham, H. D. 1983. Visual phenomenology of the LSD flashback. *Archives of General Psychiatry* 40: 884–889.

Adovasio, J. M., and G. F. Fry. 1976. Prehistoric psychotropic drug use in northeastern Mexico and Trans-Pecos Texas. *Economic Botany* 30: 94–96.

Alger, N. 1976. Pharmacology, ethnopharmacology, and the use of drugs in curanderismo. In *Actas del XLI Congreso Internacional de Americanistas* (Mexico). 3: 262–268.

Allen, J. R. F., and B. Holmstedt. 1980. The simple β-carboline alkaloids. *Phytochemistry* 19: 1573–1582.

Brimblecombe, R. W., and R. M. Pinder. 1975. *Hallucinogenic Agents*. Bristol: Wright-Scien-technica.

Bruhn, J. G., and B. Holmstedt. 1981. Ethnopharmacology: Objectives, principles, and perspectives. In *Natural Products as Medicinal Agents*. Eds. E. Reinhard and J. L. Beal. Stuttgart: Hippokrates. 405–430.

Buckholtz, N. S. 1979. Neurochemical and behavioral effects of β-carbolines. *Psychopharmacology Bulletin* 15: 56–57.

Buckley, J. P., R. J. Theobald, Jr., I. Cavero, B. A. Krukoff, A. P. Leighton, and S. M. Kupchan. 1973. Preliminary pharmacological evaluation of extracts of takini: *Helicostylis tomentosa* and *Helicostylis pedunculata*. *Lloydia* 36: 341–345.

Cassady, J. M., G. E. Blair, R. F. Raffauf, and V. E. Tyler. 1971. The isolation of 6-methoxyharmalan and 6-methoxyharman from *Virola cuspidata*. *Lloydia* 34: 161–162.

Castaneda, C. 1970. *The Teachings of Don Juan: A Yaqui Way of Knowledge*. Harmondsworth: Penguin Books.

———. 1972. *A Separate Reality—Further Conversations with Don Juan*. New York: Pocket Books.

De Carvalho, J. E., and A. P. Lapa. 1990. Pharmacology of an Indian-snuff obtained from Amazonian *Maquira sclerophylla*. *Journal of Ethnopharmacology* 30: 43–54.

De Mille, R. 1980. Allegory is not ethnobotany: Analyzing Castaneda's letter to R. Gordon Wasson and Carlos's Spanish fieldnotes. In *The Don Juan Papers—Further Castaneda Controversies*. Santa Barbara, CA: Ross-Erikson Publishers. 319–332.

De Smet, P. A. G. M. 1983a. Ritual plants and reputed botanical intoxicants of New Guinea natives. *Farmaceutisch Tijdschrift voor België* 60: 291–300.

———. 1983b. A multidisciplinary overview of intoxicating enema rituals in the Western Hemisphere. *Journal of Ethnopharmacology* 9: 129–166.

———. 1985a. A multidisciplinary overview of intoxicating snuff rituals in the Western Hemisphere. *Journal of Ethnopharmacology* 13: 3–49, 235.

———. 1985b. Ethnopharmacological table on some reputedly psychoactive fumigatories among Middle and South American natives. *Pharmaceutisch Weekblad Scientific Edition* 7: 212–218.

De Smet, P. A. G. M., and F. J. Lipp, Jr. 1987. Supplementary approach to ritual enemas and snuffs in the Western Hemisphere. *Journal of Ethnopharmacology* 19: 327–331.

De Smet, P. A. G. M., and L. Rivier. 1985. Intoxicating snuffs of the Venezuelan Piaroa Indians. *Journal of Psychoactive Drugs* 17: 93–103.

———. 1987. Intoxicating paricá seeds of the Brazilian Maué Indians. *Economic Botany* 41: 12–16.

———. 1989. A general outlook on ethnopharmacology. *Journal of Ethnopharmacology* 25: 127–138.

Díaz, J. L. 1979. Ethnopharmacology and taxonomy of Mexican psychodysleptic plants. *Journal of Psychedelic Drugs* 11: 71–101.

Efron, D. H., B. Holmstedt, and N. S. Kline, eds. 1967. *Ethnopharmacologic Search for Psychoactive Drugs*. U.S. Department of Health, Education and Welfare Publication no. 1645. Washington, DC: Government Printing Office.

Faillace, L. A., and S. Szára. 1968. Hallucinogenic drugs: Influence of mental set and setting. *Diseases of the Nervous System* 29: 124–126.

Falch, E., V. Christensen, P. Jacobsen, J. Byberg, and P. Krogsgaard-Larsen. 1984. *Amanita muscaria* in medicinal chemistry. I. Muscimol and related GABA agonists with anticonvulsant and central non-opioid analgesic effects. In *Natural Products and Drug Development*. Eds. P. Krogsgaard-Larsen, S. Brøgger Christensen, and H. Kofod. Proceedings of the Alfred Benzon symposium 20, Munskgaard, Copenhagen. 504–522.

Fanchamps, A. 1978. Some compounds with hallucinogenic activity. In *Ergot Alkaloids and Related Compounds*. Eds. B. Berde and H. O. Schild. Vol. 49, *Handbuch der Experimentellen Pharmakologie*. Berlin: Springer-Verlag. 567–614.

Fuller, R. W., S. K. Hemrick-Luecke, and K. W. Perry. 1981. Lowering of dopamine metabolites in rat brain by harmaline. *Journal of Pharmacy and Pharmacology* 33: 255–256.

Furst, P. T. 1976. *Hallucinogens and Culture*. San Francisco: Chandler and Sharp.

Gelfand, M. and C. S. Mitchell. 1952. Buphanine poisoning in man. *South African Medical Journal* 26: 573–574.

Gleye, H. J. J., C. Moulis, L. J. Mensah, C. Roussakis, and C. Gratas. 1988. Justicidin B, a cytotoxic principle from *Justicia pectoralis*. *Journal of Natural Products* 51: 599–600.

Glick, L. B. 1967. Medicine as an ethnographic category: The Gimi of the New Guinea highlands. *Ethnology* 6: 31–56.

Goosen, A., and F. L. Warren. 1960. The alkaloids of the Amaryllidaceae. Part VII. The alkaloids from *Boophone disticha* Herb., buphanitine ("crystalline" haemanthine), buphanamine, and buphanidrine (distichine). *Journal of the Chemical Society*. 1094–1096.

Gordon, I. 1947. A case of fatal buphanine poisoning. *Clinical Proceedings* 6: 90–93.

Grieco, A., and R. Bloom. 1981. Psychotherapy with hallucinogenic adjuncts from a learning perspective. *International Journal of the Addictions* 16: 801–827.

Griffin, W. J. 1976. Agronomic evaluation of *Datura candida*—a new source of hyoscine. *Economic Botany* 30: 361–369.

Grinspoon, L., and J. B. Bakalar. 1981. *Psychedelic Drugs Reconsidered*. New York: Basic Books.

Hall, R. C. W., B. Pfefferbaum, E. R. Gardner, S. K. Stickney, and M. Perl. 1978. Intoxication with angel's trumpet: Anticholinergic delirium and hallucinosis. *Journal of Psychedelic Drugs* 10: 251–253.

Haut, H., and D. Stauffacher. 1961. Die Alkaloide von *Boophone disticha* (L.f.) Herb. *Helvetica Chimica Acta* 44: 491–502.

Hegnauer, R. 1966, 1973. *Chemotaxonomie der Pflanzen—Eine Übersicht über die Verbreitung und die systematische Bedeutung der Pflanzenstoffe*. Vol. 4, *Dicotyledoneae: Daphniphyllaceae–Lythraceae*. Vol. 6, *Dicotyledoneae: Rafflesiaceae–Zygophyllaceae*. Basel: Birkhäuser Verlag.

Heimann, H. 1965. Die Wirkung von Ololiuqui im Unterschied zu Psilocybin. In *Neuro-Psychopharmacology*, vol. 4. Eds. D. Bente and P. B. Bradley. Amsterdam: Elsevier. 474–477.

Hochman, J. S., and N. Q. Brill. 1971. Marijuana intoxication: Pharmacological and psychological factors. *Diseases of the Nervous System* 32: 676–679.

Hoffer, A., and H. Osmond. 1967. *The Hallucinogens*. New York: Academic Press.

Hofmann, A. 1982. *LSD—Mein Sorgenkind*. Frankfurt: Klett-Cotta im Verlag Ullstein.

Hollister, L. 1976. Interactions of ^{9}-tetrahydrocannabinol with other drugs. *Annuals of the New York Academy of Sciences* 281: 212–218.

Holmstedt, B., J. E. Lindgren, T. Plowman, L. Rivier, R. E. Schultes, and O. Tova. 1980. Indole alkaloids in Amazoniana Myristicaceae. Field and laboratory research. *Botanical Museum Leaflets* (Harvard University) 28: 215–234.

Holmstedt, B., J. E. Lindgren, L. Rivier, and T. Plowman. 1979. Cocaine in blood of coca chewers. *Journal of Ethnopharmacology* 1: 69–78.

Jaffe, J. H. 1980. Drug addiction and drug abuse. In *The Pharmacological Basis of Therapeutics*. Eds. A. G. Gilman, L. S. Goodman, and A. Gilman. 6th ed. New York: Macmillan. 535–584.

Jaffe, J. H., and W. R. Martin. 1980. Opioid analgesics and antagonists. In *The Pharmacological Basis of Therapeutics*. Eds. A. G. Gilman, L. S. Goodman, and A. Gilman. 6th ed. New York: Macmillan. 494–534.

Janiger, O., and M. Dobkin de Rios. 1976. *Nicotiana* an hallucinogen? *Economic Botany* 30: 149–151.

Joseph, H. J. Gleye, C. Moulis, I. Fouraste, and E. Stanislas. 1988. O-methoxylated C-glycosylflavones from *Justicia pectoralis*. *Journal of Natural Products* 51: 804–805.

Laing, R. O. 1979. Three cases of poisoning by *Boophone disticha*. *Central African Journal of Medicine* 25: 265–266.

Li, H. L. 1977. Hallucinogenic plants in Chinese herbals. *Botanical Museum Leaflets* (Harvard University) 25: 161–181.

Lipp, F. J. 1989. Methods for ethnopharmacological field work. *Journal of Ethnopharmacology* 25: 139–150.

List, P. H., W. Schmid, and E. Weil. 1969. Reinsubstanz oder Galenische Zubereitung? Versuch einer Klärung am Beispiel des Belladonna-Extraktes. *Arzneimittel-Forschung* 19: 181–185.

Manno, J. E., B. R. Manno, G. F. Kiplinger, and R. B. Forney. 1974. Motor and mental performance with marijuana: Relationship to administered dose of delta-9-tetrahydrocannabinol
and its interaction with alcohol. In *Marijuana—Effects on Human Behavior*. Ed. L. L. Miller.
New York: Academic Press. 45–72.

Ohlsson, A., J. E. Lindgren, A. Wahlen, S. Agurell, L. E. Hollister, and H. K. Gillespie. 1980.
Plasma delta ⁹-tetrahydrocannabinol concentrations and clinical effects after oral and intravenous administration and smoking. *Clinical Pharmacology and Therapeutics* 28: 409–416.

Paly, D., P. Jatlow, C. Van Dyke, F. Cabieses, R. Byck. 1980. Plasma levels of cocaine in native
Peruvian coca chewers. In *Cocaine 1980*. Ed. F. R. Jeri. Proceedings of the Inter-American
Seminar on Medical and Sociological Aspects of Coca and Cocaine. Lima, Peru: Pan Amer-
ican Health Office/World Health Organization. 86–89.

Plowman, T. 1981. Amazonian coca. *Journal of Ethnopharmacology* 3: 195–225.

Resnick, O., D. M. Krus, M. Raskin. 1967. LSD-25—drug interactions in man. In *Neuro-psychopharmacology*. Ed. H. Brill. International Congress series no. 129. Amsterdam: Excerpta
Medica Foundation. 1103–1107.

Rivier, L. 1981. Analysis of alkaloids in leaves of cultivated *Erythroxylum* and characterization
of alkaline substances used during coca chewing. *Journal of Ethnopharmacology* 3: 313–335.

Routledge, P. A., and D. G. Shand. 1979. Presystemic drug elimination. *Annual Reviews of Pharmacology and Toxicology* 19: 447–468.

Rovinski, J. M., G. L. Tewalt, and A. T. Sneden. 1987. Maquiroside A, a new cytotoxic cardiac
glycoside from *Maquira calophylla*. *Journal of Natural Products* 50: 211–216.

Rowland, M., and T. N. Tozer. 1980. *Clinical Pharmacokinetics—Concepts and Applications*. Philadelphia: Lea & Febiger.

Schultes, R. E., and A. Hofmann. 1980. *The Botany and Chemistry of Hallucinogens*. 2nd ed.
Springfield, IL: Charles C. Thomas.

Schultes, R. E., B. Holmstedt, J. E. Lindgren, and L. Rivier. 1977. Phytochemical examination
of Spruce's ethnobotanical collection of *Anadenanthera peregrina*. De plantis toxicariis e
Mundo Novo tropicale commentationes. XVIII. *Botanical Museum Leaflets* (Harvard University) 25: 273–287.

Siegel, R. K. 1981. Castanedas Privatapotheke. In *Rausch und Realität—Drogen im Kulturvergle-
ich*, Teil 2. Eds. G. Völger, K. von Welck, and A. Legnaro. Cologne: Rautenstrauch-Joest-
Museum für Völkerkunde. 716–723.

Siegel, R. K., P. R. Collings, and J. L. Díaz. 1977. On the use of *Tagetes lucida* and *Nicotiana rustica* as a Huichol smoking mixture: The Aztec 'Yahutli' with suggestive hallucinogenic
effects. *Economic Botany* 31: 16–23.

Stark-Adamec, C., R. E. Adamec, and R. O. Phil. 1981. The subjective marijuana experience:
Great expectations. *International Journal of Addictions* 16: 1169–1181.

Stienstra, R., J. F. Van Poorten, F. A. Vermaas, and J. Rupreht. 1981. Psilocybine-vergiftiging
door het eten van paddestoeien. *Nederlands Tijdschrift voor Geneeskunde* 125: 833–835.

Szára, S. 1961. Hallucinogenic effects and metabolism of tryptamine derivatives in man. *Fed-
eration Proceedings* 20: 885–888.

Turner, C. E. 1980. Marijuana research and problems: An overview. *Pharmacy International* 1:
93–96.

Turner, C. E., M. A. Elsohly, and E. G. Boeren. 1980. Constituents of *Cannabis sativa* L. XVI. A
review of the natural constituents. *Journal of Natural Products* 43: 169–234.

Van Dyke, C., J. Ungerer, P. Jatlow, P. Barash, and R. Byck. 1982. Intranasal cocaine: Dose relationships of psychological effects and plasma levels. *International Journal of Psychiatry in
Medicine* 12: 1–13.

Völger, G., K. von Welck, and A. Legnaro, eds. 1981. *Rausch und Realität—Drogen im Kul-
turvergleich*, Teil 1 and 2. Cologne: Rautenstrauch-Joest-Museum für Völkerkunde.

Wilbert, J. 1987. *Tobacco and Shamanism in South America*. New Haven, CT: Yale University
Press.

Young, R. E., S. Hutchinson, R. Milroy, and C. M. Kesson. 1982. The rising price of mushrooms. *Lancet* i: 213–215.

PART 9

Ethnomycology

Undoubtedly the youngest of our sections of ethnobotany is now known as ethnomycology, the study of fungi and their use and influence in the development of cultures, religions, and mythology. Credit for the birth of this subdivision must go to Dr. R. Gordon Wasson and his wife, Valentina Pavlovna, a Russian medical doctor, in the publication in 1957 of their two-volume book, *Mushrooms, Russia and History* (Pantheon Books, New York, 1957). This extraordinary book rapidly became the genesis in scientific circles of what grew into a fast-developing interest in fungi as an integral part of cultural, religious, and mythological aspects of numerous civilizations in various parts of the world. Shortly thereafter, Wasson published in May 1957 an excellent popular article in *Life* entitled "Seeking the Magic Mushroom." This article immediately attracted attention to the religious use of hallucinogenic fungi in primitive societies and also led, unfortunately, to the use of these mushrooms in the public domain, including for hedonic purposes.

Following the early death of Valentina, Gordon, a banker, retired and devoted the rest of his time and energy almost uniquely to reading, traveling, field work, and writing on fungal significance in various aspects of history and culture. His earliest outstanding research centered on several groups of Indians in Oaxaca, Mexico, and the historical background and origins of the use of fungi in pre-Conquest Mexico. He proposed the identification of the enigmatic hallucinogen of ancient India as *soma*, a plant which was so sacred that it became a god. Use of this hallucinogen died out many centuries ago, and the source plant was unknown. Wasson's theory, now widely accepted, was that the narcotic was *Amanita muscaria*, the fly agaric mushroom.

Wasson realized that he was an amateur in the true sense of that word. One characteristic of his research is that he always collaborated with specialists—anthropologists, linguists, botanists, chemists, students of religion, and others, often taking them on his many expeditions. His collaboration with the famous French mycologist, the late Roger Heim, led to the book *Les Champignons Hallucinogènes du Mexique* (Muséum National d'Histoire Naturelle, Paris, 1959) with additional chapters by nine specialists in a variety of tangential fields of research.

During his years of research, Wasson created a large library and collected many art and archaeological objects, which he donated to the Botanical Museum of Harvard where he had held an academic appointment for 22 years. This collection, known as the Tina and Gordon Wasson Mycological Collection, represents the only ethnomycological library in the world.

The one contribution in this section is the last paper written by Wasson. In 1990, a Festschrift was published in his honor: *The Sacred Mushroom Seeker—Essays for R. Gordon Wasson* (ed. T. Riedlinger, Dioscorides Press, Portland, Oregon, 1990).

The late Bernard Lowy of the Louisiana State University in Baton Rouge carried out research on hallucinogenic mushrooms, particularly *Amanita muscaria*, in Guatemala. Lowy published several technical articles ("Amanita and the Thunderbolt Legend in Guatemala and Mexico" and "Hallucinogenic Mushrooms in Guatemala") and, at the time of his death in 1992, was cowriting with chemist Michael Montagne a comprehensive book on hallucinogenic mushrooms. Rolf Singer of the Field Museum of Natural History has done extensive research on the taxonomy of mushrooms and has carried out ethnobotanical field work on the hallucinogenic species of Oaxaca. The outstanding Mexican taxonomic mycologist, Gastón Guzmán of the Instituto de Ecología in Jalapa, Mexico, has collected and described a number of new species of hallucinogenic mushrooms, and his taxonomic research on the species employed by indigenous peoples has greatly contributed to ethnomycology. Chemist Jonathan Ott and photographer Jeremy Bigwood have, primarily through excellent publications on hallucinogenic plants, helped to enrich understanding of ethnomycology. *Teonanacatl—Hallucinogenic Mushrooms of North America* (Madronas Publishers, Seattle, 1978), with three chapters by Ott and Bigwood and four other specialists, summarized what was known of the sacred hallucinogenic mushrooms of Mexico in 1978.

There is no question that ethnomycological research will continue to develop, even though probably slower than other fields of ethnobotany. The reason is the extremely limited number of trained mycologists. The fungi, especially in the wet tropics of the world, are probably much richer in species than the higher plants. The extraordinary success in the field of antibiotics has convinced the scientific world of the untapped wealth, chemical and otherwise, awaiting study in this group of plants. The use of mushrooms in traditional medicine and as hallucinogens in native magico-religious contexts is sporadically appearing in numerous far away parts of the world as researchers continue to intensify their efforts.

Ethnomycology:
Discoveries About *Amanita muscaria*
Point to Fresh Perspectives

R. GORDON WASSON

In the 1970s there have been two major discoveries in ethnomycology involving the psychoactive mushroom *Amanita muscaria*.

The Discoveries

The first major discovery was made by Carl A. P. Ruck, outstanding classical scholar of Boston University, who showed that soma, the long-mysterious plant permeating the *Rig Veda* (one of four sacred writings in Hinduism), was known to the ancient Greeks and may well have been consumed ceremonially on the island of Delos, the plants coming from the far north, probably in Asia (Wasson et al. 1978, 1986).

In regions where there is a mushroom cult, it is the invariable custom among local believers not to consider the sacred mushroom as a mushroom. Instead, it may have a distinctive name, like soma, and is set apart from all other mushrooms in the thoughts and feelings of the people. Thus, among the Greeks, as well as among the Aryans of the Indus Valley, the words designating the holy mushroom occur seldom, being subject to a religious taboo. It would be a mistake, however, to assume the Greeks borrowed their terms for the holy mushroom from the Aryans. Both groups drew their manner of referring to it from their ancestors, from a common fund of ideas that expressed the thoughts of humankind about the miraculous fungus, the entheogen,[1] on which they had focused their adoration from far, far back.

Perhaps from the dawn of the human race the supernal plant had been linked with the lightning bolt. According to the *Rig Veda* (IX 82.3, 83.3) nothing less than the lightning bolt with its accompanying ear-splitting thunder inseminated the rain-softened earth at the base of the Tree of Life with the miraculous plant, whose paternity was confirmed in the marvelous red garb with which it came into existence, set off by patches of a virgin white veil with which it was born.

One of the words in Greek originally designating soma is *cyclops*, "single eye," referring to a race of one-eyed giants that played a vigorous role in Greek mythology. In *Soma*

(Wasson 1967), I assembled passages from the *Rig Veda* that spoke of soma as a single eye and pointed out how appropriately this mushroom could be spoken of as a single eye, showing a photograph of *Amanita muscaria*. This figure of speech could conceivably have been taken by the Greeks from Aryans in the Indus Valley, but it is far more likely the Greeks brought it down from Asia whence they emigrated in the second millennium B.C.

A second concept that finds expression in Greek literature is "one-leg," "shade-foot," or "cover-foot." Soma was the only divinity known to early humans that grew a single "leg," the stipe, and this singularity impressed itself forcefully on the imagination of the adorers.

The third figure in Greek literature is "tongue-in-belly": the divinity soma had only his torso to live in—no arms, head, neck, eyes, nose, mouth, or ears. These organs must therefore reside in the torso, the pileus of the mushroom. Ruck provided us with the Greek effigies of this figure and also citations of Iambe-Baubo in the *Homeric Hymn to Demeter* (Wasson et al. 1978, lines 203–205), where we are told that he/she played a minor role in the Eleusinian rites.

The Greeks, Aryans, and other Asiatic peoples strove to reconcile the mushroom shape of *Amanita muscaria* with a divine being! The Greek efforts were repeated in Pliny the Elder's *Historia Naturalis*, a text that was certainly responsible for the reappearance of the mushroom/divine being in manuscripts of the high Middle Ages and in the early generations of printed books, when the vogue for monsters living beyond the limits of the known world was fashionable science fiction. Until Ruck's discovery, however, no one had known what those bizarre figures meant.

The second major discovery in ethnomycology was made by Bernard Lowy (1974), retired professor of mycology at Louisiana State University, who explored the Maya world in Guatemala. Knowing the emphasis I placed on the words for *lightning bolt* in the study of ethnomycology, Lowy learned those words and other thunderstorm vocabulary in two Mayan languages—Quiche and Tzutuhil. He then explored the mountains adjacent to the Quiche town of Chichicastenango, found specimens of the scarlet variety of *Amanita muscaria*, and returned to Chichicastenango's marketplace, where mushroom vendors immediately identified Lowy's specimens as *kakuljá*, the precise word that is used in Quiche for a blinding flash of lightning followed by a deafening clap of thunder.[2]

The identity of one word for lightning bolt and *Amanita muscaria* in Quiche is breathtaking. For almost fifty years I had been pursuing the link between the two, and finally Lowy found it preserved in the Quiche highlands of Guatemala in its pristine freshness. The Maya must have brought the lightning bolt meaning from Asia and used the same word as a most powerful trope for *A. muscaria*.

Fresh Perspectives

What are the implications of these two discoveries for future study in ethnomycology? Several areas are indicated.

NEW TRANSLATION OF THE *POPOL VUH*

A long poem in Quiche, running to 8544 lines in Edmonson's English translation, survives to this day, having been preserved by human memory for untold millennia until a respected Quiche wrote it out in the middle of the sixteenth century. Called *Popol Vuh*, this poem is a kind of *Book of Genesis*. In it, the word *kakuljá* had been translated,[3] until now, in its primary sense—lightning or lightning bolt. Yet the word is obviously used in the poem in its secondary sense—*Amanita muscaria*, the mighty entheogen.

This entheogen is also called *hurakan*, "one-leg," in the *Popol Vuh*, or *ekapād*, "single foot," in the *Rig Veda*. In the Vedic texts, the latter term is coupled with the epithet *aja* in the phrase *aja ekapād*, "not-born single foot." Vedic exegetes have astutely arrived at the meaning of the Sanskrit *aja*, "not born," a name given to the mushroom because it has no visible seed, as do seed-bearing plants and animals, and because in that prehistoric age it was believed the entheogen was miraculously conceived in the rain-softened earth by the single lightning bolt of Indra. What these scholars failed to see was that soma was a cryptogram pointing to the holy plant *Amanita muscaria*, the spores of which were invisible to the Brahmans.

The great Vedist Louis Renou once wrote that all the themes of the *Rig Veda* are summed up *in nuce* in the single word *soma*; such was this entheogen's all-pervading influence on the Aryan Brahmans. Renou's utterance concerning soma is equally true for *kakuljá* in the *Popol Vuh*:[4] all the Quiche poem's problems are summed up in the words *kakuljá hurakan*, referring to *Amanita muscaria* one-leg. The *Popol Vuh*, however, shows also the influence of the New World; while *A. muscaria* is still the number-one entheogen and is expressly given this honor when it is first introduced, there are two other categories of entheogens: (1) *ch'ipi* or "dwarf" *kakuljá*, the various entheogenic *Psilocybe* species that today are used in many parts of Mesoamerica, and (2) *raxi* or "green" *kakuljá*, the various phanerogamic entheogens that have been discovered by Mesoamericans (e.g., the two species of morning glory seeds and the peyote cactus in northern Mexico). The *Popol Vuh* cites these three classes of entheogens periodically with formality, but *hurakan* enjoys the distinction of being cited a half dozen times by itself. Now that we see the vital role entheogens play in the *Popol Vuh*, perhaps we are taking our first big step in understanding this complex poem.[5] Someone with a real knowledge of Quiche Maya and a sense of poetry should give us a good translation of the *Popol Vuh* that considers the botanic meaning of *kakuljá*.

I have a personal reason for delight at Ruck's and Lowy's discoveries: though my wife and I felt we were on the track of something important, it is gratifying to see our view confirmed by scholars who have embarked on our subject with success. The mighty entheogen of the Old World was *Amanita muscaria*, and this entheogen underlines not only the *Rig Veda* but also, half-way around the world, inspires a second body of prehistoric verse, the Quiche *Popol Vuh*. The identical visceral reaction from the entheogen expresses itself not only in Quiche in the powerful metaphor "lightning bolt" and in its second name "one-leg" but also in the *Rig Veda* as "single-foot." These figures go back far, perhaps to the dawn of the human race. In the Hellenic tradition of the soma surrogates and even in medieval manuscripts we see the one-leg figure, whose meaning no one knew for centuries.

I must add a note on a basic difference in the emphasis given to the parallels between the *Rig Veda* and the *Popol Vuh*. The Hindu writing (*RV* IX 82.3) says that Parjanya, the god of thunder, was the father of soma, and that the gods, "those fathers with a commanding glance, laid the Somic germ" (*RV* IX 92.3). All 1028 hymns in the *Rig Veda* deal directly or indirectly with the provenance of soma, and no alternative or conflicting explanation is given. Is it not a little odd that the Vedic poets do not say more about the origins of soma?

Again, the phrase *aja ekapād*, "not born single-foot," occurs six times in the *Rig Veda* but all in a small body of hymns to the divinities known as the Vi'svadevas. These passages sound like archaic survivals. By contrast, in the *Popol Vuh*, the lightning bolt and accompanying thunder are repeated loud and clear at intervals, sounding like a living belief. In the New World, time marched on more slowly than in the Old, although in both the *Rig Veda* and the *Popol Vuh* we are confronting a prehistoric age.

IDENTIFICATION OF THE TREE OF LIFE

Through the centuries we have heard references to the Tree of Life, the World Tree, the Tree of the Knowledge of Good and Evil, and perhaps other names, always pointing to a tree that is revered as holy, the focus of the religious feeling of the people thereabouts. The tree is never concealed, but it is a pity that seldom has the genus or species of tree been noted. I now make bold to suggest that it is always the mycorrhizal host of *Amanita muscaria*. The "fruit" of that tree is subject to taboo, spoken about only one-to-one, most frequently in the evening by candlelight; it is never mentioned in the marketplace or in mixed company. We must assemble all references to this tree and prepare a map showing where it has been worshipped and by whom. Does the religion of that tree still survive?

IDENTIFICATION OF THE ONE-SIDED FIGURE

Rodney Needham (1980) of All Souls, Oxford, has published a significant essay on the "one-sided man" in prehistory. Needham calls the drawing of this man the "unilateral figure," a man with a single leg, a single arm, a single eye—in short, a monster. Although Needham never mentions the entheogen, it is clear to me that the unilateral figure stands for the entheogen from earliest times. Someone should check the sources Needham gives in his bibliography to see how far they go in consolidating this thesis. To the extent these sources confirm this hypothesis, they expand the use of *Amanita muscaria* in humankind's past, perhaps vastly.

USE OF *AMANITA MUSCARIA* AMONG PALEOSIBERIAN PEOPLES

Unfortunately, in the West we do not know what Russian scholars have done on the study of the *Amanita muscaria* religion among the northern tribes of Siberia and also perhaps in Europe. Toporov, a leading Indo-Europeanist in the world, published a major treatise in Russian in 1985. Since my digest in *Soma* (Wasson 1967) of what the eighteenth- and nineteenth-century writers said about the mushroom cult in Siberia, we have heard little.

Someone (Toporov?) should inspire a concerted effort to capture the recollections of the former practices among the Paleosiberian tribes, especially all the taboo words for *Amanita muscaria* and the practices attending the ritual eating of it. Similar studies should be done for the other peoples who consumed the spectacular mushroom—the Samoyeds, the Finnics, and the Ob-Ugrians. Also, among the Finns, it would be worthwhile to learn whether the various documents used in compiling the *Kalevala* included references to *A. muscaria*. It might be that nineteenth-century editors thought readers would be offended by such texts and perhaps they themselves did not understand the texts. The Lapps, especially the Inari Lapps, should be questioned about their use of *A. muscaria*.

MUSHROOM ARTIFACTS IN LATIN AMERICA

Of course, there remains much work to be done in southern Mexico. I hope my observations there will prompt others to inquire discreetly in places that I have not visited, as well as in places where I made observations. In the Quiche area of the highlands of Guatemala, every linguistic group should be explored, tactfully and carefully, to learn whether there are memories of use of *Amanita muscaria*, *ch'ipi kakuljá*, and *raxi kakuljá*.

Certainly Colombia should be included in the field. Work should start seriously in the Museo del Oro. Since the authorities are naturally and rightly alert to the possibilities of theft in the Museo del Oro, anyone who goes down there should be well introduced. Armed, for example, with a letter from Professor Richard Evans Schultes, a member of the

museum's staff, an investigator should photograph every artifact suggestive of mushrooms, even though repetitive. The following data should be attached to each photograph:
the date the museum obtained the artifact, who found the artifact and where, which Indians lived in the area, and, if that group still lives there, whether individuals can be questioned tactfully and by whom. A mine of data about the use of the fungal entheogen
remains to be discovered in Colombia, and perhaps also in the Museo de Lima, Peru.

MYCOPHOBIA IN ENGLISH SPEAKERS

A final area for future research in ethnomycology relates to the English-speaking
world. In Chapter 2 of *Mushrooms, Russia and History* (Wasson and Wasson 1957),
Valentina Pavlovna and I tracked down in detail the mycophobia that has marked English herbals through the centuries. We also pointed to pertinent passages in the works of
poets, playwrights, and prose writers. In dramatic contrast with the Russians, English-
speaking people have been mycophobes. I will not repeat the material here but will continue the inquiry further.

The mushroom manuals published in increasing numbers from the mid-nineteenth
century were written by amateur and professional mycologists to help beginners identify
mushrooms in the field. These manuals grossly exaggerated the perils of eating *Amanita
muscaria*; a goodly number of them even led readers to believe that those who ate this
mushroom risked death. This is not a scientific finding but rather a deep-seated prejudice,
a folk belief from long, long ago.

Let us suppose that a small child has nibbled a red mushroom (in North America,
perhaps a yellow one) with white spots. When the parents learn of that fact, they take
alarm, whisking off the child to the hospital where the doctor orders a stomach pump applied to the child, who is probably kept in the hospital overnight. The child is discharged
the next morning, saved! The doctor modestly and silently accepts the parents' heartfelt
expressions of gratitude for "curing" the child. How often do we read such accounts in
small-town newspapers with limited circulation! If the child eats green apples and gets
sick, we never hear of it, yet the sickness from *Amanita muscaria* is no more serious.

Camille Fauvel, a retired French policeman who in the First World War had been in
charge of the Mata Hari portfolio, was a formidable mycophage. When living in Gascony
he ate vast quantities of *Amanita muscaria* because he enjoyed the taste. Locally the plant
was called *crapaudin*, "toad mushroom," but city folk called it *tue-mouche*, a translation
from the German name, or *fausse orange*. These ugly neologisms of the nineteenth century are nothing more than the invention of city folk who are unfamiliar with local words
and go outside the country for another name.

The false belief in the toxicity of *Amanita muscaria* must go back for thousands of years.
Dorothy Sayers, an excellent and conscientious writer of mystery stories, used this mushroom in *The Documents in the Case*. Involving herself in a complicated entanglement based
on muscarine, a poisonous alkaloid that exists in *A. muscaria* but only in trace amounts,
Sayers had her villain administer the "lethal" mushroom to the victim. When Sayers discovered her mistake, she apologized publicly in an article in *The Listener* in 1931.

The deep-seated prejudice against *Amanita muscaria* has affected our vocabulary.
Although a common mushroom, it has no name in English.[6] To speak of it, one must be
familiar with the Linnean binomial. Its original name must have been *toadstool*, but
because a taboo lay on *A. muscaria*, the name came to mean any mushroom that is not
recognized; and since most people recognize no wild mushrooms, toadstool became the
name of all of them.

Anglo-Saxon discrimination against this particular mushroom extends further and
may be responsible for the poverty of the English language when it comes to expressing

the blinding and ear-splitting lightning flash that (as was once believed) marked the conception of the mighty entheogen in the rain-soaked earth. Yet other languages have words for the two categories of lightning bolts:

ENGLISH	FRENCH	SPANISH	QUICHE MAYA
Lightning	Éclair	Relámpago	Xkoyopá
—	Foudre	Rayo	Kakuljá

Has no one ever called attention to this shortcoming of the English language? What brought it about? Is it possible the lack of a common word for *Amanita muscaria* gives us a clue? Has the deep-rooted fear of the rejected belief in the divine *coup de foudre*,[7] as the inseminator of our entheogen, driven the word for the blinding and deafening flash of lightning from our language? If I were younger, I should test this hypothesis by polling the Indo-European languages to see how the presence or absence of this word can be correlated with mycophobia and mycophilia.

NOTES

[1]The word *entheogen* signifies a plant that was and/or is being served in holy agapes and that affords the celebrants what they consider supernatural insights.

[2]Lowy's discovery was confirmed in the field by Dennis Tedlock (1985).

[3]Prior to Tedlock's translation in 1985, there have been twelve translations of the *Popol Vuh* into Spanish, English, French, German, and Russian.

[4]Renou died before it was learned that soma was the same plant called *kakuljá* by the Quiche Maya.

[5]Although Tedlock knew of Lowy's discovery in 1985, in his translation of the *Popol Vuh* he rendered *ch'ipi kakuljá* as "newborn thunderbolt" and *raxi kakuljá* as "raw thunderbolt." Both of these phrases make perfect nonsense, and I rather doubt lightning is even mentioned in the *Popol Vuh*.

[6]In *Soma* (Wasson 1967) I tried out a name for it, fly-agaric, but this name has not taken hold.

[7]Do Christians not perceive in the Annunciation a reappearance in new habiliments of the age-old belief? In many old paintings (and perhaps prints) of the Virgin receiving the wonderful news from an angel of Jehovah, there is at the same moment a visible ray from heaven that comes down and enters the Virgin. That ray discharges in the New Incarnation the duty of the *coup de foudre* in the Old.

LITERATURE CITED

Lowy, B. 1974. *Mycologia*. 141–151.

Needham, R. 1980. *Reconnaissances*. University of Toronto Press.

Sayers, D., and R. Eustace. 1930. *The Documents in the Case*.

Tedlock, D. 1986. *Popol Vuh: The Definitive Edition of the Mayan Book of the Dawn of Life and the Glories of Gods and Kings*. New York: Simon & Schuster.

Toporov, N. V. 1985. On the semiotics of mythological conceptions about mushrooms. *Semiotica* 53/54. Trans. from Russian by S. Rudy.

Wasson, R. G. 1967. *Soma, Divine Mushroom of Immortality*. New York: Harcourt, Brace & World.

Wasson, R. G., A. Hofmann, and C. A. P. Ruck. 1978. *The Road to Eleusis: Unveiling the Secret of the Mysteries*. With a new translation of the *Homeric Hymn to Demeter* by D. Staples. New York: Harcourt Brace Jovanovich.

Wasson, R. G., S. Kramrisch, J. Ott, and C. A. P. Ruck. 1986. *Persephone's Quest: Entheogens and the Origins of Religion*. New Haven, CT: Yale University Press.

Wasson, V. P., and R. G. Wasson. 1957. *Mushrooms, Russia and History*. New York: Pantheon Books.

PART 10

Archaeoethnobotany

Archaeoethnobotany (often called paleoethnobotany) is the study of archaeological remains of wild and cultivated plants, and their relationship to the life and development of primitive societies and often the evolution of agricultural plants. It is a discipline of interest to anthropologists, botanists, geneticists, agriculturists, evolutionists, and historians. Furthermore, it often can provide valuable information on the environment in which ancient peoples lived.

As early as 1826, C. Kunth examined plant remains from ancient Egypt, studies of which are still continuing. Forty years later, O. Heer and his colleagues stimulated collection and study of Swiss Lake Dweller materials. Towards the end of the 1800s and in the early 1900s, L. Wittmack and others began research on plant remains in Germany and other parts of Europe; important publications on this research started to appear and were summarized by K. and F. Bertsch and by Tackholms in the 1940s. Archaeobotanical studies have continued on European material, particularly in England by G. Dimbleby and J. M. Renfrew. One of the most active students of archaeological plant remains of Europe and the Near East and their relationship to the evolution of cultivated species was the Danish scientist H. Helbaek, whose prolific publications greatly advanced and encouraged archaeoethnobotanical research. A magnificent example of the value of archaeoethnobotany is evident in the book *The Origins of Agriculture and Settled Life* by the archaeoethnobotanist Richard S. MacNeish (1992).

Research soon spread to material of the New World, particularly southwestern United States, Mexico, and Peru. It was stimulated by the early work of M. R. Gilmore and

V. Jones on plant remains in the American Southwest and on areas still inhabited by local tribes. The number of investigators now active in this field in the United States is greatly increased. Outstanding is the work of botanists P. C. Mangelsdorf, who, with his students and colleagues, researched in depth the rich archaeological finding of R. MacNeish in Tehuacan and southern Mexico. These studies have greatly increased our knowledge of the age of incipient agriculture in the Americas, and of the origin and evolution of maize and other food plants.

A comprehensive summary of the ethnobotany of Columbian Peru by the late M. Towle appeared in 1961 (*The Ethnobotany of Pre-Columbian Peru*, Aldine Publishing Co., Chicago, 1961). Another extremely important paper was written in 1976 by J. M. Adovasio and C. F. Fry and describes, from dry caves of northern Mexico and Texas, the remains of two hallucinogens (peyote and the mescal bean) in contexts suggesting their use as psychoactive drugs as early as 8000 years ago ("Prehistoric Psychotropic Drug Use in Northeastern Mexico and Trans-Pecos Texas," *Economic Botany* 20: 94–96).

In 1991, the use in Chile of the hallucinogen snuff prepared from *Anadenanthera* was established from archaeological material dated 750 A.D. by C. M. Torres, D. B. Repke, K. Chan, et al. ("Snuff Powders from Pre-Hispanic San Pedro de Atacama: Chemical and Contextual Analysis," *Current Anthropology* 32: 640–649). An exciting discovery by Anna Roosevelt in the Amazon has, on the basis mainly of archaeological studies, suggested changes in our former ideas on the history and life of Amazonian Indians in ancient times.

A number of very active archaeobotanists are working in the United States. The late E. C. Smith carried out much research on primitive cottons; Stephen Williams and his students are studying the archaeology of the Mississippi Valley; L. Kaplan is concerned with the archaeoethnobotany and evolution of beans; and students at the ethnobotany laboratory of the University of Michigan are interested in plant remains from the Southwest. Among the numerous American archaeologists interested in plant material from foreign areas are Gordon Willey (Peru) and Wilma E. Wetterstrom (Egypt).

A number of diverse specialists have used archaeological material in studies of the origin of cultivated plants, including A. P. de Candolle (*Origin of Cultivated Plants*, Hafner Publishing Co., New York, 1959), N. I. Vavilov (*The Origin, Variation, Immunity, and Breeding of Cultivated Plants*, *Chronica Botanica*, Waltham, Massachusetts, 1951), and F. Schwanitz (*The Origin of Cultivated Plants*, Harvard University Press, Cambridge, Massachusetts, 1966). An interesting aspect of archaeobotanical research is that it has adapted numerous relatively new techniques to its study of plant remains, among which are radiocarbon dating and pollen analysis.

Archaeoethnobotany is a discipline destined to advance appreciably in the future, particularly in Europe, the United States, and parts of southwestern Asia. More universities are paying attention to this type of study, and our anthropological and botanical journals are increasingly publishing articles of ancient plant remains. All these indications suggest a international recognition of the great value of this field of research.

Two contributions are included in this section. Plutarco Naranjo discusses how ceramics from the Valdivia and other periods provide evidence of the ancient use of psychoactive plants. Gordon R. Willey summarizes the importance of archaeobotany to prehistoric and culture historical studies and to botanical sciences. Readers are also referred to "Guideposts in Ethnobotany" by Vorsila L. Bohrer (*Journal of Ethnobiology* vol. 6, no. 1 (1986), pp. 27–43), which discusses maize classification from prehistoric plant remains via archaeobotanical site reports.

Archaeology and Psychoactive Plants

PLUTARCO NARANJO

How long has the human race used psychoactive plants, and why? Probably the use of psychoactive plants goes back to earliest times. *Homo sapiens* and ancestors, faced with the absolute need to feed themselves, must have tried all plants and animals within reach. In a long and empirical process, they must have learned which plants were useful for food and other needs, and psychoactive species must have been easily recognized as the result of cause and effect.

In most primitive cultures, psychoactive plants were associated with myths, rituals, and ceremonies. Thus these plants acquired positions in the culture as magical or sacred objects. These positions were denoted in many vernacular names, including *teonanacatl*, which in Nahuatl of Mexico signifies "divine flesh," and *ayahuasca*, which in Quechua signifies "vine of the soul or spirit" and refers to a drug from *Banisteriopsis caapi* (Spruce ex Griseb.) Morton that made possible communication with the ancestors.

If we restrict our discussion to the Western Hemisphere, archaeology allows us to go back at least 10,000 years. In various caves in a semi-arid region in northern Mexico and southern Texas (Taylor 1956; Adovasio and Fry 1976, p. 108), seeds of *Sophora secundiflora* Lag. ex DC., now known as mescal bean, red bean, or coralillo, have been found. Seeds found in the deepest strata are dated by radioactive carbon to be from 8440 to 8120 B.C., and successive strata have yielded seed up to 1000 B.C. Even in the deepest strata, the mescal bean is associated with crafted objects, such as arrowheads, indicating the bean was not used as a food.

What use could a non-food bean have for nearly ten millennia? When the first Spaniards came to this region in Mexico, they observed the use of the mescal bean in certain ceremonies, similar to the ceremonies that survive today to a very limited extent. Unfortunately, plant remains, wooden objects, bones, or other artifacts are preserved for centuries or millennia only in exceptional soil and climatic conditions. Our knowledge of the use of psychoactive plants, such as the mescal bean and coca, in coastal deserts of Peru and belonging to Inca or pre-Inca cultures is thus based not on discovery of plant remains but on other archaeological discoveries, like ceramics, that give us clues as to how a certain plant may have been used.

Ceramics and Plant Use

The discovery of ceramics gave primitive people a marvelous plastic material, which at first served to make objects of immediate utility: plates, pots, and a variety of cups. It allowed primitive people to cook for the first time; previously they had eaten raw or flame-roasted plant and animal foods.

The use of the ceramic pot considerably extended the use of grains, tubers, and other plant parts in human nutrition. For example, the bean (*Phaseolus vulgaris* L.), when ripened and dried, is not only very hard to chew, but it is also toxic. When cooked, however, the poisonous compounds are often destroyed and the bean becomes soft and takes on an agreeable flavor. Thus, the discovery of ceramics made it possible for primitive people to give their young children soft cooked foods, which provided good nutrition and in turn resulted in a population increase.

Simultaneously, the pot took on an importance in agriculture beyond the possibility of cooking hard, ripe grain. Pots could be used to store maize and beans for long periods for later consumption. Furthermore, this development of agriculture left time available for primitive people to work on other activities, including those of artistic or sumptuary character. Consequently, the development of ceramics shows that after the phase of utilitarian objects, other objects appeared related to adornment and the practice of rituals, ceremonies, and magic medicine. In some cases, a rich paraphernalia connected with ceremonial use of psychedelic plants was created. Accordingly, ceramics allow us, in many cases, to know the sequence of use of psychoactive plants.

Valdivia Ceramics

According to archaeological findings made during two decades, the earliest ceramic culture in the Western Hemisphere is that known as Valdivian, so named for the coastal area of Ecuador (Lathrap 1963; Meggers et al. 1965; Estrada 1976). The oldest ceramic pieces date from 4000 B.C. and are utilitarian objects, but from 3000 B.C. on they begin to diversify. Many female figures known today as the Venus of Valdivia appeared, as well as other objects connected with the use of psychedelic plants:

1. Miniature representations of the ceremonial stool used by shamans when presiding over certain rituals, ceremonies, or sessions of magical curing (Figure 1).

2. Small cups 2 to 3 centimeters (0.8 to 1.2 inches) in diameter (Figure 1). Known as *lliptas* or *poporos*, ashes from certain shells or snails were kept in these containers. In some of them it has been possible to identify the ashes of *Certhedea pulchra* (Lathrap and Marcos 1975) and bone spatulas or shell fragments used to transfer the ashes to the mouth.

3. Small but rare anthropomorphic figurines with a prominent face, a feature denoting the presence of a ball of leaves between the teeth and the cheek (Lathrap 1975).

4. Small figurines of the Venus of Valdivia in which the head is flattened like a platform or even excavated (Figure 2). Generally this female figure appears with hairdressing or high headdress; the absence of this feature indicates some of the figurines were idols and that psychedelic powders were placed in the platform or cavities of the head, a practice the Spaniards found in the Caribbean Islands where these idols were called *Cemis*.

Figure 1. Miniature ceramic reproductions of ceremonial stools and ash containers. Three types of stools used by shamans in the Valdivia culture (about 3000 B.C.) are shown and two containers with shell ashes. One of the vessels has a bone spatula probably used to take the ashes, used in the chewing of leaves from psychotropic plants, to the mouth.

Figure 2. Venus of Valdivia with concave head. The smallest figurines could be only miniature or symbolic representations of the larger heads that held sufficient quantities of psychedelic powders for inhaling.

5. Two-headed figurines of the Venus of Valdivia (Figure 3). The two- or three-headed figures found in various New World cultures are related to the use of potent psychedelic plants. Under the effect of these drugs, a psychological phenomenon known as "depersonalization" or "impersonalization" is produced in which the individual feels like himself/herself and at the same time like other persons or personalities. Frequently, one of these persons/personalities escapes the body, migrates, and goes on a journey. This psychological phenomenon is represented in ceramic figurines by the presence of two or three heads.

The types of archaeological pieces mentioned above reveal that Valdivian culture used a plant that could be chewed and one, perhaps the same plant, that could be inhaled through the nose as snuff. Which plant(s) was(were) used? The Valdivia region is semi-dry and, although six thousand years ago it could have been more humid than now, there is no evidence that coca had been domesticated and cultivated; instead, the coca plant reached the current territory of Ecuador probably one or more millennia later, following the course of rivers that flow towards the Amazon. A common plant by the name *florón*, however, grows in the region of Valdivia. Known by the scientific name *Ipomoea carnea* Jacq., this plant contains lysergic acid ethylamide in all its organs (although in different proportions). In periods of prolonged drought, it is almost the only herbaceous plant that survives. It is probable, therefore, that *I. carnea* was the psychoactive plant used by the Valdivians.

We should also consider the possibility that other psychedelic plants were used, among them *Anadenanthera peregrina* (Benth.) Speg., known in the Caribbean as *niopo* or *yopo*. The use of powdered seeds of this plant was widespread from the Caribbean to Peru and Argentina at the time the New World was discovered. This leguminous tree is not found currently in the region of Valdivia, but it is well known that all forests near Guayaquil were cut down during the Spanish conquest and the timber used to build ships. In fact, Guayaquil and Puná became the most important ship-building regions of South America's Pacific coast.

Another psychedelic powder has been used in South American humid tropics from time immemorial. It is obtained from a resin extracted from the bark of trees of the genus *Virola*. Although this tree, too, is not found in Valdivia, it is found in the northern coast of Ecuador, where it is currently exploited for its wood.

In conclusion, it is possible that more than one psychedelic plant was used as a snuff or in chewable form in the Valdivia period.

Post-Valdivian Ceramics

In approximately 2300 B.C. the Valdivian culture disappeared and was replaced by the Machalillian, which in turn was replaced by the Chorreran. In the latter two cultures, as well as in subsequent ones and in others that arose along the coast and in the Ecuadorian Andes, there appeared many ceramic pieces connected with the inhalation of psychedelic powders (Naranjo 1984a, 1984b).

In the Jama-Coaque culture (500 B.C. to 500 A.D.), a great variety of pipes appeared (Figure 4). Many are anthropomorphic and others, zoomorphic. They are made up of a container 2 to 4 centimeters (5 to 10 inches) in diameter that extends into a tube through which powder was inhaled. In none of the pieces studied is there evidence of combustion. Accordingly, these pipes must not have been used to smoke but rather to inhale powders. This coincides with the observations of Spanish explorers, who, when they arrived in the New World, found that smoking tobacco already existed in North America and

Figure 3. Two-headed figurines of Venus of Valdivia. The psychological phenomenon of feeling like two people, produced while under the influence of potent psychedelic drugs, may be represented in ceramic figures by the presence of two or more heads.

the Caribbean but not in South America. Instead, the Spaniards found the technique of inhaling powders was widespread among South American cultures.

At least a thousand years after the Valdivian culture, paraphernalia connected with the use of psychedelic plants appeared in the ceramics of Peruvian and Mesoamerican cultures. In the ceramics of pre-Inca and Inca Peru, many images relate to the use of coca and the cactus known as San Pedro (*Trichocereus pachanoi* Britton & Rose), and in the ceramics of Ecuador, many figures of the *cacique*, or shaman, chewing coca appeared in the Carchi culture (500 B.C. to 500 A.D.). The latter figures are known in archaeological slang as *coqueros*.

Another psychedelic plant with widespread use in the Amazon basin is ayahuasca or caapi, known as *Banisteriopsis caapi* (Naranjo 1985a, 1985b). Due to the trans-Andean trade between the coast and the Amazon region of present-day Ecuador, first coca and later ayahuasca reached the coast. *Banisteriopsis caapi* is still found on the northern coast of Ecuador, where it is known as *nepi* and *pilde*. In ceramics of the Milagro-Quevedo culture (500 B.C. to 500 A.D.), certain ceremonial vessels appeared, richly adorned with zoomorphic and anthropomorphic figures (Figure 5). Known today as *vasos de brujo* or *cocinas de brujo*, these containers must have been used to cook or boil psychotropic plants.

Summary

In conclusion, the careful study of archaeological materials, especially ceramics, has yielded, and will continue to yield, new insights into the use of psychoactive plants in ancient cultures.

Figure 4. Anthropomorphic pipe from the Jama-Coaque culture (500 B.C. to 500 A.D.), representing a flute player and his instrument. Like many other ceramic pipes of the culture, this one was not used to smoke but rather to inhale the snuff of psychotropic plants such as tobacco and perhaps others.

Figure 5. A *vaso* or *cocina de brujo*. This ceramic container of the Milagro-Quevedo culture (500 B.C. to 500 A.D.) perhaps served as a collective vessel so that each participating member of the ceremony could drink part of the liquid.

ACKNOWLEDGMENTS

All the pieces in the photographs belong to the Museum of Archaeology of the Central Bank of Ecuador.

LITERATURE CITED

Adovasio, J. M., and G. F. Fry. 1976. Prehistoric psychotropic drug use in northeastern Mexico and Trans-Pecos Texas. *Economic Botany* 30: 94–96.

Estrada, E. 1976. *Las culturas Pre-clásicas, Formativas o Arcaicas del Ecuador. Museo Estrada.* Guayaquil: Edit. Vida.

Lathrap, D. W. 1963. Possible affiliations of the Machalilla complex of coastal Ecuador. *American Antiquity* 29: 239.

———. 1975. *Ancient Ecuador: Culture, Clay and Creativity 3000–300 B.C.* Chicago: Field Museum of Natural History.

Lathrap, D. W., and J. Marcos. 1975. Informe preliminar sobre las excavaciones del sitio Real Alto por la Misión Antropológica de la Universidad de Illinois. *Rev. Universidad Católica* 3: 85.

Meggers, B. J., C. Evans, and E. Estrada. 1965. *The Early Formative Period of Coastal Ecuador: The Valdivia and Machalilla Phases.* Smithsonian Contributions to Anthropology, vol. 1. Washington, DC: Smithsonian.

Naranjo, P. 1984a. *Ayahuasca: Etnomedicina y Mitología.* Quito: Ediciones Libri Mundi.

———. 1984b. La Medicina en el Ecuador Preincaico. *Rev. Ecuat. de Medicina* 20: 93.

———. 1985a. La ayahuasca en la Arqueología ecuatoriana. 45th International Congress of the Americas. Bogota: University of Andes. 438.

———. 1985b. Plantas alimenticias del Ecuador precolombino. *Interciencia* 10: 227.

Taylor, W. W. 1956. Some implications of the carbon-14 dates from a cave in Coahuila, Mexico. *Bulletin of the Texas Archaeological Society* 27: 215.

Archaeobotany:
Scope and Significance

GORDON R. WILLEY

This chapter is a summary statement on archaeobotany and its importance to prehistoric and culture historical studies, as well as to botanical sciences. "Archaeobotany," as the ligature implies, refers to "old botany" in the sense of the study of ancient plant remains—those found in archaeological contexts. The name is obviously a parallel to and a derivative from "ethnobotany." In both archaeobotany and ethnobotany the focus of attention is upon the uses of plants by, and their associations with, people.

In the present volume, Nancy J. Turner (see "Ethnobotany Today in Northwestern North America") has described ethnobotany as the story or the study of "peoples' interaction with plants." Her immediate field of interest is the aboriginal peoples of northwestern North America, so that her work is closely linked to ethnography and ethnology; but, as she notes, there is an earlier, archaeological chapter to the ethnobotanical story in that area. Similarly, ethnobotanist S. K. Jain (1986) emphasized the present-day peoples and plants of India, although he made it clear that the subject in India has deep historical and archaeological roots. These conditions are to be found almost anywhere in the world—at least where there is an archaeological record.

As of the late 1980s, a substantial literature has accumulated in archaeobotany. Much of this literature relates to specific archaeological discoveries; however, there are some general works on archaeobotany. A classic one is G. W. Dimbleby's (1967) *Plants and Archaeology*, which remains a basic source on the nature of plant evidence and the history of plants as revealed by archaeology. Richard I. Ford's (1978) *The Nature and Status of Ethnobotany* also offers a valuable introduction to the subject, inasmuch as his "ethnobotany" is broadly projected and addresses archaeobotanical data as well. Still more recently, a collection of essays on *The Analysis of Prehistoric Diets* (Gilbert and Mielke 1985) provides a number of articles on or tangential to the field of archaeobotany.

The subject matter of archaeobotany includes all former living plant remains. Such may be recovered macroscopically, as wood particles, leaves, grasses, mosses, lichens, fruits, seeds, pods, husks, fibers, or other residues (Carbone and Keel 1985). In addition, some herbaceous plants, although largely lost to decay, may leave visible traces, or phytoliths, in the soil. These phytoliths (siliceous fragments or opaline traces of plant cell walls) may allow for species identification on the bases of anatomically distinct features.

As an example, maize plant phytoliths have been identified in the soils of Ecuadorian archaeological sites dating back some four thousand years (Pearsall 1978). Other plant remains in archaeological and geological deposits are recoverable only microscopically, as pollen and spores. Thus, palynologists (Dimbleby 1967, 1985) studying such pollens have been able to identify plant species from such diverse contexts as house floors in archaeological sites or geologic soil cores from deep sedimentary deposits.

Plants are, and have been, used by people in countless ways: as foodstuffs, for medicines and narcotics, in dwellings, for clothing and containers, for ornamental and ritual purposes, and so on. We are aware of this through observation of the life around us; ethnologists make such observations in the societies which they study; and, obviously, plants were so utilized in the ancient past, as we know from archaeological records.

The preservation of plant remains in archaeological sites varies greatly, depending upon the environmental setting. Thus, dry cave deposits, usually in arid or semi-arid areas, are ideal for the recovery of plant materials. Wooden objects, textiles, seeds of wheat, barley, and maize, and the husks, rinds, pods, and stems of other plants are found in these contexts. While such items are desiccated, they are often more or less intact and readily identifiable. There are also regions where, because of extreme aridity, vegetal materials are preserved in ancient sites. The rainless coastal desert of Peru is such a region. Here, it is not uncommon to find caches of dry maize cobs, maize leaves, maize husks, peanut shells, various withered but intact fruits, and abundant cotton clothing—all preserved in middens or graves of archaeological sites dating back hundreds or even thousands of years into pre-Columbian times.

Archaeological sites in swampy, boggy, or extremely wet environments often present conditions that are as favorable as those of extreme dryness for the preservation of organic and plant materials. The constant wetness prevents the aeration of specimens that is necessary for decay. A famous example is Key Marco, a site off the coast of southwest Florida, where F. H. Cushing (1896) found an amazing set of wooden sculptures preserved in muck debris. Plant remains other than wood have been found in excavations in this same Florida Keys area, both as preserved macroscopic specimens and as pollen (see Sears 1982; Widmer 1988).

Most archaeological sites in temperate or tropical areas of moderate or abundant annual rainfall present a more difficult situation for the archaeologist-archaeobotanist. Here, reliance usually must be placed upon charred or burned wood and plants. Fireplaces or fire-hearths are obvious provenances for such charred plant debris; however, within the last twenty-five years or so archaeologists have become more alert to the possibilities of finding smaller but still identifiable charred plant bits in routine refuse digging. To recover these, flotation techniques have been developed. Excavated soil is carefully sifted and then washed or dumped into water so that fine plant residues are either screened out and saved or skimmed from the surface of the flotation tub (Smith 1985). Such techniques have been employed with notable success, particularly in the eastern United States (Struever 1968).

There are also what might be referred to as "fortuitous" examples of archaeobotanical recovery. For example, wood and textile portions will sometimes be preserved through contact with eroding copper or bronze, or imprints of cereal grains will occasionally be found on pottery surfaces or in the plasterwork of buildings where they had been accidentally impressed before the firing of the pottery or during the construction of the house (Dimbleby 1967, pp. 98–100). The point to be made is that plant remains—organic items that generally are thought of as rapidly perishable with the passage of time—have been preserved in a variety of ways in archaeological sites, and modern scientific aids and recovery techniques increasingly enhance the archaeologist's chances of recovering them for study.

In that study, botanical identification of the remains is only the beginning. From this point, research can move in several directions. The emphasis may be on the reconstruction of the ancient environment of the area of the archaeological site, or sites, in question. For this, the archaeobotanical remains, together with geological and paleoecological information, are the keys to reconstruction (Smith 1985). Or, to follow an ethnographic or cultural line of inquiry, the archaeologist may collaborate with an ethnobotanist who is familiar with present-day peoples and cultures of the area who are assumed to be continuing in a cultural tradition similar to those of the archaeological past. By visiting native households and markets and asking questions about plant collection, cultivation, and usage, the investigator may better understand past plant use.

A more strictly archaeological line of inquiry focuses within the site itself. It attempts to find out how recovered plant specimens correlate with site and cultural stratigraphy. It asks questions about the changes of food and other plant remains through time and seeks to find out how these might have related to social or cultural changes. Or, the emphasis might be on plant genetics and genetic change through time. In brief, archaeobotanical study moves forward with the aid of both botanical and culture historical perspectives.

An extraordinarily difficult but crucial procedure in archaeobotanical study is the quantifying of plant remains—especially food plants—with an eye toward judging their importance in archaeological contexts (Smith 1985). In other words, how may we count or estimate the relative abundance and importance of a food plant in archaeological household or domestic refuse? And how can these counts or estimates be translated into subsistence, caloric, or nutritional values for the people who were once involved? Such computations are obviously necessary if we want to take the next step of correlating food changes and the increases or decreases of certain species with such phenomena as population size and growth or decline, and pulsations in cultural creativity. Does one count individual plant specimens per level or provenance unit, or are there other and better ways to tabulate archaeological plant residues and translate these into potential caloric intake? Estimates and computations on these and related questions have been attempted in various ways (MacNeish 1958, 1964, 1967; Flannery 1986) and are subject to debate.

With this sketch of archaeobotany—its nature, scope, and procedural relationships to archaeology—behind us, let us view the subject in its wider significance. What is its potential as a means of understanding culture history? In giving a general answer to this question, it is not an exaggeration to say that the archaeology of plants has served as one of the primary guidelines in tracing the development of human societies from small, foraging communities to agriculturally based civilizations.

This story of social, cultural,and agricultural evolution is a long one. In the Old World, it begins at least as early as 9000 B.C. with the first cultivation of cereals, quite probably in the hill country of the Middle East—an area that extends from Israel and Jordan into southern Turkey and from there eastward and southward to flank the Tigris-Euphrates rivers and the Persian Gulf (Dimbleby 1967, ch. 6). In the early post-Pleistocene (9000–8000 B.C.) this "nuclear area" for agriculture was forested terrain and also was the native habitat of wild einkorn wheat (*Triticum monococcum*), wild emmer wheat (*Triticum dicoccum*), and wild barley (*Hordeum spontaneum*). Soils and rainfall would have been such as to have favored the early experimental growing of these grains, and this, over the centuries, resulted in domestication. The domestication process involved genetic modifications of the plants, which led to their improvement as food sources and to an economic dependence upon them.

The time span, between 9000/8000 B.C. and 5000 B.C., or between the earliest beginnings of cereal cultivation and the subsequent intensive irrigation agriculture of the Mesopotamian lowlands, has been divided into two main periods by Middle Eastern archaeologists (Braidwood 1975, pp. 106–107). The earlier period has been designated as an era of "incipient cultivation and domestication"; referring to plants and animals, it lasted from 9000/8000 to 7000 B.C. The term "incipient" is well applied in this context, for it was during this time that plants were undergoing the genetic changes that were to transform them into fully domesticated farm products; and it also was during this period that human groups were going through those incipient changes that were to transform them from semi-sedentary foragers to settled village farmers. The later period, from 7000 to 5000 B.C., is designated, by contrast, as a time of "full village-farming." This latter lifestyle was characterized by the establishment of economically successful food production through agriculture and animal husbandry, with these activities based in or near permanent villages.

After 5000 B.C., such cereal crops were carried down into the arid lowlands of the Tigris-Euphrates, where they were grown more intensively and productively through irrigation. This shift may be said to mark a third stage in the Middle Eastern story of agriculture. It was further characterized by the founding of sizable towns and the construction of small temples.

A fourth stage, beginning at 3500/3000 B.C., was accompanied by a continued expansion of irrigation farming and by the founding of the first large cities replete with their monumental architecture. In brief, the first city-states of Mesopotamia came into being in this fourth stage. Thus, people and plants were linked together dynamically in Southwest Asia, in an interactive series of relationships that involved plant domestication, agriculture, demographic growth, and social and cultural change. All of this lasted for some 6000 years, or from 9000 to 3000 B.C. To be sure, after about 5000 B.C. changes in food plants and in agricultural technology were less dramatic than they had been earlier; still, the nature and fortunes of agricultural production remained clearly linked to the histories of the Mesopotamian city-states in later times, and, indeed, these linkages continue to today.

Besides the Southwest Asian-Mesopotamian area, there were two other "nuclear areas" for Old World plant cultivation. One of these was the Hwang-ho River Valley of China, the apparent setting for the first cultivation of millets (*Setaria italica, Panicum miliaceum*) (Flannery 1973; Braidwood 1975, p. 101). The other "nuclear area" was in the tropics of Southeast Asia where there was a very ancient "incipient agriculture" featuring root crops like taro and yam. Indeed, Carl Sauer (1952) has argued that the very first plant cultivation anywhere was with root crops, through techniques of vegeculture, and that such beginnings were made in tropical forests, both in the Old and New Worlds. The Southeast Asian setting also may have been the "nuclear area" for the cultivation of rice (*Oryza sativa*) (Chang 1970; Flannery 1973). Rice, known as early as the fourth millennium B.C. on the Chinese mainland, was to the rise of civilization in the Far East what wheat and barley were to the rise of civilization in Southwest Asia.

In the Americas there were three main "nuclear areas" of early plant cultivation, comparable in their roles in the development of agriculture to the Old World areas of Southwest Asia, China, and Southeast Asia. The American areas were Mesoamerica, the Andes, and the South American tropical forest. In all these areas the plants were local New World cultigens. For Mesoamerica, these were maize (*Zea mays*), beans (*Phaseolus vulgaris, P. coccineus, P. acutifolius*), and the squashes (*Cucurbita pepo, C. mixta*); for the Andes, they were quinoa (*Chenopodium quinoa*), potato (*Solanum tuberosum*), lima bean (*Phaseolus lunatus*), another strain of *P. vulgaris*, and the squashes (*Cucurbita moschata, C. ficifolia*); and in the tropical forest, manioc (*Manihot esculenta*), peanuts (*Arachis hypogaea*), and other root crops were the main staples.

In all three major American areas of early plant domestication, there was undoubtedly a long period of incipient cultivation. For Mesoamerica, the story of wild plant collecting by semi-nomadic bands living in the semi-arid valleys of southern Mexico can be traced in the archaeological record back to the ninth millennium B.C. (Willey 1966, ch. 3; Flannery 1973; Flannery et al. 1981). Between then and about 5000 B.C., plant cultivation slowly got under way. This incipient cultivation was accompanied by social and cultural changes, including a greater degree of sedentariness, especially after 5000 B.C. with the domestication of maize as an important food crop (MacNeish 1964, 1967, 1981; Flannery 1973, 1986). Maize underwent rapid genetic changes over the next thirty-five hundred years, and the plant was carried into a number of different environmental niches in Mesoamerica as well as farther afield in the Americas.

The farming of maize, beans, and squashes transformed the seasonal campsites of the earlier hunting-collecting and incipient cultivation era into permanent villages. With this step, the peoples of Mesoamerica were well on their way toward complex society. Monumental architecture, great art, and urban living followed, so that the term "civilization" could be applied appropriately to Mesoamerican cultures by the beginning of the Christian Era.

In the Central Andes (MacNeish et al. 1970; Willey 1971, ch. 3), an incipient cultivation era lasted from the end of the Pleistocene (about 9000/8000 B.C.) until about 3000/2000 B.C. The potato (*Solanum tuberosum*) was the major staple in the Andean highlands, although maize appears to have been introduced from Mesoamerica during this period. On the Peruvian coast, the lima bean was an important early food source prior to the introduction of maize; but here an early sedentariness was also significantly supported by a maritime-based subsistence (Moseley 1975). Large-scale public building appeared in many Peruvian coastal sites as early as 2000 B.C., substantially in advance of similar phenomena in Mesoamerica. From then until the arrival of the Spanish, Peruvian civilization increased in socio-political complexity and technological sophistication, arriving at a stage of urbanism and of regional states in the first millennium B.C. and going from these to large territorial empires. The last of these empires, that of the Inca, was in full swing in A.D. 1532 when Pizarro reached Peru. As in Mesoamerica, the Peruvian food and textile (cotton) plant inventory was enormous, including numerous fruits and vegetables not mentioned in this brief summary. Together with the food plants first domesticated in Mesoamerica and the South American tropical forest, it has been estimated that Native Americans are responsible for almost one-half of the agricultural plants known to the world today.

The third great "nuclear area" for New World plant cultivation, the South American tropical forests, probably had a long period of incipient cultivation (Lathrap 1977; see also Flannery 1973), but this must be hypothesized inasmuch as there is little in the way of preserved archaeological evidence to document it. Manioc and the peanut are two of the most important domesticates of this area, and there were many other fruits and vegetables originally native to, and subsequently domesticated in, the tropical forests. Large villages and ceramics date to at least 2000 B.C. in the area, and they may be much older than this. Archaeological research is still not so advanced here as it is in either Mesoamerica or the Andes. We know, though, that the South American tropical forest had significant cultural and agricultural interchange with the Andes. This included not only Peru but the highland regions of Colombia and Ecuador, where settled village agricultural life appears to be as old as, if not older than, that in Mesoamerica or Peru (Lathrap 1975).

In summary, for those of us in archaeology, one thing is certain. As long as we are interested in tracing out the past and in understanding the way human culture has grown and diversified, archaeobotany will have a very significant place in our studies.

LITERATURE CITED

Beadle, G. W. 1972. The mystery of maize. Field Museum of Natural History Bulletin 43 (10): 2–11.

Braidwood, R. J. 1975. *Prehistoric Men*. 8th ed. Glenview, IL: Scott, Foresman and Company.

Carbone, V. A., and B. C. Keel. 1985. Preservation of plant and animal remains. In *The Analysis of Prehistoric Diets*. Eds. R. I. Gilbert, Jr., and J. H. Mielke. New York: Academic Press. 1–20.

Chang, K. C. 1970. The beginning of agriculture in the Far East. *Antiquity* 44: 175–185.

Cushing, F. H. 1896. Exploration of ancient key dwellers' remains on the Gulf Coast of Florida. *Proceedings of the American Philosophical Society*, Philadelphia, 35: 329–448.

Dimbleby, G. W. 1967. *Plants and Archaeology*. London: John Baker Publishers.

———. 1985. *The Palynology of Archaeological Sites*. New York: Academic Press.

Flannery, K. V. 1973. The origins of agriculture. *Annual Review of Anthropology* 2: 271–310.

———, ed. 1986. *Guila Naquitz, Archaic Foraging and early Agriculture in Oaxaca, Mexico*. New York: Academic Press.

Flannery, K. V., J. Marcus, and S. A. Kowalewski. 1981. The preceramic and formative of the valley of Oaxaca. In *Supplement to the Handbook of Middle American Indians*, vol. 1., *Archaeology*. Eds. V. R. Bricker and J. A. Sabloff. Austin: University of Texas Press. 48–93.

Ford, R. I., ed. 1978. *The Nature and Status of Ethnobotany*. Anthropological Papers, no. 67. Ann Arbor, MI: University of Michigan Museum of Anthropology.

Gilbert, R. I., Jr., and J. H. Mielke, eds. 1985. *The Analysis of Prehistoric Diets*. New York: Academic Press.

Jain, S. K. 1986. Ethnobotany—with special reference to India. *Interdisciplinary Science Reviews* 11(3): 285–292.

Lathrap, D. W. 1975. *Ancient Ecuador: Culture, Clay, and Creativity, 3000–300 B.C.* Chicago: Field Museum of Natural History.

———. 1977. Our father the Cayman, our mother the gourd: Spinden revisited or a unitary model for the emergence of agriculture in the New World. In *Origins of Agriculture*. Ed. C. A. Reed. The Hague: Mouton. 713–751.

MacNeish, R. S. 1958. *Preliminary Archaeological Investigations in the Sierra de Tamaulipas, Mexico*. Transactions of the American Philosophical Society, vol. 48, no. 6. Philadelphia.

———. 1964. The food-gathering and incipient agriculture stage of prehistoric Middle America. In *Handbook of Middle American Indians*, vol. 1. Ed. R. Wauchope. Austin: University of Texas Press.

———. 1967. A summary of the subsistence. In *Prehistory of the Tehuacan Valley*, vol. 1, *Environment and Subsistence*. Ed. D. S. Byers. Austin: University of Texas Press. 290–309.

———. 1981. Tehuacan's accomplishments. *Supplement to the Handbook of Middle American Indians*, vol. 1., *Archaeology*. Eds. V. R. Bricker and J. A. Sabloff. Austin: University of Texas Press. 31–47.

MacNeish, R. S., A. Nelken-Turner, and A. G. Cook. 1970. *Second Annual Report of the Ayacucho Botanical-Archaeological Project*. Andover, MA: R. S. Peabody Foundation.

Moseley, M. E. 1975. *The Maritime Foundations of Andean Civilization*. Menlo Park, CA: Cummings Publishing Company.

Pearsall, D. 1978. Phytolith analysis of archaeological soils: Evidence of maize cultivation in formative Ecuador. *Science* 199: 177–178.

Sauer, C. O. 1952. *Agricultural Origins and Dispersals*. New York.

Sears, W. H. 1982. *Fort Center: An archaeological site in the Lake Okeechobee basin*. Gainesville: University Presses of Florida.

Smith, C. E., Jr. 1985. Recovery and processing of botanical remains. In *The Analysis of Prehistoric Diets*. Eds. R. I. Gilbert, Jr., and J. H. Mielke. New York: Academic Press. 97–126.

Struever, S. 1968. Flotation techniques for the recovery of small-scale archaeological remains. *American Antiquity* 33: 353–362.

Widmer, R. J. 1988. *The Evolution of the Calusa*. Tuscaloosa: University of Alabama Press.

Willey, G. R. 1966. *An Introduction to American Archaeology*, vol. 1, *North and Middle America*. Englewood Cliffs, NJ: Prentice-Hall.

———. 1971. *An Introduction to American Archaeology*, vol. 2, *South America*. Englewood Cliffs, NJ: Prentice-Hall.

Index of Scientific Names